AF701261

Handbook of Quenchants and Quenching Technology

G.E. Totten

C.E. Bates

N.A. Clinton

Acquisitions Editor - Mary Thomas Haddad

Production Project Manager - Linda Kacprzak

Production/Design - Dawn R. Levicki

ASM INTERNATIONAL®

The Materials Information Society

Copyright 1993
by
ASM International®
All rights reserved

No part of this book may be reproduced, stored in a retrieval system, or transmitted, in any form or by any means, electronic, mechanical, photocopying, recording, or otherwise, without the written permission of the copyright owner.

First printing, January 1993

This book is a collective effort involving hundreds of technical specialists. It brings together a wealth of information from worldwide sources to help scientists, engineers, and technicians solve current and long-range problems.

Great care is taken in the compilation and production of this Volume, but it should be made clear that NO WARRANTIES, EXPRESS OR IMPLIED, INCLUDING, WITHOUT LIMITATION, WARRANTIES OF MERCHANTABILITY OR FITNESS FOR A PARTICULAR PURPOSE, ARE GIVEN IN CONNECTION WITH THIS PUBLICATION. Although this information is believed to be accurate by ASM, ASM cannot guarantee that favorable results will be obtained from the use of this publication alone. This publication is intended for use by persons having technical skill, at their sole discretion and risk. Since the conditions of product or material use are outside of ASM's control, ASM assumes no liability or obligation in connection with any use of this information. No claim of any kind, whether as to products or information in this publication, and whether or not based on negligence, shall be greater in amount than the purchase price of this product or publication in respect of which damages are claimed. THE REMEDY HEREBY PROVIDED SHALL BE THE EXCLUSIVE AND SOLE REMEDY OF BUYER, AND IN NO EVENT SHALL EITHER PARTY BE LIABLE FOR SPECIAL, INDIRECT OR CONSEQUENTIAL DAMAGES WHETHER OR NOT CAUSED BY OR RESULTING FROM THE NEGLIGENCE OF SUCH PARTY. As with any material, evaluation of the material under end-use conditions prior to specification is essential. Therefore, specific testing under actual conditions is recommended.

Nothing contained in this book shall be construed as a grant of any right of manufacture, sale, use, or reproduction, in connection with any method, process, apparatus, product, composition, or system, whether or not covered by letters patent, copyright, or trademark, and nothing contained in this book shall be construed as a defense against any alleged infringement of letters patent, copyright, or trademark, or as a defense against liability for such infringement.

Comments, criticisms, and suggestions are invited, and should be forwarded to ASM International.

Library of Congress Cataloging Card Number: 92-73857
ISBN: 0-87170-448-X

ASM International®
Materials Park, OH 44073-0002

Printed in the United States of America

PREFACE

The ability of mankind to control and use properties of materials is so central to the technology of civilization that historians describe various levels of technical development by reference to the materials used to fabricate tools and weapons—for example, the Stone Age, the Bronze Age, and the Iron Age. The High-Strength Alloy Age has been suggested by G. Beck as the appropriate title for the modern era.

The first iron implements were produced from meteoric iron and must have been of great rarity and value. Iron implements made from the smelting of iron ores have been identified as dating to about 2500 B.C. Early iron would have been wrought and would have been inferior to bronze because of its susceptibility to rusting, its inability to be cast (due to the low temperatures available to the ancient metalsmith), and its inability to hold a sharp edge. By about 1400 B.C. the Hittites developed what they called "good iron." They had found that iron could be alloyed with carbon (probably the charcoal used in the furnace) to produce steel. They also discovered the process of quenching a hot part to produce a hard steel. They kept this technology secret until 1300 B.C., when they burst upon the world's stage and, after conquering Assyria, challenged and fought to a stalemate the armies of the Pharaohs. Steelmaking technology soon became widely disseminated. By 1200 B.C. the metalsmiths of the ancient world had realized that they did not need to alloy and harden the full body of a blade, but could selectively carburize and harden just the cutting edge.

By the time of Homer (circa 800 B.C.), the use and purpose of quenching was well known. In fact, in *The Odyssey*, Book Nine (using the translation of R. Fitzgerald), Homer writes:

... *In a smithy*
one sees a white-hot axehead or an adze
plunged and wrung in a cold tub, screeching steam—
the way they make soft iron hale and hard—

It was another 400 years before the discovery that tempering after quenching allowed the production of parts that were hard but not brittle.

The production and quenching of steel has a 3500-year history. Even after all this time, the comment made in 1538 (*The Pirotechnia of Vannoccio Biringussio*) that this was an “art comprising great knowledge” is still accurate. It is to this proposition that this book is addressed. We have included examples, tables, charts, illustrations, and references to assist the heat treater in extending his practical art. For the student and researcher, we have also included references and discussions of the direction of current research. It is our belief that more systematic knowledge should be applied to heat treating to transform it from an art into a science. Much work will be needed for this to occur, and we hope this book will help to stimulate the effort.

Heat treating of steel (Chapter 1) is the initial topic of the book. This chapter discusses the transformations that occur during both heating and cooling and the importance of the kinetics of these transformations. The various types of heat treating and the relationships between composition and transformation kinetics and temperatures are presented, as are the various types of steel structures produced. Hardenability (Chapter 2) is reviewed in terms of how it is quantified, how quenchant variables impact on part hardness, and how quench severity is measured.

Chapter 3 focuses on cooling curve analysis. Included are methods of measuring and interpreting cooling curves, critical cooling curve parameters, and the calculation of heat-transfer rates and metallurgical properties from cooling curves.

Chapter 4 deals with the types and performance characteristics of quench oils, and with oil classification by quench severity. The contribution of quench oil additives and the impact of bath variables on cooling rates are discussed. The impact of oil degradation on bath performance is also presented.

Polymer quenchants are examined in Chapter 5. Polymers are classified by chemical structure, and the relationship between structure and cooling mechanism is developed. Bath temperature, polymer concentration, and agitation all have an important impact on quench severity. Polymer dragout and chemical stability are considered.

Quenchant bath maintenance for both oil and polymer quenchants is critical, and is the subject of Chapter 6. The mechanisms of degradation of oils and polymers are presented, and standard tests to track bath quality are described. The importance, interpretation, and impact of the measured properties on quench bath performance are discussed, as are general guidelines for proper bath maintenance.

The topic of spray quenching is covered in Chapter 7. The fundamentals of cooling by the impact of droplets of a volatile liquid are reviewed, as are the mathematics used to describe the heat transfer from the sprayed hot surface, the influence of spray variables on heat transfer, and the results achieved with spray quenching of steel and aluminum parts. Induction heating as well as new techniques such as laser heating are included. Other quenching methods (Chapter 8) include the use of air, water, brine, molten salt, fluidized bed, gas, HIP, and Aus-Bay quenching. The use of ultrasonic, electric, or magnetic fields also can have an impact on the quenching process.

The optimization of bath design (Chapter 9) is critical. General guidelines are given, along with formulas for calculating the proper size of baths, heat exchangers, and pumps. The importance of agitation is discussed, followed by a section on the selection and use of recirculation pumps, jet mixers, forced-air sparging, and impellers. The use of membrane separators in aqueous quench systems is examined. An in-depth discussion of propeller agitation follows (Chapter 10), including agitation design criteria for quench tanks with either open or draft-tube systems. Scaleup correlations for flow and power are given, as are methods for comparing different impeller configurations. The principles of agitation are presented, with guidance in selection of mixer size and type.

We conclude (Chapter 11) with an analysis of the metallurgical basis of residual stress. This includes a discussion of the generation of thermal and transformational stresses. Methods for measuring residual stress and an overview of the classical tests for cracking and distortion are discussed.

G.E. Totten
C.E. Bates
N.A. Clinton

To Alice Totten,
for her untiring dedication
in typing the manuscript

Table of Contents

CHAPTER 1

Introduction to Heat Treating of Steel

Most carbon, low-alloy, and high-alloy steel parts produced today are heat treated before being put into service. Parts are heat treated to enhance particular properties, such as hardness, toughness, and corrosion resistance, and to improve uniformity of properties. The exact heat treatment applied depends on both the type of alloy and the intended service conditions.

In general, heat treatment is an operation involving heating to an elevated temperature followed by controlled cooling to obtain particular microstructures and combinations of properties. The utility of steel lies largely in the relative ease with which properties can be altered by controlling the way parts are heated and then cooled. The changes in properties occur principally because different microstructural constituents are achieved with different cooling rates.

The essential elements of any heat treatment are the heating cycle, holding temperature and time, and the cooling cycle. A heating cycle and three different types of cooling cycles are schematically illustrated in Fig. 1.1. The length of time that a part is held at temperature and the rate at which it is then cooled are important factors. The holding time at temperature should be long enough to complete the desired elevated-temperature microstructural transformations.

Carbon and Low-Alloy Steels

Transformations During Heating

Most carbon and low-alloy steels contain less than 0.8% C. As a consequence, the microstructure is a two-phase mixture consisting of α iron (ferrite) and pearlite. Ferrite is a body-centered cubic (bcc) phase, and pearlite is a mixture of ferrite and cementite (Fe_3C) existing as alternating lamellae. If such a steel is heated to a moderate temperature of about 425 to 650 °C (800 to 1200 °F), internal stresses may be relieved without substantially altering the microstructure or physical properties.

A portion of the iron-carbon phase diagram is illustrated in Fig. 1.2. The iron-carbon diagram is a map that depicts the phases present as a function of carbon concentration and temperature. At carbon concentrations below about 0.83% and temperatures below the eutectoid value of about 720 °C (1330 °F), the stable phases are ferrite and pearlite. At carbon concentrations above 0.83%, the stable phases below 720 °C (1330 °F) are pearlite and cementite.

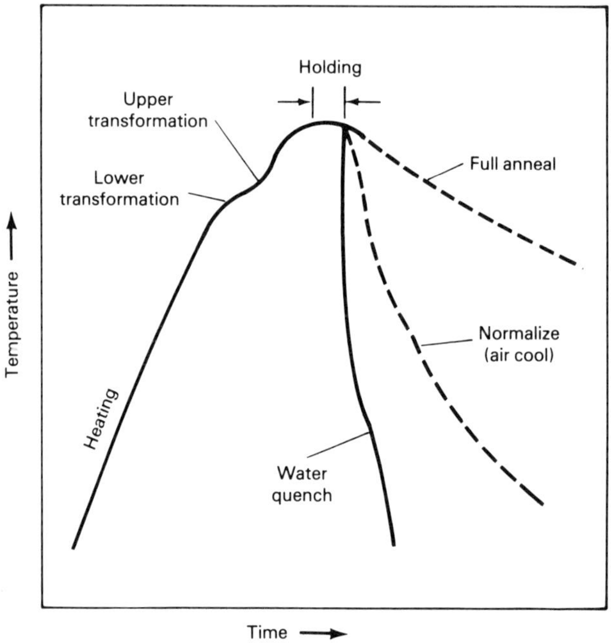

Fig. 1.1 A generalized schematic of the steps in the heat treating of steel

If a steel containing less than the eutectoid carbon concentration is heated to a temperature just above the eutectoid temperature, pearlite decomposes to produce face-centered cubic (fcc) austenite (γ iron), resulting in a mixture of ferrite and austenite phases. If heating is continued to higher temperatures, the ferrite progressively decomposes to form more austenite. At a sufficiently high temperature, the ferrite and any carbides present are completely decomposed to produce a homogeneous austenitic microstructure; this heating process is called solution treating or austenitizing. The upper critical temperature for a steel containing less than 0.83% C is represented by the line G-S in Fig. 1.2.

Heating a carbon steel containing more than about 0.83% C to a temperature above the eutectoid value again results in decomposition of the pearlite to form austenite and produces a structure containing both austenite and cementite. At higher temperatures, the cementite is progressively decomposed until, at a sufficiently high temperature, the structure consists of homogeneous austenite. The upper critical temperature for a steel containing greater than 0.83% C is represented by the line S-E in Fig. 1.2. The lower critical temperature for both hypoeutectoid and hypereutectoid steels is represented by the line P-S-K.

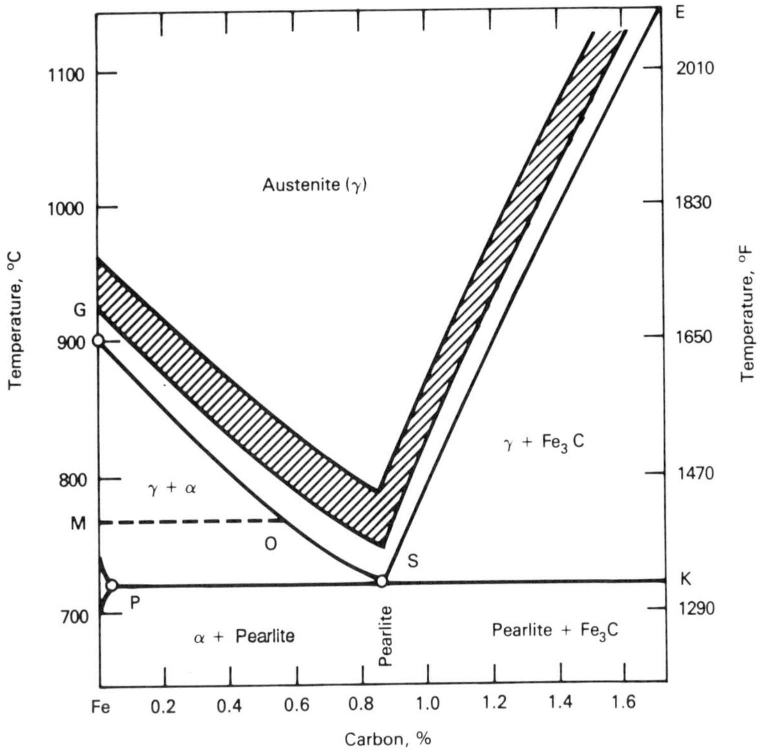

Fig. 1.2 Simplified iron-carbon phase diagram. Shaded band is the annealing or normalizing temperature range.

Transformations During Cooling

If a solution-treated steel containing less than about 0.8% C is slowly cooled from the austenite temperature region, the first phase to precipitate from the austenite is ferrite. This transformation results in a change in the atomic arrangement from the fcc austenite to bcc ferrite. Because ferrite can dissolve only a small amount of carbon, the concentration of carbon in austenite increases during cooling, until at the eutectoid temperature of about 720 °C (1330 °F), the remaining austenite transforms to pearlite. The appearance of ferrite formed on austenite grain boundaries during cooling of a 1045 sheet steel is illustrated in Fig. 1.3. This steel had been normalized at 1095 °C (2000 °F) and cooled in air.

Slow cooling a hypereutectoid steel containing more than 0.8% C is similar to cooling a lower-carbon steel, except that the first constituent to precipitate from austenite is cementite. Cementite usually forms at austenite grain boundaries and forms a network structure that can be observed in the steel at room temperature after the transformation is complete. The appearance of cementite at austenite grain boundaries is illustrated in Fig. 1.4 at a magnification of 1000×. The microstructure illustrated is that of a 52100 steel bar hot rolled between 1175 and 925 °C (2150 and 1700 °F) and air cooled to room temperature. At exactly the eutectoid composition, austenite decomposes into pearlite without the formation of either primary ferrite or primary cementite.

Fig. 1.3 1045 steel sheet, 3 mm (0.13 in.) thick, normalized by austenitizing at 1095 °C (2000 °F) and cooling in air. The structure consists of pearlite (gray) with a network of grain-boundary ferrite (white) and a few plates of ferrite. Picral etch. 500×

Fig. 1.4 Microstructure of 124 mm (4.8 in.) 52100 steel bar, hot rolled between 1175 and 925 °C (2150 and 1700 °F) and air cooled to room temperature. The grain-boundary carbide outlines the primary austenite grain boundaries. The matrix is pearlite. Etched in equal parts of 4% picral and 4% nital. 1000×

Because both ferrite and pearlite formation requires some diffusion of carbon, higher cooling rates reduce the thickness of the ferrite and cementite phases in pearlite. Pearlite in a ferritic matrix of 1045 steel is illustrated in Fig. 1.5. In general, higher cooling rates produce thinner lamellae, which in turn produce a progressive increase in hardness and strength.

At even higher cooling rates, it is possible to avoid the transformation of austenite to ferrite and pearlite in many carbon and low-alloy steels. The transformation of austenite at temperatures below 540 °C (1000 °F), but above about 200 °C (400 °F), produces a structure composed of ferrite and cementite, but the phases are not arranged in a lamellar form. Instead, the ferrite and cementite have a feathery or acicular appearance, and neither constituent can be clearly resolved with an optical microscope. These structures are known as bainite.

When the transformation occurs in the range of about 370 to 540 °C (700 to 1000 °F), bainite has a feathery appearance and is called upper bainite. When the transformation occurs at temperatures in the range of about 200 to 370 °C (400 to 700 °F), the bainite has a more acicular appearance and is called lower bainite. The hardness of bainite depends on the exact structure achieved, but typically ranges from about 45 to 60 HRC. The appearance of bainite in a 1045 steel austenitized for 10 min at 1205 °C (2200 °F) and held for 10 min at 340 °C (640 °F) for partial isothermal transformation is illustrated in Fig. 1.6. After partial isothermal transformation, the steel was cooled in air at room temperature to produce a martensite matrix.

Transformation of austenite at even lower temperatures, below about 200 °C (400 °F), produces a body-centered tetragonal (bct) structure called martensite, which is found in fully

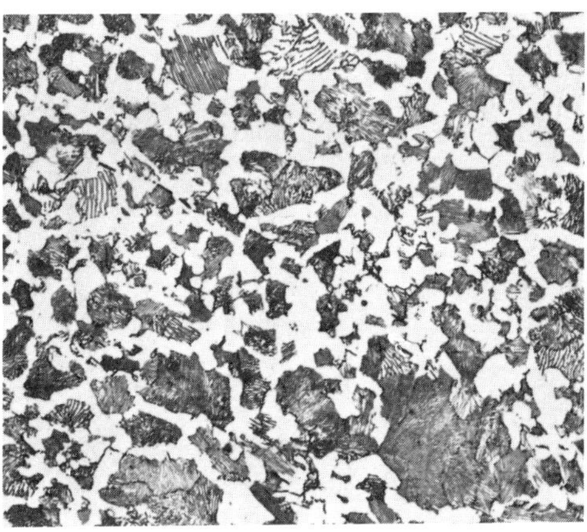

Fig. 1.5 1045 steel sheet, 3 mm (0.13 in.) thick, normalized by austenitizing at 1095 °C (2000 °F) and cooling in air. The structure consists of pearlite (dark gray) and ferrite (light). Picral etch. 500×

hardened steels. Microstructurally, martensite has a needlelike appearance, as illustrated in Fig. 1.7. This 10B35 steel was solution treated at 850 °C (1560 °F) and quenched in agitated water. It was then tempered for 1 h at 230 °C (450 °F). The severe water quench suppressed the formation of ferrite, pearlite, and bainite and produced a fully martensitic structure. Because martensite occupies a greater volume than austenite, steel expands

Fig. 1.6 1045 steel austenitized 10 min at 1205 °C (2200 °F), held 10 min at 340 °C (640 °F) for partial isothermal transformation, and cooled in air to room temperature. Lower bainite (dark) in a matrix of martensite (white). Picral etch. 500×

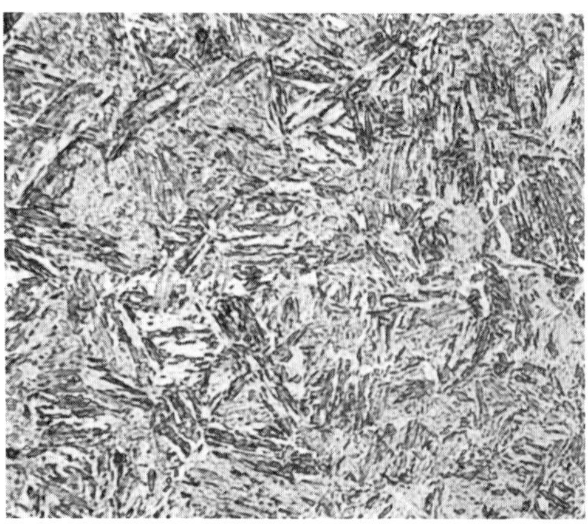

Fig. 1.7 10B35 steel austenitized 1 h at 850 °C (1560 °F) and quenched in agitated water. Tempered 1 h at 230 °C (450 °F). The structure is tempered martensite.

when martensite forms and results in dimensional changes that may cause distortion. In severe situations, distortion may cause parts to crack during quenching.

In some cases, all austenite in a severely quenched steel does not transform to martensite during cooling. When this occurs, martensite needles appear in a matrix containing some austenite, as illustrated in Fig. 1.8 for an AMS 6419 grade steel solution treated for 1.5 h at 860 °C (1575 °F), salt quenched 15 min at 290 °C (550 °F), and then air cooled. The structure consists of tempered martensite, with small amounts of retained (untransformed) austenite. Retained austenite is soft and detracts from hardness when a steel requiring full hardening is desired. Retained austenite can subsequently transform to martensite during machining or after a part is put into service. The expansion that occurs during this transformation creates internal stresses that may cause distortion and loss of tolerance in machined parts. The resulting untempered martensite is brittle and may lead to cracking in service. Consequently, the presence of retained austenite is usually objectionable.

The transformation of austenite to martensite begins at a temperature called the martensite start (M_s) temperature and continues with decreasing temperature to the martensite finish (M_f) temperature. At the M_f, the reaction of austenite to martensite ceases, even though some retained austenite may be present. The M_s and M_f temperatures are influenced by the concentrations of alloying elements present, by carbon concentration, and by alloy segregation that occurred during solidification. The influence of carbon concentration on M_s and M_f is illustrated in Fig. 1.9 (Ref 1).

The M_f temperature in some alloy steels is below room temperature. More complete transformation of austenite in such steels may be obtained by cooling to the M_f temperature, followed by reheating to room temperature and cooling to M_f a second or

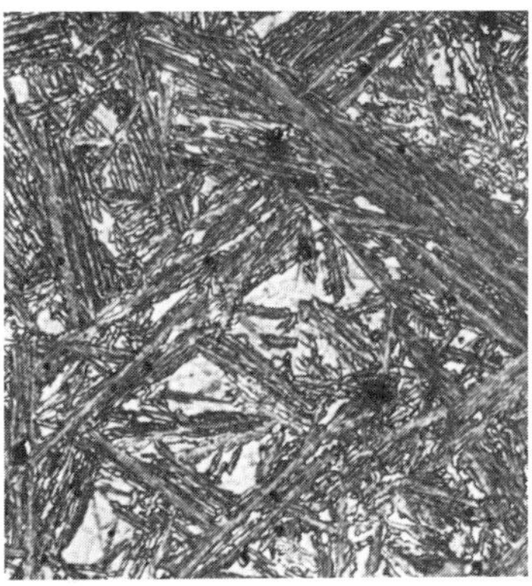

Fig. 1.8 AMS 6419 steel austenitized 1.5 h at 860 °C (1575 °F), salt quenched 15 min at 290 °C (550 °F), then air cooled to room temperature. The structure is tempered martensite, with small amounts of retained austenite. 2% nital etch. 500×

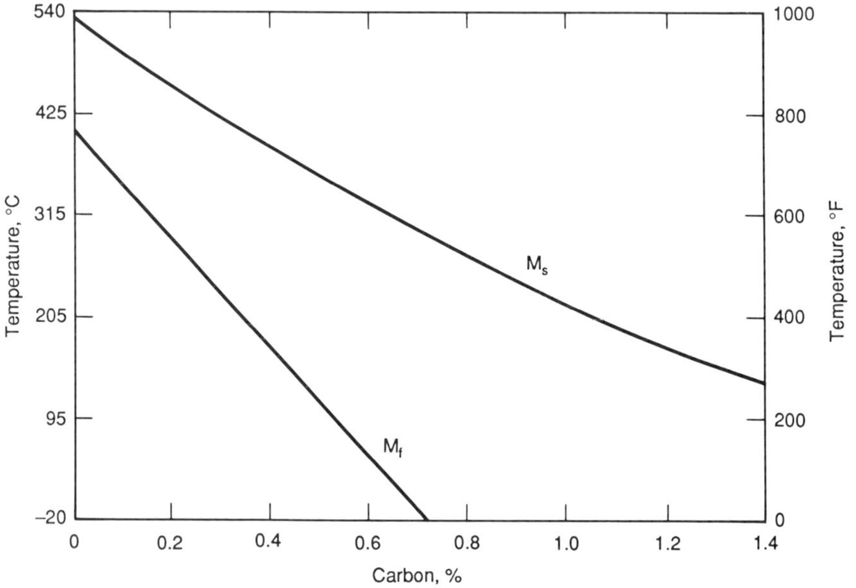

Fig. 1.9 Influence of carbon on M_s and M_f temperatures

third time. With each cooling cycle, a smaller percentage of retained austenite is converted to martensite.

The M_s temperatures of many steels have been determined experimentally and have been approximated using several empirical formulas. One such formula, developed by Grange and Stewart (Ref 2) is:

$$M_s\ (°F) = 1000 - 650 \times \%\ C - 70 \times \%\ Mn - 35 \times \%\ Ni - 70 \times \%\ Cr - 50 \times \%\ Mo \quad \text{(Eq 1.1)}$$

All elemental concentrations are expressed in weight percent, and Eq 1.1 applies when all carbides are dissolved in the austenite. Another empirical formula was reported by Atkins (Ref 3):

$$M_s\ (°C) = 539 - 432 \times \%\ C - 30.4 \times \%\ Mn - 17.7 \times \%\ Ni - 12.1 \times \%\ Cr - 7.5 \times \%\ Mo \quad \text{(Eq 1.2)}$$

The M_f temperature is important when attempting to minimize retained austenite in quenched steels. The M_f temperature ranges from about 165 to 245 °C (300 to 475 °F) below the corresponding M_s temperature. The influence of carbon concentration on the proportion of martensite formed at a given temperature below the M_s is illustrated in Fig. 1.10 (Ref 2).

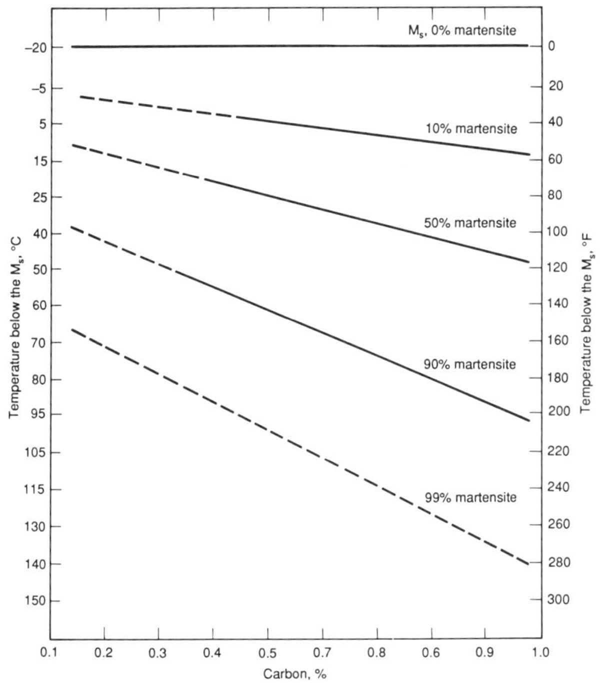

Fig. 1.10 Relationship between carbon concentration in steel and the proportion of martensite formed at a given temperature

The most common method of hardening steel involves continuously cooling from the solution treating (austenitizing) temperature in a liquid that rapidly extracts heat. The rate of cooling in the part depends on the heat-removal characteristics of the cooling medium, the thermal characteristics of the alloy, and the section thickness of the part. Cooling curves superimposed on C curves representing the beginning and end of transformation are illustrated in Fig. 1.11. The points where the cooling curves cross the beginning and ending curves represent the beginning and end of transformations for the indicated cooling rates.

Curve A in Fig. 1.11 represents a slow cooling condition such as in a furnace, with transformation taking place at high temperatures and producing thick, lamellar pearlite. Curve B represents conditions that might be encountered during air cooling or fan cooling that would result in the formation of thinner pearlite lamellae than those produced by cooling condition A. The cooling characteristics illustrated by curve C represent an initial formation of fine pearlite, but with insufficient time in the upper temperature region to allow complete transformation to pearlite. As a consequence, some of the austenite transforms to bainite or martensite during continued cooling.

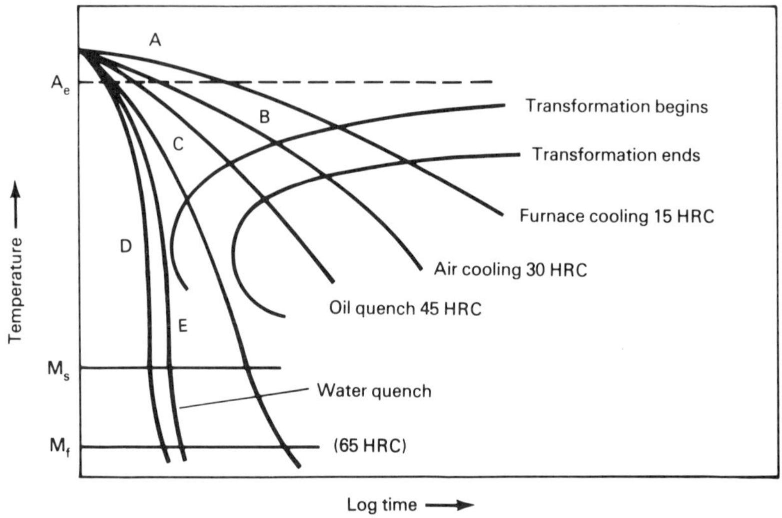

Fig. 1.11 Typical continuous cooling transformation curves. Hardness values are approximate for a 0.55% C steel.

Under this condition, the transformation produces two or more types of phases, such as pearlite and bainite or bainite and martensite, and is called a split transformation.

At a sufficiently high rate of cooling, such as indicated by curve D in Fig. 1.11, transformation does not occur until the Ms temperature is reached and martensite formation begins. Some cooling conditions, such as that indicated by curve E, will produce a fully hardened steel; such a cooling rate is called the critical cooling rate. A cooling rate slower than the critical value will allow formation of some bainite or pearlite and thus will not produce full hardening.

Transformation Diagrams

Transformation diagrams were developed to illustrate the time dependence of microstructural transformations. There are two widely used diagrams for presenting transformation data: the time-temperature-transformation (TTT) diagram, sometimes called an isothermal transformation (IT) diagram, and the continuous cooling transformation (CCT) diagram.

Isothermal Transformation Diagrams. TTT diagrams are developed by solution treating small samples of steel at the proper temperature and then rapidly cooling from the austenitizing temperature to a temperature between the A_3 and the M_s temperatures, holding for a period of time, completing the quench, and analyzing the resulting microstructure for the transformation products.

TTT diagrams for carbon steels containing about 0.20% and 0.30% C are illustrated in Fig. 1.12 and 1.13, respectively (Ref 4). A TTT diagram for a steel containing approximately 0.30% C plus some chromium and molybdenum is illustrated in Fig. 1.14. Although carbon shifts the start of transformation curve to longer times at a given temperature (as can be seen by comparing Fig. 1.12 and 1.13), the unalloyed carbon steels are very difficult to quench to fully hard microstructures because the transformation of austenite to ferrite and pearlite begins at such short times over the temperature range of 425 to 650 °C (800 to 1200 °F). The quick transformation of austenite to ferrite and pearlite makes it virtually impossible to achieve fully martensitic structures except in very thin sections.

The addition of chromium and molybdenum, however, retards the transformation and allows hardening to occur with a less severe quench. The TTT curve for 4130 steel containing chromium and molybdenum, illustrated in Fig. 1.14, is usually interpreted to mean that the steel must be cooled past 540 °C (1000 °F) and to the start of the martensitic transformation (M_S) in less than about 1.5 s to produce a fully martensitic structure and to the M_S in about 10 s to produce a 50% martensitic structure. Slower

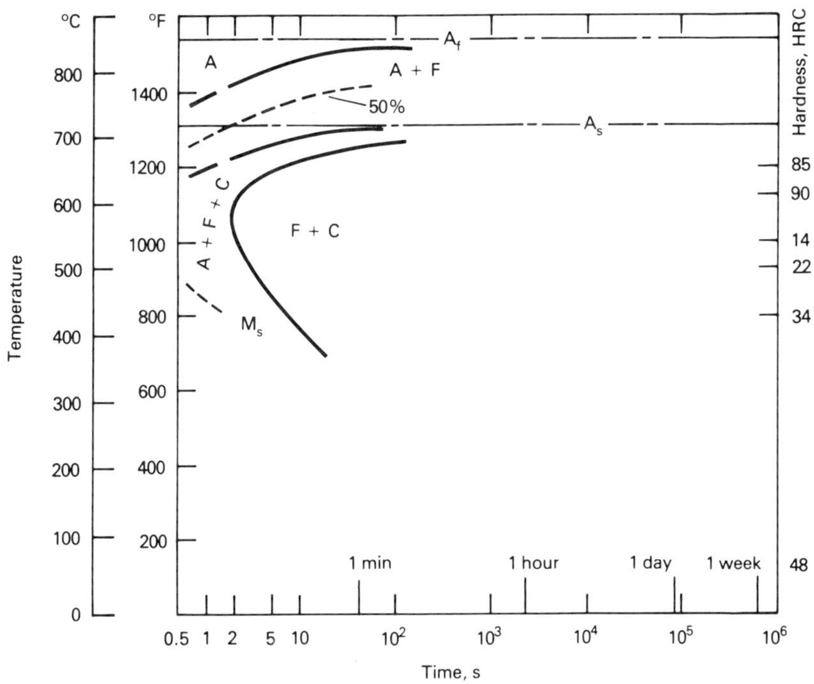

Fig. 1.12 TTT diagram for 1021 carbon steel

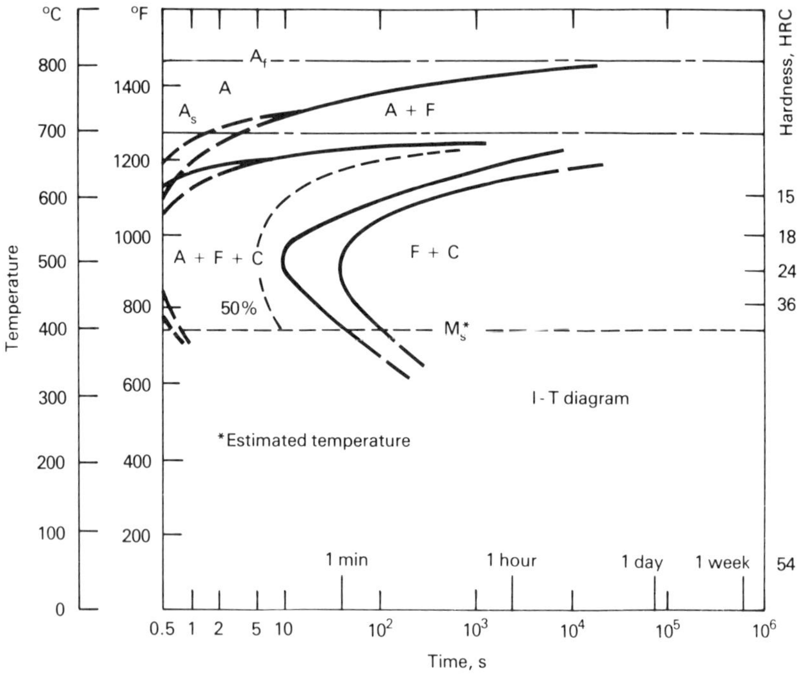

Fig. 1.13 TTT diagram for 1027 carbon steel

cooling rates may result in less than 50% hardening. Several books containing TTT diagrams are available (Ref 4, 5).

Continuous Cooling Diagrams. The second method of studying and reporting transformations involves the examination of samples of steel that are continuously cooled at specified rates. CCT curves provide data on the temperature regime of each phase formation, the amount of transformation occurring at a given cooling rate with time, and the cooling rate necessary to produce martensite.

A TTT diagram with a superimposed CCT diagram is illustrated in Fig. 1.15 (Ref 6). Cooling curves associated with selected positions on a Jominy bar are also superimposed.

CCT diagrams have several characteristics not seen in TTT diagrams, such as a depressed M_S temperature at slow cooling rates and the prevalence of bainite in the microstructure. CCT diagrams are particularly helpful when devising heat treatments for high-hardenability, thick-section steels (Ref 6-10).

The CCT diagram for a carbon steel containing approximately 0.18% C is illustrated in Fig. 1.16 (Ref 3). This diagram also indicates the difficulty of achieving full hardening

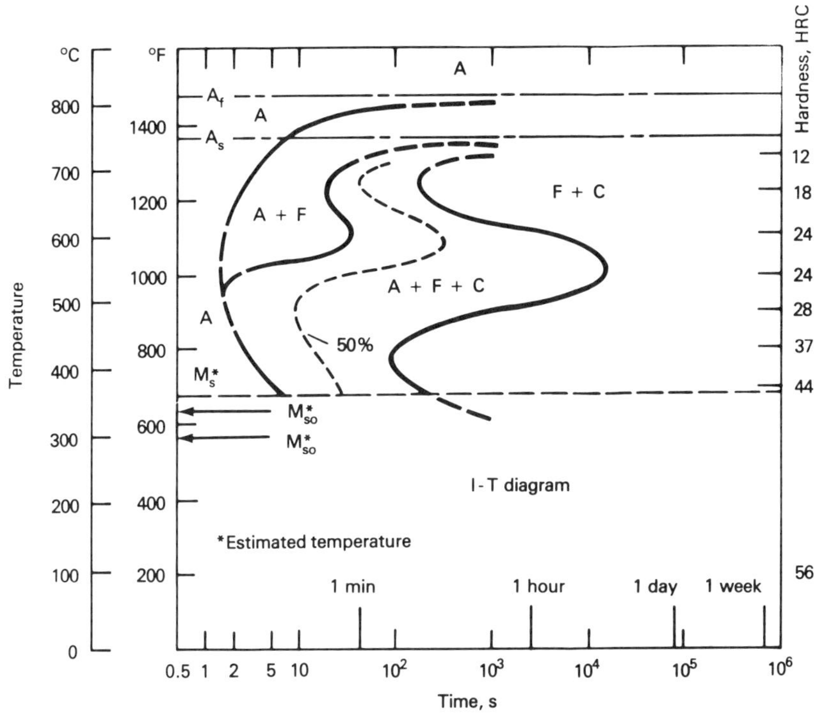

Fig. 1.14 TTT diagram for 4130 carbon steel

in an unalloyed low-carbon steel and suggests that cooling rates in excess of 10,000 °C/min (18,000 °F/min) are required to achieve full hardening.

The CCT diagram for a chromium-molybdenum (AISI 4130) steel, illustrated in Fig. 1.17, shows how the required cooling rates for hardening are reduced with these alloy additions (Ref 3). Achieving a 50% martensitic microstructure requires a cooling rate of about 70 °C/s at 700 °C (120 °F/s at 1300 °F), and a fully martensitic microstructure requires a cooling rate above 170 °C/s at 700 °C (300 °F/s at 1300 °F).

The discrepancies in the required cooling rates needed to produce 50% martensite in 4130 steel, as determined from the TTT diagram (Fig. 1.14) and from the CCT diagram (Fig. 1.17), are principally a result of the different methodologies involved and possibly differences in composition and grain size of the steels used to develop the transformation curves.

These and other published diagrams present approximate cooling rates needed during the quenching of various alloys to obtain desired microstructures (Ref 4, 11). The actual rates required in parts made of different alloys depend on the specific compositions and grain sizes of the steels being quenched.

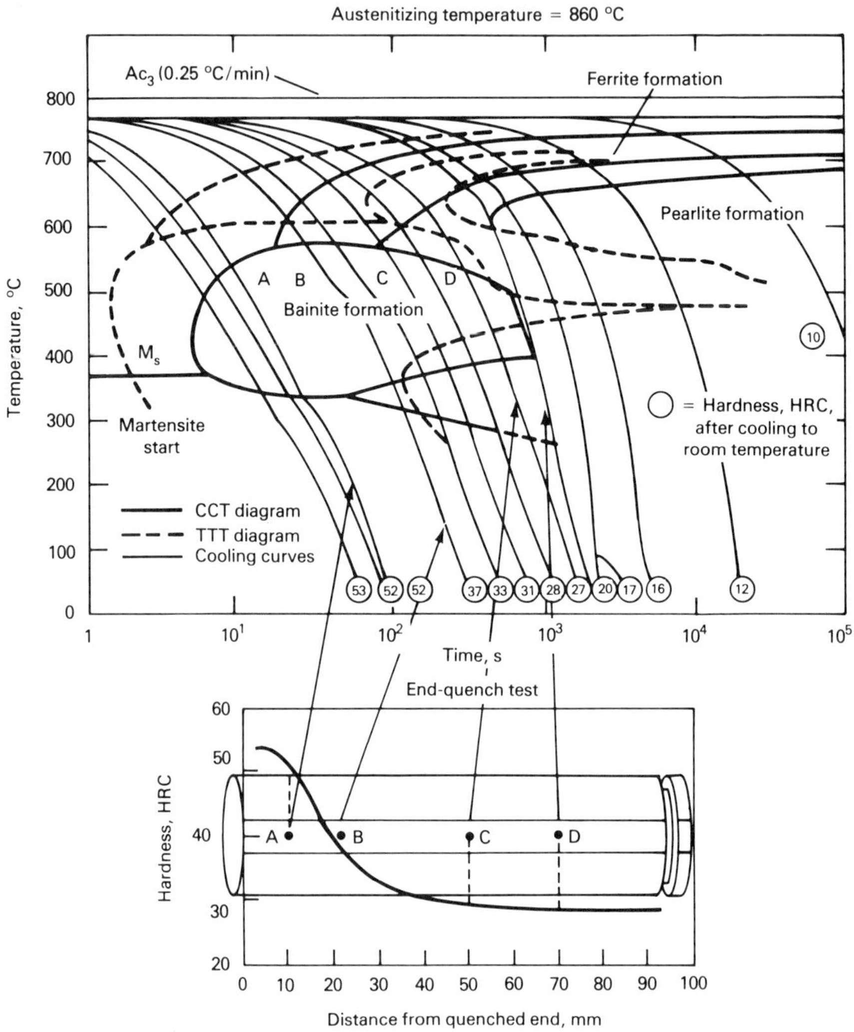

Fig. 1.15 Experimentally determined CCT diagram (solid lines) for steel with German designation 42 CrMo 4 (0.38% C, 0.99% Cr, and 0.16% Mo). TTT diagram is also shown (dashed lines).

Types of Heat Treatments

It is apparent from the foregoing discussion that a range of structures and properties can be achieved in steel by controlling the composition, solution-treating temperature, and conditions used to cool the steel from the austenitizing temperature back to room

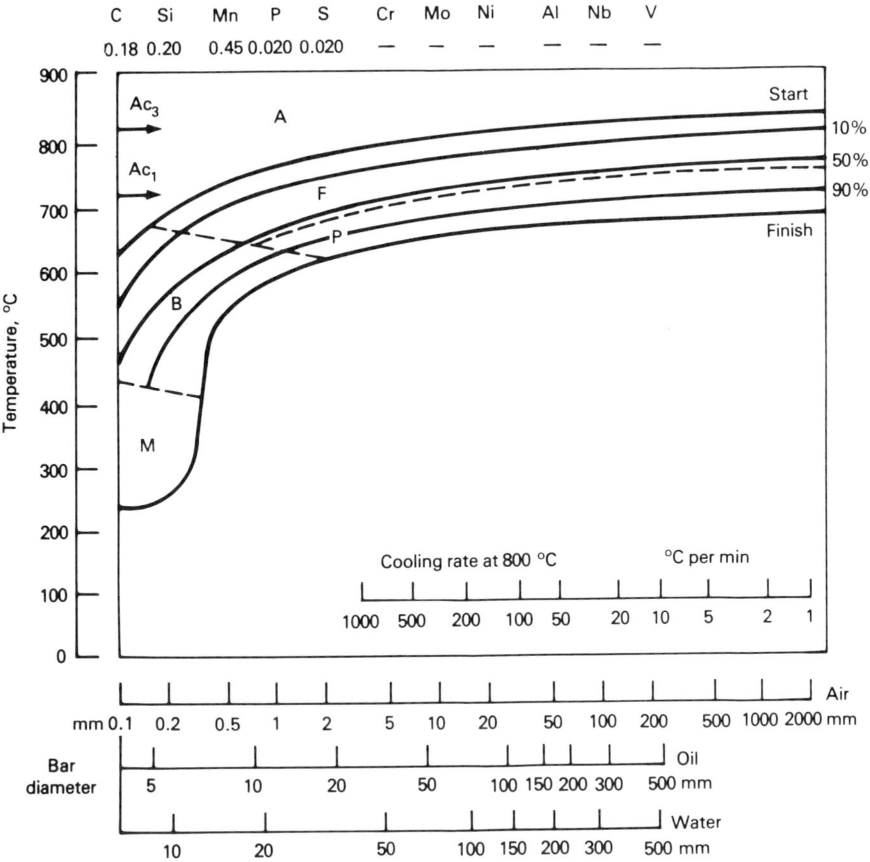

Fig. 1.16 CCT diagram for 1020 steel

temperature. This section will describe several heat treatment cycles for achieving particular microstructures and properties.

Stress relief refers to the process of heating steel to a temperature in the range of 425 to 650 °C (800 to 1200 °F) to relieve internal stresses without substantially altering the microstructure or properties of the part. Little or no stress relief occurs below 260 °C (500 °F), as illustrated in Fig. 1.18, but about 90% of the stress is removed at 540 °C (1000 °F) (Ref 12). Sometimes a part is heat treated and later stress relieved after operations such as welding, induction hardening, or grinding. The maximum temperature for stress relief is limited to about 30 °C (50 °F) below the tempering temperature used after quenching. Higher stress-relief temperatures cause softening.

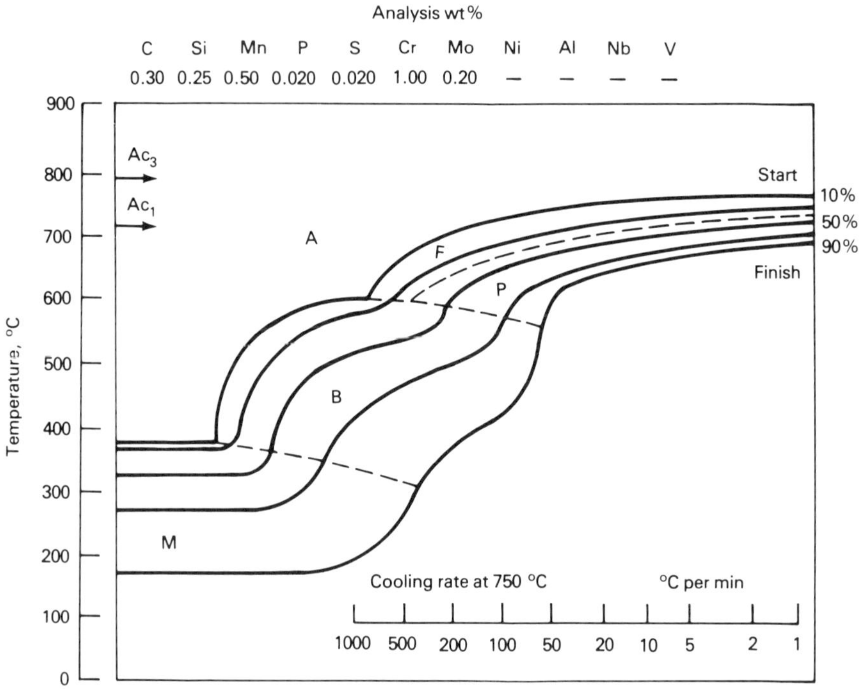

Fig. 1.17 CCT diagram for 4130 steel

Solution treating, or austenitizing, is an elevated-temperature process intended to achieve a uniform carbon distribution in the austenite. The objective of the heat treatment is to dissolve second phases, such as carbides, and produce a homogeneous structure with uniformly distributed carbon.

Homogenization treatments are usually the same as solution treatments and are intended to achieve a uniform distribution of carbon in austenite. In some circumstances, homogenization refers to efforts to achieve a homogeneous distribution of all elements in the austenite. Homogenization temperatures may range from 1095 to 1260 °C (2000 to 2300 °F) or higher in efforts to achieve uniform distributions of nickel, chromium, molybdenum, and carbon in the matrix. However, the diffusion coefficients for these elements are so low that the time required is long. The results do not generally warrant the costs associated with extended homogenizing treatments.

Annealing refers to heating and cooling operations applied to induce softening. The parts are generally solution treated and cooled slowly, usually in a furnace, to produce soft microstructural constituents. The purpose of annealing is to soften the metal, increase its ductility, relieve internal stresses caused by previous treatments, and enhance machinability. In full annealing, a steel is heated to at least 55 °C (100 °F) above the

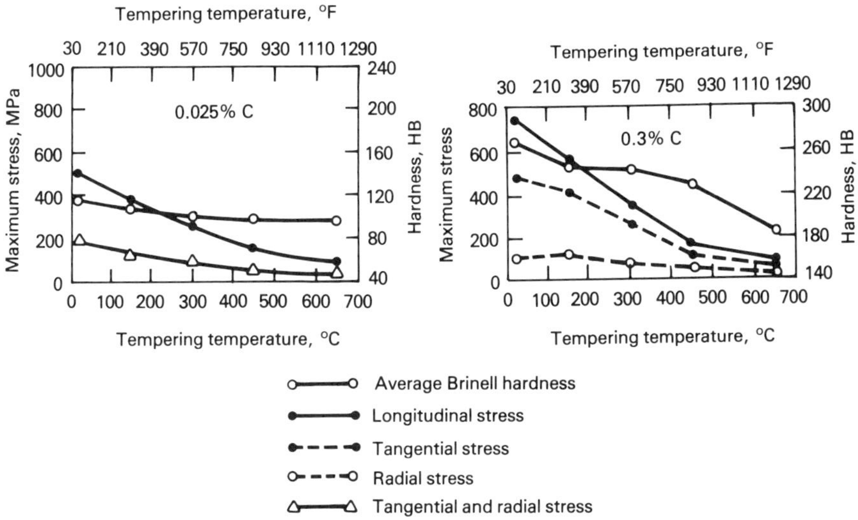

Fig. 1.18 Effect of tempering on residual stresses in quenched cylinders

upper critical temperature, as illustrated in Fig. 1.2, and held for sufficient time to achieve a uniform distribution of carbon in austenite. This is followed by slow cooling, usually in a furnace. When parts have cooled to below the lower critical temperature, the transformation of austenite is usually complete. The parts can then be removed from the furnace and air cooled. Furnace cooling to lower temperatures wastes furnace time and requires that the furnace be reheated for the next load.

The strength of annealed parts is low, but the ductility is high. Some low-alloy steels are annealed for machinability and given a final heat treatment after rough machining.

Annealing treatments can be modified for specialized purposes—for example, to achieve spheroidized pearlite structures. Spheroidized pearlite has excellent machinability, but the thermal cycle is long and expensive and should not be specified unless the last increment in ductility and machinability is required.

Normalizing refers to solution treating a steel at a temperature at least 55 °C (100 °F) above the upper critical temperature, followed by cooling in air. The normalizing temperature depends on the carbon concentration, as illustrated in Fig. 1.2. The purpose of normalizing is usually to refine the grain structure and possibly harden the steel slightly.

Normalized parts must be placed on trays so air can circulate freely around every part as the load is cooled. If air flow is restricted, the operation will be closer to annealing. Accelerated cooling by fans or forced airflow may produce a result more like quenching.

The microstructure produced by normalizing is a mixture of ferrite and pearlite, usually resulting in low residual stresses and almost no distortion. Some parts are tempered after being normalized to achieve a slight additional softening and stress relief. Machinability is usually good because of the uniform structure obtained.

Quench Hardening. It is possible to accelerate cooling from the solution-treating temperature and control the transformation of austenite to bainite and martensite to achieve higher strengths and hardnesses than obtained with annealing or normalizing treatments. In hardening by quenching, the work is austenitized at the temperatures illustrated in Fig. 1.19 and then cooled fast enough that pearlite and ferrite do not have time to form. A quenching cycle is illustrated schematically in Fig. 1.20(a).

The maximum attainable hardness of a quenched steel is controlled almost exclusively by the carbon concentration and is obtained by cooling at a rate equal to or greater than the critical cooling rate for the alloy. Quenching media include water, brine solutions, oil, water-polymer solutions, and, in some cases, inert gases.

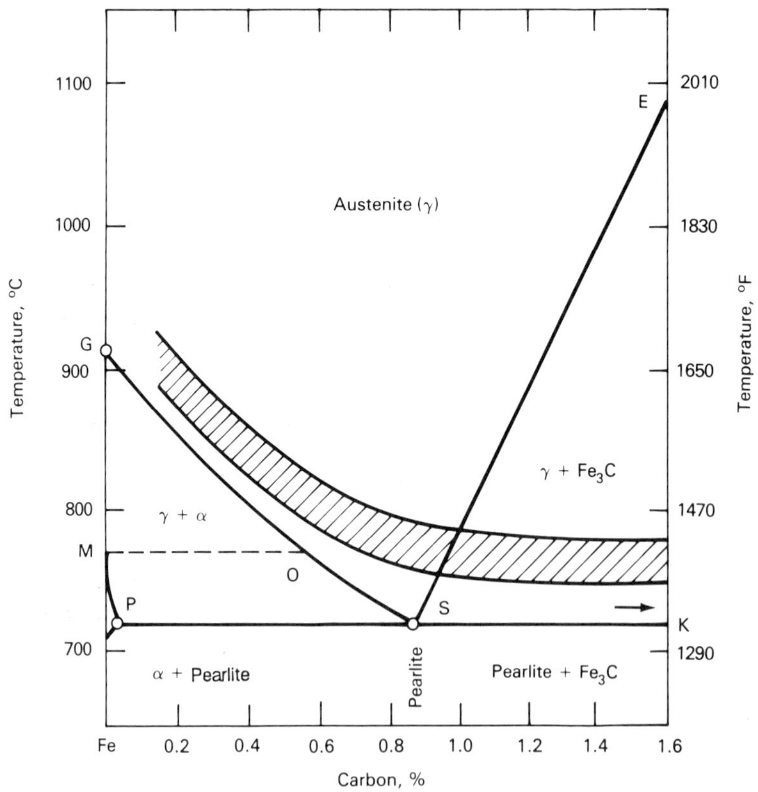

Fig. 1.19 Simplified iron-carbon diagram. The shaded band represents the hardening temperature range for carbon steels.

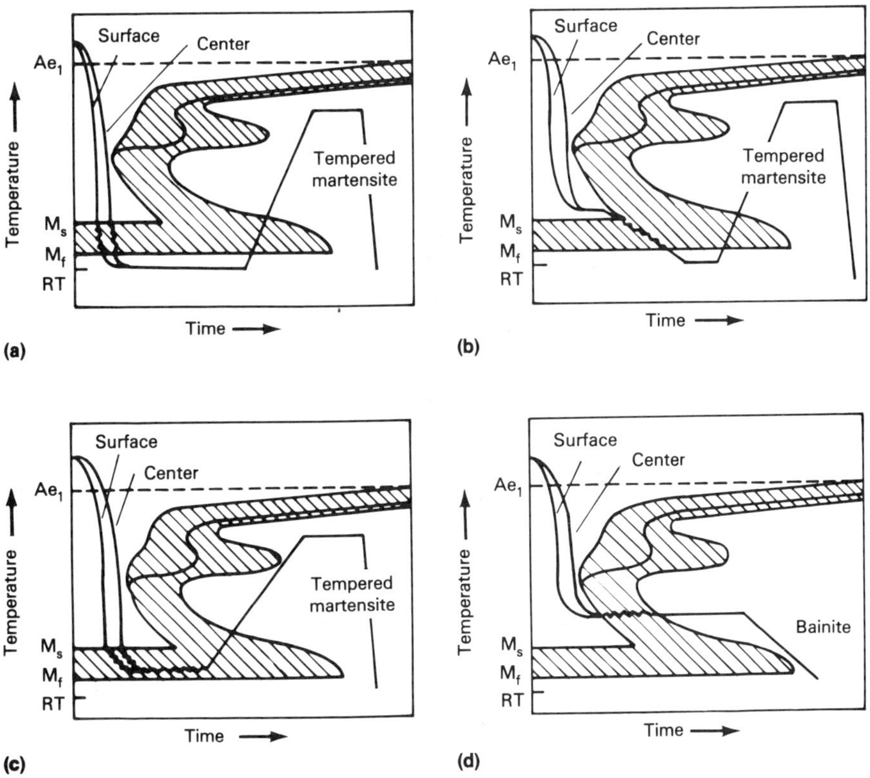

Fig. 1.20 Types of quenching cycles. **(a)** Conventional quenching and tempering. **(b)** Martempering. **(c)** Isothermal quenching and tempering. **(d)** Austempering

Water and oil are the media most commonly used for quenching steel. Water is used whenever possible, but high-carbon and deep-hardening steels require oil quenching. Some complicated shapes require oil quenching to minimize quench cracking. Oil cools steel more slowly than water under almost all conditions.

Certain organic polymers can be added to water to provide a quenching solution that resembles oil in its heat-removal characteristics. The main advantage of these solutions is that they remove heat more slowly than water, without the fire hazard of oil. The disadvantage is that they require careful concentration, temperature, and agitation control to achieve consistent quench behavior. The quench severity provided by these baths can vary widely with the type of polymer, polymer concentration, bath temperature, and extent of agitation during quenching.

Interrupted or Slack Quenching. In some situations it is desirable to quench steel in water or brine to produce a high surface hardness. However, water or brine quenching may cause cracking or excessive distortion if used to harden parts fully. If it is not necessary to harden the part throughout the cross section, an interrupted quench may be used. Interrupted or slack quenching usually involves quenching in water for a predetermined time and then transferring the part to an oil bath to complete the transformation. This is not a common procedure, but it can be used under certain circumstances where surface hardness is desired and complete hardening is not required. Quenching of carbon and low-alloy steels is always followed by tempering.

Tempering is the process of heating a hardened steel to below the lower critical temperature to achieve some softening, and then cooling the steel back to room temperature. The objective of a tempering treatment is to relieve some internal stress and to reduce hardness, thereby creating higher ductility than is possible in as-quenched parts. Tempering slightly modifies the structure of martensite, and this change can be used to adjust strength, hardness, toughness, and other mechanical properties to specified levels.

Carbon and low-alloy steels are tempered in the temperature range of 175 to 700 °C (350 to 1300 °F). The holding time may vary from 30 min to several hours. A longer tempering time at a given temperature, or a higher temperature for a given time, increases the extent of softening. Martensite softens more than pearlite at a given tempering temperature, and the steel composition affects the rate of tempering. Some effects of temperature are illustrated in Fig. 1.21, where hardness is used to measure the response to tempering. The presence of carbide-forming elements causes steel to be more resistant to tempering.

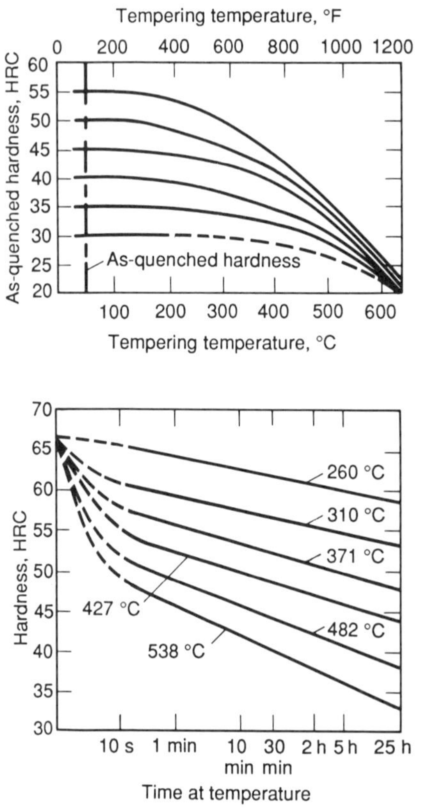

Fig. 1.21 Effect of tempering temperature and time on the softening characteristics of steel

Tempering below 595 °C (1100 °F) may cause temper embrittlement in certain steels. The decrease in notch toughness caused by temper embrittlement is illustrated in Fig. 1.22 for a Ni-Cr-Mo AISI 4320 steel tempered in the range of about 260 to 370 °C (500 to 700 °F) (Ref 13). Tempering in the temper-embrittlement range usually is avoided, and when higher tempering temperatures are used, the work may be quenched from the tempering temperature to minimize the time in the embrittling zone.

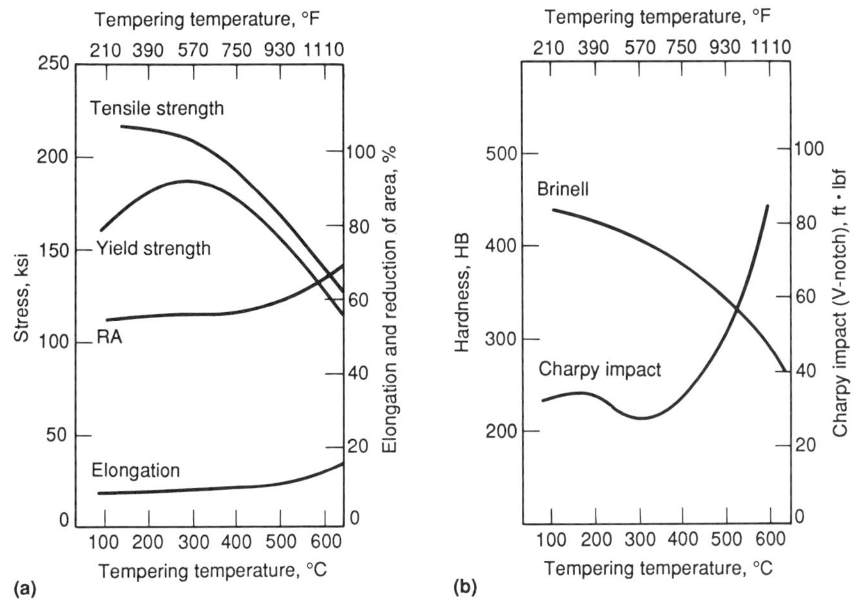

Fig. 1.22 Loss in room-temperature toughness caused by temper embrittlement of oil-quenched wrought Ni-Cr-Mo (4320) steel (13 mm, or ½ in., section)

Table 1.1 Temperature range of martensite formation for several carbon and low-alloy steels

Steel SAE No.	Martensite start (M_s) °C	Martensite start (M_s) °F	50% martensite formed °C	50% martensite formed °F	99% martensite finish (M_f) °C	99% martensite finish (M_f) °F
1030	343	650	293	560	232	450
1065	277	530	219	425	149	300
1090	219	425	157	315	82	180
1335	343	650	293	560	232	450
2340	304	580	293	560	207	405
3140	335	635	288	550	221	430
4130	380	715	343	650	288	550
4140	343	650	299	570	227	440
4340	288	550	249	480	188	370
4640	343	650	299	570	249	480
5140	343	650	299	570	232	450
6140	327	620	293	560	232	450
8630	366	690	332	630	277	530
9440	330	625	282	540	207	405

Martempering. In the usual quench hardening process, parts are cooled at a rate greater than the critical rate to form martensite. In some parts prone to cracking or distortion, however, it is desirable to cool faster than the critical rate, and then interrupt cooling above the M_S temperature until the temperature becomes uniform across the section. The parts are then cooled to room temperature in air. This cycle is called martempering and is schematically illustrated in Fig. 1.20(b). The martemper cooling cycle favors uniform martensite formation throughout the section and helps minimize distortion and cracking.

In martempering, the work is usually quenched in a molten salt bath to a temperature just above the M_s and held long enough for the temperature to equalize. The work is then removed from the salt and air cooled. Martensite forms during cooling from the M_s to the M_f. Table 1.1 lists M_s and M_f temperatures for several carbon and low-alloy steels (Ref 14).

In modified martempering, the work is quenched into a bath at a temperature between the M_S and M_f temperatures. When the temperatures at the surface and the center of the work have equalized, the work is removed from the bath and either air cooled or placed in a tempering furnace. This thermal cycle is schematically illustrated in Fig. 1.20(c).

Most steels that can be oil quenched can be martempered. The maximum section size that can be through-hardened by martempering is appreciably less, however, than the section size that can be hardened by conventional quenching.

Austempering. The austempering process requires that a steel be cooled from the solution-treating temperature sufficiently quickly to miss the nose of the C curve. Cooling is stopped, and the temperature is held below the nose of the C curve and above the M_S to produce a bainitic structure. This cooling cycle eliminates the formation of pearlite as well as of martensite. Minimizing the volume expansion associated with martensite formation helps eliminate cracking, but some sacrifices are made in hardness and toughness compared with martensitic structures. The austempering cycle is schematically illustrated in Fig. 1.20(d). Upper or lower bainite can be formed, depending on the temperature at which the part is transformed.

Age Hardening. The age hardening reaction results from the precipitation of copper from a supersaturated solid solution or from the formation of certain intermetallic compounds, such as nickel aluminide and nickel titanium compounds. These reactions are favored by holding a steel containing these elements in the temperature range of about 480 to 595 °C (900 to 1100 °F) after parts have been solution heat treated and quenched.

Age hardening has some advantages, because the precipitation occurs uniformly across the section and results in a minimum amount of distortion. All age hardening steels require an initial solution treatment to dissolve the hardening compounds and elements prior to quenching and aging. Age hardening temperatures are sufficiently low that a minimum amount of surface discoloration or oxidation occurs during the hardening process.

Surface Hardening Processes

Several heat treating processes, known as surface hardening or case hardening processes, are used for producing hard, wear-resistant surfaces on steel parts while maintaining a tough core. The processes include carburizing, cyaniding, carbonitriding, nitriding, laser hardening, electron beam (EB) hardening, and flame hardening. The surface hardness obtained during carburizing, carbonitriding, and cyaniding depends on heat treatment after the composition of the case has been changed by a high-temperature diffusion process. Nitriding alters the composition of the case to form compounds that are inherently hard. Surface hardening by lasers, electron beams, or flames depends entirely on heat treating the surface of a hardenable steel by the local application of heat followed by quenching.

Quenching processes for surface-hardened materials are similar to those for the base steels. The hardenability of the surface is increased by the addition of carbon and nitrogen. The core, which represents the majority of the material in the part, is unaffected.

The process of carburizing involves heating steel samples to an elevated temperature in an atmosphere having a higher carbon potential than the surface of the steel. At the elevated temperatures, carbon passes from the atmosphere into the steel surface. This produces a carbon concentration gradient, with the carbon concentration decreasing with distance from the surface.

There are several methods of carburizing, but gas carburization employing hydrocarbon gases is by far the most common. Pack carburization employing solid carburizing materials and liquid baths employing fused salts are also used under selected circumstances. The carbon solubility of steel is higher above the Ac_3 temperature, and, as a consequence, carburization takes place most rapidly above that temperature. It is common to select a carburization temperature at least 55 °C (100 °F) above the Ac_3 temperature to promote the absorption of carbon into the surface and increase the rate of diffusion of carbon from the surface toward the core.

After carburization, the outer surface usually consists of hypereutectoid steel. Heat treatment of the resulting carburized steels depends on the actual temperature used to carburize, the carbon concentration in the surface, and the required properties in the resulting part. If the carburization temperature has been just above the Ac_3 temperature, the heat treatment may consist of simply quenching the steel directly from the carburization temperature into a bath. This treatment is common when the surface carbon concentration is below the eutectoid composition. If the carbon concentration equals or exceeds the eutectoid concentration, the steel may be treated by cooling to room temperature and then reheating to a temperature slightly above the Ac_3, quenching, and then tempering. In general, the second cycle produces the best microstructural uniformity and combination of properties and is commonly applied when the carbon concentration in the case is above the eutectoid concentration.

Oil quenchants are commonly used with carburized parts. However, water-base polymer quenchants are being increasingly used where tight door seals between the quench chamber and the carburizing chamber can be maintained.

Carbonitriding is a variation on the gas carburization process in which an addition of anhydrous ammonia is made to the furnace atmosphere to cause both carbon and nitrogen to be absorbed into the surface of the steel. Nitrogen and carbon both increase the hardenability of the steel at the surface and lower the critical temperature. The increased surface hardenability sometimes lowers the quench severity needed to harden the steel.

Nitriding is a case hardening process in which nitrogen is diffused into the surface to form hard, wear-resistant nitrogen compounds by reactions with elements contained in the steel. The required core properties must be developed prior to the application of the nitride layer. The desired core structure is usually quenched and tempered martensite.

Cyanide treatments are used both to carburize and to add nitrogen to the surface of steel parts. The steel is immersed in a bath of molten sodium cyanide for periods of up to 3 h. During this time, nitrogen and carbon diffuse from the bath into the steel surface. It is necessary to quench directly from the cyanide bath into oil or water to obtain high surface hardnesses.

Several other processes are used to achieve localized high surface hardnesses, including flame hardening, induction hardening, laser surface hardening, and electron beam hardening. Each of these processes increases the surface hardness by elevating the temperature to produce austenite and then requenching the surface to produce martensite. No attempt is made to increase the concentration of the hardening elements during these treatments. The hardenability is controlled by the hardenability of the base metal.

Flame hardening consists of heating the surface of the steel with an oxyacetylene or oxyhydrogen torch and then immediately quenching in water. The result is a surface layer consisting of martensite over a softer interior core structure.

Induction hardening is done by placing a steel part in a magnetic field generated by a high-frequency alternating current passing through the inductor. The depth of heating produced by the induction field is related to the frequency of the field and the duration of the cycle. When the surface of the steel has reached the appropriate temperature, spray jets are usually used to quench the surface to produce a fully martensitic structure.

Laser surface heat treatment is used to harden localized areas of steel. The heat is generated by the absorption of laser energy at the surface. The localized high temperatures transform the steel to austenite, which then self-quenches to martensite to produce the hardened surface. It is important that the steel not be melted during laser treatment if high surface hardnesses are to be achieved.

Electron beam hardening of steel is similar to laser hardening, except that a high-energy electron beam is used to produce locally high surface temperatures. Like laser beam hardening, the EB process eliminates the need for quenchants, but requires a sufficient workpiece mass to produce self-quenching.

High-Strength Low-Alloy (HSLA) Steels

This family of materials has good strength and low-temperature toughness, combined with excellent weldability. The carbon concentration in HSLA steels is kept low to promote toughness and weldability. Strength is maintained by the addition of manganese for solid-solution strengthening and by the addition of microalloying elements, such as vanadium and/or niobium, that form carbonitride precipitates during heat treatment.

During the heat treatment of HSLA cast steels, both the conventional phase transformations and the precipitation of carbonitrides take place. During the initial austenitizing heat treatment, the material is transformed to austenite and carbonitride precipitates are dissolved. Solution-treating temperatures as high as 1040 °C (1900 °F) may be necessary to dissolve niobium carbonitrides.

During air cooling or quenching from the austenitizing temperature, a ferrite plus pearlite matrix or a martensite matrix is developed, but the precipitation of carbonitrides is prevented by the rapid cooling. During tempering (aging), the matrix hardness decreases slightly because of conventional tempering reactions, but strengthening occurs as carbonitride precipitates form. The net result is that the tempering results in a lesser decrease in hardness than occurs in conventional steels. The hardness changes observed during the tempering (aging) of an HSLA steel are shown in Fig. 1.23 for different heat treatments and tempering temperatures (Ref 15). Good impact toughness

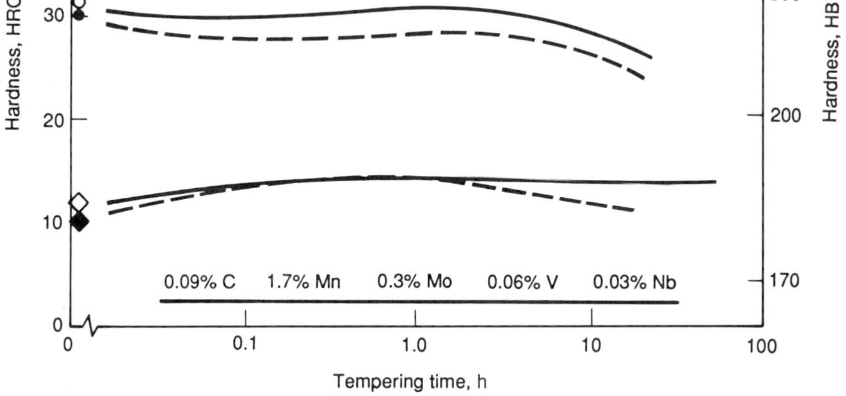

Fig. 1.23 Hardness changes observed as a function of tempering time and temperature for normalized and tempered (N & T) and quenched and tempered (Q & T) heat treatments after austenitizing at 950 °C (1740 °F)

is achieved for these materials when the material is tempered to a slightly overaged condition. In particular, special intercritical heat treatments have been observed to increase the toughness of the HSLA cast steels without sacrificing strength (Ref 16).

Austenitic Manganese Steels

The purpose of heat treating manganese steel is to produce a uniform austenite structure supersaturated with dissolved carbon. In this condition, the steel is wear resistant because it is very work hardenable on the wear surface, but maintains a ductile and tough core. Austenitic manganese steels are commonly solution treated at temperatures between 1050 and 1075 °C (1925 and 1975 °F) for best toughness, although higher temperatures are necessary for the higher-carbon and higher-alloy grades (Ref 17).

Manganese steel should be held at the austenitizing temperature long enough to ensure complete carbide dissolution. After complete solution treatment, parts are rapidly cooled in highly agitated quenchants until the parts are near room temperature.

TTT curves for two manganese steel compositions are shown in Fig. 1.24 (Ref 18). Manganese steels are hardenable if properly solution treated, but they must be rapidly quenched, primarily to minimize grain-boundary carbide precipitation, which can reduce toughness.

Martensitic Stainless Steels

Martensitic and higher-alloy stainless steels are solution treated similar to carbon, low-alloy, and manganese steels. The solution treatment achieves a uniform carbon distribution in the austenite. Producing a "stainless" steel with the ability to resist corrosion depends on obtaining a uniform distribution of chromium in the matrix. Chromium concentrations above about 12% result in a passive surface that maximizes corrosion resistance. However, precipitation of chromium carbides at grain boundaries can lead to a significant loss in corrosion resistance because of the associated chromium depletion along the grain boundaries.

Whether the microstructure of a stainless steel is ferritic, martensitic, or austenitic is determined principally by the chromium, nickel, and carbon concentrations. The chromium steels are either martensitic or ferritic, depending on the chromium and carbon concentrations.

The chromium steels are predominantly martensitic when the following inequality is met:

$$\%\mathrm{Cr} - (17 \times \%\mathrm{C}) < 12.5\% \quad \text{(Eq 1.3)}$$

The amount of ferrite in chromium steels increases as the chromium concentration increases or as the carbon concentration decreases.

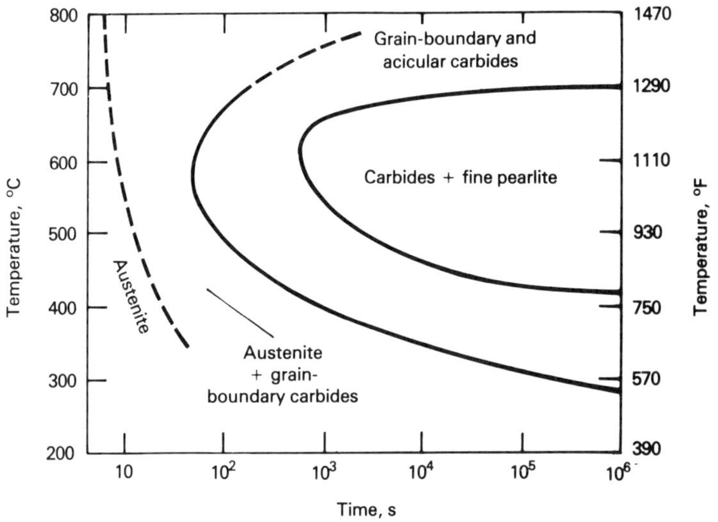

(a)

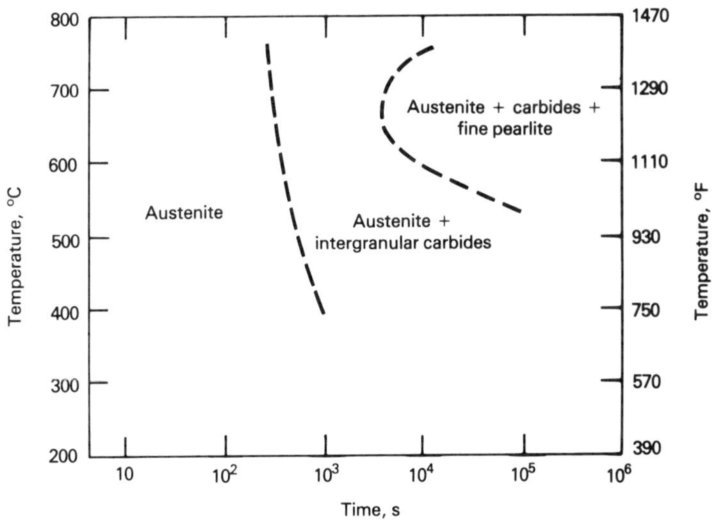

(b)

Fig. 1.24 TTT curves for manganese steel. (**a**) Steel contained 1.28% C, 12.4% Mn, 0.35% Si, 0.009% S, 0.030% P, 0.28% Ni, and 0.23% Cu. Austenitized 30 min at 1050 °C (1920 °F). Grain size 5 to 6. (**b**) Steel contained 1.2% C, 14.0% Mn, 0.07% S, and 2.0% Mo. Austenitized at 1100 °C (2010 °F)

It is desirable to solution treat welded components to eliminate grain-boundary carbides that may have formed either during welding or during cooling of parts to room temperature. The object is to dissolve second phases and produce homogeneous structures. A quench or other accelerated cooling process is usually required to retain carbon in solution and prevent precipitation of carbides.

The hardenability of the 12% Cr martensitic stainless steels is high, as illustrated by the TTT curve shown in Fig. 1.25 (Ref 19). The presence of excessive amounts of ferrite in the microstructure can result in some loss of hardenability, as illustrated in Fig. 1.26 (Ref 20, 21).

Austenitic Stainless Steels

Austenitic stainless steels are also solution heat treated to optimize corrosion resistance. The objective of the heat treatment is to dissolve carbides and produce a homogeneous chromium solid solution. Accelerated cooling is usually required to retain

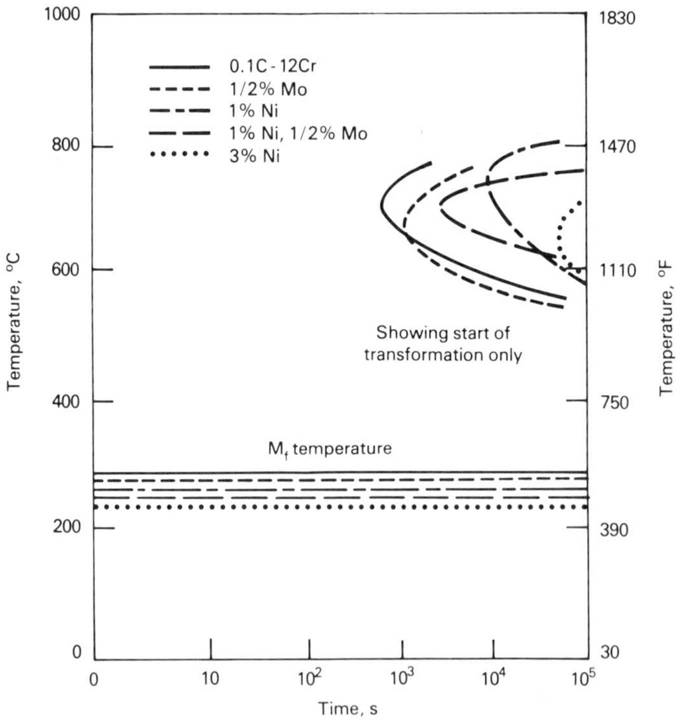

Fig. 1.25 TTT diagram for 12% Cr steels

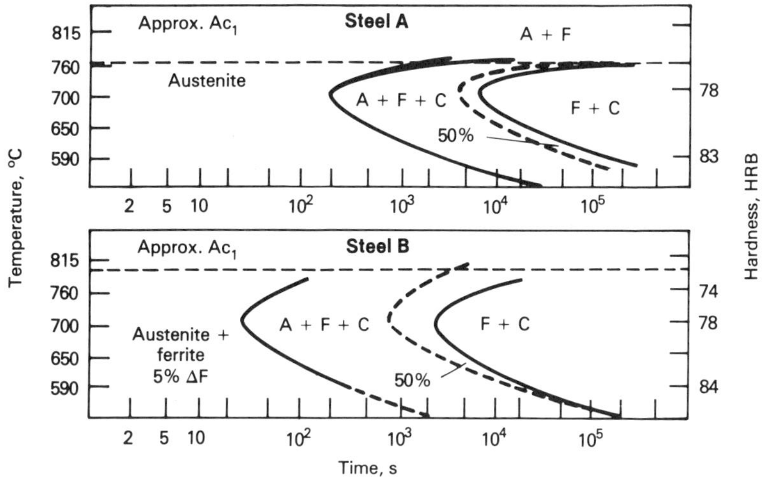

Fig. 1.26 Reduction of hardenability by the presence of ferrite. Steel B contained more ferrite at the time of quenching than steel A.

carbon in solution and prevent precipitation of chromium carbides, which results in chromium depletion and localized loss of corrosion resistance.

Corrosion-resistant stainless steels usually have a low carbon concentration, normally less than 0.08% and often less than 0.03%. For best corrosion resistance, alloys should be solution treated to dissolve grain-boundary carbides before being put into service. Solution-treating temperatures range from 1040 to 1150 °C (1900 to 2100 °F), depending on alloy content, section size, and corrosion resistance requirements.

The solubility of chromium carbides in an 18Cr-8Ni steel is illustrated in Fig. 1.27 (Ref 22). The solubility of carbon in austenite is about 0.03% at 650 °C (1200 °F), increases to about 0.1% at 870 °C (1600 °F), and then sharply increases to a value of about 0.5% at 1370 °C (2500 °F).

Because of higher carbon solubility at elevated temperatures, it is possible to solution treat weldments that may contain small amounts of grain-boundary carbides, dissolve the carbon in the matrix, release the chromium, and produce a uniform chromium concentration throughout the structure, even at grain boundaries. This condition promotes maximum corrosion resistance.

Several precipitates can form during fabrication or during use of stainless steel at moderately elevated temperatures. Time-temperature precipitation curves for grain-boundary carbides and several other phases in U50 stainless steel have been developed by Solomon and Devine (Ref 23) and are illustrated in Fig. 1.28 (Ref 23). U50 contains about 21% Cr and 7% Ni. The formation of $M_{23}C_6$ carbides can occur quickly as parts are cooled to room temperature after welding. Carbides can be avoided by solution treating followed by rapid cooling, often by water quenching, to room temperature.

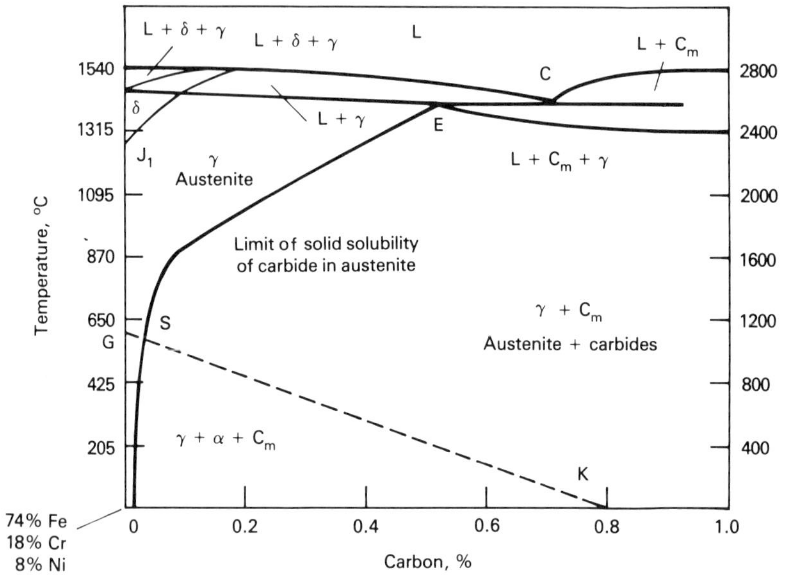

Fig. 1.27 Phases in 18Cr-8Ni steel for carbon content between 0 and 1%

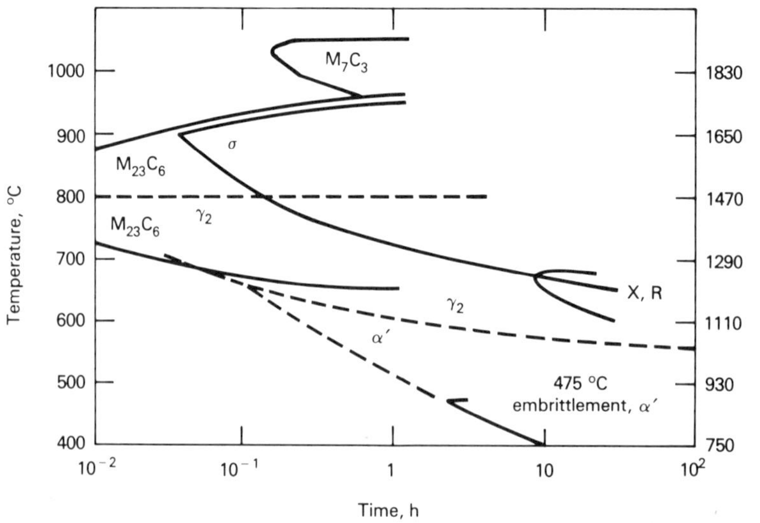

Fig. 1.28 Time-temperature precipitation curves for phases in U50 stainless steel

Precipitation Hardening Steels

It has been known for many years that precipitation hardening can occur in low-carbon iron alloys containing titanium, molybdenum, nickel, copper, and tungsten. However, because it is possible to strengthen low-alloy iron-base materials containing carbon with a simple quench and temper treatment, the possibility of strengthening alloys by precipitation hardening has not been developed to any great extent, probably because of the higher cost of the elements required compared with carbon. The development of maraging steels, however, and the possibility of obtaining high strengths without distortion that often accompanies quench and temper heat treatments, has generated some interest in precipitation hardening alloys.

A portion of the binary phase diagram of the iron-copper system is shown in Fig. 1.29 (Ref 24). The system has a simple eutectoid transformation at 850 °C (1560 °F) and a copper solubility in austenite just above the eutectoid temperature of approximately 4%. On transformation to ferrite during cooling, the solubility of copper decreases to about 1.9% and continues to decrease with decreasing temperature. The solubility at room temperature is not precisely known, although it is low. Because of the increased solubility of copper in austenite, more copper can be retained in solid solution by quenching from this phase field than can be retained on quenching from the ferrite phase field (Ref 25).

Solution treating creates a soft steel that can be machined more easily than a fully hardened aged steel. The cooling rate from the solution-treating temperature must be fast enough to prevent the precipitation hardening reaction from taking place during quenching.

Age-hardenable steels can be aged to harden at a controlled rate. One advantage of this type of hardening process is the absence of distortion compared with that which occurs in steels hardened by the martensite reaction. Another advantage is the ability to machine components in the soft, solution-treated condition, followed by aging at relatively low temperatures, without significant scaling.

A fine dispersion of copper spherical precipitates formed during age hardening produces the maximum internal stress and, consequently, the highest strength attainable in copper-bearing irons. Hardness measurements indicate that by the time the spheres begin to appreciably increase in size, the hardness decreases. If rod-shaped precipitates begin to form, the material is well overaged (Ref 24).

The manganese-copper steels represent a family of low-alloy steels that can be precipitation hardened. These low-carbon steels contain about 0.90 to 1.50% Mn and 1.50 to 1.80% Cu. The usual heat treating practice is to normalize from 925 °C (1700 °F) and precipitation harden for about 2 h in the range of 480 to 510 °C (900 to 950 °F). This hardening reaction produces an increase of 138 MPa (20 ksi) in tensile strength, with a small decrease in ductility.

7-14 PH and 15-5 PH are stainless precipitation-hardenable steels. These steels can be hardened by heating to 1050 °C (1925 °F), followed by cooling to room temperature

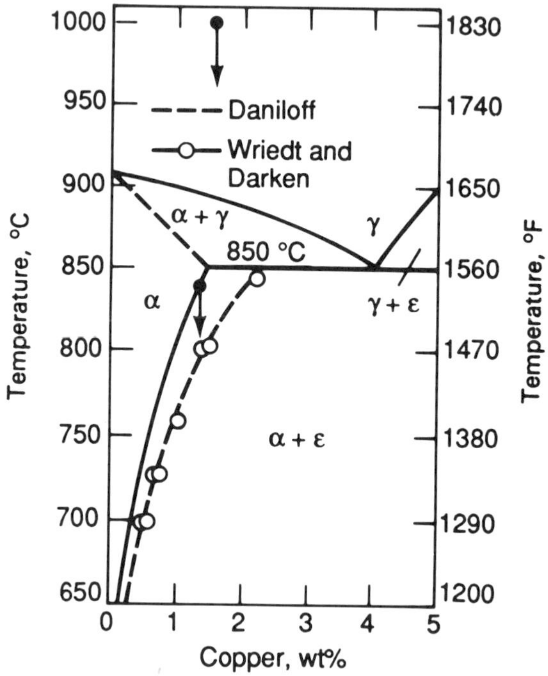

Fig. 1.29 Iron-copper phase diagram

and aging at temperatures in the range of 480 to 620 °C (900 to 1150 °F), depending on the desired properties.

References

1. A.R. Troiano and A.B. Greninger, *Met. Prog.*, Vol 50, 1946, p 303
2. R.A. Grange and N.M. Stewart, The Temperature Range of Martensite Formation, *Trans. AIME*, Vol 167, 1946, p 467
3. M. Atkins, *Atlas of Continuous Cooling Transformation Diagrams for Engineering Steels*, American Society for Metals, 1980
4. *Atlas of Isothermal Transformation and Cooling Transformation Diagrams*, American Society for Metals, 1979
5. *Atlas of Isothermal Transformation Diagrams*, 3rd ed., U.S. Steel, 1963
6. G. Krauss, *Principles of Heat Treatment of Steels*, American Society for Metals, 1980
7. *Source Book on Heat Treating*, Vol 1, *Materials and Processes*, American Society for Metals, 1975
8. C.A. Siebert, D.V. Doane, and D.H. Breen, *The Hardenability of Steels*, American Society for Metals, 1977

9. K. Sachs, Presentation of Transformation Data, *Heat Treat. Met.*, Vol 1, 1974, p 43-57
10. K.E. Thelning, CCT Continuous Cooling Transformation Diagrams With Natural Cooling, *Scand. J. Metall.*, 1978, p 252-263
11. *Austenite Transformation Kinetics of Ferrous Alloys*, Vol 1 and 2, Climax Molybdenum Co.
12. *The Making, Shaping, and Treating of Steel*, 7th ed., U.S. Steel Corp., 1957
13. *Nickel Alloy Steels Data Book*, 3rd ed., Bull. A, Section 2, International Nickel Co.
14. R.A. Grange and N.M. Stewart, The Temperature Range of Martensite Formation, *Trans. AIME*, Vol 167, 1946, p 467
15. R.C. Voigt, M. Blair, and J. Rassizadehghani, High Strength Low Alloy Cast Steels, *Proc. 1990 ASME Pressure Vessels and Piping Conf.*, Nashville, 17-21 June 1990, p 147-154
16. R.C. Voigt, Analysis of Intercritical Heat Treatment of Cast Steels, *J. Heat Treat.*, Vol 7, 1989, p 95-105
17. R. Molnar, "A Literature Survey on Certain Aspects of Austenitic Manganese Steels," Steel Castings Institute of Canada, June 1974
18. G. Coletta, *et al.*, Contributions a l'Etude des Transformations des Austenites a 12 pour cent Mn, *Rev. Metall.*, Vol 54 (No. 6), June 1957, p 433-486
19. K.J. Irvine, D.J. Crow, and F.B. Pickering, The Physical Metallurgy of 12% Chromium Steels, *J. Iron Steel Inst.*, Aug 1960, p 386
20. J. Beech, Metallurgy of 13% Cr Steel Castings, *Foundry Trade J.*, 3 Nov 1960, p 569
21. P. Shahinian and J.R. Lane, Modifying 12% Cr Steel for Better Properties, *Iron Age*, Oct 1956, p 106
22. E.E. Thum, Ed., *The Book of Stainless Steels*, 2nd ed., American Society for Metals, 1935
23. H.D. Solomon and T.M. Devine, "Duplex Stainless Steels—A Tale of Two Phases," ASM Metals/Materials Technology Series 8201-089, 1984
24. E. Hornbogen and R.C. Glenn, A Metallographic Study of Precipitation of Copper from Alpha Iron, *Trans. AIME*, Vol 218, Dec 1960, p 1064-1069
25. E. Hornbogen, Precipitation From Binary Substitutional Solid Solutions of Alpha Iron, *Proc. Metall. Soc. AIME*, Vol 28, Oct 1965, p 1-69

CHAPTER 2

Measuring Hardenability and Quench Severity

The best combination of properties in carbon and low-alloy steel can be obtained by quenching to a martensitic microstructure, followed by tempering. A considerably higher yield strength can be obtained at a given tensile strength by quenching and tempering than by normalizing and tempering, as illustrated in Fig. 2.1.

Successful hardening depends on the geometry of the part, the hardenability of the steel, and the quenching practices employed. Quenching to develop a martensitic structure without warping or cracking requires good part design, a steel with adequate hardenability, properly maintained heat treating furnaces and quench facilities, the proper quenchant, knowledgeable technicians, and experience.

Some heat treating problems are related to poor part design. Long parts with small cross sections, those lacking symmetry, and those with holes or recesses are difficult to heat treat without distortion or cracking. Rapid or nonuniform cooling of parts can also cause distortion or cracking (Ref 1).

The principal method of hardening carbon and low-alloy steels consists of quenching the steel from the austenitizing temperature. Steels vary in their response to the quenching operation, because the depth below the surface that can be hardened depends on the composition of the steel and the severity of the quench. Because the hardening reaction is composition dependent, hardenability is an important consideration in alloy selection for the production of high-strength parts.

Improved quenching practices permit the use of a less hardenable, less expensive steel to achieve the desired properties. The quenching practices employed, however, are usually a compromise between a practice that could harden to the greatest possible depth and the requirement that quenching be accomplished without excessive distortion or cracking (Ref 2).

Various aspects of steel hardenability will be discussed in this chapter, including quantifying hardenability, effect of hardenability on as-quenched martensite, effect of quenching variables on hardenability.

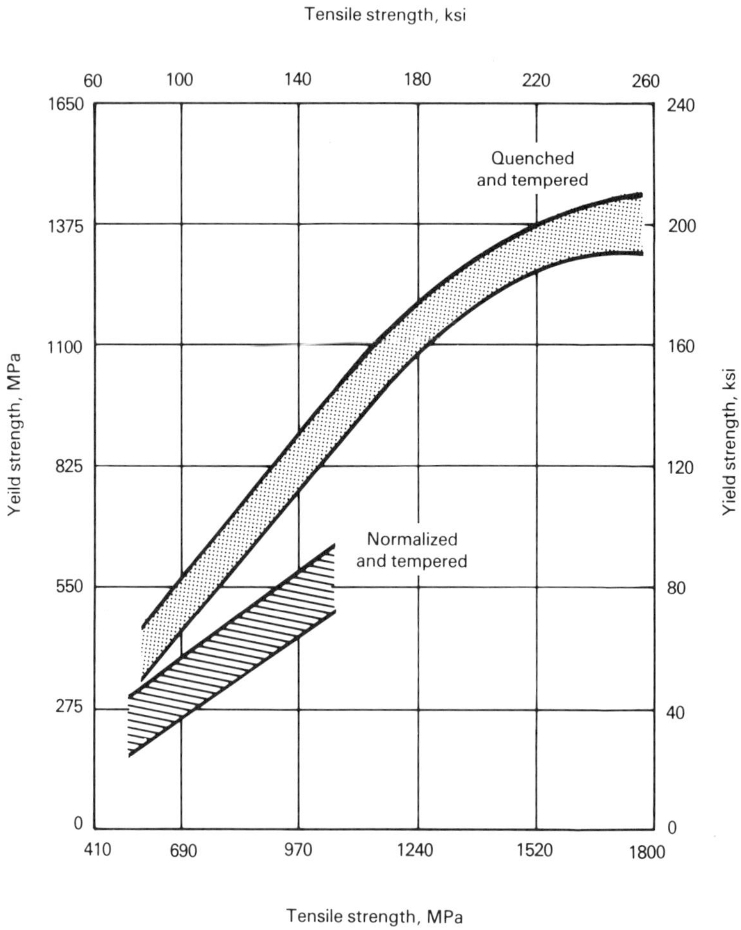

Fig. 2.1 Tensile and yield strengths of normalized and tempered and of quenched and tempered steels

Hardenability

Hardenability refers to the ability of a steel to be transformed partially or completely from austenite to martensite at a given depth when cooled under prescribed conditions. Hardenability should not be confused with hardness. Hardness refers to the resistance of a material to penetration as measured with a Rockwell, Brinell, or other indentation test. Hardenability reflects the ability of a steel to be hardened when cooled in a particular way. Hardenability is empirically determined, and several types of experiments have been devised to measure it.

Jominy Procedure

The most familiar and commonly used procedure for measuring hardenability is the Jominy bar end-quench test. This test has been standardized and is described in SAE J406 and ASTM A255 specifications. A 100 mm (4 in.) long by 25 mm (1 in.) diam bar is austenitized at the proper temperature, dropped into a fixture, and one end rapidly cooled with 24 °C (75 °F) water issuing from a 13 mm (½ in.) orifice under prescribed conditions. The austenitizing temperature is selected according to the specific steel and may be as low as 815 °C (1500 °F) or as high as 925 °C (1700 °F). However, most steels are heated in the range of 870 to 900 °C (1600 to 1650 °F). Special precautions are taken to minimize scaling during solution treating.

The end-quenched bar is subjected to a continuous stream of cold water and is rapidly quenched at the end contacted by the water stream. The cooling rate is slower and approximates air cooling at the opposite end of the bar. Rates comparable to still-water quenching, oil quenching, and forced-air cooling are represented at locations between the water-quenched and air-cooled ends of the bar.

After quenching, parallel flats are ground on opposite sides of the bar, and hardness measurements are made at $\frac{1}{16}$ in. (1.6 mm) intervals along the bar length. The standard specimen, specimen support, water quench requirements, and a composite of hardness-distance results obtained on a particular steel in nine laboratories are illustrated in Fig. 2.2.

The Jominy test is considered to provide valid data on steels having an ideal diameter (D_I) from about 25 to 150 mm (1 to 6 in.). The D_I can be less than 25 mm (1 in.), but this usually requires that Vickers hardness readings be taken closer to the quenched end and closer together than possible using standard Rockwell hardness measuring equipment.

The austenitizing time and temperature, extent of oxidation or surface decarburization during austenitizing, care and consistency of surface flat preparation, and bar positioning prior to making hardness measurements are important procedural factors that influence test results. Water temperature, water jet height, orifice diameter, and transfer time from the furnace to the quench fixture are also important, but not as critical, in obtaining consistent data. However, all tests should be conducted in compliance with recognized standards (Ref 3, 4).

The hardness versus distance representation of Jominy data provides hardness data on the steel at various cooling rates. The hardness readings are plotted against distance from the quenched end to produce the hardenability or end-quench curve. Figure 2.3 illustrates schematic end-quench curves for four steels to show various shapes assumed by different alloys. The carbon steel is shallow hardening, and the chromium-molybdenum steel is deep hardening.

Some variation in composition occurs between heats of a given grade. Therefore, the end-quench curves can (within limits) vary from heat to heat. When enough heats within a given grade have been tested, the end-quench curves may be plotted to form a hardenability band. Hardenability bands and specific hardness limits for AISI-SAE 1045H, 8630H, 4130H, and 4340H are illustrated in Fig. 2.4 to 2.7, respectively.

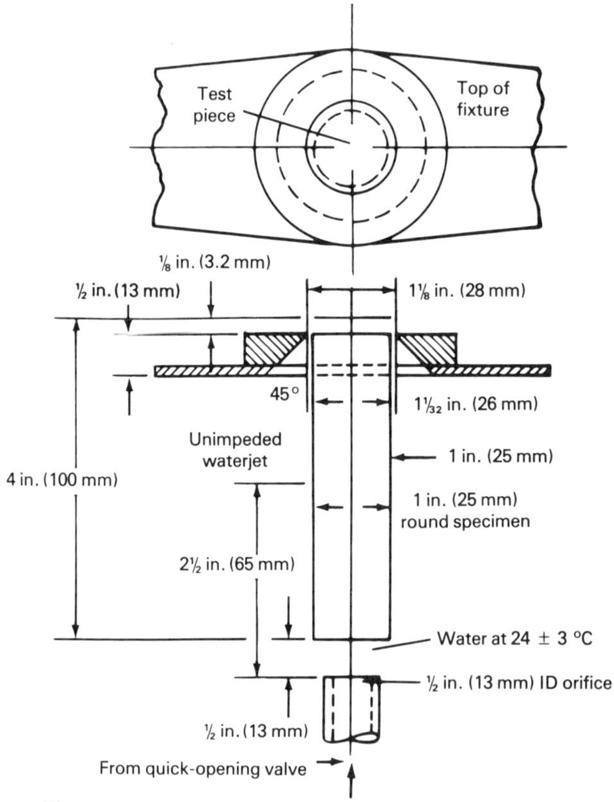

(a)

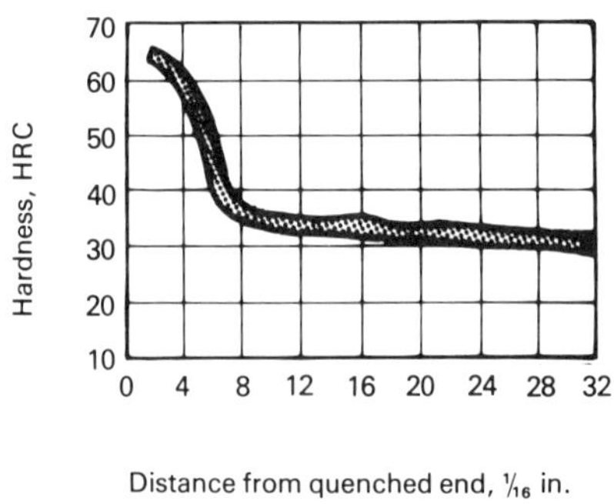

(b)

Fig. 2.2 **(a)** Standard form of end-quench test specimen (Jominy), support, and water-quench requirements. **(b)** Hardenability range, based on tests by nine laboratories on an SAE 406B steel

The standard end-quench procedure employs 27 °C (80 °F) water, but some investigators use oils or polymers to perform the quench and then compare the results obtained with the alternate quenchant with the results produced by water.

Ideal Critical Diameter

Jominy hardness curves can be used to estimate the ideal critical diameter of an alloy. The critical diameter (D) refers to the largest bar diameter that contains 50% martensite at the center after being quenched. The ideal critical diameter (D_I) is the largest bar diameter that can be quenched to produce 50% martensite at its center after being given an "infinite" or "ideal" quench. The ideal quench is one that lowers the surface temperature of an austenitized steel to the bath temperature instantaneously. Under these conditions, the cooling rate at the bar center depends only on the thermal diffusivity of the steel.

The ideal critical diameter can be estimated from a Jominy curve using Fig. 2.8. For example, a steel with a 50% martensite point $^8/_{16}$ in. (13 mm) from the end of the Jominy specimen can be quenched to 50% martensite at the center of a 68.6 mm (2.7 in.) diam bar under ideal quenching conditions. The ideal critical diameter has been correlated with the 50% martensite point in many alloys to develop this figure.

Calculated End-Quench Curves

There are situations where end-quench curves are not available, such as in alloy development programs, or where the end-quench test is not practical, such as on thin-section parts or stock. In these cases, the end-quench curve can be computed from the chemical analysis and grain size.

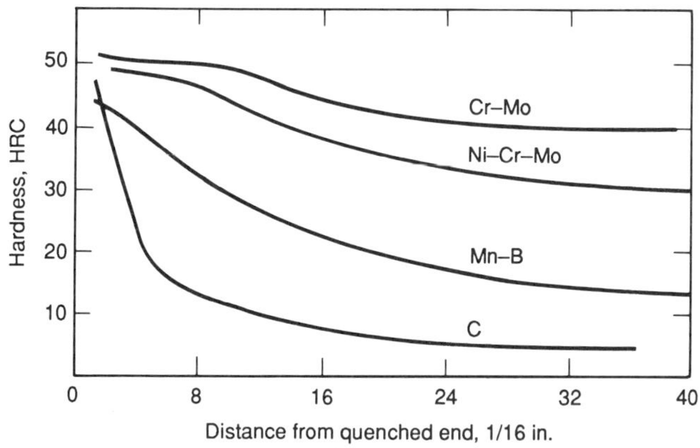

Fig. 2.3 Schematic end-quench (hardenability) curves for carbon, manganese-boron, nickel-chromium-molybdenum, and chromium-molybdenum steels

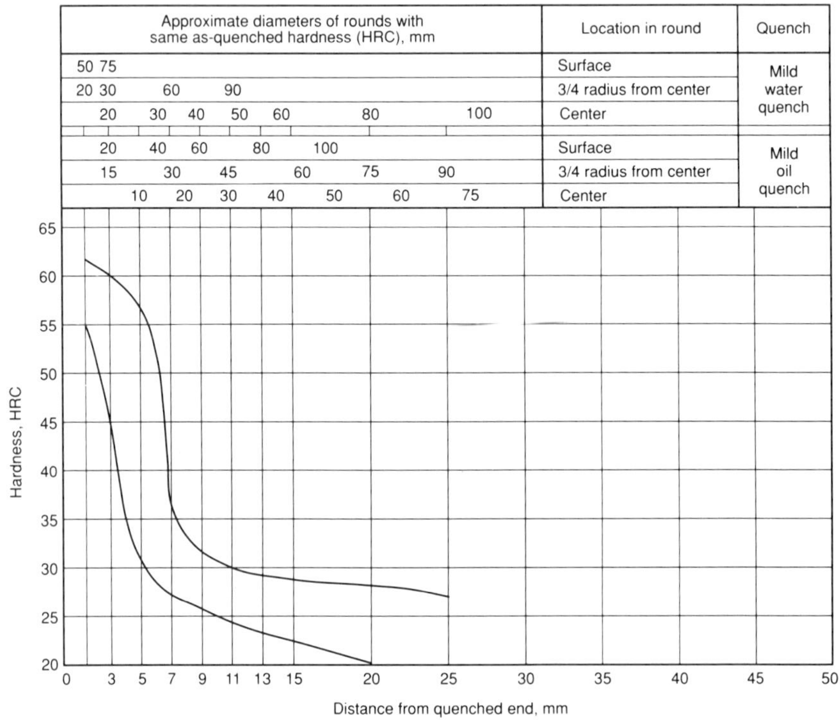

Hardness limits for specification purposes

J distance, mm	Hardness, HRC Maximum	Hardness, HRC Minimum
1.5	62	55
3	60	45
5	53	31
7	36	27
9	32	25
11	31	24
13	30	23
15	29	22
20	28	20
25	27	...
30	...	...

Fig. 2.4 Jominy hardenability band and hardness limits for AISI-SAE 1045H

The hardenability of a steel is largely controlled by the carbon content, grain size, and alloy element concentrations (Ref 3). For most steels, the grain size is small and does not vary widely. Hardenability of steels, therefore, is determined largely by the chemical analysis.

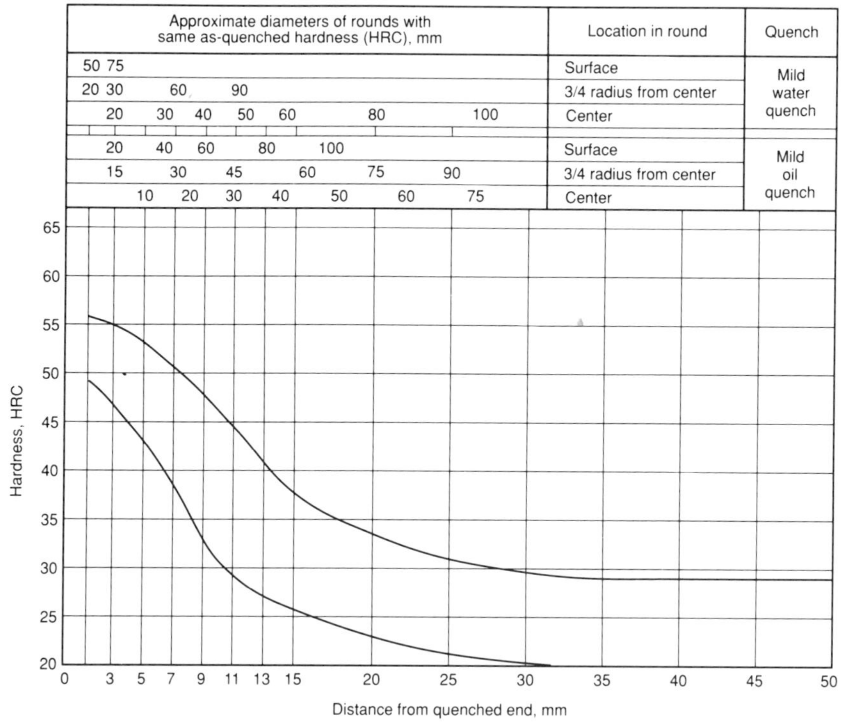

Hardness limits for specification purposes

J distance, mm	Hardness, HRC Maximum	Hardness, HRC Minimum
1.5	56	49
3	55	46
5	54	42
7	51	38
9	48	33
11	44	29
13	41	27
15	38	26
20	34	23
25	31	21
30	30	20
35	29	...
40	29	...
45	29	...
50	29	...

Fig. 2.5 Jominy hardenability band and hardness limits for AISI-SAE 8630H

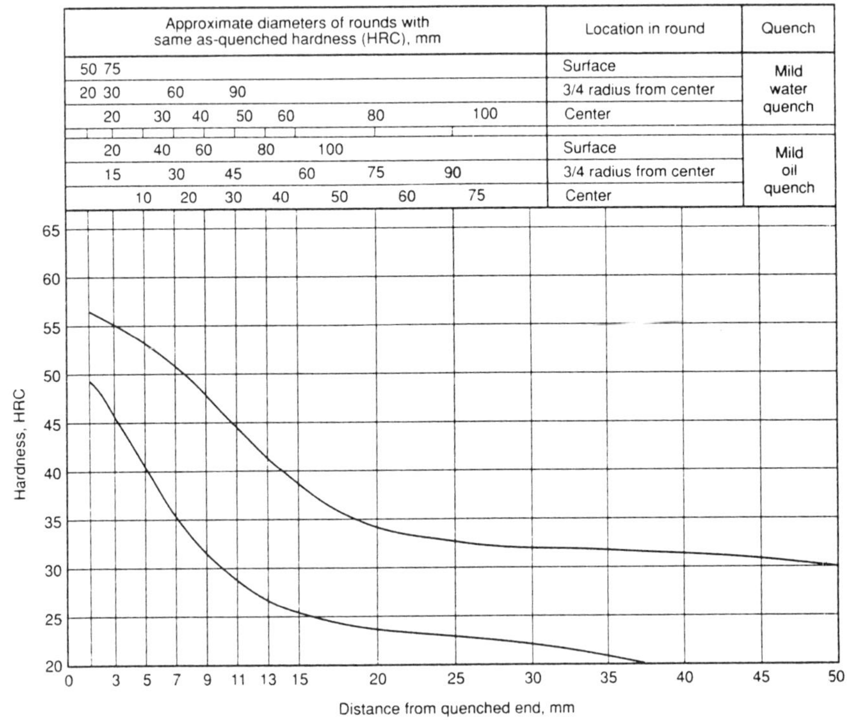

Hardness limits for specification purposes

J distance, mm	Hardness, HRC Maximum	Minimum
1.5	56	49
3	55	46
5	53	40
7	51	36
9	48	32
11	44	28
13	41	26
15	39	25
20	34	24
25	33	23
30	33	22
35	32	20
40	31	...
45	31	...
50	30	...

Fig. 2.6 Jominy hardenability band and hardness limits for AISI-SAE 4130H

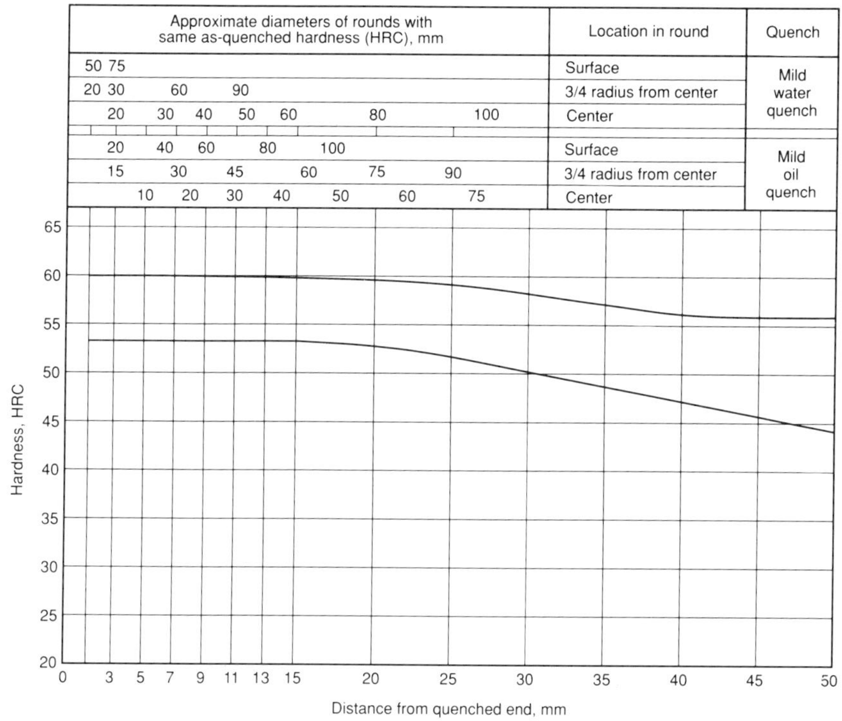

Hardness limits for specification purposes

J distance, mm	Hardness, HRC Maximum	Minimum
1.5	60	53
3	60	53
5	60	53
7	60	53
9	60	53
11	60	53
13	60	53
15	60	53
20	60	52
25	59	51
30	58	50
35	58	49
40	57	47
45	57	46
50	57	44

Fig. 2.7 Jominy hardenability band and hardness limits for AISI-SAE 4340H

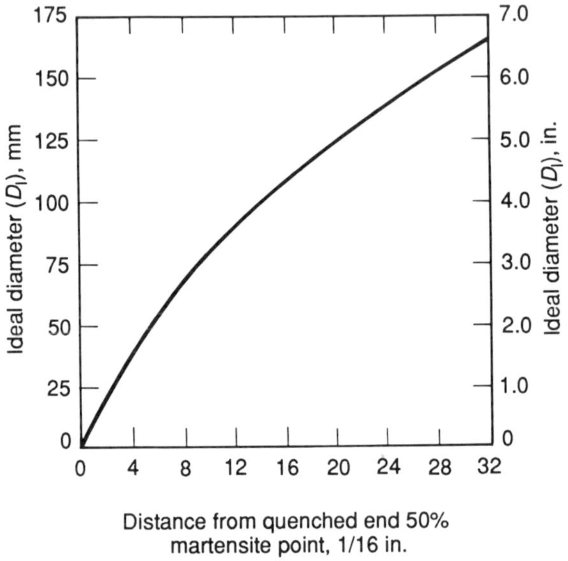

Fig. 2.8 Relationship between end-quench distance and hardenability (D_I)

Hardenability is calculated in terms of the ideal critical diameter, and equations most often used for this prediction contain multiplicative factors for the alloy elements. The base D_I is determined by the carbon concentration and grain size. The base D_I is then multiplied by the alloy factors that account for the hardenability effects of the elements present. Equation 2.1 is one expression for calculating the ideal critical diameter:

$$D_I = D_{I_{base}}(\text{carbon and grain size}) \cdot f_{Mn} \cdot f_{Si} \cdot f_{Cr} \cdot f_{Mo} \cdot f_V \cdot f_{Cu} \quad \text{(Eq 2.1)}$$

where f_x is the multiplicative factor for the particular alloying element.

The base D_I and one set of alloy factors are presented in Table 2.1 (Ref 5). The alloy factors were developed based on data from medium-carbon steels of medium hardenability. These factors and others for steels of particular composition ranges have been incorporated into calculators for quickly making calculations and plots of Jominy end-quench data (Ref 6, 7).

The procedure for calculating the hardenability of a steel from the composition includes the following steps:

- Determine the ASTM grain size.
- Determine the chemical composition.
- Determine base D_I from carbon content and grain size (see Table 2.1) (Ref 5).

- Determine alloy factors (see Table 2.1).
- Multiply the factors according to Eq 2.1 to calculate the ideal diameter.

Table 2.1 does not provide alloy factors for calculating the effects of vanadium or boron. There is some inconsistency regarding the behavior of these elements. Vanadium up to about 0.1% increases hardenability, but higher concentrations produce vanadium-rich carbides that are difficult to dissolve at standard austenitizing temperatures.

Boron increases hardenability of steels with carbon contents below about 0.3%, but has little effect in higher-carbon steels. It is appropriate to test heats from particular suppliers to determine the effects of boron and vanadium in the alloys of interest.

Once D_I has been determined, the end-quench curve can be determined using Table 2.2 and Fig. 2.9 (Ref 8, 9). The second column of Table 2.2 gives the maximum hardness of quenched steel as a function of the carbon content. Subsequent columns provide the hardness at reduced martensite contents of 95, 80, and 50%. The maximum hardness is the initial hardness value often plotted $1/16$ in. (1.6 mm) from the quenched end of a Jominy specimen.

The maximum hardness is then used in the ratio of initial hardness to distance hardness in Fig. 2.9. The numbers on the curves of Fig. 2.9 indicate the number of $1/16$ in. (1.6 mm) increments from the quenched end—for example, $4/16$, $8/16$, $12/16$ in. (6.4, 13, 19 mm), etc. Using the D_I calculated from Eq 2.1, proceed horizontally across Fig. 2.9 and note the ratio

Table 2.1 Hardenability factors for carbon grain size and selected alloying elements in steel

	Carbon grain size No.			Alloying element				
Percent	6	7	8	Mn	Si	Ni	Cr	Mo
0.05	0.0814	0.0750	0.0697	1.167	1.035	1.018	1.1080	1.15
0.10	0.1153	0.1065	0.0995	1.333	1.070	1.036	1.2160	1.30
0.15	0.1413	0.1315	0.1212	1.500	1.105	1.055	1.3240	1.45
0.20	0.1623	0.1509	0.1400	1.667	1.140	1.073	1.4320	1.60
0.25	0.1820	0.1678	0.1560	1.833	1.175	1.091	1.54	1.75
0.30	0.1991	0.1849	0.1700	2.000	1.210	1.109	1.6480	1.90
0.35	0.2154	0.2000	0.1842	2.167	1.245	1.128	1.7560	2.05
0.40	0.2300	0.2130	0.1976	2.333	1.280	1.146	1.8640	2.20
0.45	0.2440	0.2259	0.2090	2.500	1.315	1.164	1.9720	2.35
0.50	0.2580	0.2380	0.2200	2.667	1.350	1.182	2.0800	2.50
0.55	0.273	0.251	0.231	2.833	1.385	1.201	2.1880	2.65
0.60	0.284	0.262	0.241	3.000	1.420	1.219	2.2960	2.80
0.65	0.295	0.273	0.251	3.167	1.455	1.237	2.4040	2.95
0.70	0.306	0.283	0.260	3.333	1.490	1.255	2.5120	3.10
0.75	0.316	0.293	0.270	3.500	1.525	1.273	2.62	3.25
0.80	0.326	0.303	0.278	3.667	1.560	1.291	2.7280	3.40
0.85	0.336	0.312	0.287	3.833	1.595	1.309	2.8360	3.55
0.90	0.346	0.321	0.296	4.000	1.630	1.321	2.9440	3.70
0.95	...	...	...	4.167	1.665	1.345	3.0520	...
1.00	...	...	...	4.333	1.700	1.364	3.1600	...

Table 2.2 Effect of carbon concentration and martensite on the as-quenched hardness of steel

Carbon, %	Hardness, HRC				
	99% M	95% M	90% M	80% M	50% M
0.10	38.5	32.9	30.7	27.8	26.2
0.12	39.5	34.5	32.3	29.3	27.3
0.14	40.6	36.1	33.9	30.8	28.4
0.16	41.8	37.6	35.3	32.3	29.5
0.18	42.9	39.1	36.8	33.7	30.7
0.20	44.2	40.5	38.2	35.0	31.8
0.22	45.4	41.9	39.6	36.3	33.0
0.24	46.6	43.2	40.9	37.6	34.2
0.26	47.9	44.5	42.2	38.8	35.3
0.28	49.1	45.8	43.4	40.0	36.4
0.30	50.3	47.0	44.6	41.2	37.5
0.32	51.5	48.2	45.8	42.3	38.5
0.34	52.7	49.3	46.9	43.4	39.5
0.36	53.9	50.4	47.9	44.4	40.5
0.38	55.0	51.4	49.0	45.4	41.5
0.40	56.1	52.4	50.0	46.4	42.4
0.42	57.1	53.4	50.9	47.3	43.4
0.44	58.1	54.3	51.8	48.2	44.3
0.46	59.1	55.2	52.7	49.0	45.1
0.48	60.0	56.0	53.5	49.8	46.0
0.50	60.9	56.8	54.3	50.6	46.8
0.52	61.7	57.5	55.0	51.3	47.7
0.54	62.5	58.2	55.7	52.0	48.5
0.56	63.2	58.9	56.3	52.6	49.3
0.58	63.8	59.5	57.0	53.2	50.0
0.60	64.3	60.0	57.5	53.8	50.7

M, martensite

given by the intersection with each curve. The hardnesses at the indicated distances from the quenched end can then be calculated by dividing the initial hardness by the ratio given on the abscissa and plotting the resulting values to draw the end-quench curve. Because the curve is drawn based on carbon content and D_I, tables have been published that give the hardness ratio at various distances from the quenched end for various combinations of D_I and carbon content (Ref 10, 11).

Application of Ideal Critical Diameter

The definition of D_I assumes that parts are cooled in an ideal quench. Real quenching operations can only approach ideal conditions. The depth of hardening obtained in practice depends not only on the hardenability of the particular steel but also on the quench severity.

Quench severity is often expressed by a number that is designated as the H value. Larger H values are associated with more severe quenching. The H value depends

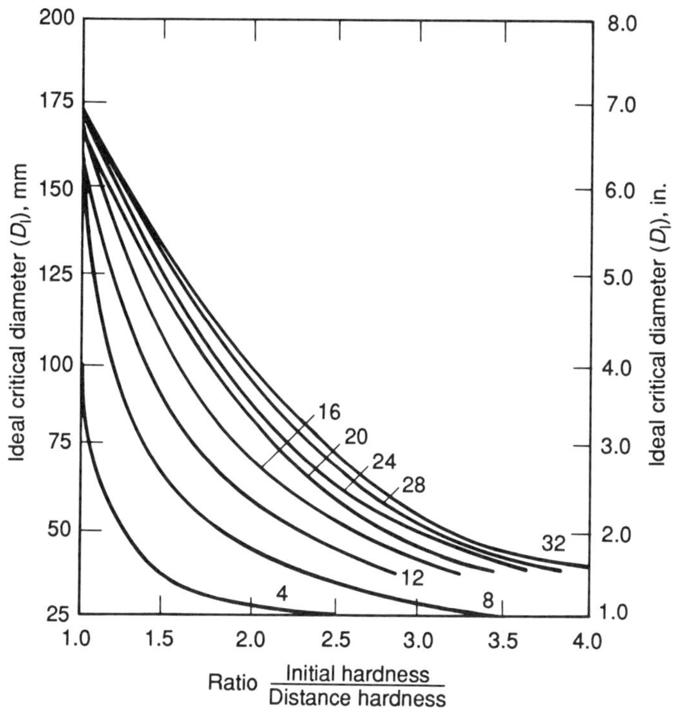

Fig. 2.9 Curves used for the determination of hardness for certain distances from the quenched end

primarily on the quenching medium, its temperature, and the degree of quenchant agitation. Table 2.3 lists H values for several quenchants and quenching conditions (Ref 12).

The diameter of a cylinder that will quench to produce 50% martensite at the center (D) in commercial quenching operations is always less than the ideal critical diameter (D_I). The relation between D and D_I for quenches of various H values is illustrated in Fig. 2.10. For example, if a steel has $D_I = 100$ mm (4 in.) and is quenched in water where $H = 1.5$, an 89 mm (3.5 in.) diam bar can be expected to quench to 50% martensite at the center. If the same steel is quenched in oil with $H = 0.4$, a cylinder with a diameter of only 57 mm (2.25 in.) can be expected to harden to produce 50% martensite at the center.

The relationship between ideal critical diameter, quench severity, and the thickness of plates that can be hardened is illustrated in Fig. 2.11. For example, assume that a plate of 4340 steel with a D_I of 163 mm (6.42 in.) is quenched in still water ($H = 1.1$ from Table 2.3). The corresponding critical plate thickness is 76 mm (3.0 in.). This is the thickness of a plate that can be hardened to 50% martensite. A 150 mm (6 in.) diam rod or cylinder will harden to 50% martensite in the center under the same conditions. The difference between plates and cylinders arises from the slower rate of cooling in plates, because plates have lower ratios of surface area to mass than cylinders.

Table 2.3 Grossmann numbers and film coefficients for selected quenchants

Quenchant	Quenchant temperature °C	Quenchant temperature °F	Quenchant velocity m/s	Quenchant velocity ft/min	Grossmann number (H)	Effective film coefficient W/m^2 · K	Effective film coefficient Btu/ft^2 · h · °F
Water	32	90	0.00	0	1.1	5000	880
			0.25	50	2.1	9000	1600
			0.51	100	2.7	12,000	2100
			0.76	150	2.8	12,000	2100
	55	130	0.00	0	0.2	1000	180
			0.25	50	0.6	2500	440
			0.51	100	1.5	6500	1100
			0.76	150	2.4	10,500	1850
Fast oil	60	140	0.00	0	0.5	2000	350
			0.25	50	1.0	4500	790
			0.51	100	1.1	5000	880
			0.76	150	1.5	6500	1200
25% polyvinyl pyrrolidone	43	110	0.00	0	0.8	3500	620
			0.25	50	1.3	6000	1100
			0.51	100	1.5	6500	1200
			0.76	150	1.8	7500	1300
Conventional oil	65	150	0.51	100	0.7	3000	530
Martempering oil	150	300	0.51	100	1.2	5000	880
Air	27	80	0.00	0	0.05	200	35
			2.54	500	0.06	250	44
			5.08	1000	0.08	350	62

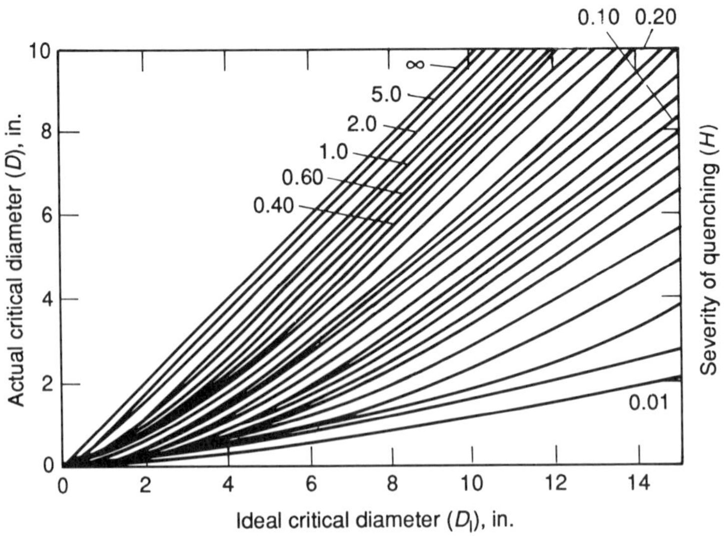

Fig. 2.10 Relationships among ideal critical diameter, actual critical diameter, and severity of quenching

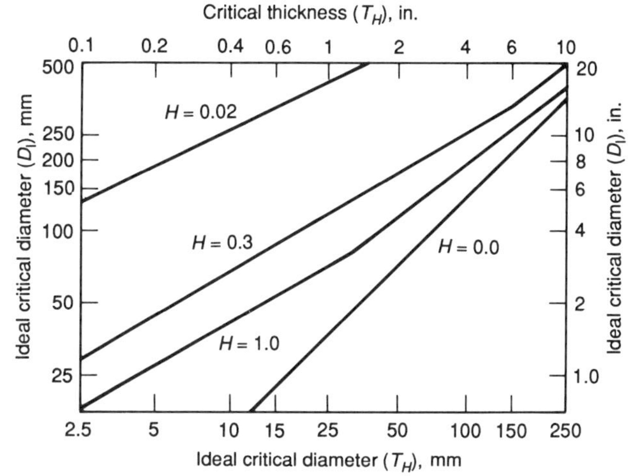

Fig. 2.11 Relationship between ideal critical diameter and the critical thickness that can be fully hardened using a quenching medium with severity *H*

Cooling Rate Equivalence

Different parts of a section cool at different rates when being quenched. The cooling rates are affected by the severity of quench and section thickness, as indicated in Fig. 2.12 and 2.13 (Ref 13). The cooling rates along the length of a Jominy end-quench bar also vary, depending on the distance from the quenched end. Consequently, there should be a cooling rate somewhere along an end-quench bar that corresponds to the cooling rate at a point of interest in a part being quenched.

To determine the point of correspondence on the end-quench bar, the cooling rates of interest in the actual part are measured over some temperature interval, typically from 730 °C (1350 °F) down to 315 or 370 °C (600 or 700 °F). This is the range where the greatest tendency of transformation to pearlite or bainite occurs.

Figure 2.12 provides the end-quench equivalent at the surface, midradius, and center of cylindrical bars quenched in solutions providing different quench severities (Ref 14). These curves allow the determination of hardness at a particular distance from the end of a Jominy bar and prediction of hardness at the surface, midradius, and center of bars quenched in oil, water, and brine. Similar data are also presented in Table 2.4.

The curves in Fig. 2.13 and 2.14 can be used for shapes having other cross sections, such as square or hexagonal bars, by determining equivalent diameters. The equivalent diameter is considered to be the diameter of a circle that can be inscribed in the section of interest.

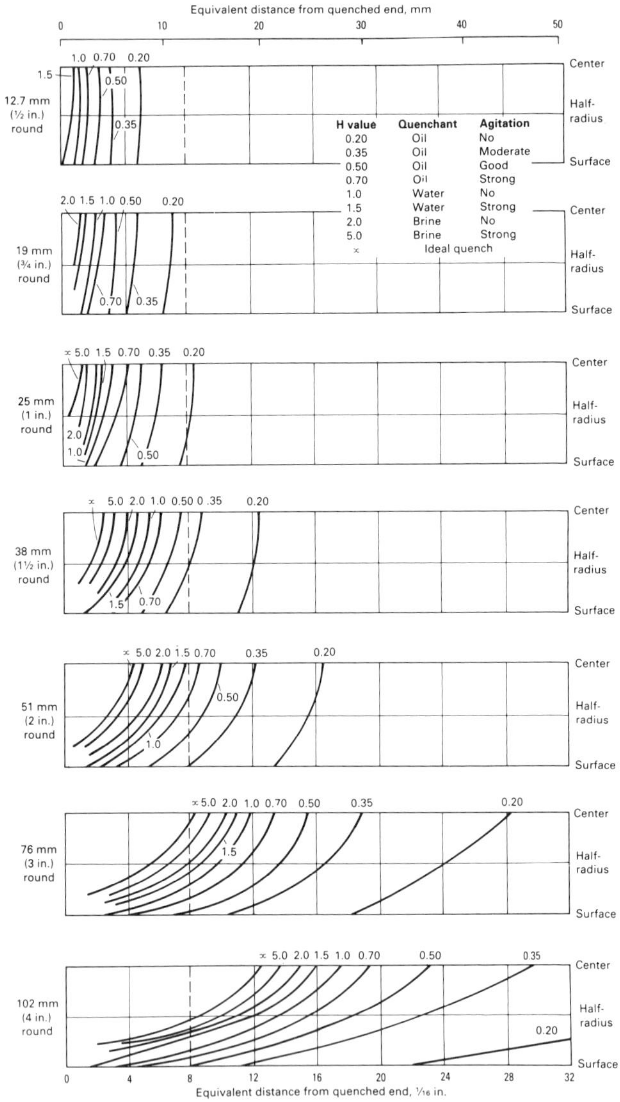

Fig. 2.12 Correlation of equivalent hardness positions in end-quench hardenability specimen and various locations in round bars quenched in oil, water, and brine. The dashed line shows the various positions in ½ to 4 in. (13 to 100 mm) diam rounds that are equivalent to the 8/16 in. distance on the end-quench bar. To determine cross-section hardnesses from results of end-quench tests, pick out the end-quench hardness at an appropriate point on the bottom line and extend an imaginary line upward to the curved line that corresponds to the quenching severity needed to obtain that hardness for the given diameter of round.

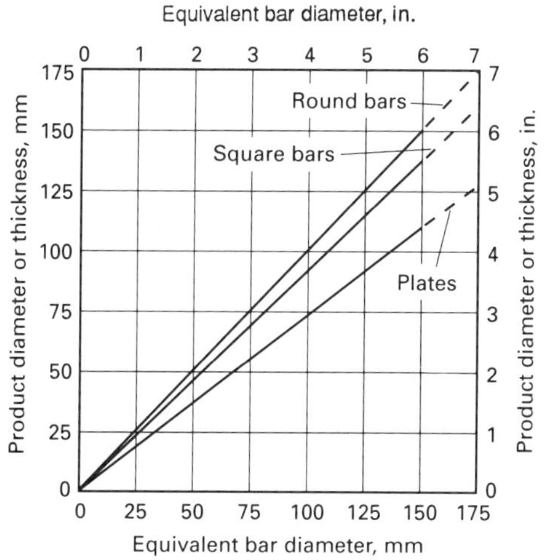

Fig. 2.13 Relationship between through-quenched bar diameter and through-quenched thickness of plates and square bars

Data on end-quench equivalence of plates are rather scarce. Figure 2.13 provides equivalence among round bars, square bars, and plates, and Fig. 2.14 provides a correlation between the Jominy equivalent distance and the center cooling rates in plates of various thicknesses quenched with different severities. The result provides a correlation between distance from the quenched end of a Jominy bar and points in the cross section of the quenched shape having equivalent hardness.

Specifying Hardenability

Hardenability is specified in several different ways. SAE J435a covers three grades of 0.3% C steel. The hardenability bands for the three grades are illustrated in Fig. 2.15. The choice of analysis to fit the chosen grade is usually made by the manufacturer.

When the hardenability band has been defined for the steel, it can be accepted by both producer and customer. The end-quench curve for each heat made to this specification is required to fall within the band. The specification may call for a range of hardness at one or more distances from the quenched end—for example, 38 to 53 HRC at $^{6}/_{16}$ in. (9.5 mm) from the quenched end. The distance is chosen to be equivalent to a critical region in the part being heat treated. Other specifications may set maximum and minimum distances from the quenched end for a given hardness—for example, 40 HRC at $^{3}/_{16}$ to $^{14}/_{16}$ in. (4.8 to 22.2 mm) from the quenched end.

Still other hardenability specifications may set lower and upper limits on the ideal critical diameter. This range is not calculated from the extreme composition limits of the steel grade, but from the distribution of D_I values determined from several production heats.

Hardenability Bands

Hardenability bands for steels have been determined for many steels, including carbon and many of the frequently used low-alloy steels. Several bands are illustrated in Fig. 2.4 to 2.7. Hardenability bands are useful in selecting alloys for particular applications.

Table 2.4 End-quench distance versus bar diameter

	Equivalent bar diameter when quenched, in.					
Distance from quenched end, in.	Still oil ($H = 0.25$)	Circulated oil ($H = 0.45$)	Still water ($H = 1.0$)	Circulated water ($H = 1.5$)	Still brine ($H = 2.0$)	Infinite or idealized quench ($H = \infty$)
1/16	0.1	0.15	0.3	0.35	0.4	0.7
2/16	0.2	0.3	0.5	0.65	0.75	1.15
3/16	0.35	0.55	0.85	1.0	1.25	1.6
4/16	0.5	0.80	1.15	1.3	1.5	1.9
5/16	0.6	0.95	1.4	1.6	1.75	2.2
6/16	0.8	1.2	1.6	1.8	2.0	2.4
7/16	1.0	1.4	1.8	2.0	2.3	2.7
8/16	1.1	1.5	2.1	2.3	2.5	2.9
9/16	1.3	1.7	2.3	2.5	2.7	3.2
10/16	1.4	1.9	2.5	2.7	2.9	3.4
11/16	1.6	2.1	2.8	3.0	3.2	3.6
12/16	1.7	2.2	3.0	3.2	3.4	3.8
13/16	1.9	2.4	3.2	3.4	3.5	4.0
14/16	2.0	2.5	3.3	3.5	3.7	4.2
15/16	2.1	2.7	3.5	3.7	3.9	4.4
16/16	2.3	2.8	3.7	3.9	4.1	4.6
17/16	2.4	3.0	3.9	4.1	4.2	4.7
18/16	2.5	3.1	4.0	4.2	4.4	4.9
19/16	2.6	3.3	4.1	4.4	4.5	5.0
20/16	2.7	3.4	4.3	4.5	4.7	5.1
21/16	2.8	3.5	4.4	4.7	4.8	5.3
22/16	2.9	3.6	4.5	4.8	4.9	5.4
23/16	3.0	3.7	4.7	5.0	5.1	5.5
24/16	3.1	3.8	4.8	5.1	5.2	5.6
25/16	3.2	4.0	4.9	5.2	5.3	5.8
26/16	3.3	4.0	5.0	5.3	5.4	5.9
27/16	3.4	4.1	5.1	5.4	5.5	6.0
28/16	3.5	4.2	5.2	5.5	5.6	6.1
29/16	3.6	4.3	5.3	5.6	5.6	6.2
30/16	3.6	4.4	5.4	5.7	5.7	6.2
31/16	3.7	4.5	5.5	5.8	5.8	6.3
32/16	3.8	4.5	5.5	5.8	5.9	6.4

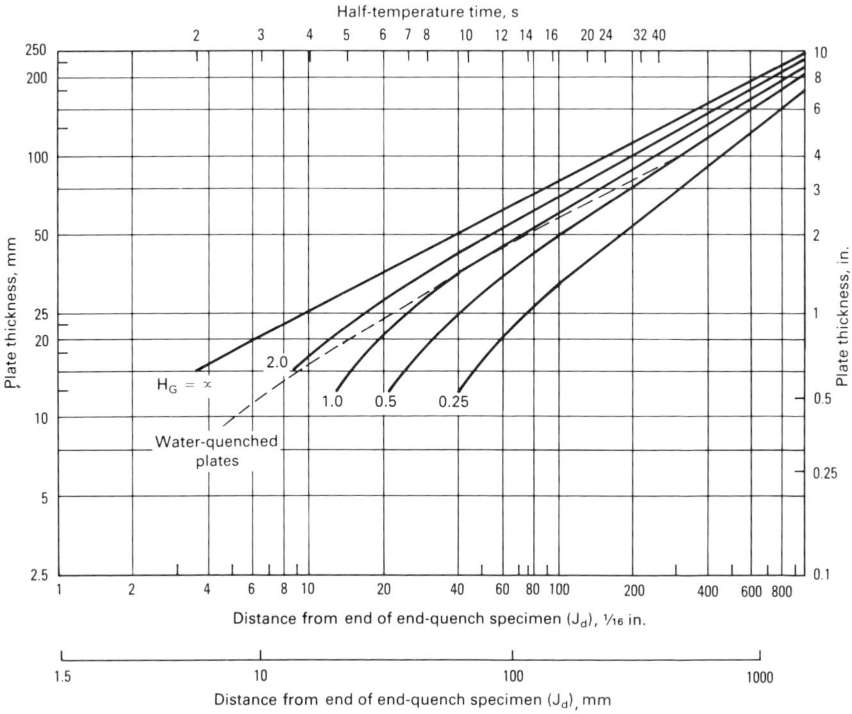

Fig. 2.14 Correlation between Jominy equivalent distance and center cooling rates in plates quenched at various severities

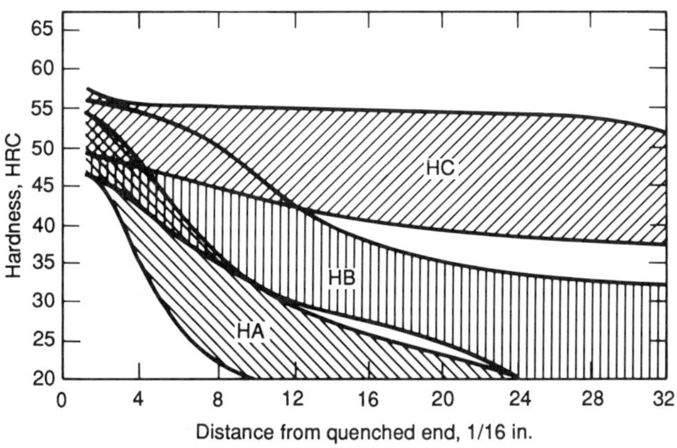

Fig. 2.15 Hardenability bands specified by SAE J435a

Part of the width of a band for a given alloy is caused by variations in analysis, including residual elements, from heat to heat. Some elements (chromium and molybdenum, for example) have an appreciable effect on hardenability even when present in small quantities. As a consequence, residual alloy element concentrations must be monitored and controlled as closely as specified elemental concentrations if hardness values within the hardenability band are to be maintained.

Quenching and Quenchants To Produce Martensite

The term quenching is sometimes used to refer to the rapid heat extraction required when parts are being austempered or martempered in hot oil or salt baths, but its present use is restricted to the process of removing heat to form martensite. Just how rapidly the steel must be cooled to miss the nose of the transformation curve depends on the hardenability of the particular alloy. More hardenable materials do not require fast cooling rates, but achieving martensitic microstructures in most grades of steel requires faster cooling than can be obtained in air. It is usually necessary for steel to be sprayed or immersed in a liquid quenchant to increase the rate of heat extraction; in a few cases, it is possible to use gas quenching, especially agitated gas under pressure.

Three stages of heat extraction are associated with quenching. Duplicate temperature-time curves in a part being quenched and corresponding cooling rate curves are illustrated in Fig. 2.16, which also shows the various stages of cooling. Slow cooling occurs in stage A as a result of a vapor barrier around the part. Heat transfer during this period occurs primarily by radiation through the vapor blanket. Stage B quenching is the interval where the rate of heat removal is highest. During this stage, the vapor blanket collapses and high heat-extraction rates associated with nucleate boiling on the part surface are achieved. The final stage of quenching (stage C) is associated with the end of boiling and slow heat removal by conduction and convection from the steel into the liquid quenchant (Ref 3, 12).

Agitating the quenchant accelerates heat removal by limiting the duration of stage A and accelerating heat removal in stages B and C. An increase in the quenchant temperature usually slows the rate of heat removal if other conditions are held constant.

The cooling rates and temperature ranges of the three stages vary with the type of quenchant and the mass of the parts being quenched. The fastest overall cooling rates are achieved with brine solutions, generally followed by water, polymer solutions in water, oils, inert gases, and air.

Transformation diagrams provide data on various steels and the cooling rates necessary to form martensite and to avoid ferrite or pearlite formation. The required cooling rate is determined by the hardenability of the steel. The actual rate that can be achieved is a function of the quenchant and facility, the thickness and geometry of the part, and the heat capacity and thermal conductivity of the alloy. The quenching capacity

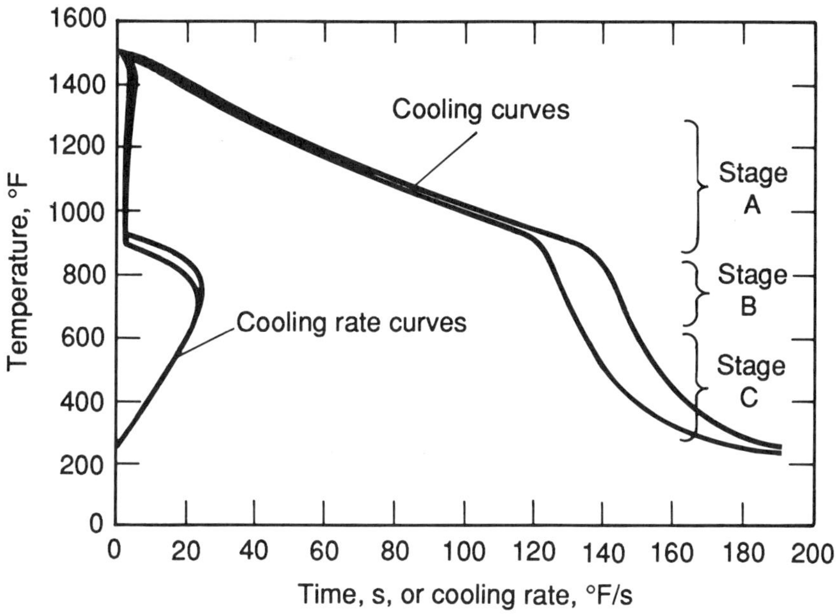

Fig. 2.16 Cooling curves and cooling rate curves at center of a 38 mm (1.5 in.) diam probe quenched in unagitated hot water

of a given facility depends on the quenchant used, its temperature, and the extent of agitation.

Quenchants

Many materials and solutions have been used as quenchants for steels. The fastest quench rates are obtained using water or brine (water-salt solutions). Slower, less drastic quenchants include the water-base polymers, oils, and salts. Very slow rates can be obtained by cooling in air, inert gas atmospheres, or vacuum. These various quenching media are discussed in Chapters 4, 5, 7, and 8.

Quenching Variables

Quench facility operation involves the careful analysis of two intimately interrelated factors. The first is the type of quenchant and the operating conditions in use, and the second is the hardenability of the particular alloy(s) being heat treated. The hardenability is a function of the chemical composition and grain size of the specific heat.

Water is an excellent quenchant, but its heat-removal characteristics are variable, depending on the bath temperature and the relative velocity between the water and the parts being quenched, as illustrated in Fig. 2.17 and by the data in Table 2.5. Figure 2.17 illustrates cooling curves obtained while quenching 38 mm (1.5 in.) diam probes in water at 27, 32, 60, and 70 °C (80, 90, 140, and 160 °F). The water velocity past the probe during quenching was a constant 15 m/min (50 ft/min). The cooling rates decreased rather dramatically, from a value of about 33 °C/s (60 °F/s) with 27 °C (80 °F) water to about 20 °C/s (35 °F/s) when the water temperature was 70 °C (160 °F).

Data on the effect of quenchant type, temperature, and quenchant velocity on the Grossmann number and film coefficient produced by water, a fast oil, a conventional oil, a martemper oil, a 25% water-soluble polymer solution, and air are presented in Table 2.3 (Ref 12). Water at 27 to 32 °C (80 to 90 °F) can produce film coefficients greater than 5700 W/m^2 · K (1000 Btu/ft^2 · h · °F), depending on the extent of agitation. The film coefficient decreases rather rapidly, however, as the water temperature rises or as the water velocity past the part decreases.

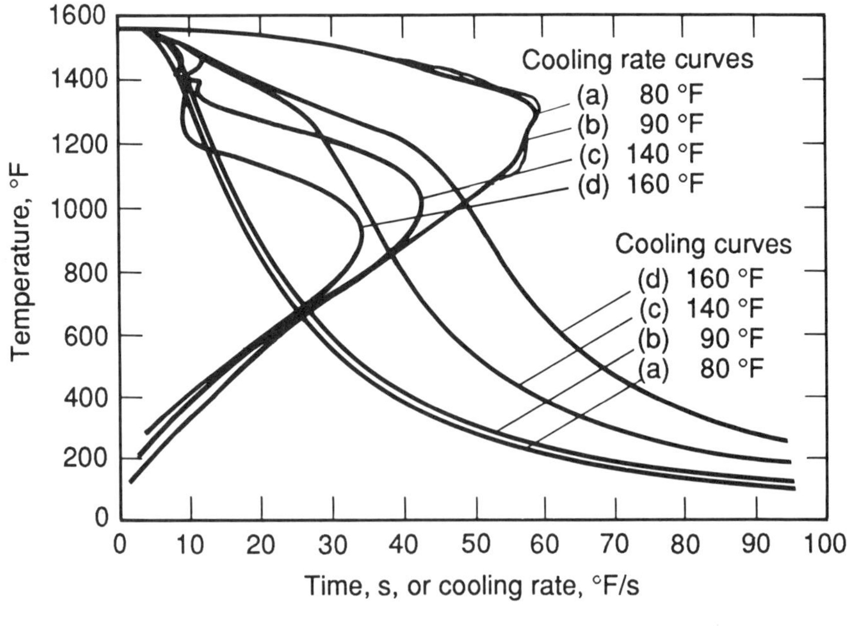

Fig. 2.17 Cooling curves and cooling rate curves produced by **(a)** 27 °C (80 °F), **(b)** 32 °C (90 °F), **(c)** 60 °C (140 °F), and **(d)** 70 °C (160 °F) water flowing at 15 m/min (50 ft/min) past a 38 mm (1.5 in.) diam bar

The heat-removal characteristics of water at 55 °C (130 °F) are substantially reduced, especially at lower velocities, compared with 27 °C (80 °F) water. The film coefficient produced by 55 °C (130 °F) water without agitation was only 340 W/m^2 · K (60 Btu/ft^2 h · °F), compared with a value of approximately 4150 W/m^2 · K (730 Btu/ft^2 · h · °F when used at a velocity of 45 m/min (150 ft/min). The decrease in cooling rate with higher water temperature is evident from the rate curves in Fig. 2.17.

Without agitation, water at 32 °C (90 °F) can remove heat approximately five times faster than water at 55 °C (130 °F) without agitation. At 15 m/min (50 ft/min), 55 °C (130 °F) water has less than one-third the ability to remove heat compared with 32 °C (90 °F) water at the same velocity. At a velocity of 30 m/min (100 ft/min), the heat-removal ability of 55 °C (130 °F) water is approximately half that of 32 °C (90 °F) water.

These data are sufficient to illustrate the variability of water and water-base quenchants as heat-removal media. When large parts are being quenched, the localized temperature and velocity of water past parts significantly affect heat-removal rates. The hardenability of the specific alloy and the heat-removal rate significantly affect the hardness that can be expected in quenched parts.

Table 2.5 Effect of water temperature on cooling rates and film coefficients in 38 mm (1.5 in.) diam 4130 steel bars quenched from 845 °C (1550 °F)

Water bath temperature		Velocity		Cooling rate, °C/s (°F/s), at :			Film coefficient	
				705 °C	345 °C	205 °C		
°C	°F	m/s	ft/min	(1300 °F)	(650 °F)	(400 °F)	W/m^2 · K	Btu/ft^2 · h · °F
25	80	0.25	50	32.5 (58.6)	14.5 (26.2)	7.3 (13.2)	9420	1652.5
30	90	0.25	50	32.2 (58.1)	14.6 (26.3)	7.2 (13.0)	9060	1589.4
40	100	0.25	50	31.0 (55.8)	14.3 (25.8)	6.9 (12.5)	7675	1346.5
50	120	0.25	50	24.3 (43.7)	14.4 (25.9)	6.8 (12.2)	3520	617.1
60	140	0.25	50	9.3 (16.8)	14.3 (25.7)	6.3 (11.3)	760	132.8
70	160	0.25	50	5.6 (10.1)	13.7 (24.6)	5.9 (10.6)	420	73.5
80	180	0.25	50	4.7 (8.5)	13.4 (24.2)	5.4 (9.7)	340	60.1
95	200	0.25	50	3.5 (6.3)	13.2 (23.7)	5.1 (9.1)	240	42.5
100	212	0.25	50	3.0 (5.4)	12.5 (22.6)	5.5 (9.9)	205	35.8

Measuring Quench Severity

Methods for determining the ability of a quenching solution or facility to extract heat can be divided into two broad categories: mechanical and thermal. The mechanical procedures usually employ hardness measurements or etching procedures applied to quenched parts. These methods determine the depth of martensite formation in a particular alloy after quenching under standardized conditions. Thermal techniques involve measuring the time required to cool between specified temperatures, determining the rate of cooling in a standard specimen, or defining by some other thermal means the ability of a quenchant to extract heat.

Grossmann Technique

The Grossmann technique has been used for many years to evaluate the quench severity (H) provided by a quenchant. This procedure is based on the observation that with increasing bar diameter, the unhardened core diameter increases with the use of less severe quenchants. Determining the Grossmann H value involves quenching a series of bars of different diameters (D) in a particular quenchant or quench facility. The bars are then sectioned and the cross section of the unhardened diameter (D_u) determined by either hardness measurements or by chemical etching.

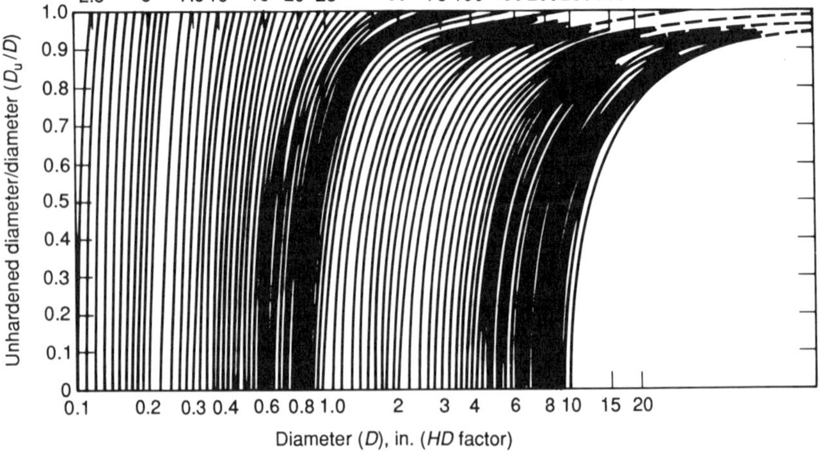

Fig. 2.18 Curves for estimating the cooling power (severity of quench) of quenching baths

A graph on the appropriate scale is made of D_u/D for the several quenched bar sizes (Ref 3). This D_u/D curve can be superimposed on Fig. 2.18 and matched with the best-fit curve. The *HD* factor on the abscissa is the product of the Grossmann *H* factor and the bar diameter *D*. This value is read from the scale and divided by the actual bar diameter to obtain the quench severity factor, *H*. One limitation of this procedure is that it may not be possible to match a curve if there is a large difference between the bar diameter and the depth of hardening (Ref 3, 15, 16).

The Grossmann chart illustrated in Fig. 2.19 can also be used to obtain *H* values for quench severity under operating conditions (Ref 14). As an example of the use of the chart, the hardness at the center of a 30.5 mm (1.2 in.) diam bar quenched in still water ($H = 1.0$) will be the same as the hardness a distance of 4⁄16 in. (6.4 mm) from the quenched end of a Jominy bar of the same alloy.

To determine the *H* value of a commercial quench, an end-quench bar and cylinders of two different diameters are needed. The cylinder diameters are chosen so that the center hardnesses fall on the sloping part of the end-quench curve. The analysis is selected to give a usable end-quench curve.

The two cylinders are quenched with a load of parts. In the example in Fig. 2.19 for a still-water quench, the hardness at the center of the 38 mm (1.5 in.) diam bar equaled that at 5⁄16 in. (8 mm) from the end of the end-quench bar. The hardness at the center of the 25 mm (1 in.) diam bar was equivalent to 3⁄16 in. (4.8 mm) from the quenched end of the Jominy bar. These two points and a line joining them are plotted in Fig. 2.19. The *H* value of this quench is slightly over 1.0. The curves for oil and hot salt were determined in the same manner.

Rushman Technique

The Rushman method is a variation of the bar quenching technique. Two bars having the same composition, intermediate hardenability (typically 1141), and different diameters (typically 13 and 25 mm, or ½ and 1 in.) are quenched in the medium of interest. The bars are then cut in two and the hardness at the bar center measured. Using the bar diameter and the equivalent Jominy distance based on hardness achieved when the two bars were quenched, the data points are placed on a Lamont diagram, illustrated in Fig. 2.20 (Ref 17, 18), or compared with Table 2.3 to assign an *H* value for the quenchant.

Cooling Curves

Data obtained by measuring the cooling rates in a test specimen also can be used to characterize quenchant behavior under a variety of conditions. Typical cooling curves are illustrated in Fig. 2.17. A specimen is usually machined from a nonscaling, nontransforming material, such as silver, nickel, or an austenitic stainless steel. The test specimen is usually cylindrical, with a diameter ranging from 13 to 50 mm (½ to 2 in.) or longer and a length-to-diameter ratio greater than 4. Two positions in the specimen, one near the surface and one in the center, are instrumented with thermocouples to provide information on the cooling rates achieved. The thermocouple and recording system used

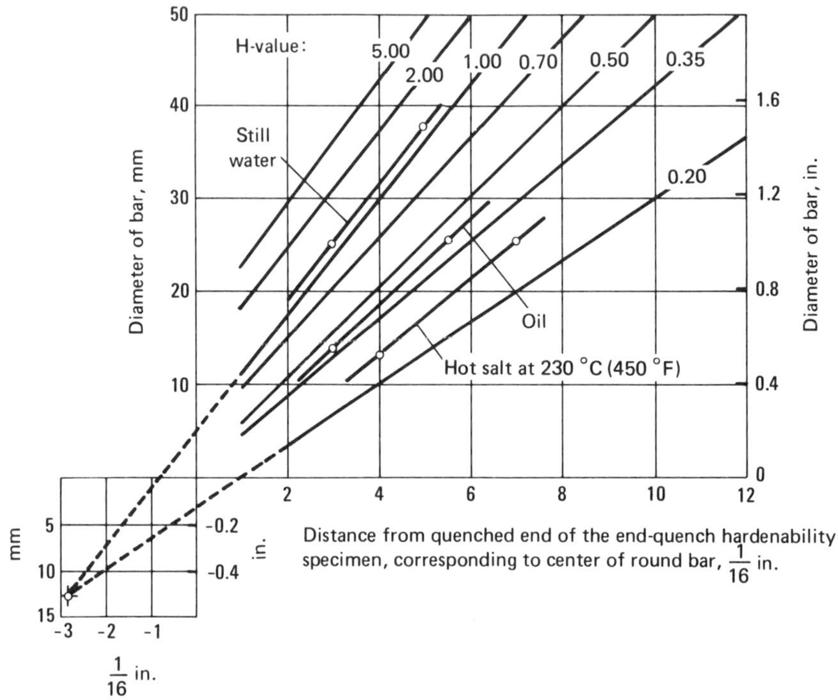

Fig. 2.19 Grossmann chart relating bar diameter, hardenability of steel, and severity of quench (H). H-values: 5.00, strong brine quench, violent agitation; 2.00, poor brine quench, no agitation; 1.00, poor water quench, no agitation; 0.70, strong oil quench, good agitation; 0.50, good oil quench, good agitation; 0.35, good oil quench, moderate agitation; 0.20, poor oil quench, no agitation

must be capable of measuring rapid temperature changes, especially if a thermocouple is placed near the surface of a probe. This type of equipment has been used to study quenchants, spray quenching, agitation, bath temperature, and the effects of surface condition on heat extraction from metal.

Quench Factors

Jominy curves have long been used to represent hardenability of various grades of steel, as has the Grossmann H value to provide information about the ability of a quenchant to extract heat from parts. More recently, quench factors have been introduced for quantifying quenchant and steel interactions (Ref 19-22).

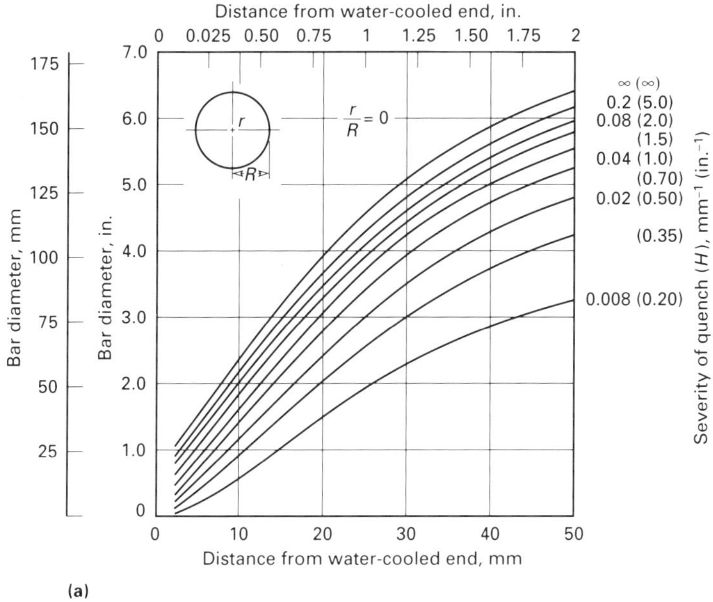

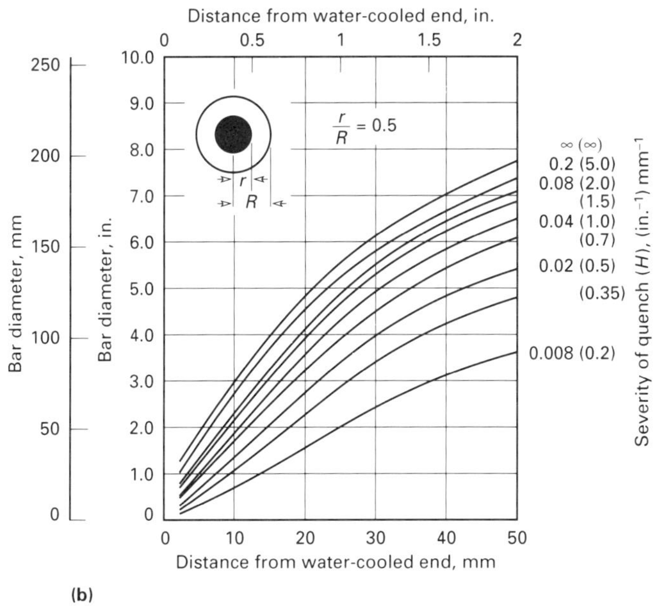

Fig. 2.20 Location in end-quench Jominy hardenability round bar specimen corresponding to **(a)** center position and **(b)** midradius position

Quench factor analysis (QFA) provides a single number that interrelates the cooling rate as produced by the quenchant, the section thickness of the part, and the transformation rate of the alloy as reflected by the time-temperature-property (TTP) curve. The TTP curve is the mathematical representation of the start of transformation curve and reflects the amount of transformation required to reduce the as-quenched hardness or strength from the maximum value by a fixed percentage. The alloy composition and quench conditions must be controlled so that the maximum quench factor corresponding to the minimum acceptable as-quenched hardness is not exceeded.

Lower quench factor values are associated with higher as-quenched hardness values. The critical value of the quench factor (Q) is the maximum value that will result in the desired hardness value and can be defined in terms of the desired martensitc content, that is, 99%, 95%, 90%, etc.

Quench Factor Calculation. Quench factors can be calculated from time-temperature (cooling curve) data and the equation describing the transformation kinetics of the alloy of interest (C_T function).

The C_T function is usually described by an equation of the form:

$$C_T = -K_1K_2 \exp\left[\frac{(K_3K_4^2)}{RT(K_4 - T)^2}\right] \exp\left[\frac{K_5}{RT}\right] \qquad \text{(Eq 2.2)}$$

where C_T is the critical time required to form a constant amount of a new phase or to reduce the hardness by a specified amount (the locus of the critical time values as a function of temperature forms the TTP curve); K_1 is a constant that equals the natural logarithm of the fraction untransformed during quenching, that is, the fraction defined by the TTP curve; K_2 is a constant related to the reciprocal of the number of nucleation sites; K_3 is a constant related to the energy required to form a nucleus; K_4 is a constant related to the solvus temperature; K_5 is a constant related to the activation energy for diffusion; R = 8.3143 J/K · mol; and T is temperature in degrees Kelvin. The constants K_1, K_2, K_3, K_4, and K_5 define the shape on the TTP curve.

An incremental quench factor, q, for each time step in the cooling curve is calculated using Eq 2.3:

$$q = \frac{\Delta t_i}{C_{Ti}} \qquad \text{(Eq 2.3)}$$

where t is the time step used in cooling curve data acquisition. The incremental quench factor, q, represents the ratio of the amount of time a steel was at a particular temperature divided by the time required for a specified amount of transformation, typically 1%, at that temperature.

The incremental quench values are summed over the entire transformation range between Ar3 and M_s to produce the cumulative quench factor, Q, according to:

$$Q = \sum q = \sum_{T=M_s}^{T=Ar_3} \frac{\Delta t_i}{C_{Ti}} \qquad \text{(Eq 2.4)}$$

The cumulative quench factor reflects the heat-extraction characteristics of the quenchant as a function of the quenchant type, velocity, and temperature over the transformation range of the steel being quenched. This factor includes section thickness effects on cooling rate and the transformation rate of the alloy as a function of temperature.

The calculated quench factor can be used to predict the as-quenched hardness in steel using the following:

$$H_p = H_{min} + (H_{max} - H_{min}) \exp(K_1 Q) \qquad \text{(Eq 2.5)}$$

where H_p is the predicted hardness, H_{min} is the minimum hardness for the alloy, H_{max} is the maximum hardness for the alloy, $K_1 = \ln 0.995 = -0.00501$, and Q is the quench factor.

Hardenability of Cast 4130 Steel. Calculated hardenability data for cast 4130 steel are presented in Table 2.6 (Ref 20). The first column provides composition and hardenability data, assuming all alloy elements to be at the low end of the specified composition range. Although such an assumption is unrealistic, the calculation is made to illustrate the potential extremes that can be encountered. The carbon range has been expanded from the normal AISI range of 0.28 to 0.33% to cover the range acceptable in castings: 0.25 to 0.35% C. The second column in Table 2.6 provides data for an actual cast steel composition tested, and the third column provides data assuming all elements to be at the high end of the composition range. The composition data are followed by calculated alloy factors and ideal diameters for the low, actual, and high-end compositions.

The next data group in Table 2.6 contains calculated martensite hardness data. At 99% martensite, a 4130 steel containing 0.25% C can be expected to have a maximum hardness of 47.3 HRC, and steel containing 0.31% C is expected to have a maximum martensite hardness of 51 HRC. As the martensite content decreases, the as-quenched hardness decreases to a value of 38.3 HRC at 50% martensite. The M_s temperature for the steel was calculated to be 367 °C (692 °F), and the temperature at which the steel had transformed to 90% martensite was 263 °C (506 °F), using equations published by Atkins (Ref 23).

Calculated and experimental Jominy hardness values for the low-specification composition, the actual composition, and the high-specification composition 4130 are shown in Fig. 2.21 (Ref 20). The solid lines represent the calculated (predicted) Jominy hardness distribution using the methods, procedures, and coefficients described in

Table 2.6 Hardenability data for 4130 steel

	Low specification	Actual composition	High specification
Chemical composition, %			
Carbon	0.250	0.315	0.350
Silicon	0.150	0.379	0.400
Manganese	0.400	0.576	0.700
Nickel	0.000	0.103	0.103
Chromium	0.800	1.140	1.500
Molybdenum	0.150	0.244	0.250
Copper	0.000	0.125	0.125
Vanadium	0.000	0.014	0.014
Boron	0.000	0.000	0.000
Alloy factors			
Carbon	0.135	0.170	0.189
Silicon	1.105	1.265	1.280
Manganese	2.333	2.920	3.333
Nickel	1.000	1.035	1.035
Chromium	2.728	3.462	4.240
Molybdenum	1.450	1.732	1.750
Copper	1.000	1.046	1.046
Vanadium	1.000	1.024	1.024
Ideal diameter, mm (in.)	35.1 (1.38)	106.2 (4.18)	168.4 (6.63)
Martensite hardness, HRC			
M_{99}	47.3	51.2	53.3
M_{95}	43.9	47.9	49.9
M_{90}	41.5	45.5	47.4
M_{80}	38.2	42.0	43.9
M_{50}	34.7	38.3	40.0
Transformation temperature, °C (°F)			
A_3	819.9 (1507.9)	820.6 (1509.1)	815.5 (1500.0)
A_1	732.6 (1350.7)	739.6 (1363.3)	743.8 (1370.8)
B_5	658.1 (1216.5)	589.2 (1092.6)	542.9 (1009.3)
M_5	402.2 (756.0)	366.6 (691.9)	345.4 (653.7)
M_{90}	299.2 (570.6)	263.6 (506.5)	242.4 (468.3)

Caterpillar Tractor Company Specification 1E38. The data points illustrate the experimental Jominy hardness measurements on the steel after austenitizing and quenching from 860 °C (1575 °F). There was reasonably good agreement between the predicted Jominy hardness and the measured hardness in this alloy.

TTP Curves for Cast 4130 Steel. Most published TTT curves were developed for specific steel compositions and do not take into account the possible variations in composition and shifts in TTT curve position that can occur within the allowable composition range. Using quench factor analysis, it is possible to calculate the start of the transformation curve for a steel where a modest amount of data exists.

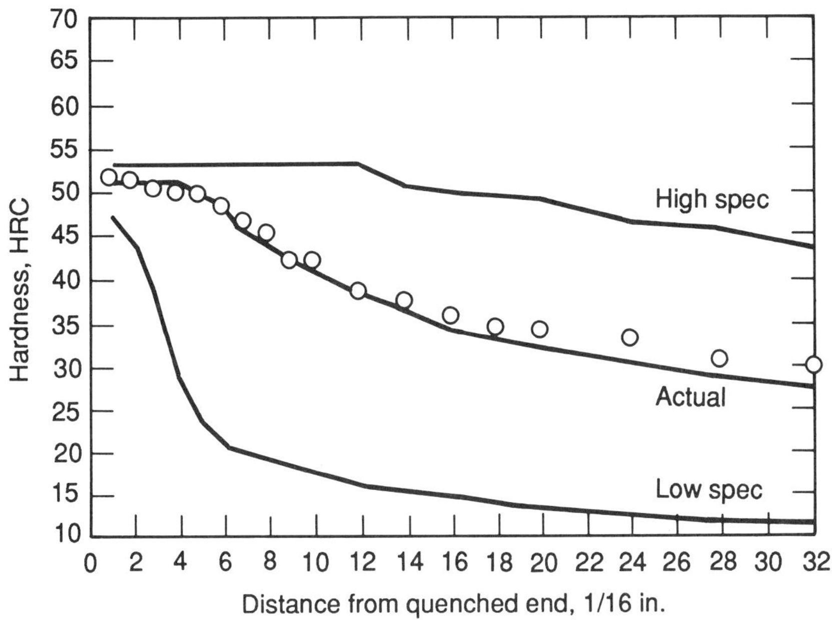

Fig. 2.21 Calculated and experimental Jominy data for cast 4130 steel

The approximate TTP curves have been developed for the upper- and lower-limit compositions of cast 4130 steel given in Table 2.6. The shift in C curve with composition is clearly evident in Fig. 2.22 (Ref 20). These C curves are plotted for times to 1000 s (16.7 min) rather than the usual 1,000,000 s (20 days) used in many published diagrams. Transformations that occur over a 20-day period are of little interest in heat treating operations. At 595 °C (1100 °F), the transformation begins in about 0.15 s in the low-specification composition, in 1 s in the actual composition, and after about 3 s in the high-specification composition.

The C curves illustrate the shift in the start of transformation with alloy content, and the mathematical expression describing the curves allows hardness predictions to be made under a variety of quenching conditions. Several correlations between quench factor and predicted hardness and measured hardness in cast 4130 steel are presented in Table 2.7 and are graphically illustrated in Fig. 2.23. The solid line in Fig. 2.23 represents the predicted hardness as a function of quench factor, and the data points are measured hardness values at locations in a quenched part where cooling curves were available.

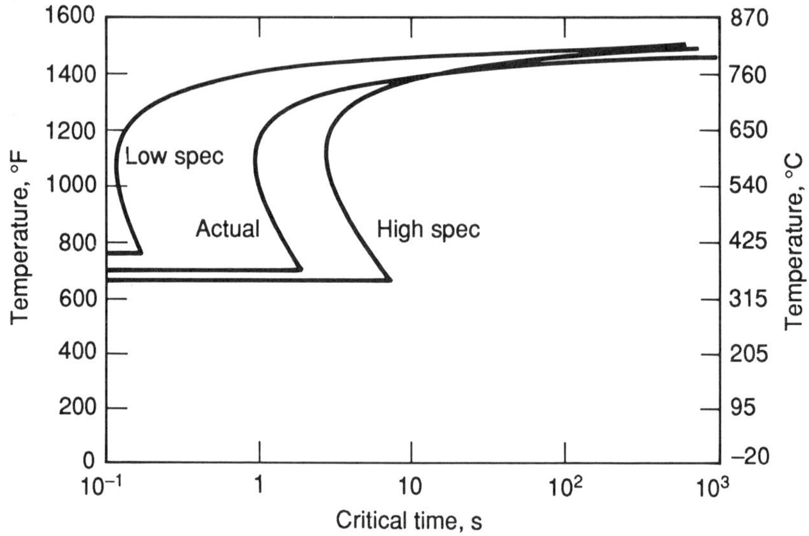

Fig. 2.22 TTP curves for cast 4130 steel

Table 2.7 Measured and predicted hardnesses at 17 positions in water-quenched 4130 steel

Casting position	Cooling rate at 705 °C (1300 °F), °C/s	Cooling rate at 705 °C (1300 °F), °F/s	Quench factor	Predicted hardness, HRC	Measured hardness, HRC
1	114.0	205.4	7.2	51.8	52
2	67.5	121.7	12.4	50.9	51
3	40.0	72.1	18.5	49.9	50
4	29.4	52.9	25.6	48.8	50
5	21.5	38.8	33.1	47.7	49
6	17.4	31.3	41.5	46.4	47
7	13.2	23.8	50.1	45.2	45
8	11.0	19.9	58.7	44.1	42
9	9.4	17.0	67.3	43.0	42
10	7.0	12.7	86.8	40.7	39
11	5.4	9.8	103.1	38.9	38
12	4.5	8.1	119.1	37.3	36
13	3.8	6.8	134.8	35.8	35
14	3.3	5.9	150.1	34.5	34
15	2.6	4.6	175.3	32.5	33
16	2.2	3.9	197.6	31.0	31
17	1.8	3.3	215.6	29.9	30

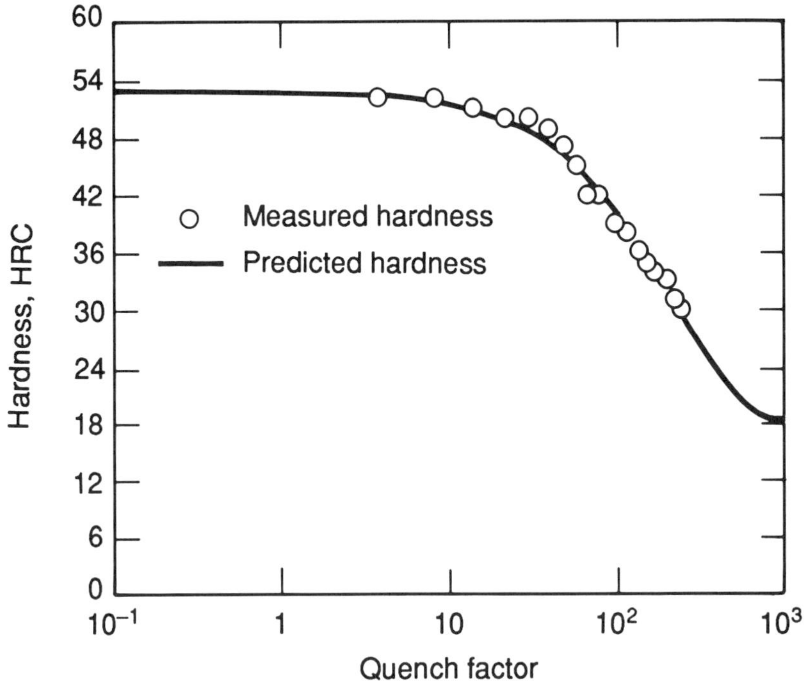

Fig. 2.23 Correlation between predicted hardness and measured hardness at 18 locations in a quenched part

Advantages and Disadvantages of QFA

Quench factor analysis has several advantages over other techniques used to evaluate quenchants:

- The quench factor provides a single number describing quench severity.
- The quench factor is related to the hardness in the quenched product.
- Intermediate manual interpretations are not required.

Disadvantages of the technique include the fact that very few TTP curve coefficients have been published for steel alloys. Modeling is difficult for the rapid transition from soft ferrite and pearlite structures to martensite in low-hardenability steels.

References

1. R.F. Kern, Thinking Through to Successful Heat Treatment, *Met. Eng. Q.*, 1971, p 1-4
2. *Source Book on Heat Treating*, Vol II, *Production and Engineering Practices*, American Society for Metals, 1975

3. C.A. Siebert, D.V. Doane, and D.H. Breen, *The Hardenability of Steels*, American Society for Metals, 1977
4. ASTM A 255, "Jominy Test, Standard Method for End-Quench Test for Hardenability of Steel," *Annual Book of ASTM Standards*, ASTM
5. "Hardenability Slide Rule," United States Steel Corp., 1970
6. STECAL computer program, ASM International
7. S.E. Feldman, The Minitech Computerized Alloy Steel Information System, *Proc. Metall. Soc. AIME Heat Treatment Comm.*/ASM Activity on Phase Transformation Symp., Chicago, 24-26 Oct 1977, p 546-567
8. J.M. Hodge and M.A. Orehoski, Relationship Between Hardenability and Percentage of Martensite in Some Low-Alloy Steels, *Trans. AIME*, Vol 167, 1946, p 627
9. M.A. Grossmann and E.C. Bain, *Principles of Heat Treatment*, 5th ed., American Society for Metals, 1964
10. "Hardenability Calculations and Data," Steel Casting Design, Engineering Data File, Section 12, Steel Founders' Society of America
11. Calculating Hardenability Curves From Chemical Composition, *Met. Prog.*, June 1977, p 114
12. C.E. Bates, Predicting Properties and Minimizing Residual Stress in Quenched Steel Parts, *J. Heat Treat.*, Vol 6 (No. 1), 1988, p 27-45
13. M.A. Grossmann, *Elements of Hardenability*, American Society for Metals, 1952
14. Quenching of Steel, *Metals Handbook*, 8th ed., Vol 2, American Society for Metals, 1964, p 15
15. M.A. Grossmann, *Principles of Heat Treatment*, American Society for Metals, 1935
16. B.D. Wakefield, Agitation Quenching: A Good Way to Avoid High Alloying Costs, *Iron Age*, 1 Oct 1970, p 54-55
17. J.L. Lamont, How to Estimate Hardening Depth in Bars, *Iron Age*, 14 Oct 1943, p 64-70
18. W.F. Rushman, How to Determine the Quench Severity in Oil and Salt Baths, *Met. Prog.*, Dec 1963, p 91
19. C.E. Bates, G.E. Totten, and R.L. Brennan, Quenching of Steel, *ASM Handbook*, 10th ed., Vol 4, ASM International, 1991, p 67-120
20. C.E. Bates, Effects of Composition and Quenching Practice on the As-Quenched Hardness of Cast 4130 Steel, *Proc. 44th Steel Founders' Society of America Technical and Operating Conf.*, Chicago, 8-10 Nov 1990, p 1-38
21. C.E. Bates, T. Landig, and G. Seitanakis, Quench Factor Analysis: A Powerful Tool Comes of Age, *Heat Treat.*, Vol 85, Dec 1985, p 13-17
22. C.E. Bates, G.E. Totten, and K.B. Orszak, Quench Severity Effects on the Properties of Selected Steel Alloys, *IFAT Conf. Proc.*, Melbourne, 2-5 Sept 1991
23. M. Atkins, *Atlas of Continuous Cooling Transformation Diagrams for Engineering Steels*, American Society for Metals, 1980

CHAPTER 3

Cooling Curve Analysis

The primary function of a quenchant is to control the heat-transfer rates from parts during the quenching process and produce the desired hardness, strength, and toughness. Simultaneously, the quenchant should minimize distortion and residual stresses by enhancing the quenching uniformity over the total surface of the part. The physical properties obtained are directly related to cooling rates in the part.

As the heat treating industry becomes more competitive, monitoring of quenching by heat treaters becomes more important. Consistency checks are needed to identify variations in quench severity and to qualify new quenchants or quenchant processes.

It has long been known that the severity of a quenchant can be measured by "hardening power" tests, such as cross-sectional hardness surveys (U curves) (Ref 1) and Jominy end-quench tests (Ref 2). However, these traditional tests are time consuming. Furthermore, metallurgical responses (Rockwell hardness, for example) vary with test specimen chemistry. As a consequence, development of alternative procedures for predicting the ability of a quench bath to produce the desired physical properties is of considerable interest.

A quenchant bath can be characterized by methods other than direct hardness measurements on test specimens or parts. A common alternative procedure is to measure the "cooling power" of the quenchant. Various methods have been developed to simplify the measurement of cooling power, including the General Motors (GM) quenchometer or "nickel ball" test (Ref 3), the hot wire test (Ref 4), the 5-second interval test, and the cooling curve test. Of these procedures, the cooling curve test has been generally accepted as the most useful means of describing the mechanisms of quenching (Ref 5). Cooling curves are particularly sensitive to factors that affect the ability of quenchants to extract heat, including quenchant type and physical properties, bath temperature, and bath agitation.

This chapter reviews the development of the cooling curve analysis and discusses in detail the methods used for cooling curve interpretation. Some of the methods reviewed include recently published studies conducted to correlate cooling curves with metallurgical transformations and with physical properties. Throughout this chapter, the notations of the original authors have been retained where possible.

Cooling Mechanisms

The shape of a cooling curve is indicative of the various cooling mechanisms that occur during the quenching process. For example, when austenitized steel is quenched in a volatile liquid medium such as water, oil, or an aqueous polymer, at least three stages of heat transfer occur (Fig. 3.1) (Ref 1, 6). These stages reflect three different cooling mechanisms that occur during the quenching of hot metals. These three cooling zones are known as stages A, B, and C.

Stage A, or the vapor blanket stage, is characterized by the formation of an unbroken vapor blanket around the hot metal. Upon further cooling, stage B, or the nucleate boiling stage, begins. This cooling mechanism is characterized by violent boiling at the metal interface. The transition temperature from A- to B-stage cooling is classically known as the leidenfrost temperature and is independent of the initial temperature of the metal being quenched (Fig. 3.2) (Ref 7, 8).

Stage C, or the convective cooling stage, begins when the metal cools below the boiling point of the quenching fluid. The B- to C-stage transition temperature is primarily a function of the boiling point of the quenchant, and the rate of heat removal in stage C is much slower than in stage B. Heat-transfer rates in this region are affected by various process variables, such as agitation, quenchant viscosity, and bath temperature, and by the viscosity of the quenchant medium.

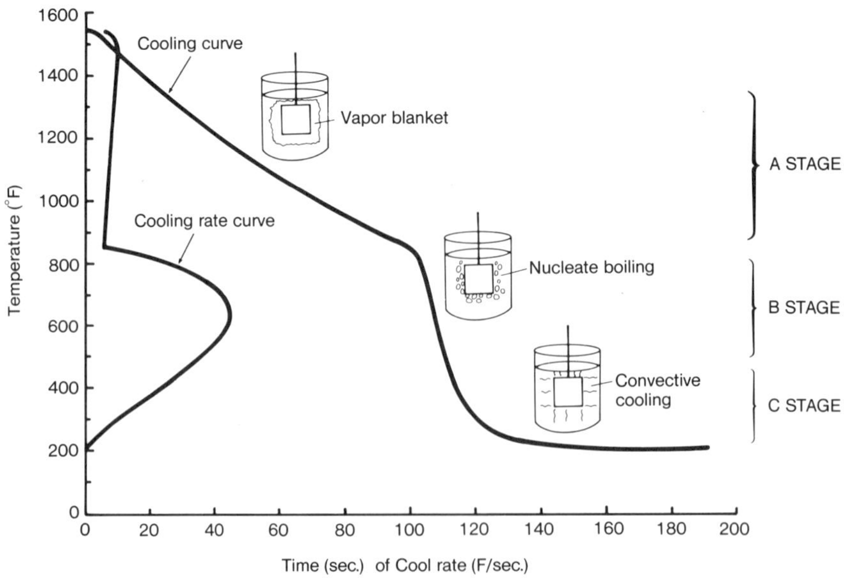

Fig. 3.1 Cooling mechanisms

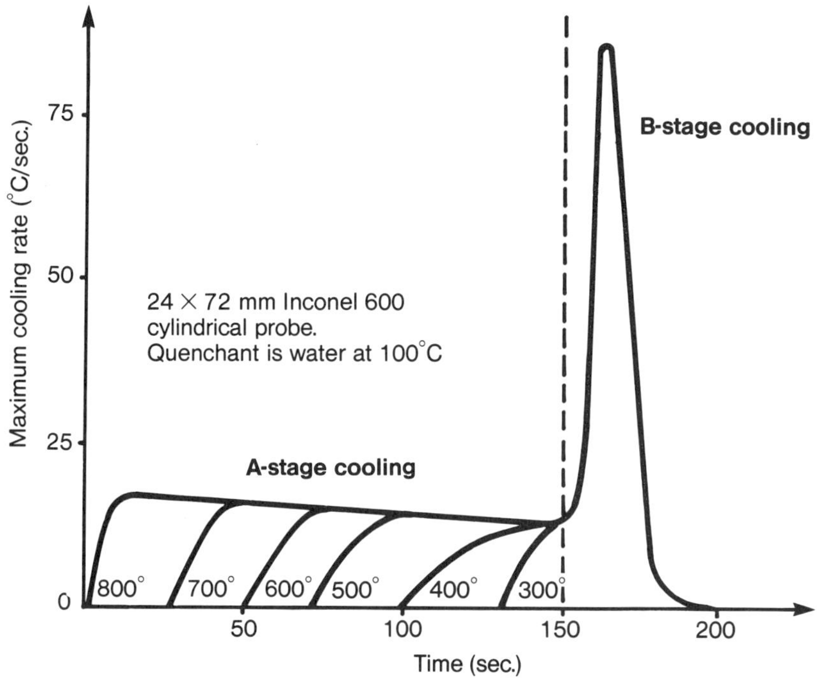

Fig. 3.2 Cooling rate as a function of metal probe temperature

Variations in film-forming properties during A-, B-, and C-stage cooling of various aqueous polymer quenchants are shown in Fig. 5.12. Similar cinematographic studies have been reported for water and for various oils and polymers (Ref 9, 10).

Cooling curves are useful because the cooling rate in a test specimen can be recorded throughout the quench cycle. It is conceptually possible, although sometimes difficult, to illustrate and quantify the cooling mechanisms encountered while quenching an actual part in a particular quenchant medium.

From the measured cooling curve, it is possible to calculate the cooling rate curve as a function of time or as a function of temperature (Fig. 3.3). The time required to cool to any temperature, the temperature after any time, and the time required to cool over a particular temperature range can be readily calculated.

Cooling Curve Acquisition

Cooling curves are relatively easy to obtain experimentally by using an apparatus that typically consists of an instrumented probe and a system for data acquisition and display. Such systems can be purchased or custom built. Either way, it is important that the user understand the basic system components. One of the most important components is the quench probe used for cooling curve acquisition.

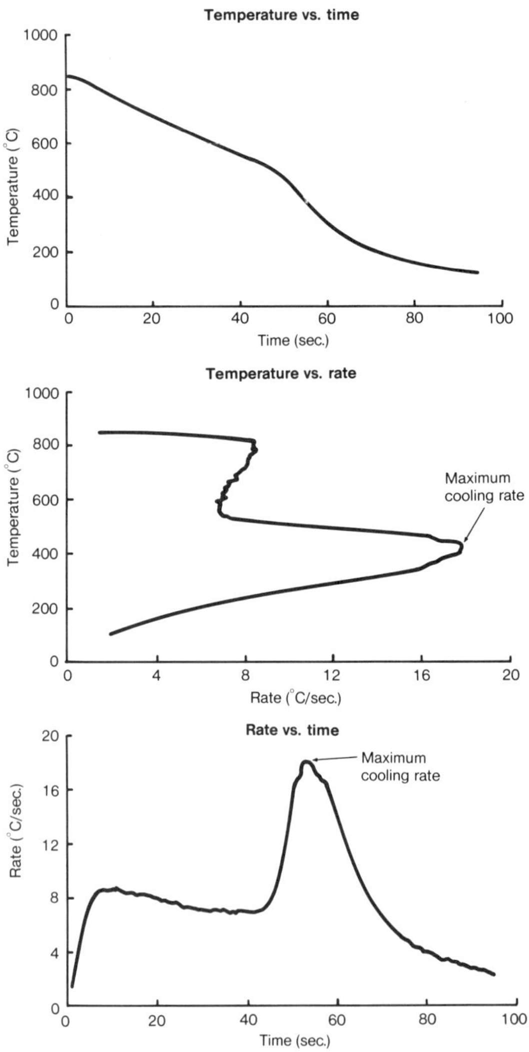

Fig. 3.3 Various representations of cooling curve data

Probes

With the advent of personal computers and other commercial systems for cooling curve acquisition, it is sometimes forgotten that cooling curves have been analyzed for more than 60 years. Some of the earliest research in this area was conducted at the Max Planck Institute in the 1930s (Ref 11). Since that time, probes have been produced in a great variety of shapes, including cylinders (Ref 12), spheres (Ref 13, 14), square bars (Ref 15), plates (Ref 16-18), rings and coils (Ref 19), round disks (Ref 20), and production parts (Ref 21). Probes have also been constructed of various materials, including alloy (Ref 12) and stainless steels (Ref 22), silver (Ref 14, 23, 24), nickel (Ref 25), copper (Ref 26, 27), gold (Ref 28), and aluminum (Ref 29).

Figures 3.4 to 3.14 illustrate various types of probes. One of the earliest probes used is shown in Fig. 3.4 (Ref 12). This probe was constructed from a 100 × 300 mm (4 × 12 in.) SAE 5145 alloy by cutting it in half, yielding two 100 × 150 mm (4 × 6 in.) sections. A chromel-alumel thermocouple was hydrogen brazed to the center of the bottom half. The thermocouple wires were passed through a 13 mm (½ in.) hole drilled through the top half of the bar. A steel tube, used as a handle, was welded to the top section, and both sections were then welded together. The cooling time-temperature data were collected using Speedomax equipment manufactured by Leeds & Northrup Company.

Figure 3.5 illustrates a smaller probe, 13 × 50 mm (½ × 2 in.), shown by French in his classic 1930 paper (Ref 30). Interestingly, cooling temperature data were obtained using an Einthoven "string" galvanometer. Except for the data acquisition equipment, the quenching equipment reported by French was not substantially different from equipment used today.

French noted some precautions that should be considered in probe construction:

- The probes should all be machined or ground by the same method to provide the same surface finish.
- The smallest practical thermocouple "bead" should be used while maximizing surface contact with the probe surface. For example, the welded bead in the thermocouple used by French was cut in half to obtain a flat contact instead of a point contact with the probe surface.

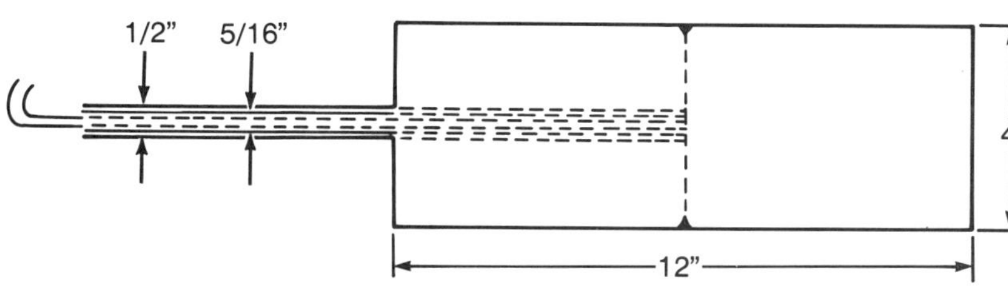

Fig. 3.4 SAE 5145 steel probe used by Grossmann

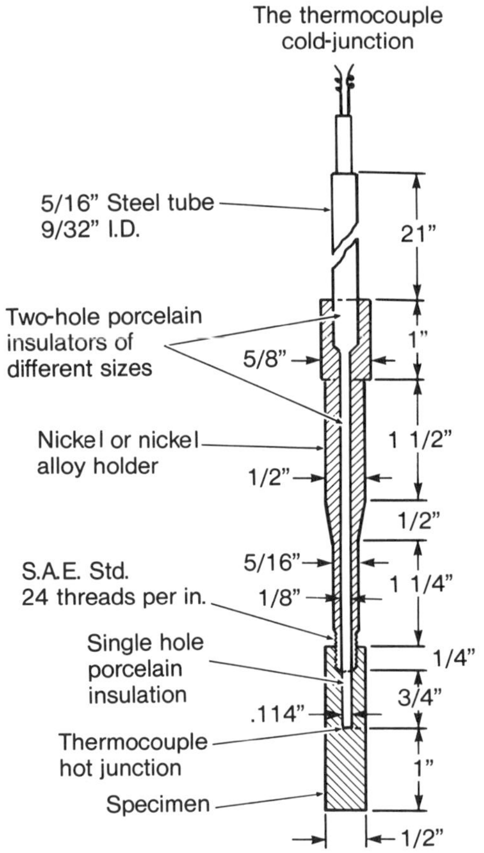

Fig. 3.5 Probe used by French

- Surface contamination between the thermocouple and the probe surface must be avoided.
- The thermocouple should be mounted to maintain uniform contact pressure during both heating and cooling of the probe.
- Leakage of the quenchant solution into the thermocouple assembly must be avoided.
- The thermocouple must be accurately calibrated.

An interesting probe construction was reported by Osman and Beck (Fig. 3.6)(Ref 27), who utilized a hollow copper sphere. High-purity copper was selected because of its high thermal conductivity relative to other metals such as steel. The spherical shape simplified heat-transfer calculations. This probe was constructed with three type E thermocouples peened into separate grooves machined in the copper after the sphere was sectioned in half. The objective was to obtain three simultaneous thermocouple readings during quenching. The three cooling curves were averaged for subsequent heat-transfer calculations in order to minimize error.

Liscic (Ref 23), Paschkis and Stolz (Ref 14), and others have reported the use of an instrumented silver ball in a testing apparatus illustrated schematically in Fig. 3.7 (Ref 31). Such probes have been relatively popular, because silver has excellent heat-transfer properties, and heat-transfer calculations from the time-temperature data can be made fairly simply because of the symmetrical shape of the spherical probe. However, several limitations have precluded their widespread use:

- It is difficult to manufacture a spherical probe with the thermocouple assembly located precisely at the geometric center. Errors in thermocouple placement result in corresponding errors in probe reproducibility.
- Although this point is somewhat controversial, many investigators believe that the high thermal conductivity of silver relative to steel does not permit realistic assessment of actual quenching conditions.
- It is difficult to maintain a consistent surface finish without affecting the diameter of the ball.

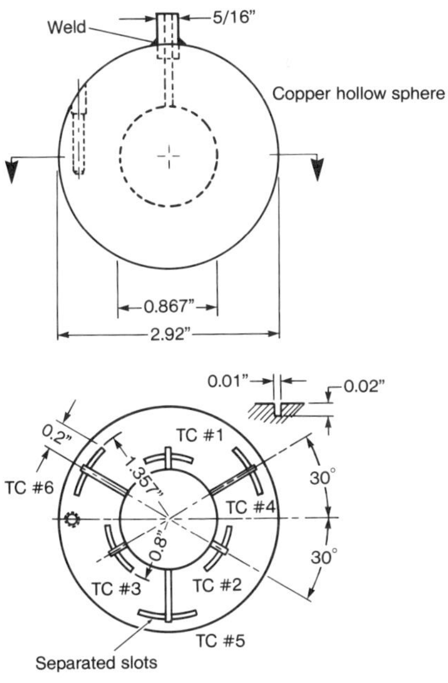

Fig. 3.6 Beck hollow spherical copper probe

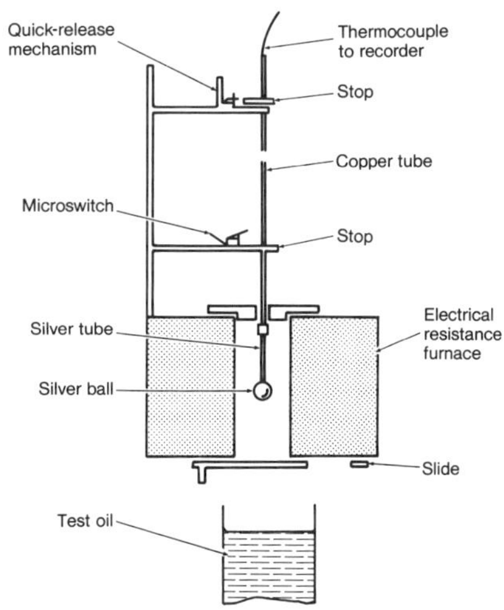

Fig. 3.7 Schematic of spherical silver probe test apparatus

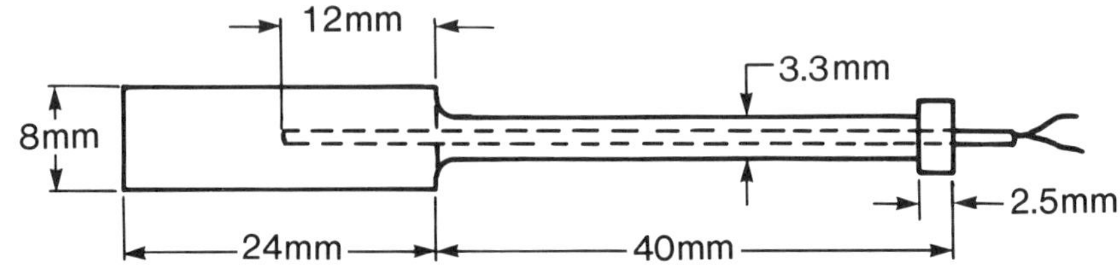

Fig. 3.8 Cylindrical silver probe

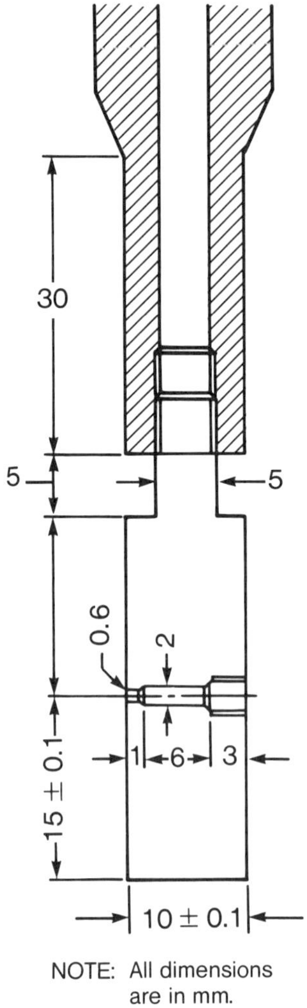

Fig. 3.9 JIS silver probe

In order to avoid the problems associated with the construction of spherical silver probes, some workers have used cylindrical silver probes (Fig. 3.8) (Ref 24). This probe assembly is being considered for an International Organization for Standardization (ISO) specification, having been recommended by the International Federation of Heat Treaters (IFHT) Quenching and Cooling Committee (Ref 25).

Luty (Ref 24) showed that satisfactory reproducibility of cooling curves from a silver probe could not be achieved by simply washing the probe after quenching. However, washing the probe and then polishing it with a fine abrasive paper, followed by another washing, did produce satisfactory reproducibility. Cooling curve reproducibility was also strongly dependent on the uniformity of agitation when aqueous polymer solutions were used as quenchants.

In order to gain insight into the quenching process, it has been proposed that it is critically important to model the heat-transfer properties that occur at the metal surface during quenching (Ref 20, 24, 32, 33). In such an analysis, the quenchant is considered to be a heat-transfer fluid that controls the rate of heat loss over the range of metal temperatures during the quenching cycle. Thus, the heat-transfer properties of the quenchant determine the metallurgical properties obtained.

Tagaya and Tamura (Ref 32) developed a Japanese Industrial Standard (JIS) (Ref 13) for cooling curve acquisition that utilizes a cylindrical silver probe with a thermocouple assembly specifically constructed to determine the surface temperature change with time during quenching (Fig. 3.9). The JIS probe was used by Tamura in his classic work describing the development of master cooling curves (Ref 34) and quench oil characterization (Ref 35), but it has not gained wide acceptance in Western industry. In addition to the general reluctance of the heat treating industry to use a probe with a thermal conductivity considerably different from that of steel, concern has been expressed regarding the cost of silver used in the probe construction, problems in maintaining a clean surface, and the comparative difficulty of preparing delicate surface thermocouple assemblies.

Liscic (Ref 23) has shown by correlations of cooling rate and hardness that it is not always possible to estimate the course of hardening over the cross section of a steel cylinder from the cooling curve obtained from a silver ball probe. His research led to the development of a steel probe (Ref 36) that utilizes a proprietary surface-temperature measurement system (Ref 37) (Fig. 3.10). The following advantages of this probe assembly have been reported:

- The response time of the thermocouple is 10^{-5} s, making it sensitive to small and rapid temperature changes.
- The surface condition of the probe can be easily maintained, because only the surface-temperature sensor needs to be polished before each measurement.

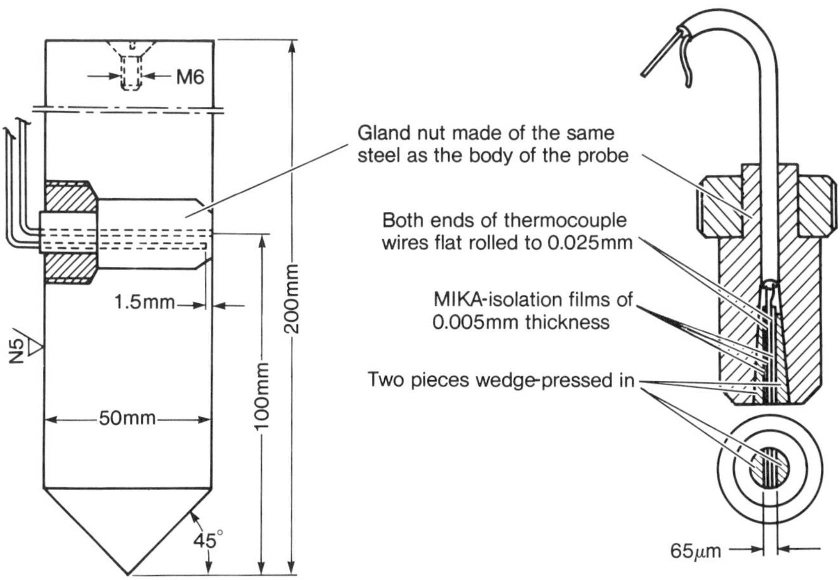

Fig. 3.10 Liscic-NANMAC steel probe

- The body of the probe is made from an austenitic stainless steel, which is not subject to volume changes during phase transformations that occur in plain carbon and alloy steels.
- The dimensions of the probe permit it to be mathematically modeled as an infinite cylinder. Therefore, it has symmetrical radial heat flow in the cross-section plane where the thermocouples are located, and end-cooling effects can be neglected in heat-transfer calculations.

Tensi and Stitzelberger-Jakob (Ref 38) have reported a probe capable of measuring surface heat-transfer properties of quenchants (Fig. 3.11). This probe is constructed of stainless steel; thermocouples are placed at the center and at three surface positions. Rewetting times can be calculated from the data obtained.

The probe currently enjoying the most rapid worldwide acceptance as a standard is probably the 12.5 × 60 mm (0.5 × 2.4 in.) Inconel 600 probe (Fig. 3.12) (Ref 25). Before use, the probe is conditioned to form a uniform surface oxide that facilitates reproducible quenching results, thus avoiding one of the most significant problems associated with silver probes. Other features include (Ref 39):

- Interlaboratory reproducibility, demonstrated by round-robin studies
- A thermal conductivity much closer to that of steel than silver
- A small probe size suited to routine use for quality control monitoring of quenchants with limited quantities of quenchant solution (approximately 2 L, or 0.5 gal)
- No phase transformations in Inconel 600 during quenching

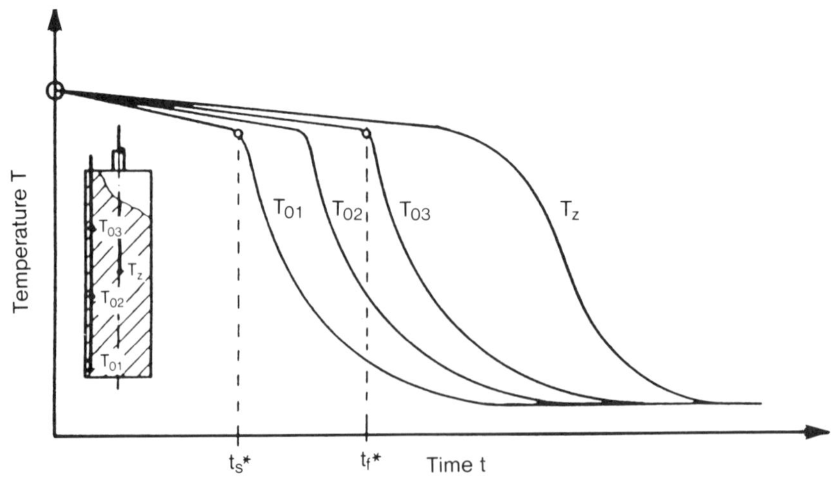

Fig. 3.11 Wetting kinematics from cooling curve measurements made by a stainless steel probe

Cooling curves analyses commonly use probes with spherical and cylindrical geometries. However, plate-shaped probes (Fig. 3.13) (Ref 18) are occasionally chosen when the primary focus is to model cooling and heat transfer from plate stock.

Probes with multiple thermocouple assemblies are also used (Fig. 3.14) (Ref 40). Multiple thermocouples allow measurement of thermal gradients. Bates and Totten (Ref 29) have used an aluminum probe with multiple thermocouples to measure temperature gradients while cooling in various quenching media. The maximum differences in temperature between the surface and the center of the probe (Fig. 3.15) can be used to estimate the propensity for stress formation and distortion during quenching.

Hines and Mueller (Ref 22) compared the performance of 25 × 100 mm (1 × 4 in.) cylindrical AISI 1546 steel and type 304 stainless steel probes. Their results showed that the low-hardenability AISI 1546 probe consistently produced longer A-stage times than the type 304 probe, both of which were quenched in the same quenchant medium (Fig. 3.16). This behavior was attributed to the exothermic phase transformation of austenite to ferrite and pearlite in the AISI 1546 steel. Similar results (Ref 22) were obtained for an AISI 5160 steel probe. Thus, use of a probe constructed of a material that does not exhibit a phase transformation during quenching, such as type 304 stainless steel (Fig. 3.17) or Inconel 600 (Fig. 3.12), allows the results obtained from various quenchant media to be compared without the complication of exothermic transformations. Furthermore, use of such materials eliminates the danger of cracking of the probe, especially in more severe quench media.

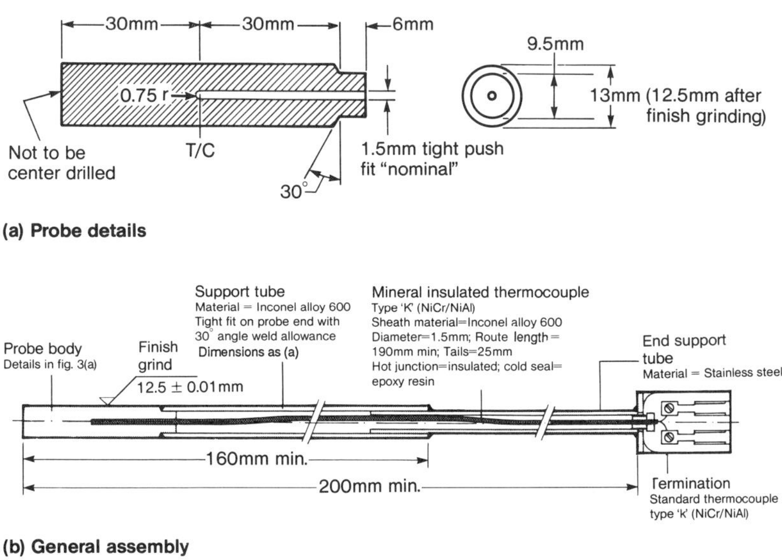

Fig. 3.12 Wolfson Inconel 600 probe

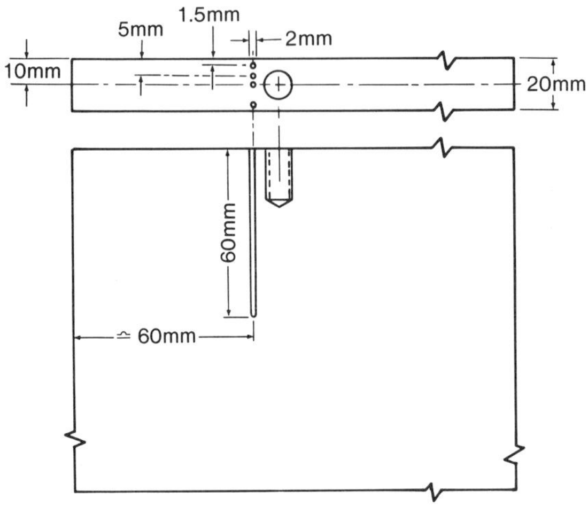

Fig. 3.13 Allen plate probe

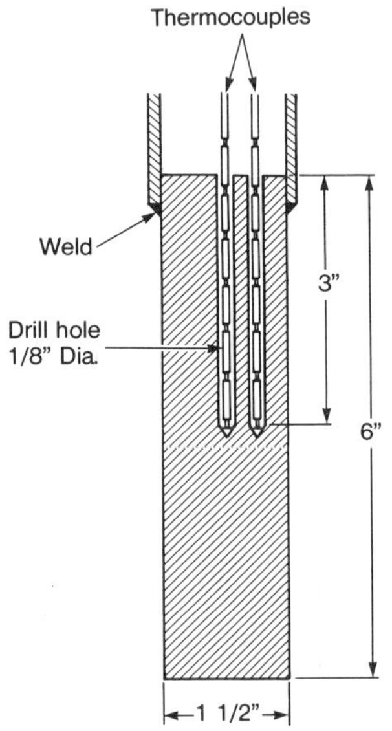

Fig. 3.14 Multiple thermocouple assembly

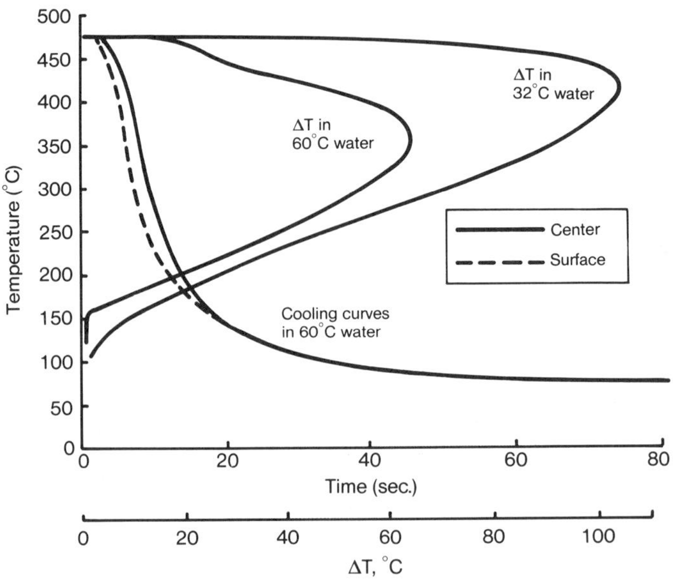

Fig. 3.15 Temperature gradients (ΔT) across an aluminum bar probe during a water quench

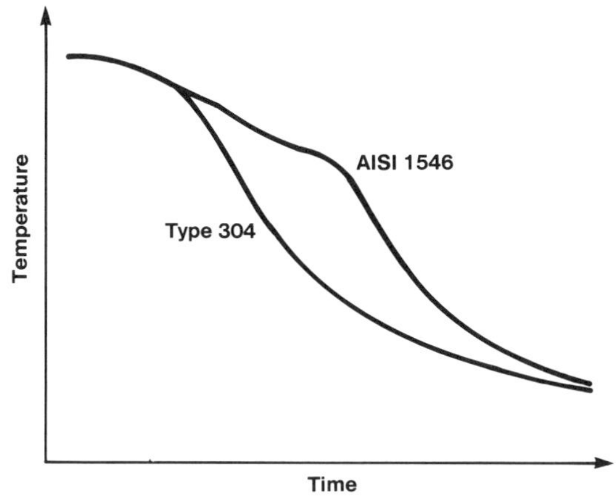

Fig. 3.16 Cooling curves for AISI 1546 steel and type 304 stainless steel test probes

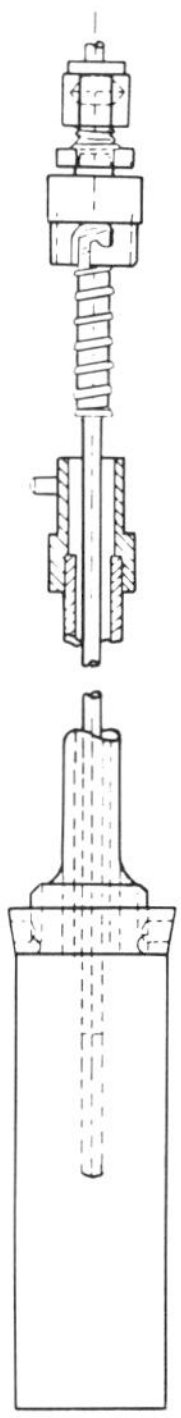

Fig. 3.17 Hines-Mueller type 304 stainless steel probe

Hines and Mueller (Ref 22) also studied the effect of cooling curve shape on as-quenched hardness. The AISI 1546 probe was quenched in selected polymer quenchant media at two agitation rates 10 and 20 m/min (35 and 70 ft/min) (Fig. 3.18). After quenching, the probe was sectioned and the hardness profile determined (Fig. 3.19). Higher agitation rates produced cooling curves with somewhat shorter A-stage times and slightly higher core hardness values.

Probe Dimensions. Most workers studying the quenching process use a cylindrical probe where the length of the cylinder is at least four times the diameter. The use of shorter probes may introduce significant end-cooling effects. This affects the recorded cooling curve, because the heat loss from the end of the probe will reduce the rate of temperature loss at the geometric center. Because the objective of cooling curve analysis is to measure the ability of the quenchant to attenuate heat transfer under various conditions, it is desirable that the experiment not be perturbed by the measurement process itself. Therefore, a probe with "semi-infinite" dimensions is usually selected.

The necessity of using a cylindrical probe with a length of at least four times the diameter (semi-infinite)

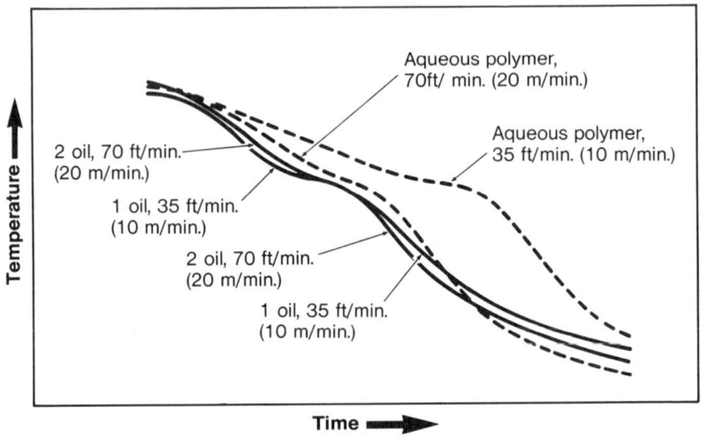

Fig. 3.18 Cooling curves for AISI 1546 steel probe showing influence of flow rate on oil and aqueous polymer quenchants

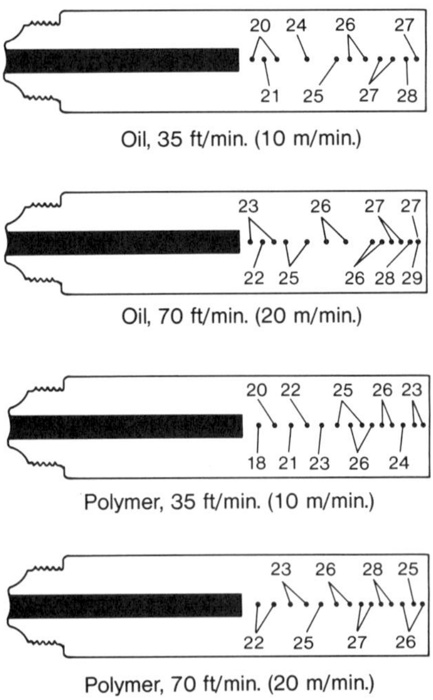

Fig. 3.19 Hardness patterns of sectioned AISI 1546 steel probes. Numbers indicate Rockwell C hardness.

can be illustrated by comparing the Fourier numbers or dimensionless time (t^+) for various solid cylinders (Ref 41). Fourier numbers less than 0.05 are considered to be acceptable, and numbers greater than 0.5 are considered to be large. The Fourier number varies linearly with time response above this value (Ref 42). The Fourier number is defined as:

$$t^+ = \frac{\alpha t}{L^2}\ \mathrm{tot}$$

where L is the length of the cylinder, t_{tot} is the test duration, α is the thermal diffusivity of the rod, and r is the probe radius.

If it is assumed that the total duration of the experiment is 30 s (since cooling is generally complete within this time for most smaller probes), it is possible to compute the relative cooling of, or heat loss from, a cylindrical probe longitudinally over the distance L versus radial heat loss from the center of the probe (with radius equal to a) to the surface. For the purposes of this calculation, it is assumed that the probe is constructed of Inconel 600 and that

α is 0.025. The relative impact of the probe size on cooling can then be computed from the dimensionless time expression:

$$t^+ = \alpha t_{tot} (a/L)^2$$

The results of this calculation for a cylindrical probe with a diameter of 12.5 mm (0.5 in.) and lengths varying from 12.5 to 125 mm (0.5 to 5.0 in.) are summarized in Table 3.1. These results show that a probe length of at least 50 mm (2.0 in.) is required to minimize the end effect on temperature measurement. A similar data analysis can be performed on plate probes to demonstrate that dimensions where the length and widths are at least four times the thickness are required to minimize edge-cooling effects.

The two most commonly used cylindrical probe sizes are 12.5 × 60 mm (0.5 × 2.4 in.) and 25 × 100 mm (1 × 4 in.). Of the two, the 12.5 × 60 mm probe is gaining the greatest acceptance, probably because it requires the least amount of quenchant, making it desirable for routine quality control tests.

Another selection criterion is probe sensitivity. The cooling time and rate curves shown in Fig. 3.20 indicate that sensitivity to quenchant characteristics increases with decreasing probe size. However, probe diameters of less than 12.5 mm (0.5 in.) are not often used because of the difficulty of temperature control during transfer from the furnace to the quench bath.

Data Acquisition and Analysis

The need to acquire sufficient data to adequately define a cooling curve for subsequent analysis has long been recognized. Special data acquisition devices, including hardware such as oscillographs, were used for work reported by Jominy (Ref 12), French (Ref 30), and others (Ref 14, 40). However, this equipment is difficult to calibrate, which has inhibited widespread use of cooling curve analysis. Currently, sufficient data acquisition rates can be achieved with personal computers equipped with analog-to-digital (A/D) converter boards.

Table 3.1 Estimation of the potential for end-cooling effects of different cylindrical probe lengths

Probe diameter		Probe length			Dimensionless
mm	in.	mm	in.	L/a	time, t^+
12.5	0.5	12.5	0.5	1.0	3.75
12.5	0.5	25.0	1.0	2.0	0.938
12.5	0.5	50.0	2.0	4.0	0.234
12.5	0.5	75.0	3.0	6.0	0.094
12.5	0.5	100.0	4.0	8.0	0.059
12.5	0.5	125.0	5.0	10.0	0.038

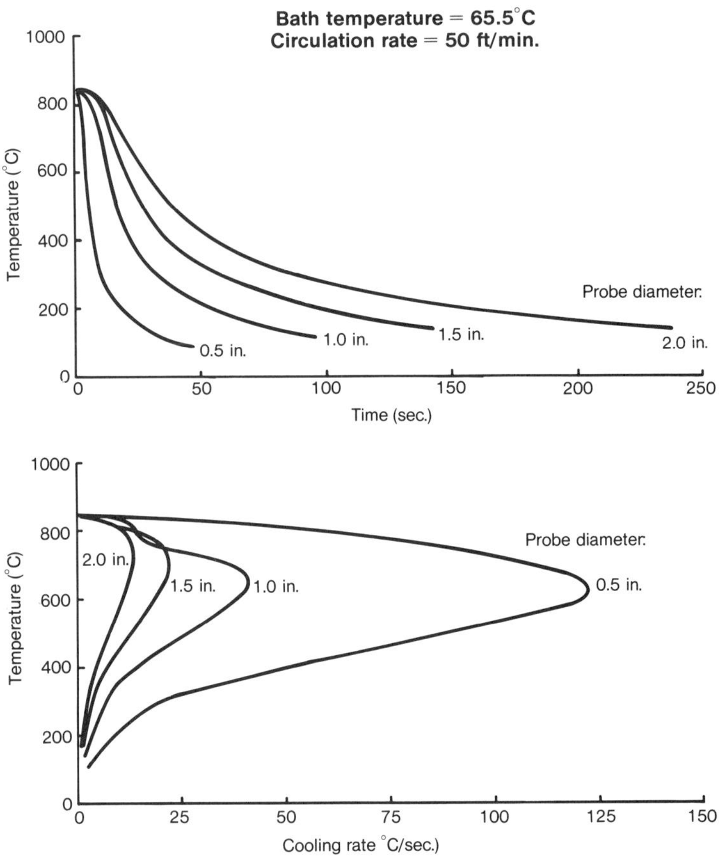

Fig. 3.20 Cooling curves for a typical accelerated quenching oil as a function of probe size

Although computer hardware is available, there are no published guidelines for selecting a data acquisition rate, which varies with probe alloy, size, and quench severity. Perhaps the best method for selecting the required acquisition rate is to determine it experimentally. This can be done by repeatedly quenching a probe in cold (25 °C, or 77 °F) water, one of the more severe quenchants, and collecting data at various acquisition rates. For example, a 13 × 100 mm (½ × 4 in.) cylindrical Inconel 600 probe with a type K thermocouple inserted at the geometric center was quenched into 25 °C (77 °F) water with data acquisition rates of 1, 2.5, and 5 Hz. (A hertz is equivalent to a data point per second.) The corresponding cooling time and rate curves are shown in Fig. 3.21. These results show that for this particular probe, a data acquisition rate of at least 5 Hz was required to obtain a smooth cooling rate curve in the maximum cooling rate region. Smooth curves are required to minimize errors in determining the maximum cooling rate

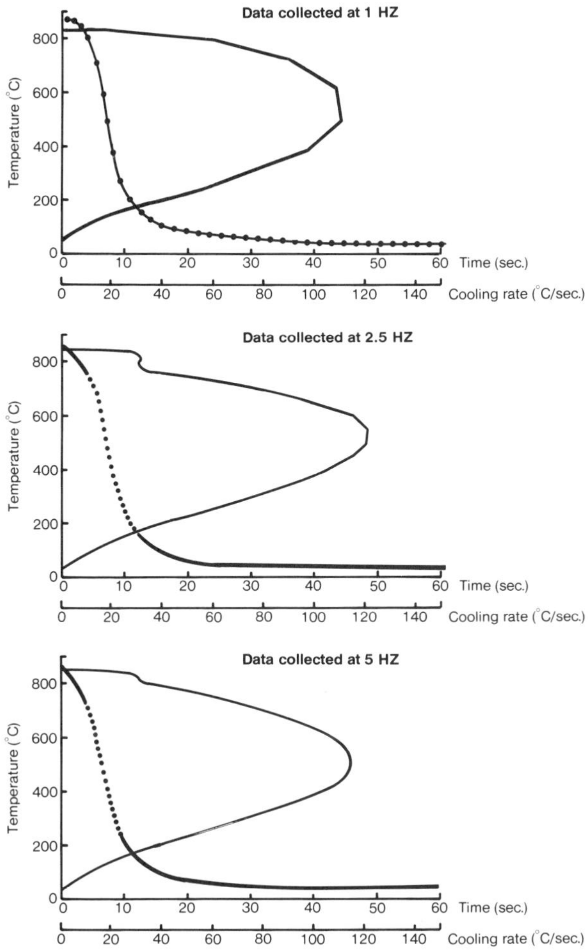

Fig. 3.21 Effect of data acquisition rate on cooling rate curve smoothness

in this critical region of the quenching process. Higher data acquisition rates can be used, although they may lead to data storage problems.

In data acquisition, the proper A/D converter card must be selected to match the thermocouple being used. The thermocouple output is nonlinear with respect to temperature. If software for data analysis is being written by the user, this nonlinearity must be taken into account. One method is to create a data file that will convert thermocouple electromotive force (emf) values to the appropriate temperature.

Zhu (Ref 43) recently analyzed computer data acquisition for cooling curve analysis and reported the following sources of error:

- Thermocouple voltage to temperature conversion
- A/D signal conversion
- Preamplifying circuit and signal transmission circuits
- Electronic component instability
- Errors caused by signal interference

Typical data acquisition errors were found to be ±0.5 °C (±0.9 °F) within the temperature range of 0 to 900 °C (32 to 1650 °F). Another source of error is the algorithm used to calculate cooling rates. In general, however, it was found that the error obtained was primarily a result of an error in the data.

Most data have some background noise; consequently, most data acquisition techniques employ data averaging or smoothing to minimize the noise. One method is a five-point "running" average (Ref 44):

$$T_n = (-\tfrac{3}{35}T_{n-2}) + (\tfrac{12}{35}T_{n-1}) + (\tfrac{17}{35}T_n) + (\tfrac{12}{35}T_{n+1}) - (\tfrac{3}{35}T_{n+2})$$

where T is the temperature at the n, $n-1$, $n+1$, $n-2$ and $n+2$ groups of points being averaged. A running average means that after T_n is calculated, the average of the next $(n+1)$ point is then averaged. This process is continued until all of the data points have been averaged.

Another feature of the A/D converter card that should be considered is ice-junction compensation. In order to obtain accurate temperature measurements, it is important to calibrate to a reference temperature, typically ice-water. The better A/D converter cards have built-in ice-junction compensation.

Interpretation of Cooling Curves

Once cooling time-temperature data are in hand, they must be interpreted. Ideally, cooling curves should be correlated with metallurgical properties of interest. Currently, there is no generally accepted method of accomplishing this task. However, several methods that have been used will be described. Selection of the "best" procedures is left to the reader. Each procedure provides some insight into the quenchant process.

Interpretation by Comparison With CCT Diagrams

One of the oldest methods of cooling curve interpretation involved taking the first derivative of the time-temperature curve obtained during the quenching of a probe with the desired cross section (Fig. 3.3). The maximum cooling rate was determined and compared with the critical cooling rate read from the continuous cooling transformation

(CCT) curve for the steel of interest (Fig. 3.22) (Ref 45). The critical cooling rate is considered to be the cooling rate at the nose of the austenite-to-pearlite transformation curve. To obtain maximum hardness in a quenched part, it would be necessary to select a quenchant that produces a maximum cooling rate equal to or greater than the critical cooling rate.

Segerberg (Ref 46) showed this procedure to be inadequate, because it did not account for the time required for the A- to B-phase cooling transition to occur. Because the cooling process involved in quenching is inherently different from the cooling process used to construct CCT diagrams, it is not rigorously correct to superimpose the data obtained by the two methods.

In order to understand the effect of the cooling process on as-quenched hardness, Segerberg instrumented a steel probe and performed simultaneous dilatometer experiments. The results showed that the A-stage cooling time affected the hardness obtained. His results also showed that the austenite-to-pearlite phase transformation occurs at a time when there is significant A-phase cooling. The transformations that occurred late in the A-stage cooling process resulted in slightly harder steels than when essentially no A-phase cooling occurred, such as obtained when quenching in cold water. This was consistent with results reported by Thelning (Ref 47). Segerberg also showed that if excessively long A-phase cooling times were observed during the austenite-to-pearlite phase transformation region, then softer steels resulted. Thus, it is important to account for the time associated with the A- to B-phase transition. Segerberg found a correlation

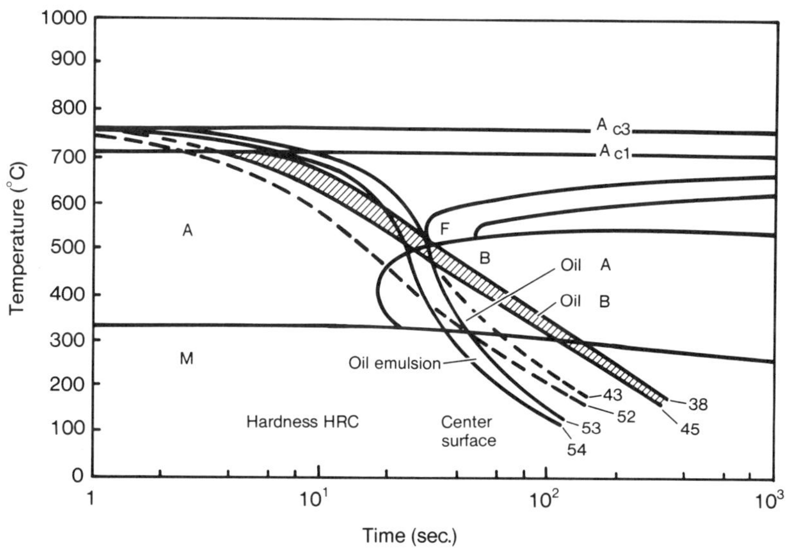

Fig. 3.22 CCT diagram for a spring steel (50M7) with superimposed cooling curves

between hardness and cooling rate at 550 °C (1020 °F), the temperature where the austenite-to-pearlite transformation occurred in the En 43 B steel under study (Fig. 3.23) (Ref 46).

Although overlaying cooling curves on various CCT curves has been utilized with some success, this procedure must be used with caution.

Cooling Time and Rate Parameters

In recent years, several rate- and time-dependent parameters have been suggested to quantify the severity of various quenchant media. Some of these parameters include the cooling rates at 705 and 205 °C (1300 and 400 °F), maximum cooling rate, and time to cool from 730 to 260 °C (1350 to 500 °F) (Ref 48). It is generally desirable to maximize the cooling rate at 705 °C (1300 °F) to avoid the pearlite transformation region. It is desirable to minimize the cooling rate at 205 °C (400 °F), which is in the region of the M_s transformation temperature for many steels, to minimize cracking tendency. It is also desirable to minimize the time to cool from 730 to 260 °C (1350 to 500 °F) in order to optimize the potential hardness by avoiding pearlite formation.

Tensi and Steffen (Ref 49) have prepared a set of critical cooling parameters for cooling curve characterization (Fig. 3.24):

- A-to B-stage transition time, temperature, and rate, $R_{\mathrm{DH}_{\min}}$. These parameters characterize the transition from A- to B-stage cooling.
- Maximize the rate of cooling, $R_{\max}$, and the temperature where this occurs, $T_{R_{\max}}$. Generally, it would be desirable to maximize $R_{\max}$ and minimize $T_{R_{\max}}$.
- Rate of cooling at temperatures such as 200 °C (R_{200}) and 300 °C, (R_{300}). In order to minimize the propensity for cracking and distortion, it is desirable to minimize cooling rates in this region.

One problem with this approach is the potential misinterpretation of low cooling rates in the M_s transformation region. Traditionally, it has been thought that minimizing the cooling rates in the Ms transformation region is essential for minimizing cracking and distortion. This is a desirable, but not necessarily a sufficient, condition for reduction of quench cracking.

Recently, Zhelokhovtseva (Ref 50) has shown that uniformity of cooling in the M_s region is a critical parameter. It seems reasonable, therefore, that minimal *and* uniform cooling rates to minimize stresses that occur due to thermal gradients are probably the necessary and sufficient conditions for optimizing quenchants in the M_s temperature region.

A seldom encountered procedure for cooling curve quantification is the acceleration index (L_D) method (Ref 51). The L_D value represents the ratio between the cooling rate in the medium being analyzed and the cooling rate in distilled water and is expressed as:

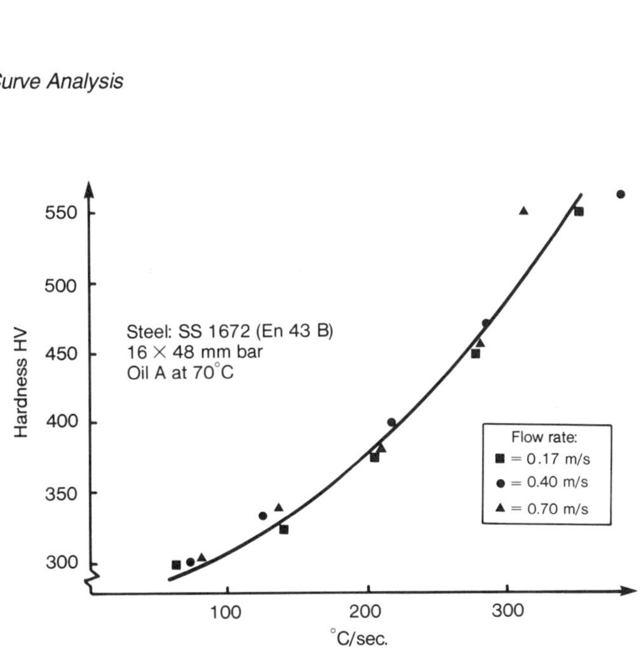

Fig. 3.23 Hardness versus cooling rate at 550 °C (1020 °F) for stainless steel 1672 (En 43 B)

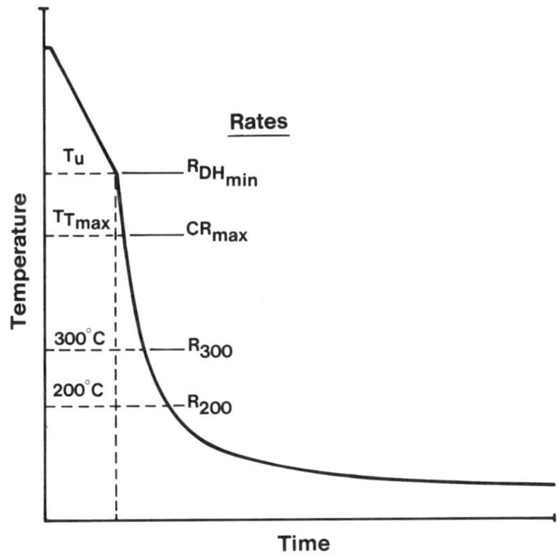

Fig. 3.24 Critical cooling curve parameters

$$L_D = \frac{t_e}{t_m}$$

where t_e is the time required for the temperature of the probe to drop below a given value in distilled water and t_m is the time required for the temperature to drop below the same limit in the medium being examined.

Area Under Cooling Time and Rate Curves

In order to simplify the analysis of cooling curves, there has been an ongoing effort to devise a characterization parameter that would provide an adequate summary of the cooling process. However, it is difficult, if not impossible, to predict the extent of hardening by utilizing only one parameter from a cooling curve, such as a single cooling rate. Instead, heat transfer must be modeled throughout the entire quenching period.

Liscic (Ref 52) demonstrated that one method of obtaining a useful correlation between cooling rates during a quench and hardness was to integrate the area under cooling rate curves. A plot of accumulated area versus time can then be used to quantify the progression of the quench cycle. An example of data obtained from this analysis is shown in Table 3.2.

Table 3.2 Characterization of quench oils using the cooling area method developed by Liscic (Ref 23)(a)

	Elapsed time from initiation of quench, s								
Quenchant	**1**	**2**	**3**	**4**	**5**	**6**	**7**	**8**	**9**
Unagitated oil	1.0	1.75	1.70	1.60	1.35	1.08	1.44	12.58	1.28
Agitated oil (1.1 m/s,or 3.6 ft/s)	9.0	15.52	26.12	3.50	1.53	0.91	0.60	0.21	0.12

(a) The cooling rate curve was obtained using a 20 mm (0.8 in.) diam silver ball probe. The areas are actually incremental areas (m^2/s) and are obtained by integrating the area under the curve with a planimeter.

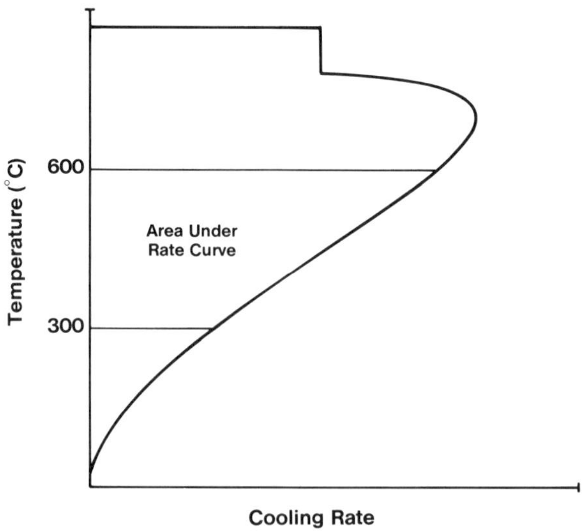

Fig. 3.25 Area under the cooling rate curve

The data in Table 3.2 show that the A- to B-phase boiling transition for the unagitated quench oil occurred 8 s after the start of quenching. The same transition occurred after 3 s in the agitated oil. The larger area under the cooling rate curve obtained for the agitated oil reflects faster heat transfer. This method allows precise determination of the cooling transition temperatures and provides a measure of the rate of heat transfer at these transitions.

Thelning (Ref 47) reported a similar method involving the integration of the area under the cooling rate curve between two temperatures, T_1 and T_2 (Fig. 3.25). Although the selection of T_1 and T_2 depends on the steel alloy, it was concluded that the best general range was 600 to 300 °C (1110 to 570 °F). Thelning found the area under the curve to increase with the average quenching speed. This was more recently confirmed by comparison of selected oil and polymer quenchants, as shown by the data in Table 3.3 (Ref 53).

Another approach to the characterization of cooling rate behavior is the utilization of the quantity called the mean reduced cooling rate, developed by Goryushin *et al.* (Ref 54). This value is expressed as:

$$v_{\text{red}}^{\text{me}} = \frac{\Delta t}{\tau_{\text{red}}}$$

where me refers to the quenching medium, Δt is the temperature range, and ($\tau_{\text{red}} = 2S/a_{\text{tot}}$ is the reduced cooling time (where S is the area under the cooling curve between the temperature range and a_{tot} is the total area under the cooling rate curve). The mean reduced cooling rate ($v_{\text{red}}^{\text{me}}$) approach appears to be similar to Thelnings' technique, except that the area integrated between the two temperatures is normalized to the total area under the curve (a_{tot}).

Table 3.3 Comparison of cooling area and rate techniques

Quenching medium	CR_{max}(a)	CR_{232}(b)	$A_{600\text{-}300}$(c), °C/°C/s
Water(d)	54.3	20.1	12,979
Conventional oil(e)	31.6	6.5	5824
Accelerated oil(e)	34.6	6.9	6663
Martempering oil(f)	33.9	3.3	4781
25% UCON® Quenchant A(g)	41.7	14.5	9121
15% UCON® Quenchant A(g)	46.8	15.1	9815
15% UCON® Quenchant E(h)	28.0	6.6	6147

(**a**) CR_{max}, maximum cooling rate. (**b**) CR_{232}, cooling rate at 232 °C (450 °F). (**c**) $A_{600\text{-}300}$, integrated area under the cooling rate curve between 600 and 300 °C (1110 and 570 °F). (**d**) Quenching conditions: 32 °C (90 °F), 30 m/min (100 ft/min). (**e**) Quenching conditions: 65 °C (150 °F), 30 m/min (100 ft/min). (**f**) Quenching conditions: 150 °C (300 °F), 30 m/min (100 ft/min). (**g**) Quenching conditions: 38 °C (100 °F), 30 m/min (100 ft/min). (**h**) Quenching conditions: 60 °C (140 °F), 15 m/min (50 ft/min).

Master Cooling Curve

Lack of standardization of probe construction (e.g., shape and alloy) has made comparison of cooling curves obtained in different laboratories difficult. Tamura (Ref 32) developed the method of master cooling curve calculation, which, at least conceptually, addresses this problem. This procedure has also been used by Wang *et al.* (Ref 55).

Tamura found that cooling curve time was the only factor affected by varying the size and alloy of the probe. The variation in the cooling curve time with respect to these variables was found to be related by:

$$t = k_1 D^n$$

where t is the time in seconds required to cool the probe from an initial temperature to any selected temperature, k_1 is a parameter independent of the specimen size, D is the diameter of the cylindrical probe in centimeters, and n is a constant related to the alloy and thermocouple location in the probe.

At each temperature during the quenching process, k_1 is determined by dividing the incremental time, t, by the probe diameter, D, raised to the nth power:

$$k_1 = \frac{t}{D^n}$$

The master cooling curve is then obtained by plotting temperature, T, as a function of k_1. Although Tamura described the procedure for precise determination of n, the values shown in Table 3.4 can be used as reasonable approximations (Ref 56).

Traditionally, the most common materials for probe construction have been silver, nickel, and steel. Each of these materials produces significantly different cooling curves

Table 3.4 Approximate values of *n* for construction of a master cooling curve

Probe material	Thermocouple placement	n
Silver	Center	1.34
	Surface	1.14
Steel	Center	1.37
	Surface	1.00

Table 3.5 Correlation factors for variation in probe construction materials

Probe material	Specific heat (S) cal/g · °C	Specific heat (S) J/kg · K	Density (ρ), g/cm^3	Density (ρ), lb/in.^{3}	$S \times \rho$
Silver	0.061	255.4	10.21	0.37	0.62
Steel	0.108	452.2	7.76	0.28	0.84
Nickel	0.140	586.2	8.80	0.32	1.23

when other variables are constant. Tamura (Ref 32) demonstrated that heat transfer within a probe can be represented by the product of specific heat, S, and density, ρ, of the probe (Table 3.5). If the product of $S \times \rho$ is known, cooling curves obtained with probes of different materials can be compared by plotting T as a function of $t/S\rho D^n$, assuming the probe is cylindrical with a thermocouple placed at the geometric center.

Various probe shapes can be utilized for cooling curve analysis, including cylindrical, spherical, and plate. Carey (Ref 56) has provided a summary of interconversion constants that permit comparison of cooling curve data obtained with different probe shapes (Table 3.6).

V-Values: Quench-Hardening Criteria

Although the master cooling curve is a rather elegant approach to the description of quenching behavior, it does have limitations, including a complicated interpretation and errors introduced during data normalization. To simplify cooling curve interpretation, Tamura *et al.* developed a quantity called the "V-value," which is proportional to the ability of an oil quenchant to harden steel (Ref 35, 57). It should be noted that Tamura *et al.* developed V-values for oil quenchants only.

The V-value is calculated from cooling curve data and the corresponding CCT diagram using the relationship:

$$V = \frac{T_c - T_d}{T_s - T_f}$$

where T_c is the characteristic temperature at the start of B-stage cooling, T_d is the temperature at the start of the C-stage and is calculated as shown in Fig. 3.26, T_s is the temperature at the start of the austenite-to-pearlite transformation, and T_f is the temperature at the start of martensite transformation. From this relationship, it is evident that the V-value represents the ratio of the nucleate boiling temperature range (B-stage cooling) to the temperature range where rapid cooling is necessary to achieve high hardness.

The use of V-values for quenchant bath characterization was originally developed for oil, where cooling rates are similar at T_c and T_d. Synthetic quenchants, however, usually exhibit different cooling rates at these temperatures. Without further develop-

Table 3.6 Calculations of Tamura's *D* factor as a function of probe shape

To convert from	To	Multiply *D* by
Cylinder	Sphere	1.33
Cylinder	Plate	0.66
Sphere	Cylinder	0.75
Sphere	Plate	0.50
Plate	Cylinder	1.50
Plate	Sphere	2.00

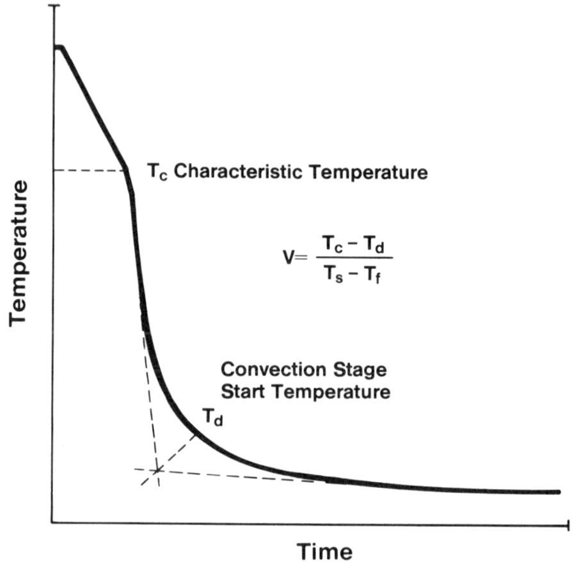

Fig. 3.26 Method for finding convection stage start temperature

mental work, this procedure cannot be readily applied to cooling curves obtained from aqueous polymer quenchants.

Grossmann Quench Severity Factor

Hardness is one of the primary quality and tensile property indicators in quenched and tempered steel. The ability of a quenchant to harden a particular type of steel under specific quenching conditions has been experimentally determined by performing cross-sectional hardness surveys on quenched bars (Ref 58, 59).

Although it is possible to determine *H* factors experimentally by quenching steel bars with the desired chemistry, it is often inconvenient because the procedure is time consuming and necessitates sacrificing test bars of different diameters. Unless substantial quantities of steel of the same composition and grain size are available in bars with different diameters, it is difficult to compare *H* factor variations from facility to facility or over time, because chemistry variations between lots within a given grade of steel introduce data scatter. Furthermore, several steels with different hardenabilities are necessary to obtain test precision when evaluating quenchants ranging from agitated water and brines to poorly agitated oils. Therefore, an alternate method for determining the quenchant *H* factor that does not involve characterizing and sacrificing steel bar specimens is desirable.

Monroe and Bates (Ref 60) have described the use of a cooling curve analysis technique to estimate *H* factors. Analytical cooling curves were first calculated using a finite-difference heat-transfer program. Thermal properties of type 304 stainless steel were put into the program, and specific quench severity values were imposed at the part surface. Cooling curves were calculated for bars with diameters of 13, 25, 38, and 50 mm (0.5, 1.0, 1.5, and 2.0 in.). The bars were of sufficient length (semi-infinite) that end-cooling effects could be neglected.

The calculated time-temperature profiles were subsequently analyzed to determine the cooling rate at 705 °C (1300 °F) as a function of imposed quench severity and bar diameter. A temperature of 705 °C (1300 °F) was selected for the cooling rate analysis because much of the metallurgical literature on transformations is related to the cooling rate at this temperature. Results of calculations that relate the *H* value to the cooling rate in 13, 25, 38, and 50 mm (0.5, 1.0, 1.5, and 2.0 in.) diam type 304 stainless steel bars are presented in Table 3.7.

A statistical model was developed to fit these data (Ref 61):

$$H = AX^{C} \exp(BX^{D})$$

Table 3.7 Cooling rate at 705 °C (1300 °F) versus *H* factor

Cooling rate, °F/s Probe diameter, mm (in.)				Approximate *H* factor
13 (0.5)	25 (1.0)	38 (1.5)	50 (2.0)	
15.2	7.5	4.9	3.6	0.10
30.0	14.6	9.5	6.9	0.20
58.3	27.5	17.1	11.9	0.40
85.2	38.5	23.1	15.7	0.60
109.8	47.8	27.9	18.6	0.80
132.6	58.8	31.9	20.9	1.00
159.5	64.4	36.0	23.3	1.25
182.7	71.8	39.4	25.1	1.50
241.3	88.7	46.7	29.0	2.25
286.9	100.1	51.5	31.5	3.00
333.7	111.2	56.0	33.8	4.00
367.9	119.0	58.9	35.1	5.00
427.2	130.9	62.7	37.0	7.00
476.2	140.0	66.6	38.5	10.00

Table 3.8 Parameters for statistical modeling of *H* factors

Probe diameter		$H = AX^{C} \exp(BX^{D})$ Curve-fitting parameters			
mm	in.	A	B	C	D
13	0.5	0.002802	0.1857×10^{-7}	1.201	2.846
25	1.0	0.002348	0.2564×10^{-9}	1.508	4.448
38	1.5	0.002309	0.5742×10^{-9}	1.749	5.076
50	2.0	0.003706	0.03456×10^{-10}	1.847	6.631

where H is the Grossmann H factor, X is the cooling rate (in °F/s) at 705 °C (1300 °F), and A, B, C, and D are statistical model parameters from Table 3.8. This model can be used to calculate the H factor from bar cooling curves if the cooling rate at 705 °C (1300 °F) and the statistical constants are known. The plot of H factor versus cooling rate at 705 °C (1300 °F) in different bar sizes shown in Fig. 3.27 indicates that the statistical model provides a good fit to the data in Table 3.7.

The use of this model and type 304 stainless steel probes in conjunction with the appropriate cooling rate curves from the quenchant media of interest permits a nondestructive and reproducible determination of the quench severity provided by various quenching media. Some representative values obtained when this technique was applied to cooling curves obtained in water, oil, air, and two aqueous polymer quenchant media under a variety of agitation conditions are listed in Table 2.3. A knowledge of H factors for quenchant media is becoming more important as computer programs such as STECAL (available from ASM International) are developed to predict hardnesses in particular steels under various quenching conditions.

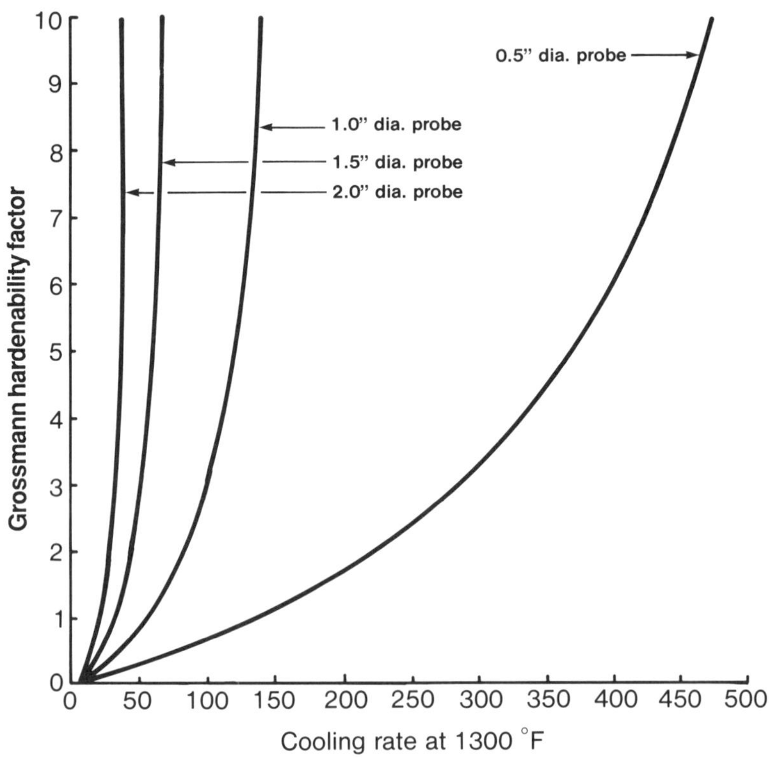

Fig. 3.27 Grossmann hardenability factor versus cooling rate at 705 °C (1300 °F)

Hardening Power

Segerberg (Ref 62) has developed two empirical approaches for the characterization of hardening power (HP). One approach was developed for determining the hardening power of quench oils without agitation and is described in Chapter 4. A second approach, developed for determining the hardening power of quenchant media with agitation, will be described here. Both of Segerberg's hardening power approaches utilized cooling curves obtained with the Wolfson probe shown in Fig. 3.12 (Ref 25). The agitation system utilized by Segerberg was developed by Tensi and is shown in Fig. 3.28 (Ref 63).

The hardening power calculation for agitated quenchants is based on two values from the cooling curve: CR_F and CR_M (Fig. 3.29). The value CR_F is the cooling rate at 550 °C/s (990 °F/s), which is the temperature at, or at least near, the nose of the pearlite transformation. The cooling rate at 330 °C/s (595 °F/s) approximates the start of the martensitic transformation for many steels.

The hardening power of a polymer quenchant was determined by first obtaining a cooling curve by using a 16 × 48 mm (0.6 × 1.9 in.) cylindrical probe of the steel alloy of interest with a thermocouple inserted at the geometric center. After the cooling curve was determined (Fig. 3.29), the probe was sectioned and the hardness at the geometric center was measured.

Regression analysis of a series of cooling curves for polymer quenchants obtained under various quenching conditions yielded this relationship between hardening power and CR_F and CR_M:

$$\text{HP} = 3.54CR_F + 12.30CR_M - 168$$

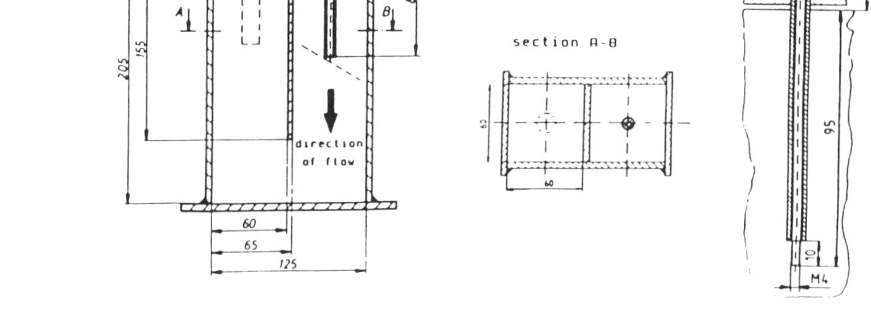

Fig. 3.28 Agitation system developed by Tensi

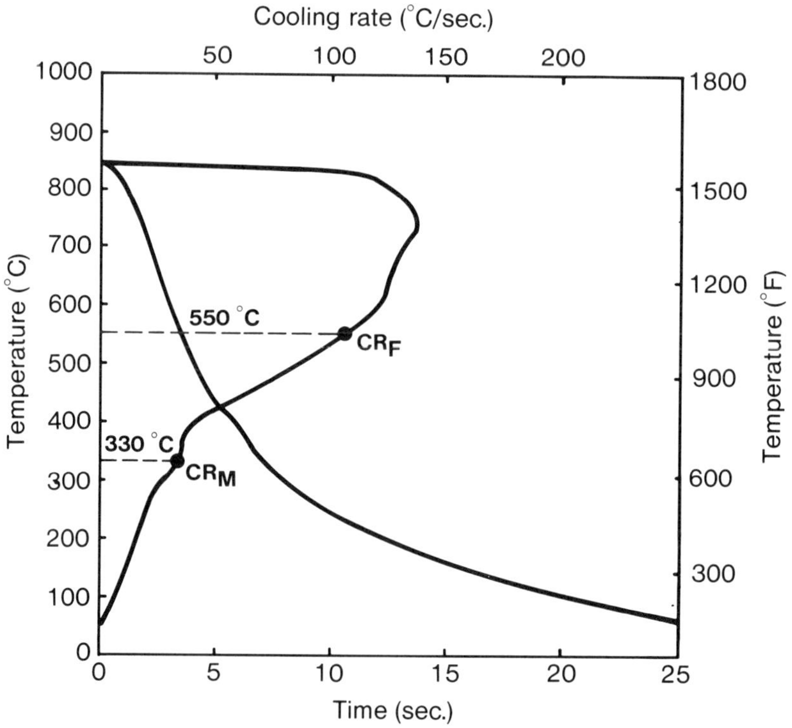

Fig. 3.29 Critical temperature determination for Segerberg's hardening power of agitated quenchants

Segerberg (Ref 62) used this procedure to predict the as-quenched hardness achievable with a commercial quench system, by strategically placing a series of instrumented steel probes throughout a typical quench load. The hardening power at these various positions was calculated from the above regression equation. Using the correlation curve for the steel of interest (e.g., the curve for the AISI 1045 steel shown in Fig. 3.30), it is possible to predict successfully the hardness for steel with a 16 mm (0.6 in.) cross section. A similar procedure can be used for other alloys and different cross section sizes.

Rewetting Times

A study of quenching is actually a study of heat transfer during cooling of hot metal parts. Numerous factors impact the transfer of heat from the metal piece during quenching, including the mass of the part, thermal conductivity of the metal, motion of the quenching medium past the part, interfacial viscosity of the film at the hot metal surface, and wetting properties of the quenchant.

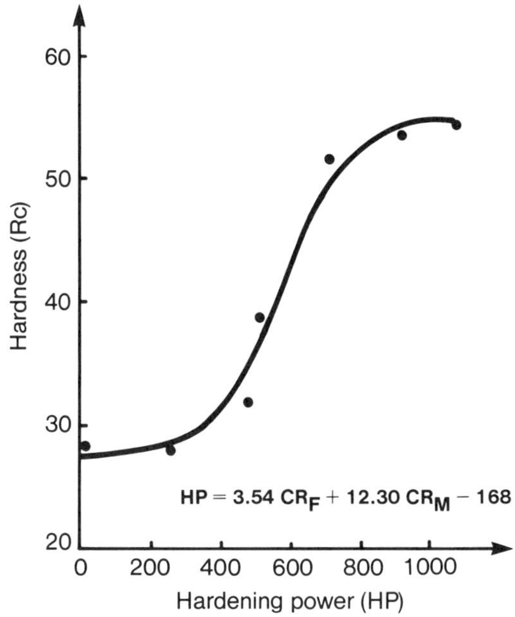

Fig. 3.30 Classification of polymer quenchant by hardening power (AISI 1045 steel calibration curve)

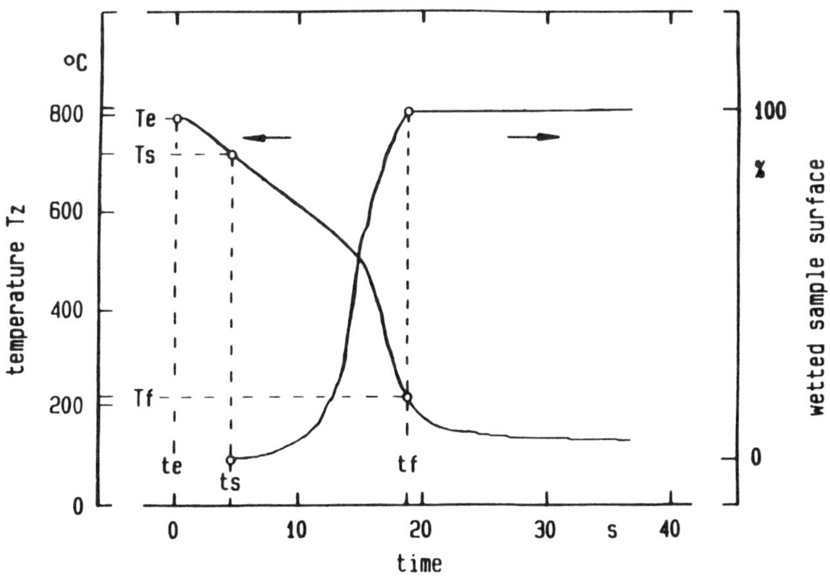

Fig. 3.31 Temperature, T_Z, in sample center and wetted sample surface as a function of time during immersion cooling of a cylindrical Cr-Ni steel sample

The highest cooling rates occur during nucleate boiling. This is the region where the vapor blanket collapses and the quenchant rewets the metal surface. Thus, to maximize hardness it is desirable, and in some cases critically important, to minimize the rewetting times.

Tensi *et al.* (Ref 38, 49, 64-69) have utilized rewetting times to characterize quenchant media. The rewetting time of a quenchant can be determined experimentally

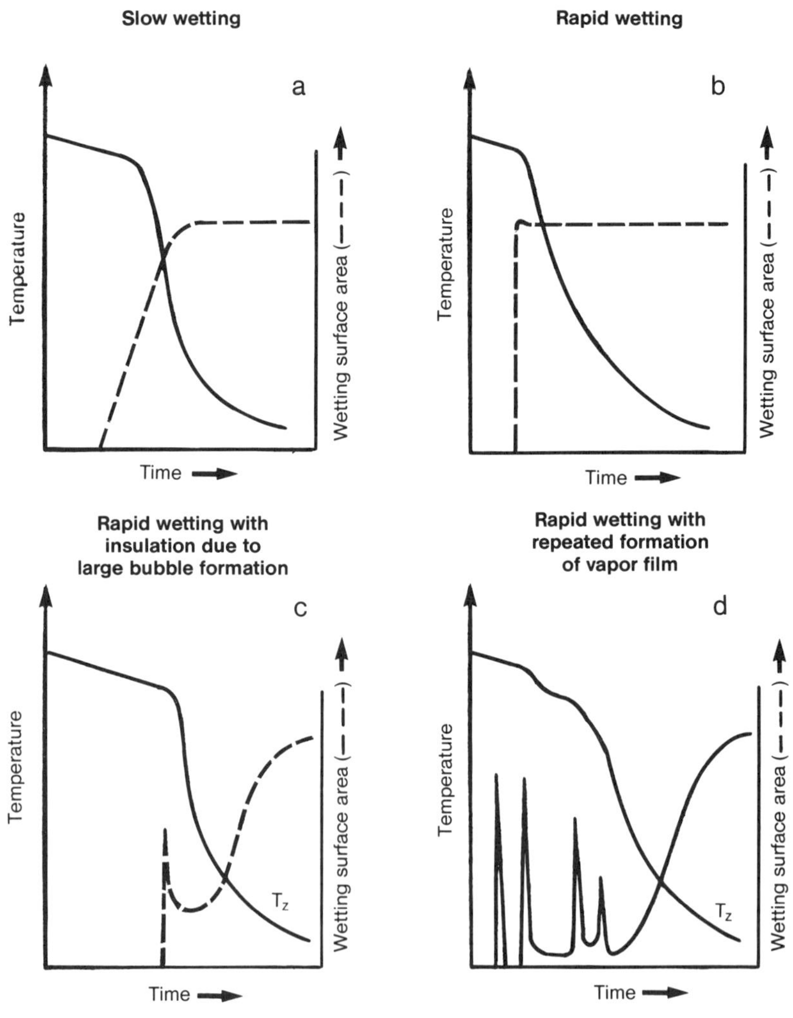

Fig. 3.32 Common rewetting profiles

by instrumenting a probe as shown in Fig. 3.11. The objective is to minimize the time required for the vapor blanket to collapse along the probe from T_1 to T_3 (Fig. 3.11). This time interval, $t_f - t_s$ (Fig. 3.31), is designated the rewetting time.

After studying numerous quenchant media, Tensi found that the rewetting mechanisms can be divided into four categories (Fig. 3.32):

(a) Slow wetting
(b) Rapid wetting
(c) Rapid wetting following prolonged A-stage cooling
(d) Rapid wetting involving repeated formation of a vapor film and subsequent wetting as in (c)

Cooling curve characterization by conductivity changes during the quench provides considerably more detailed information. Most of Tensi's rewetting time work is based on conductivity measurements at the center of a steel probe. Although not reviewed here, cooling curves can be obtained by measuring conductivity changes as a function of time, because the conductivity of steel is temperature dependent.

Tensi found a correlation between rewetting times and hardness (Fig. 3.33). This curve can be used to construct a calibration curve for hardness and rewetting time, as shown in Fig. 3.34. This calibration curve was used to predict the hardness of a series of

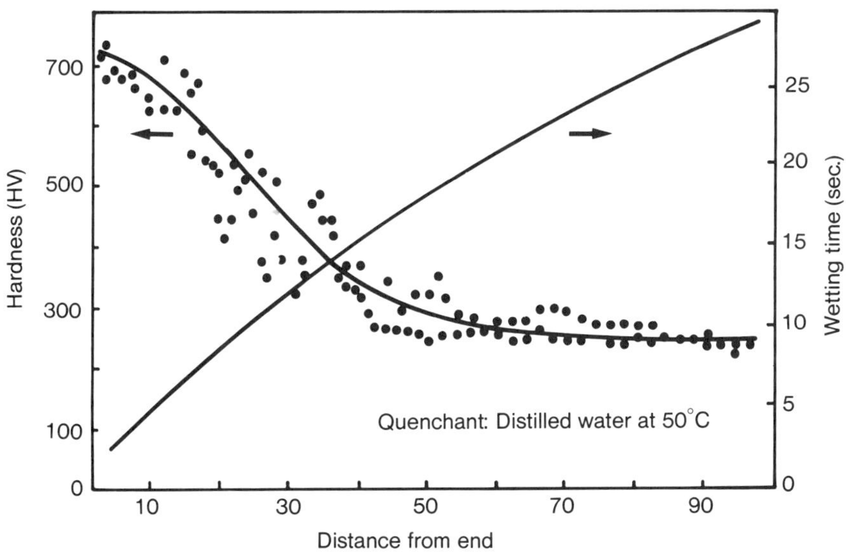

Fig. 3.33 Surface hardness versus wetting time calculation (0.45C steel)

0.45C steel bars. A reasonably good correlation between predicted and actual hardness resulted. Presumably, if appropriate calibration curves are generated for the steel alloys of interest and the hardenability bands of the alloys are accounted for, steel hardness can be correlated with rewetting times during the quenching as shown in Fig. 3.33 and 3.34.

V_S/V_C Uniformity Ratio

Russian workers have reported that the propensity for a quenchant to produce cracking and distortion is determined by quench uniformity (Ref 70). Their method of quantifying quench uniformity was to calculate the cooling rate difference between the center and surface of a steel probe at 300 and 200 °C (570 and 390 °C)—V_C and V_S, respectively. The temperatures were selected as being representative of the M_s and M_f temperatures of many steels.

The values for V_C and V_S are calculated from the respective cooling temperature time curve by dividing the temperature difference (300 – 200 °C) by the times to cool to these temperatures ($t_{300} - t_{200}$) (Fig. 3.35):

$$V_S = \frac{300 - 200\ °C}{(t_{300} - t_{200})_S}$$

The ratio of V_S and V_C is then calculated and can be shown to be V_S and $V_C = (t_{300} - t_{200})_S$. The uniformity of cooling improves as the ratio of V_S/V_C approaches 1.0.

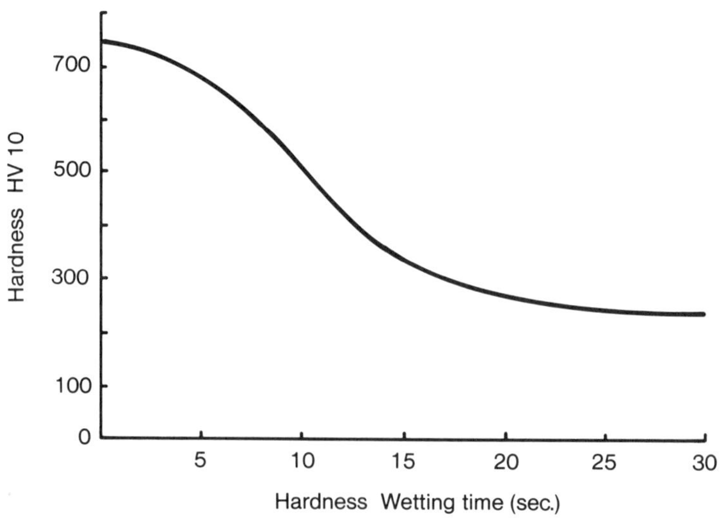

Fig. 3.34 Calibration curve for wetting time and surface hardness (0.45C steel)

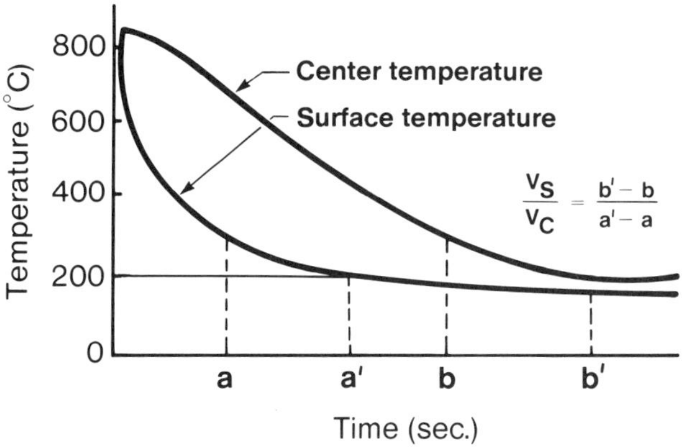

Fig. 3.35 Calculation of V_S/V_C value

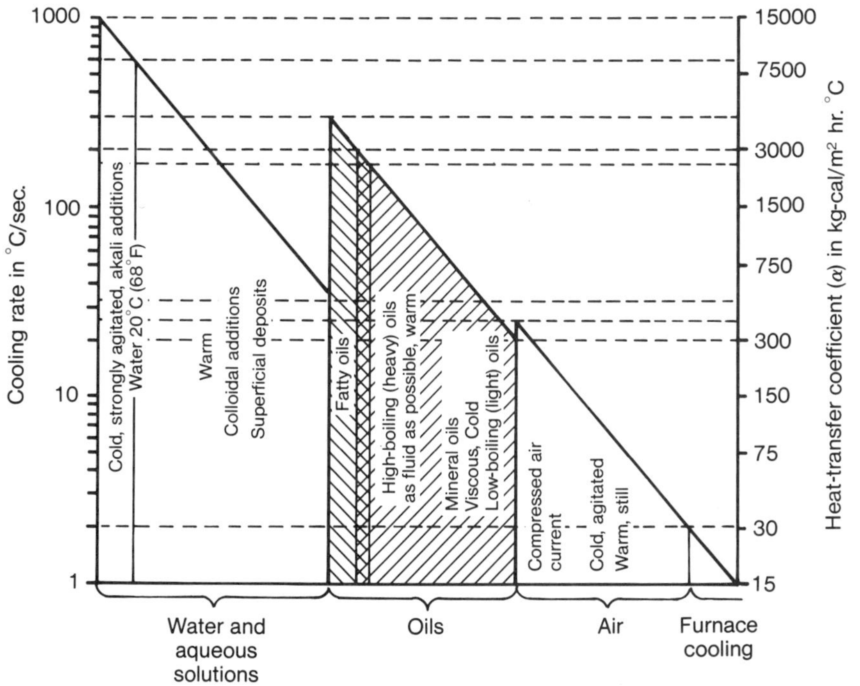

Fig. 3.36 Cooling power of various quenchants and heat-transfer coefficient at a probe temperature of 500 °C (930 °F)

Quench Factor Analysis

See the section "Quench Factors" in Chapter 2 for a detailed description of the use of quench factor analysis to obtain physical property data from cooling curves.

QTA Method

Quenching has long been recognized as an application problem in heat transfer. In 1939, Rose (Ref 11) performed cooling curve analyses using spherical silver probes and showed that the heat transfer during quenching could be correlated using cooling rates as follows:

$$\frac{dq}{dt} = h(T_p - T_k)F$$

where T_p is the temperature of the part, T_k is the temperature of the quenchant, F is the surface area of the part, dq/dt is the change in the amount of heat extracted from the steel specimens with respect to the corresponding change in time, and h is the heat-transfer coefficient. If the specific heat, C_p, is constant, and if the heat lost from a given part is solely a function of temperature, then:

$$Q = T_p GC_p$$

where G is the mass of the part. From this relationship, the change in temperature of the part with respect to a change in time, dT_p/dt, can be shown to be:

$$\frac{dT_p}{dt} = h(T_p - T_k)\left(\frac{F}{GC_p}\right)$$

or

$$h = \frac{dT_p}{dt}\frac{GC_p}{(T_p - T_k)F}$$

Rose used this equation for heat transfer to correlate the cooling power of various quenchants (Fig. 3.36).

Paschkis and Stolz (Ref 71) also examined quenching as a heat-transfer problem. Using a "heat and mass flow analyzer" (Ref 72), they derived a relationship where cooling rates are related to the surface area of the part being quenched and the temperature difference between the surface of the part and the quenching medium. This "reduced cooling rate" is called the boundary conductance ($W/m^2 \cdot K$, or $Btu/ft^2 \cdot h \cdot °F$). The objective was to correlate boundary conductance as a function of surface temperature with the quench severity (Fig. 3.37) and ultimately with hardness for any quench medium (Ref 73).

Tensi and Steffan (Ref 69) have also used heat-transfer coefficients and heat-flux calculations to characterize various quenchant media. One of the focal points of this work was to compare the heat-transfer coefficient as a function of surface temperature

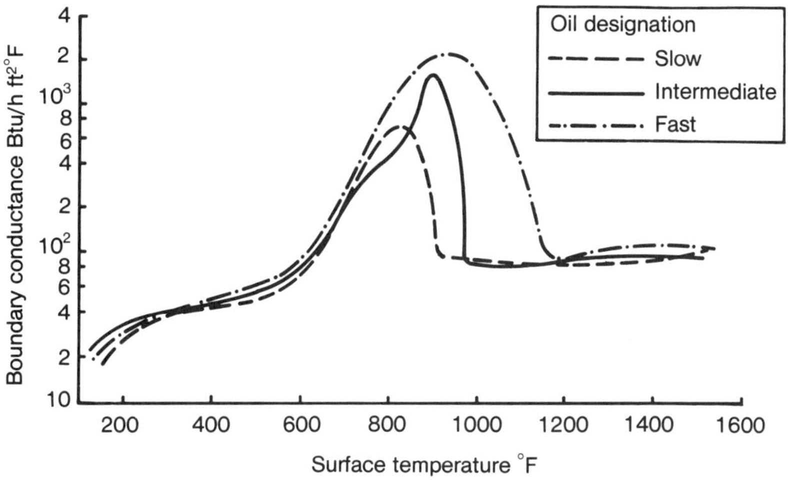

Fig. 3.37 Boundary conductance versus surface temperature for three quenching oils

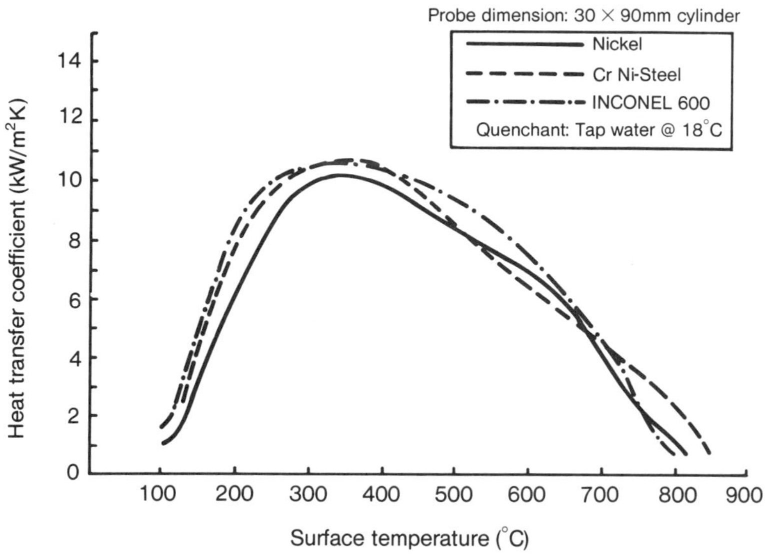

Fig. 3.38 Heat-transfer coefficient as a function of surface temperature during the quenching of a cylindrical probe

for various probe materials quenched in water (Fig. 3.38). The heat-transfer coefficients obtained with Inconel 600 and steel were very similar up to the maximum cooling rate. Above this temperature, however, differences were observed.

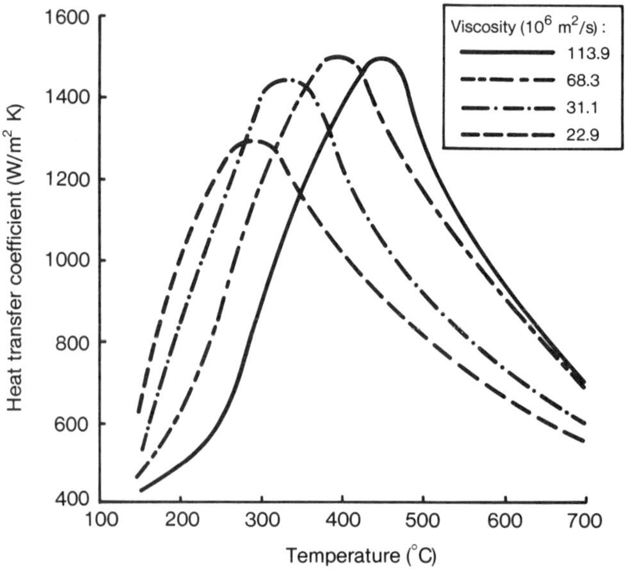

Fig. 3.39 Heat-transfer coefficients for hardening oil of various viscosities

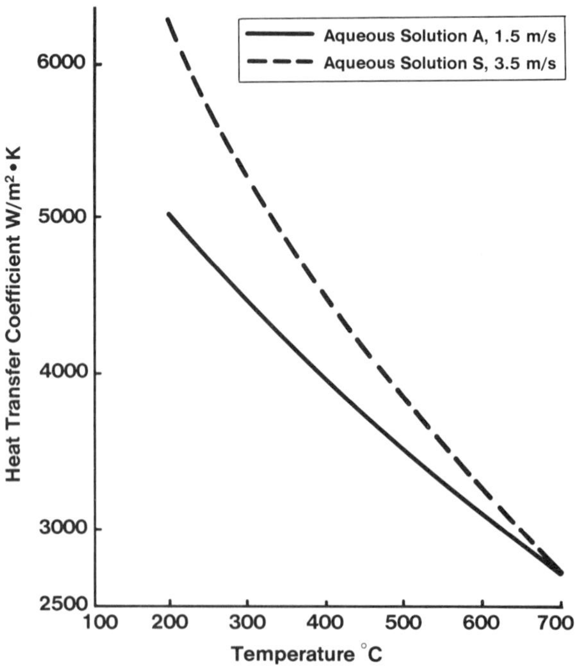

Fig. 3.40 Heat-transfer coefficients for aqueous solutions at various flow rates. Experimental conditions were the same as for Fig. 3.39.

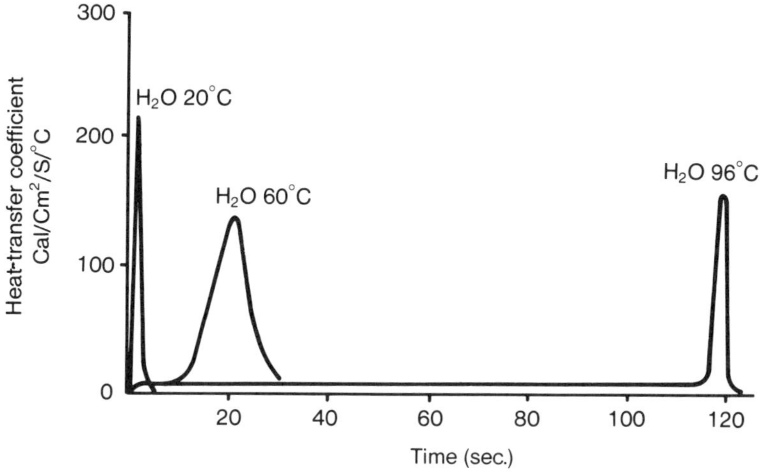

Fig. 3.41 Effect of water temperature on cooling time of a silver probe (see Fig. 3.8)

Table 3.9 Effect of water temperature on the heat-transfer coefficient

Water temperature		Heat-transfer coefficient,
°C	°F	(cal/cm^2/s/°C × 10^3)
20	68	224
60	140	142
100	212	156

Zieger (Ref 74) demonstrated that the heat-transfer properties of quenchants are dependent on the viscosity of the quenchant medium (Fig. 3.39), and on flow rate (Fig. 3.40). The heat-transfer rate decreased with increasing quenchant viscosity and increased with increased flow rate.

Chevrier and Beck (Ref 75, 76) made a detailed study of the heat-transfer properties that occurred during quenching of a silver cylinder in water at various temperatures (Fig. 3.41). The value of the film coefficient, h, goes through a minimum with increasing water temperature (Table 3.9). The effects of probe diameter (Fig. 3.42) and probe material (Fig. 3.43) on the heat flux were also studied. The data obtained for nickel and for aluminum probes were quite different, suggesting that heat transfer from a given alloy during quenching should be modeled using a probe with similar heat-transfer properties (e.g., stainless steel to model carbon and alloy steel) or, if necessary, a probe constructed of the alloy of interest.

The effect of varying probe cross section (Fig. 3.44) was studied by Kobasco (Ref 77). The heat-transfer coefficient decreased with increasing probe size. Kobasco (Ref 77) also studied the heat-transfer properties of aqueous sodium chloride and sodium

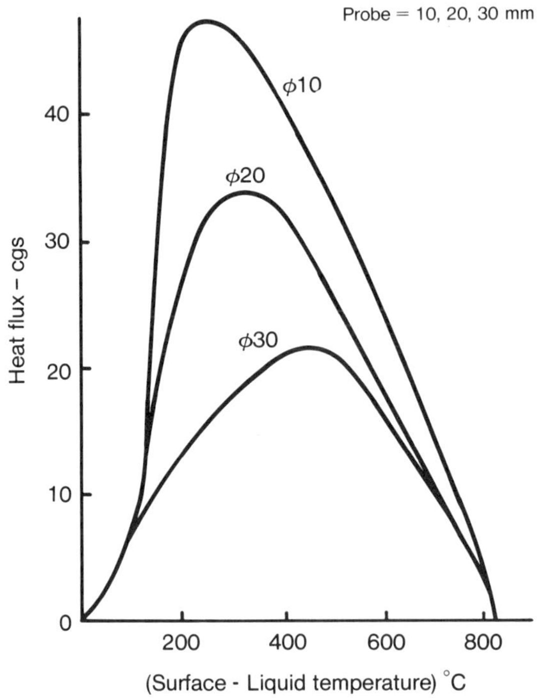

Fig. 3.42 Change in heat-flux density with probe diameter during quenching (see Fig. 3.8)

hydroxide solutions. The results of this study (Fig. 3.45) showed that effective heat transfer reaches a maximum at 10 to 12% salt concentration. Greater concentrations produce progressively lower heat-transfer rates.

Wallis (Ref 78, 79) calculated heat-transfer coefficients at various positions on a disk-shaped probe to characterize various quenchant media (Fig. 3.46). From these data, the temperature differences between different regions of an actual forged disk were predicted, which in turn were used for calculating residual stress.

The investigators discussed in the preceding paragraphs have shown the utility of heat-transfer coefficients, which are derived from cooling curves, to quantitatively describe the quenching process. However, heat-transfer data do not immediately provide information on the metallurgical properties obtained while quenching a given alloy in a particular medium. Wunning and Liedtke (Ref 80) have utilized the QTA method to provide such a correlation.

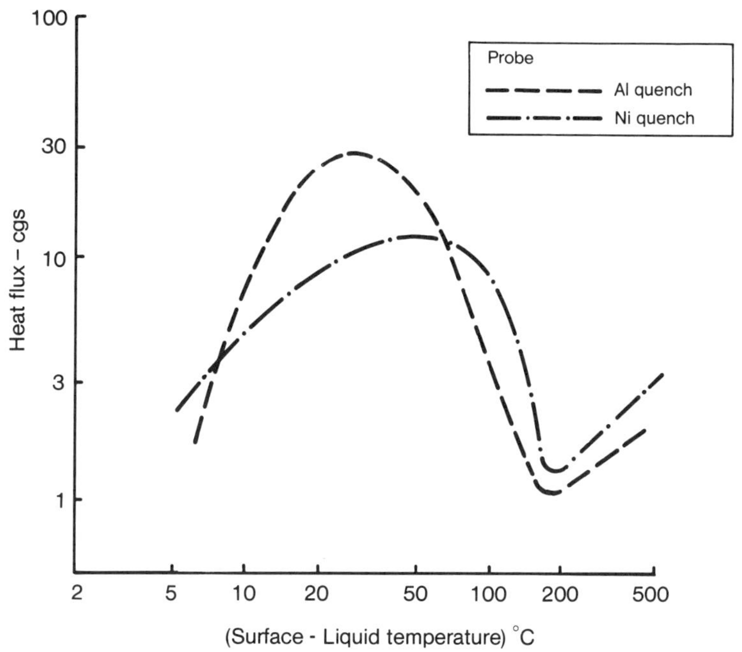

Fig. 3.43 Change in heat-flux density during quenching of nickel and aluminum test pieces in boiling water

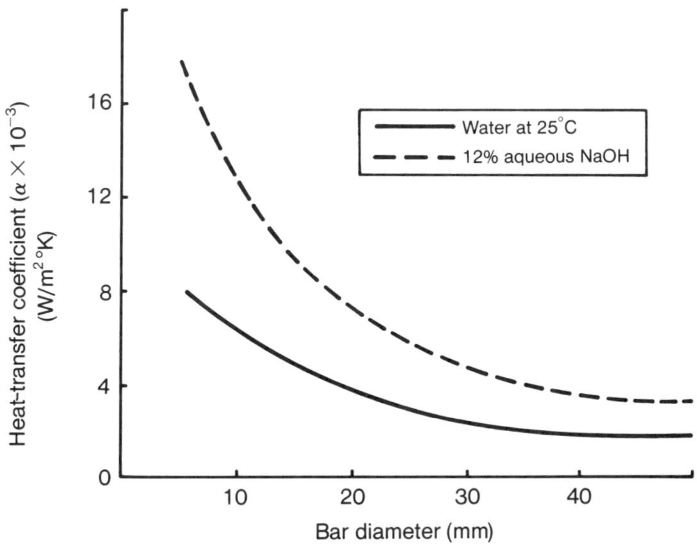

Fig. 3.44 Effect of test piece dimensions on the heat-transfer coefficient during nucleate boiling

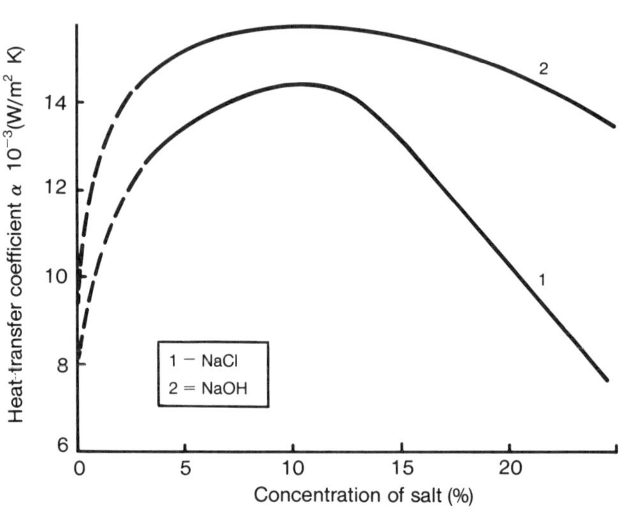

Fig. 3.45 Effect of solution concentration on the heat-transfer coefficient

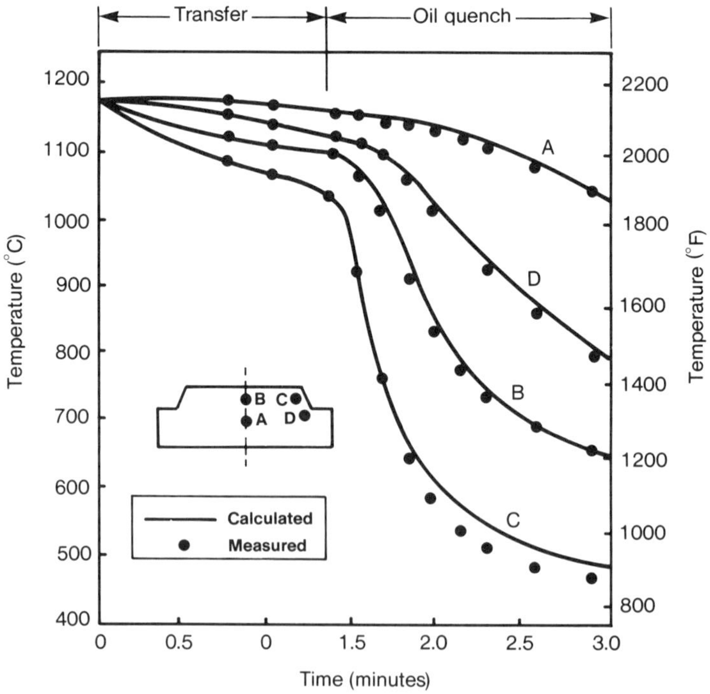

Fig. 3.46 Comparison of calculated and measured temperatures during oil quenching of a contour forging

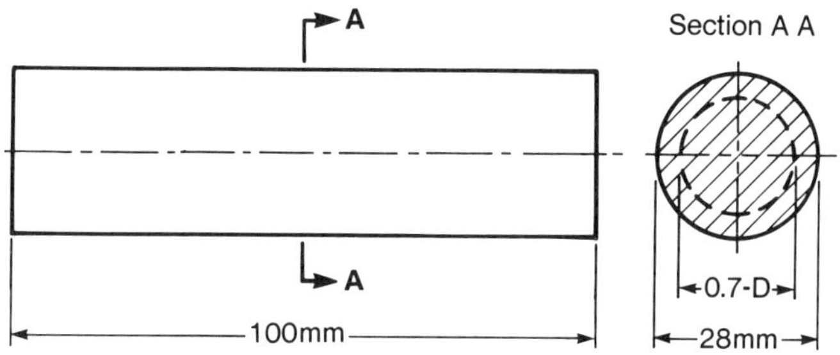

Fig. 3.47 Dimensions of the standard cylindrical specimen used in the QTA method

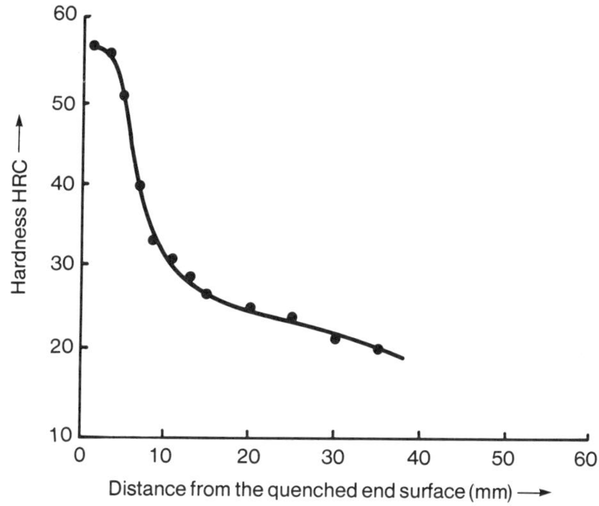

Fig. 3.48 End-quench curve for 46Cr2 steel

The first step in the QTA procedure is to quench a standard cylindrical specimen (Fig. 3.47) instrumented with thermocouples in the quenchant of interest. A cooling temperature-time curve is determined during end-quenching the test piece at the center and 0.7*D* (diameter). The Rockwell hardness is determined at the same positions, and these values are correlated with standard Jominy end-quench test results (Fig. 3.48).

The equivalent Jominy distance to produce the same hardness as obtained at 0.7D is then determined. This suggests that the same cooling process must have occurred at both positions. The cooling curves for the QTA test piece and for the equivalent Jominy distance are compared (Fig. 3.49). Similar correlations with other positions at the surface of the test bar are then obtained. These data are used to construct a *Q* value correlation (Fig. 3.50).

The correlation drawn at 0.7*D* corresponds to the *Q* value, or mean heat-flux density, and is derived from the mean cooling rate in the vapor blanket stage (Fig. 3.49). The *Q* value is inversely proportional to the square of the distance from the end surface.

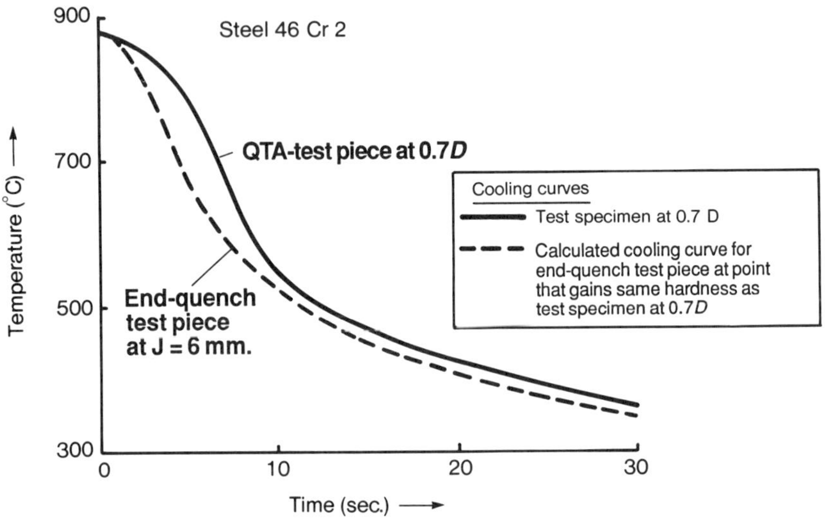

Fig. 3.49 Cooling curves at points of equal hardness (44 HRC) for QTA test pieces at 0.7D and for the end-quench test piece at J = 6 mm (0.24 in.)

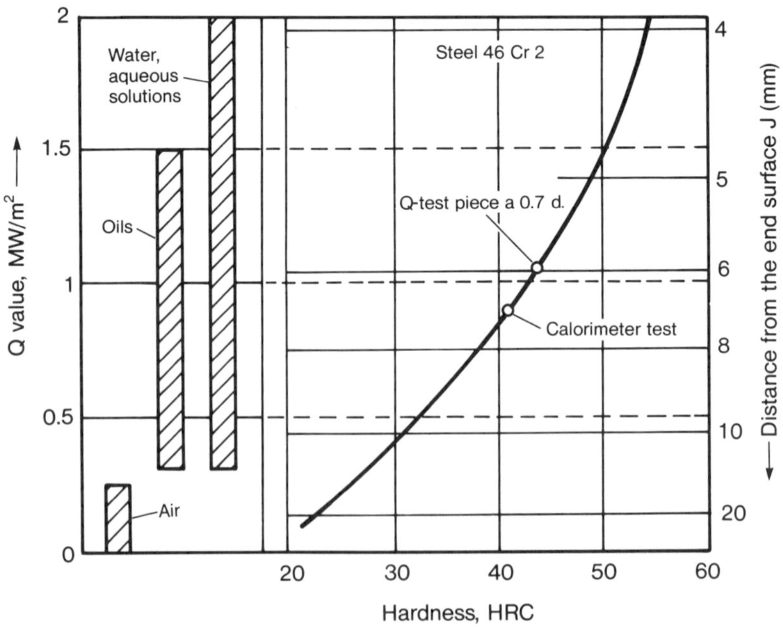

Fig. 3.50 Q value (heat flux) versus hardness

Once the Q values are known for a series of quenchants, they can be ranked with a Q-T diagram (Fig. 3.51) (where T is the boiling point of the quenchant). These Q-T plots can be used to calculate a critical cooling rate between 800 to 500 °C (1470 to 930 °F) for the surface and center of the steel being studied (Fig. 3.52).

The use of Q values to quantify quench severity is compared with Grossmann's H factor in Fig. 3.53. This figure shows that both methods are related but that the Q value allows a more rigorous characterization and differentiation in quench severity, because the boiling point of the quenchant is included in the calculation of heat flux. This provides a more quantitative model of the B- to C-phase cooling, which in turn affects the ability of a quenchant to prevent cracking.

Castrol Index

Damay and Deck (Ref 81) have described a method of calculating the thermokinetic or hardening power of a quench oil. This procedure is based on cooling curve analysis and interrelates the hardenability of the steel of interest with the hardening power of the quenchant.

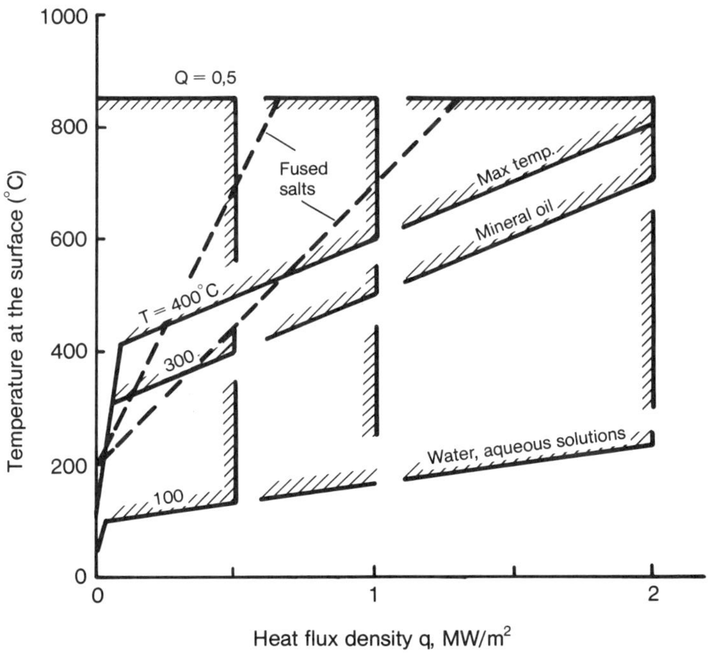

Fig. 3.51 Q-T diagram of heat-flux densities for quenching in water, oils, and fused salts

An interesting feature of this work is that it provides an interrelationship among various methods of cooling curve analysis currently used by various companies (Table 3.10). These test methods were used to characterize a series of oils (Table 3.11). The cooling curve characterization parameters used for this work are summarized in Fig. 3.54. The following

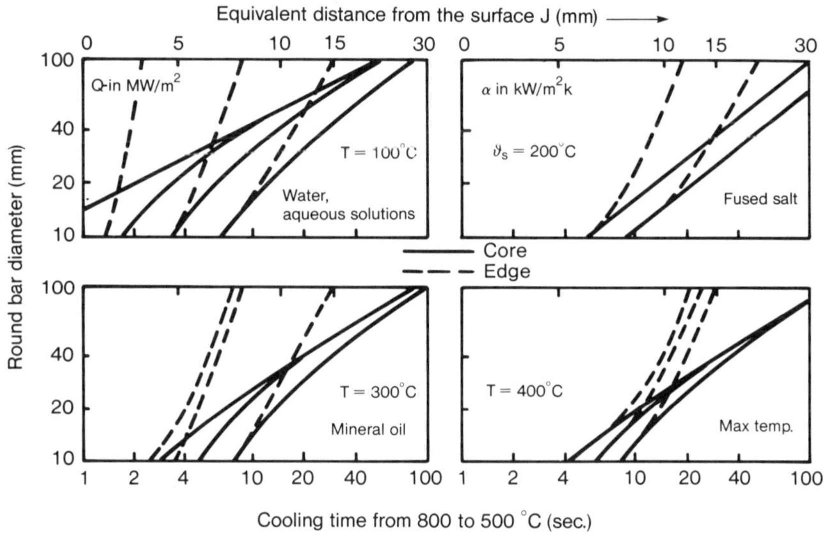

Fig. 3.52 Critical cooling time (800 to 500 °C, or 1470 to 930 °F) for the core and edge of round bars during quenching in water, oils, and fused salts as a function of Q and T

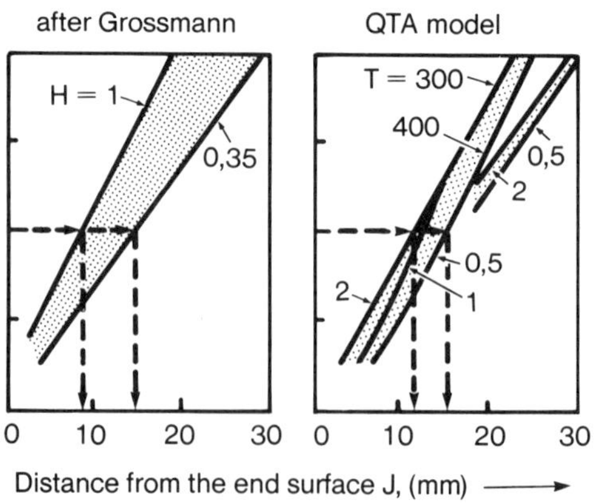

Fig. 3.53 Comparison of Grossmann's H factor and Q value correlations

Table 3.10 Experimental test methods evaluated by Damay and Deck (Ref 81)

Probe material	Dimensions (radius × length) mm	Dimensions (radius × length) in.	Heating conditions Temperature °C	Heating conditions Temperature °F	Heating conditions Atmosphere	Quenching conditions Volume L	Quenching conditions Volume gal	Quenching conditions Temperature °C	Quenching conditions Temperature °F	Method
38C2	16 × 38	0.6 × 1.5	850	1560	Nitrogen	3	0.8	140	285	RNUR(b)
Silver...............	16 × 48	0.6 × 1.9	800	1470	Air	0.7	0.2	150	300	NFT60 178(c)
Inconel (Wolfson)(a)...	12.5 × 60	0.5 × 2.4	850	1560	Air	2	0.5	150	300	ISO DIS-9950P(d)
Inconel	16 × 48	0.6 × 1.9	850	1560	Nitrogen	3	0.8	140	285	RNUR(b)

(**a**) Shown schematically in Fig. 2.12. (**b**) Specification of Regie Nationale des Usines Renault. (**c**) "Internal" specification developed by Castrol Ltd. (**d**) Quenching condition being used in the ISO DIS-9950, currently in the final balloting stage

conclusions were drawn from the results shown in Table 3.11 for the 38C2 and Inconel probes:

- The cooling curves obtained with both probes were similar, although the Inconel probe produced smaller changes in the cooling curve parameters.
- The θ_{max} value did not provide any significant correlation with hardness.
- The hardness values varied linearly with temperature at the beginning of convective cooling (θ'_2).
- Hardness was particularly sensitive to, although nonlinear with, cooling rate at 400 °C (V_{400}).

The higher thermal conductivity of the silver probe resulted in faster cooling. The cooling curve results with the silver probe showed:

- The θ_1 temperature varied with hardness; however, no significant correlation was obtained.
- The θ_2 temperatures varied linearly with hardness.

Based on this work, Damay and Deck (Ref 81) elected to use the data collected from the 38C2 probes for development of the Castrol index. The values for θ_2 and V_{400} were used because they provided a reasonable correlation with hardness. Correlation analysis of these parameters with hardness provided the regression equations for hardness prediction. For the 38C2 probe, the regression equation is (Fig. 3.55):

$$R_c\,(\text{HRC}) = 92.36 - 0.129\theta_2 - 0.147\,V_{400}$$

In order to obtain acceptable hardness versus cooling power correlations, Damay and Deck (Ref 81) developed an alternative relationship applicable to all probe materials:

$$R_c\ (HRC) = 35.7C - 5.7$$

where C is defined as:

$$C = \frac{V_{max} - (\theta_{max} - \theta_t)(oil)}{V_{max} - (\theta_{max} - \theta_t)(ref)}$$

The values for V_{max} and θ_{max} are the maximum cooling rate and the temperature at the maximum cooling rate, respectively. The θ_t value is the bath temperature.

The heat-transfer coefficient for the quench medium can be calculated from:

$$h = KV_{max}/(\theta_{max} - \theta_t)$$

Table 3.11 Thermokinetic correlation of various quench oils

Quench	Probe	θ_1		θ_2		θ'_2		V_{max}		V_{400}		θ_{max}		Hardness
oil	type	°C	°F	°C	°F	°C	°F	°C/s	°F/s	°C/s	°F/s	°C	°F	HRC
1	38C2	718	1324	565	1049	482	900	33.9	61.0	6.0	10.8	652	1206	30
	Wolfson	708	1306	493	919	409	768	56.3	101.3	6.3	11.3	605	1121	
	Silver	612	1134	442	828	355	671	135.0	243.0	9.8	17.6	548	1018	
2	38C2	712	1314	510	950	480	896	42.8	77.0	6.2	11.2	616	1141	31
	Wolfson	700	1292	500	932	411	772	59.7	107.5	6.8	12.2	612	1134	
	Silver	618	1144	440	824	351	664	139.6	251.3	10.0	18.0	553	1027	
3	38C2	736	1357	535	995	434	813	56.2	101.2	5.2	9.4	658	1216	34
	Wolfson	694	1281	455	851	374	705	71.0	127.8	6.6	11.9	622	1152	
	Silver	644	1191	410	770	330	626	165.4	297.7	17.0	30.6	574	1065	
4	38C2	758	1396	522	972	415	779	53.5	96.3	5.8	10.4	670	1238	36
	Wolfson	810	1490	461	862	357	675	71.0	127.8	14.1	25.4	626	1159	
	Silver	643	1189	387	729	308	586	180.0	324.0	36.0	64.8	560	1040	
5	38C2	729	1344	502	936	352	666	55.8	100.4	16.8	30.2	650	1202	42
	Wolfson	694	1281	441	826	354	669	78.0	140.4	11.3	20.3	613	1135	
	Silver	665	1229	382	720	306	583	199.0	358.2	10.0	18.0	577	1071	
6	38C2	778	1432	551	1024	372	702	54.2	97.6	13.0	23.4	686	1267	44
	Wolfson	735	1355	440	824	346	655	88.1	158.6	17.2	31.0	640	1184	
	Silver	678	1252	417	783	332	630	206.0	370.8	15.0	27.0	571	1060	
7	38C2	...	...	...	...	...	...	...	...	...	...	...	...	30
	Wolfson	647	1197	487	909	406	763	54.3	97.7	6.5	11.7	588	1090	
	Silver	623	1153	442	828	350	662	141.5	254.7	10.7	19.3	558	1036	

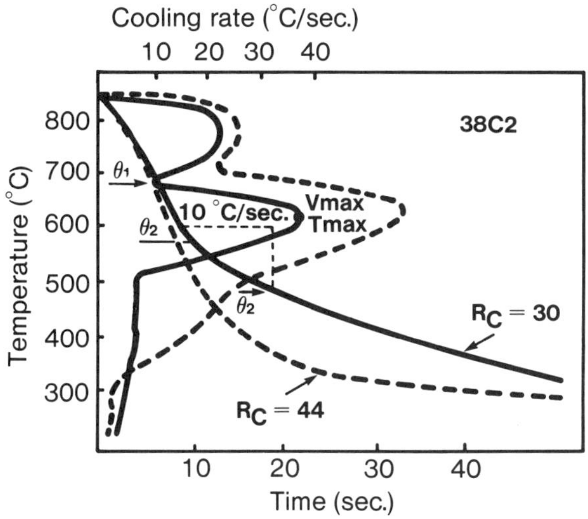

Fig. 3.54 Calculation of critical parameters from a cooling curve

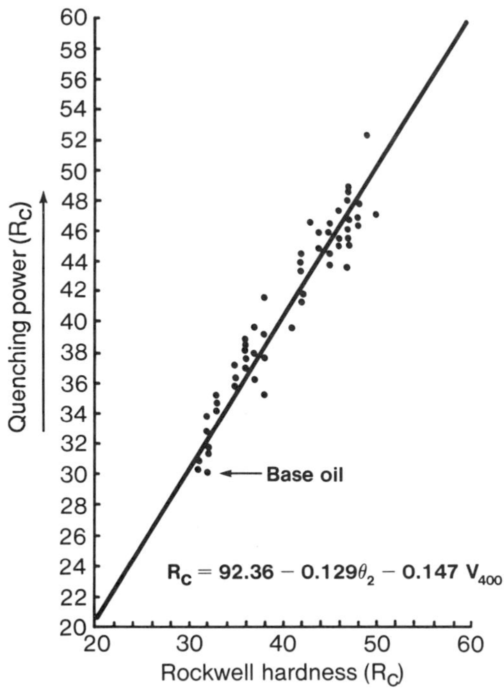

Fig. 3.55 Correlation of power and cooling power for a 38C2 steel probe

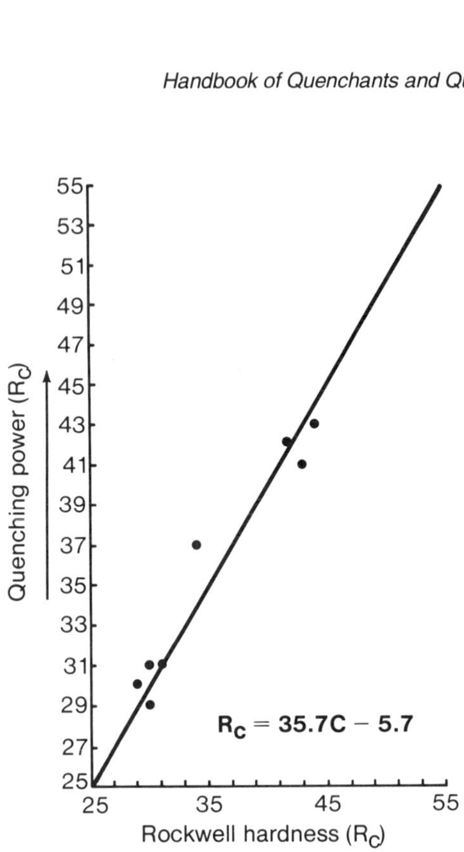

Fig. 3.56 Correlation of quenching power and cooling power for a silver probe

This relationship can be used for both the silver probe (Fig. 3.56) and the Wolfson probe (Fig. 3.57). The variation of hardness with the heat-transfer coefficient is shown in Fig. 3.58.

The Castrol index (Ci) is derived from these formulas by introducing a constant of 11.5, which corresponds to an intermediate value of H = 11 to 12, typical for an additive-free oil without agitation:

$$\mathrm{Ci} = \mathrm{K}'\left(\frac{V_{\mathrm{max}}}{\theta_{\mathrm{max}} - \theta_{\mathrm{t}}}\right)(\mathrm{oil})$$

$$\mathrm{K}' = 11.5\left(\frac{\theta_{\mathrm{max}} - \theta_{\mathrm{t}}}{V_{\mathrm{max}}}\right)(\mathrm{ref})$$

$$\mathrm{R_c\,(HRC)} = 3.1\mathrm{Ci} - 5.7$$

These same equations allow the classification of a quench oil from cooling curves obtained with standard 38C2, Wolfson, and silver probes.

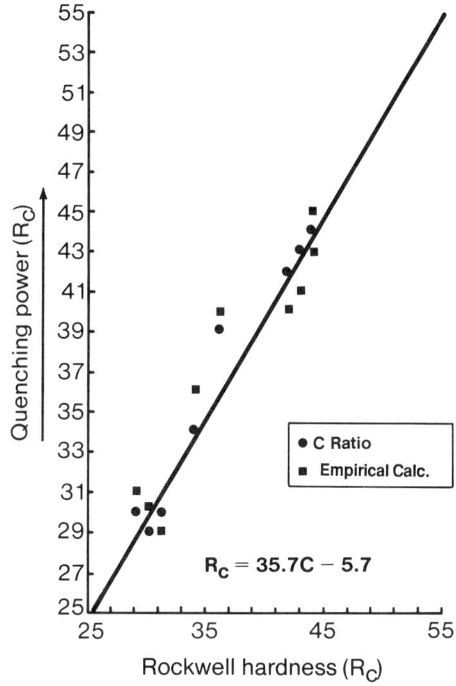

Fig. 3.57 Correlation of quenching power and cooling power for a Wolfson probe

Computer-Aided Quenchant Intensity Evaluations (Liscic Technique)

Liscic and Filetin (Ref 36) have built upon the work of Wunning and Liedtke (Ref 80), Tensi and Steffan (Ref 69), Beck and Chevrier (Ref 75, 76) and others to develop a process for measuring quenching intensity under production conditions or in the laboratory using a proprietary cylindrical probe (Fig. 3.10). This probe is reported to be particularly sensitive for measuring heat flux during quenching because it measures the temperature gradient from the surface to the center of the probe. The procedures described earlier provide near-surface temperature measurements. Near-surface temperatures may introduce error, which can sometimes be very significant, into subsequent hardness calculations.

Calculation of the cooling curve data obtained from the Liscic-NANMAC probe is based on the following algorithm:

$$\theta = \lambda \times \frac{dT}{dn}$$

where θ is the heat flux (W/m^2), λ is the thermal conductivity of the probe material (W/km), and dT/dn is the temperature gradient across the probe perpendicular to the surface (°K/m). Typical cooling curve and heat-flux output from the probe are shown in

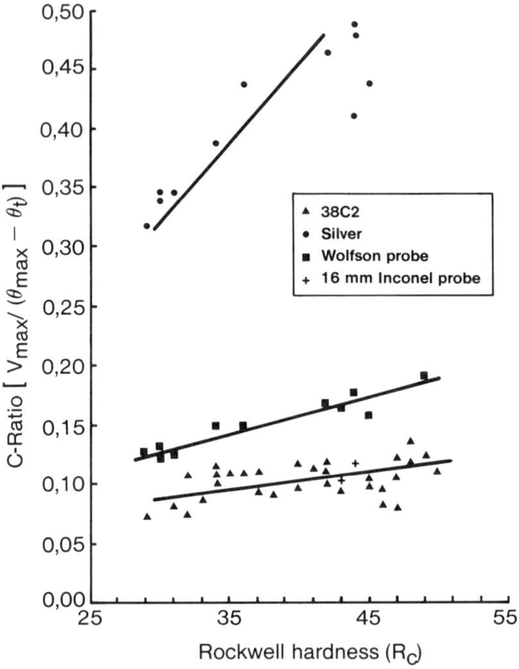

Fig. 3.58 Variation of C-ratio with hardness for 38C2 steel

Fig. 3.59 for water versus mineral oil at 20 °C (68 °F). The cooling curves were determined under laboratory conditions with the apparatus shown in Fig. 3.60.

In addition to the heat-flux data, a test specimen of the steel alloy of interest that has the same dimensions as the probe (50 × 200 mm, or 2 × 8 in.) is quenched, using the same heat treating conditions and quenchant as for cooling curve and heat-flux determination. Jominy hardenability data are also obtained. Quenching variables are selected that are representative of those encountered in the heat treating shop.

The 50 × 200 mm (2 × 8 in.) test specimen is used to determine a cross-sectional hardness survey at the surface, center, and ¼*R*, ½*R* and ¾*R* (radius). These data are used not only for building a database for various quenching conditions but also for comparison with equivalent Jominy distances.

The step-by-step data analysis process proposed by Liscic is:

1. Record quenching intensity measurements for each of the quenching conditions selected as a function of T_S, surface temperature of the probe (°C); t, time (s); θ, heat flux of the probe (MW/m^2).
2. Specify the steel grade and quenching conditions.

3. Harden a test specimen by quenching it under the quenching conditions identified in step 2.
4. Measure the cross-sectional hardness survey as described eariler.
5. From the Jominy curve for the steel alloy of interest, measure the equivalent Jominy distances, E_S, $E_{3/4R}$, $E_{1/2R}$, $E_{1/4R}$, and E_C, that yield the same hardness as determined at the positions specified for the cross-sectional hardness survey (Fig. 3.61).
6. Calculate the hypothetical quenching intensity, I, at each of the above characteristic points from the relationship:

$$E_i = \frac{AD^{B_1}}{I^{B_2}}$$

where E_i is the corresponding equivalent point on the Jominy curve, A, B_1, and B_2 are regression coefficients, D is the bar diameter, and I is the quenching intensity (or Grossmann's H factor). Regression analysis yielded the following relationships:

$$E_S = \frac{D^{0.718}}{5.11 \times I^{1.28}}, \quad I = \frac{D^{0.718}}{5.11 \times E_S^{0.78}}$$

$$E_{3/4R} = \frac{D^{1.05}}{8.62 \times I^{0.668}}, \quad I = \frac{D^{1.05}}{8.62 \times E_{3/4R}^{1.50}}$$

$$E_{1/2R} = \frac{D^{1.16}}{9.45 \times I^{0.51}}, \quad I = \frac{D^{1.16}}{8.45 \times E_{1/2R}^{1.96}}$$

$$E_{1/4R} = \frac{D^{1.14}}{7.7 \times I^{0.44}}, \quad I = \frac{D^{1.14}}{7.7 \times E_{1/4R}^{2.27}}$$

$$E_C = \frac{D^{1.18}}{8.29 \times I^{0.44}}, \quad I = \frac{D^{1.18}}{8.29 \times E_C^{2.27}}$$

These equations were reported to be valid for:

$$20\ \text{mm} < D < 90\ \text{mm}$$

$$0.2 < I < 2.0\ \text{mm}^{-1}$$

$$I < E < 40\ \text{mm}$$

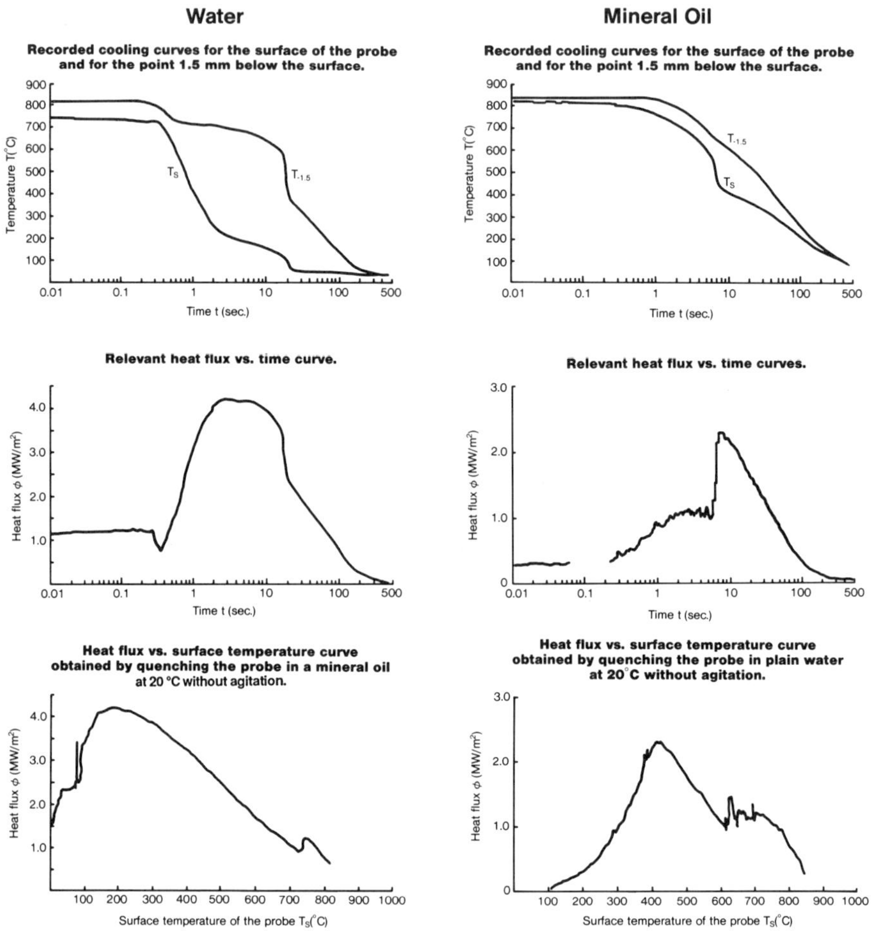

Fig. 3.59 Comparison of heat-transfer properties of water and oil quenchants (Liscic data)

7. Enter the bar diameter for which the predicted hardness is desired.
8. Calculate the equivalent Jominy distances, E_S, $E_{3/4R}$, $E_{1/2R}$, $E_{1/4R}$, and E_C, that correspond to the bar diameter D and the corresponding calculated quenching intensities.
9. Read the hardness values from the appropriate Jominy curve that corresponds to the E' values and plot the hardness distribution curve (Fig. 3.61).

Although this procedure is quite involved, it does integrate all of the classical techniques for quenchant characterization and permits the calculation of hardness data from an extensive database of experimental data obtained under shop quenching conditions.

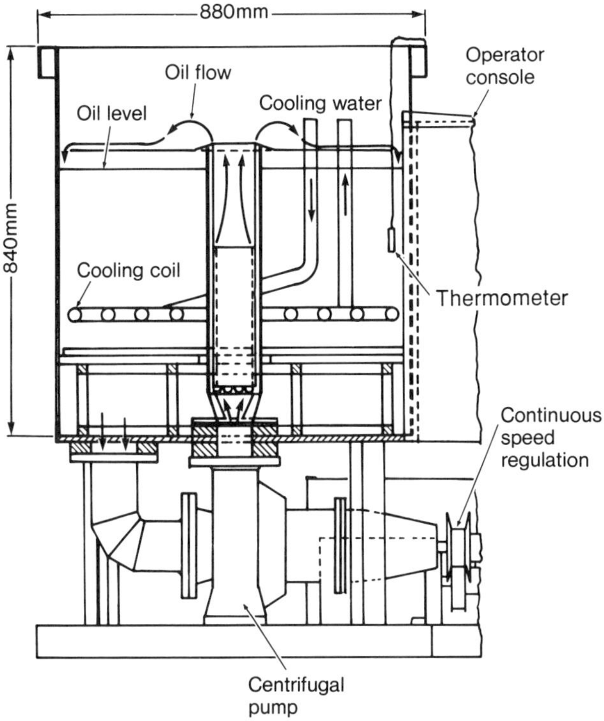

Fig. 3.60 Liscic quench tank

Predictive Analysis of Quenchant Bath Behavior

The work discussed so far has many common goals. One is to predictively relate cooling curve behavior to the metallurgical properties in a quenched part. Another is to qualitatively relate variations in quenchant bath parameters, such as bath temperature, agitation, and polymer concentration (for aqueous synthetic polymers), to cooling curve behavior.

One approach to the latter goal is to use linear regression analysis to interrelate quenchant cooling curve performance. Hilder (Ref 39) reported a set of equations describing the effects of polymer concentration, bath temperature, and linear pumping velocity for Quendila PA, an aqueous synthetic polymer quenchant based on polyalkylene glycol:

$$CR_{\max} = 244.7 - 4.3C - 1.7T + 47.2V$$

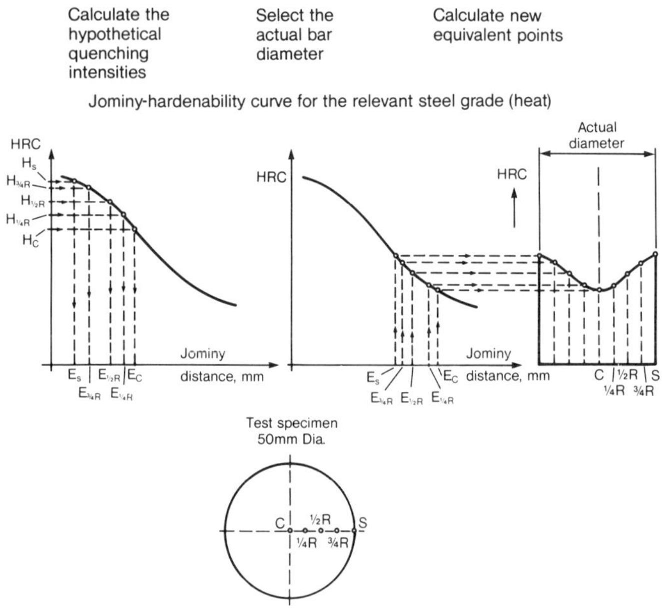

Fig. 3.61 Stepwise scheme of the process of predicting the hardness distribution

$$R_{300} = 83 - 1.68C - 0.36T + 14.1V$$

where CR_{max} is the maximum cooling rate (°C/s), R_{300} is the cooling rate at 300 °C, C is concentration (%), T is the bath temperature (°C), and V is the flow rate (m/s).

Similar expressions were developed for PARQUENCH 90, a polyvinylpyrrolidone quenchant (Ref 39):

$$CR_{max} = 223.7 - 2.94C - 1.9T + 71.9V$$

$$R_{300} = 96.3 - 1.8C - 1.02T + 37.68V$$

and for AQUAQUENCH ACR, a polysodium acrylate quenchant (Ref 39):

$$CR_{max} = 160.59 - 5.63C - 1.10T + 114.8V$$

$$R_{300} = 58.6 - 2.5C - 0.57T + 43.4V$$

Hilder's work seems to have applied linear regression analysis to each individual variable. An alternative procedure using a three-level (2^3) factorial design has been applied to a different polyalkylene glycol-based quenchant, UCON® Quenchant E (Ref 82):

$$CR_{max} = 116.2 + 0.2743X1 - 1.386X2 - 0.3634X3$$

$$R_{650} = 58.72 - 0.06680X1 - 1.035X2 - 0.09587X3 + 0.0097X1X2$$

$$H = 1.937 + 0.007150X1 - 0.02675X2 - 0.09938X3$$

where CR_{max} is the maximum cooling rate (°F/s), R_{650} is the cooling rate at 650 °F, H is the Grossmann hardenability factor (in.$^{-1}$), $X1$ is linear, turbulent flow (ft/min), $X2$ is quenchant concentration (%), and $X3$ is bath temperature (°F). It is important to note the quenching conditions used to develop this model and the use of the Hines probe (Fig. 3.17 and 5.14) are different from those used by Hilder (Fig. 3.12); therefore, the calculated results cannot be directly compared.

Park Chemical Company has published cooling rate equations for polyvinylpyrrolidone-based quenchants (Ref 83). The following model was reported, although the method of regression analysis was not:

$$CR = B_0 + B_1C + B_2T + B_3V$$

where CR is the cooling rate (°F/s), B_0, B_1, B_2, and B_3 are regression constants, C is the concentration (%), T is the temperature (°F), and V is the linear flow rate (ft/min).

Although these equations were developed under laboratory conditions and are applicable only for the quenching conditions used to perform this work, they are useful tools when used in conjunction with practical heat treating experience, because they provide guidance in selecting optimum quenching conditions. A similar strategy could easily be employed in the heat treat shop to develop a quantitative model of actual quenching conditions in the quench tank.

References

1. S. Kodama, *Netsu Shori*, Vol 25, 1985, p 23-26
2. The experimental details of a Jominy test are described in a number of specifications including: "End-Quench Test for Hardenability of Steel," ASTM A 255-67; and "Hardenability Test by End-Quenching," DIN 50, Sept 1987
3. H.G. Gilliland, *Met. Prog.*, Oct 1960, p 11
4. "Quenchability of Oils—Hot Wire Method," Aerospace Material Specification ARP 4206, 28 Sept 1990
5. *Metals Handbook*, 9th ed., Vol 4, American Society for Metals, 1981, p 32-35
6. G.E. Totten, M.E. Dakins, and R.W. Heins, *J. Heat Treat.*, Vol 6, 1988, p 87-95
7. G. Beck, *C.R. Acad. Sci. Paris*, Vol 265, 1967, p 793-796
8. J.C. Chervier, A. Simon, and G. Beck, *Heat Mass Transfer Metall. Syst. (Seminar Int. Cent. Heat Mass Transfer)*, 1979, p 535-544

9. M. Tagaya and I. Tamura, *Mem. Inst. Sci. Ind. Res., Osaka Univ.*, Vol 9, 1952, p 85-102
10. M. Schwalm and H.M. Tensi, *Heat Mass Transfer Metall. Syst. (Seminar Int. Cent. Heat Mass Transfer)*, 1981, p 563-572
11. A. Rose, *Archiv. Eisenhüttenwesen*, Vol 13, 1940, p 345-354
12. W.E. Jominy, *Hardenability of Alloy Steels*, American Society for Metals, 1939, p 73
13. "Heat Treating Oil," Japanese Industrial Standard K 2242, 1980
14. V. Paschkis and G. Stolz, *Iron Age*, Vol 22, 1956, p 95-97
15. G.P. Plekhanov and J.S. Berezhinskaya, *Sov. Teknol. Prokatki, Termoobrab. Otdelki. Tolstoloiist. Prokata, M.*, 1987, p 78-80
16. R.S. Allen, A.J. Fletcher, and A. Mills, *Steel Res.*, Vol 60, 1989, p 522-530
17. Y.I. Lipunov, K.Y. Eismondt, A.Y. Sitchenko, and A.F. Goncharov, *Freib. Forschungsh. B*, Vol B266, 1988, p 37-40
18. R.F. Price and A.J. Fletcher, *Met. Technol.*, May 1980, p 203-211
19. V. Fronek and J. Kovarik, *Hutn. Listy*, Vol 2, 1986, p 102-105
20. D.M. Trujillo and R.A. Wallis, *Ind. Heat.*, July 1989, p 22-24
21. S.D. Segerberg and J. Bodin, *Heat Treat.*, May 1988, p 26-28
22. R.W. Hines and E.R. Mueller, *Met. Prog.*, Vol 122, 1982, p 33-39
23. B. Liscic, *Härt.-Tech. Mitt.*, Vol 33, 1978, p 179-191
24. W. Luty, *Metaloz. Obrobka, Cieplna*, May-June 1984, p 12-16
25. "Laboratory Test for Assessing the Cooling Curve Characteristics of Industrial Quenching Media," Wolfson Heat Treatment Centre Engineering Group Specification, 1982 [Note: This procedure is the basis of an ISO specification which is currently being balloted for approval.]
26. B.K. Sheremeta, N.Ya. Ruddakova, G.I. Cherednichenko, A.K. Maskayev, M.R. Orazova, T.I. Tkachuk, P.S. Protsidim, Ya. Garun, Yr., E.N. Marmer, V.I. Visloboker, N.S. Gromova, and A.V. Boychenko, Soviet Patent, 247 423, 1983
27. A.M. Osman and J.V. Beck, *ASME Proc. National Heat Transfer Conf.*, Vol 3, 1988, p 109-117
28. M. Tajima, T. Maki, and K. Katayama, *Nippon Kikai Gakkai Ronbunshu (B-ken)*, Vol 54, 1988, p 3491-3496
29. C.E. Bates and G.E. Totten, *Heat Treat. Met.*, Vol 4, 1988, p 89-97
30. H.J. French, *Trans. ASST*, May 1930, p 646-727
31. J.J. Lakin, *Heat Treat. Met.*, Vol 3, 1979, p 59-62
32. M. Tagaya and I. Tamura, *Härt.-Tech. Mitt.*, Vol 18, 1963, p 63-67
33. H. Scott, *Trans. ASM*, Vol 22, 1934, p 68-98
34. M. Tagaya and I. Tamura, *Härt.-Tech. Mitt.*, Vol 18, 1963, p 63-67
35. I. Tamura, N. Shimizu, and T. Okadu, *J. Heat Treat.*, Vol 3, 1984, p 335
36. B. Liscic and T. Filetin, *J. Heat Treat.*, 1988, Vol 5, No. 2, p 115-124
37. J. Macatician and J. Nanigian, U.S. Patent 2,829,185
38. H.M. Tensi and P. Stitzelberger-Jakob, *Härt.-Tech. Mitt.*, Vol 44, 1989, p 99-105
39. N.A. Hilder, Ph.D. thesis, University of Aston (UK), Jan 1988
40. D.J. Carney and A.D. Janulionis, *Trans. ASM*, Vol 43, 1951, p 480-496
41. J.V. Beck, Michigan State University, private communication
42. J.V. Beck, B. Blackwell, and C.R. St. Clair, *Inverse Heat Conduction—Ill Posed Problems*, Wiley-Interscience, 1985, chap 1
43. H. Zhu, *Heat Treat. Met.* (China), Vol 11, 1986, p 26-31
44. S. Zhang, H. Liu, and X. Zhou, *Heat Treat. Met.* (China), Vol 4, 1986, p 35-39
45. M. Umemoto, N. Nishioka, and I. Tamura, *J. Heat Treat.*, Vol 2, 1981, p 130-138

46. S. Segerberg, *Conf. Heat Treatment*, London, 2-4 May 1984, p 19.1-19.7
47. K.E. Thelning, *Scand. J. Metall.*, Vol 12, 1983, p 189-194
48. C.E. Bates, *Heat Treat.*, Oct 1983, p 18-22
49. H.M. Tensi and E. Steffen, *Steel Res.*, Vol 56, 1985, p 489-495
50. R.K. Zhelokhovtseva, *Steel USSR*, Vol 15, 1985, p 238-239
51. M. Trusculescu, D. Trasculescu, V., Serban, and A. Raduta, *Cercet. Metal.*, Vol 27, 1986, p 223-228
52. B. Liscic, *Härt.-Tech. Mitt.*, Vol 33, 1978, p 179-191
53. G.E. Totten, M.E. Dakins, K.P. Ananthapadmanabhan, and R.W. Heins, *Heat Treat.*, Dec 1987, p 18-20
54. V.V. Goryushin, V.F. Arifmetchikov, A.K. Tsretkov, and S.N. Sinetskii, *Met. Sci. Heat Treat.*, Vol 10, 1986, p 709
55. R. Wang, Y. Wang, and L. Su, *Jinshu Rechuli Xubao*, Vol 8, 1987, p 30-40
56. P.E. Carey, *Met. Prog.*, Feb 1966, p 90
57. R.S. Wang, Y. Wang, and L.P. Su, *Jin Xu Re Chu Li Xu Bao*, Vol 8, 1987, p 30-36
58. M.A. Grossmann and M. Asimov, *Hardenability of Alloy Steels*, American Society for Metals, p 124-190
59. J.L. Lamont, *Iron Age*, Oct 1943, p 64-70
60. R.W. Monroe and C.E. Bates, *J. Heat Treat.*, Vol 3, 1983, p 833-899
61. M.E. Dakins, C.E. Bates, and G.E. Totten, *Metallurgia*, Vol 56, 1989, p 57-59 [Note: The algorithm for calculating H factor is incorrect in this paper and should be: $H = AX^c \exp(BX^D)$.]
62. S. Segerberg, paper presented at ASM Heat Treating Conf., March 1990
63. H.M. Tensi and P. Stitzelberger-Jakob, *Mater. Sci. Technol.*, Vol 5, 1989, p 718-724
64. M. Schwalm and H.M. Tensi, *Heat Mass Transfer Metall. (Seminar Int. Cent. Heat Mass Transfer)*, Vol 45, 1981, p 563-572
65. H.M. Tensi, "Technische Universitat Munchen—Lehrstahl fur Metallurgie and Metallkunde," technical brochure, 1989
66. Th. Kunzel, H.M. Tensi, and G. Wetzel, *5th Annual Congress on Heat Treatment of Materials*, Vol III, Budapest, 20-24 Oct 1986
67. H. Tensi, P. Stiltzelberger-Jakob, and T. Kunzel, *Maschinemarkt*, Vol 94, 1988, p 70-72
68. H.M. Tensi, G. Welzel, and Th. Kuenzel, *Proc. 8th Int. Heat Transfer Conf.*, San Francisco, 12-22 Aug 1986, p 3031-3035
69. H. Tensi and E. Steffan, *Warine Stoffubertragung*, Vol 19, 1985, p 279-286
70. N.N. Khavskii and R.K. Zhelokhovtseva, *Isv. VUZ Chernaya Metall.*, Vol 3, 1982, p 111-113
71. V. Paschkis and G. Stolz, *J. Met.*, Aug 1956, p 1074-1075
72. V. Paschkis, *Sci. Month.*, Vol 73, 1961, p 81-88
73. G. Stolz, V. Paschkis, C.F. Bonilla, and G. Acevedo, *J. Iron Steel Inst.*, Vol 193, 1959, p 116-123
74. H. Zieger, *Neue Hutte*, Vol 9, 1986, p 339-343
75. G. Beck and J. Chevrier, *Int. J. Heat Mass Transfer*, Vol 14, 1971, p 1731-1745
76. J.C. Chevrier and G. Beck, *Mem. Sci. Rev. Metall.*, Vol 69, 1972, p 623-632
77. N.I. Kobasco, *Metalloved. Term. Obrab. Met.*, Vol 3, 1968, p 2-6
78. R.A. Wallis, *Heat Treat.*, Dec 1989, p 26-27
79. D.M. Trujillo and R.N. Wallis, *Ind. Heat.*, July 1989, p 22-24
80. J. Wunning and D. Liedtke, *Härt.-Tech. Mitt.*, Vol 38, 1983, p 149-155
81. P. Damay and M. Deck, paper presented at Int. Heat Treating Assoc. Meet., Lamans, Sept 1990

82. G.E. Totten, M.E. Dakins, and L.M. Jarvis, *Heat Treat.*, Dec 1989, p 28-29
83. "PARQUENCH—Water Based Polymer Quenchants," Park Chemical Co.

CHAPTER 4

Quenching Oils

Water and oil are the quenchants most commonly used to harden steel. Oils used for quenching applications include various petroleum distillates and animal fats—generally mixtures of chemical structures with a range of molecular weights—and thus vary widely in composition, properties, and heat-removal characteristics, depending on the source and extent of refinement. They may also be blended with various additives. The view that oils can be used to quench any steel, including carbon, low-alloy, and high-alloy steels, is incorrect, although a variety of steels can indeed be quenched in a single quench oil.

This chapter will explore and define compositions and properties of typical quench oils, discuss the nature and effect of additives on quenching properties, outline quench oil classification, illustrate the sensitivity of quench oils to process variables such as bath temperature and agitation rate, and describe the effects of degradation and additive dragout on quenching properties.

Quench Oil Compositions

Petroleum-Derived Quench Oils. Quenching oils are analogous to other petroleum products, including engine oils, spindle oils, and industrial lubricating oils such as gear lubricants (Ref 1). Although petroleum oils are usually refined for specific applications, they remain complex mixtures with a variety of possible compositions, which may vary even when produced by a single refinery.

The complexity of a quench oil can be shown by gas chromatography, an analytical technique that separates mixtures based on differences in component volatility and adsorptivity. The complexity of the mixture, or the number of individual components, can be determined by counting the number of peaks in the chromatogram, which provides a characteristic "fingerprint" of the oil.

Figure 4.1 shows a "chromatogram" of a Wolfson reference oil (Ref 2) that has been proposed as a standard for the International Organization for Standardization (ISO) draft on quench oil (Ref 3). Figure 4.2 shows a similar chromatogram trace for an accelerated fast quenching oil. These chromatograms illustrate that both oils contain numerous components, as evidenced by the number of peaks present. In fact, the peaks are so numerous that their complete resolution is not possible unless special techniques are applied (Ref 4, 5).

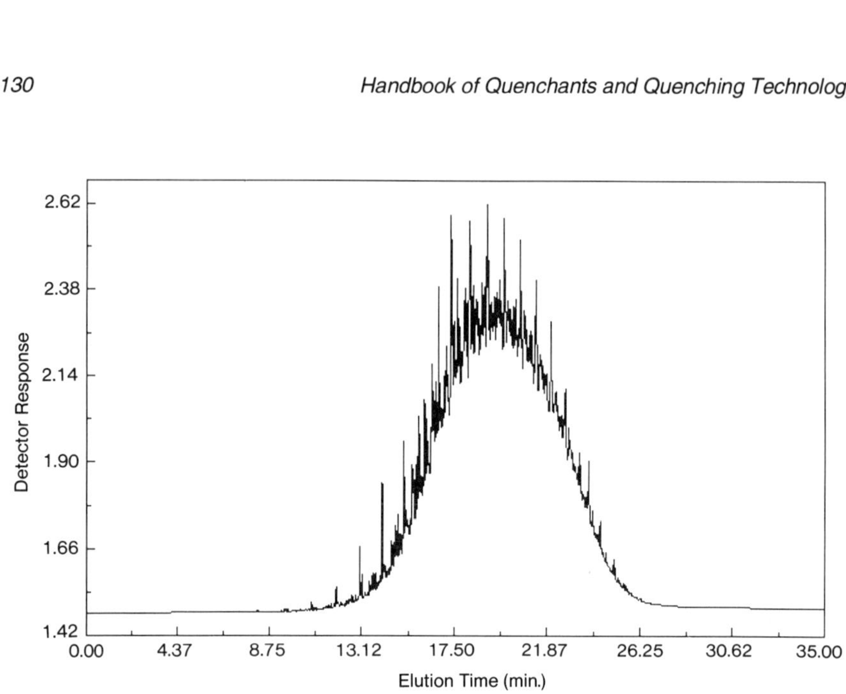

Fig. 4.1 Gas chromatogram of the Wolfson reference quench oil

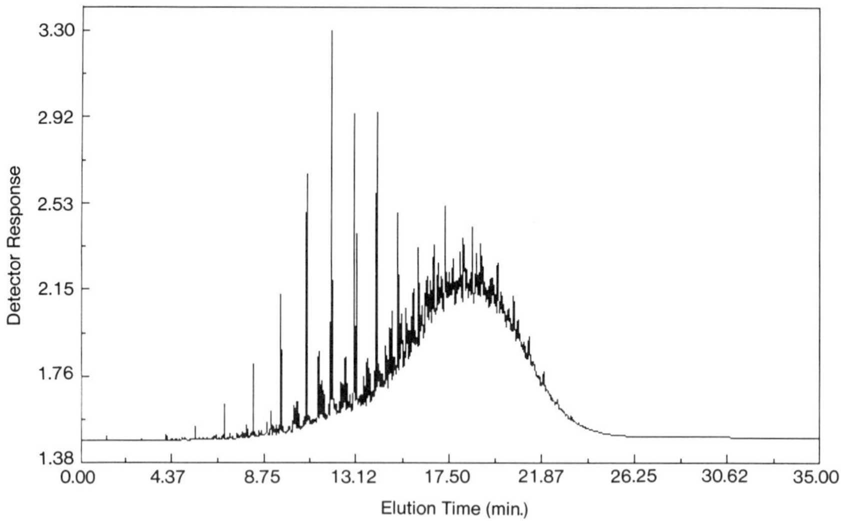

Fig. 4.2 Gas chromatogram of an accelerated quench oil

The chromatographic complexities of the two quench oils illustrated in Fig. 4.1 and 4.2 are typical of refined petroleum derivatives. The components of a petroleum oil are many, including paraffinic, naphthenic, and various oxygen-, nitrogen-, and sulfur-derived open-chain and cyclic derivatives. Examples of heterocyclic derivatives are listed in Table 4.1 (Ref 6, 7). The specific composition of a petroleum product varies with the source of the crude oil—western versus eastern United States, Middle East, Far East, Venezuela, North Sea, etc.

The volatility of components in an oil decreases as the average molecular size or carbon number of the components increases. The volatility of a quench oil component is usually inversely proportional to its "flash point," the lowest temperature at which the vapors over an oil sample (in either an open or a closed system) will ignite if exposed to a flame. The more volatile a component, the lower its flash point. Figure 4.3 lists several petroleum oil components and their relative volatilities.

The compositional complexity of quench oils affects their quenching performance. Segerberg (Ref 8) compared a series of mineral-oil-base quenchants under standard conditions and obtained a wide variety of cooling rates (Fig. 4.4). It is clear that even straight mineral oils vary in quenching performance. Formulated oils can produce an even wider range of cooling rates.

Windgassen (Ref 9) reported that quenching oils that contain substantial quantities of naphthenic derivatives usually exhibit inferior cooling characteristics, a greater deposit-forming tendency, and lower flash points than paraffinic oils. The lower flash points are particularly deleterious in heat treating applications. Protsidim *et al.* (Ref 10) also showed that small changes in the compositions of the quench oils listed in Table 4.2 resulted in significant changes in quenching properties (Fig. 4.5).

Tensi (Ref 11) has shown that the quench severity of a particular oil is directly related to its ability to wet a metal surface. The wettability of an oil is quantified by measuring "rewetting" times, as shown in Fig. 4.6.

Another method of quantifying the ability of a quench oil to wet a steel surface is to measure the contact angle of the oil on that surface. Tkachuk *et al.* (Ref 12) measured the viscosity and contact angle of four quench oils and then determined the cooling rates they produced using an instrumented silver ball probe (Table 4.3). Typical cooling curves are shown in Fig. 4.7 and 4.8.

The contact angles, expressed as cos θ, of the quenchant oils were measured after applying a droplet from a calibrated pipet (delivery tube) to a machined steel plate. After

Table 4.1 Heterocyclic derivatives that may be present in petroleum oil

Sulfur heterocycles	**Nitrogen heterocycles**
Cyclohexanethiol	Pyrrole
Thiacyclohexane	Pyridine
Benzothiophene	Quinoline
Oxygen heterocycles	Isoquinoline
Cyclopentanecarboxylic acid	Indole
Phenol	Carboxole
Fatty acids	

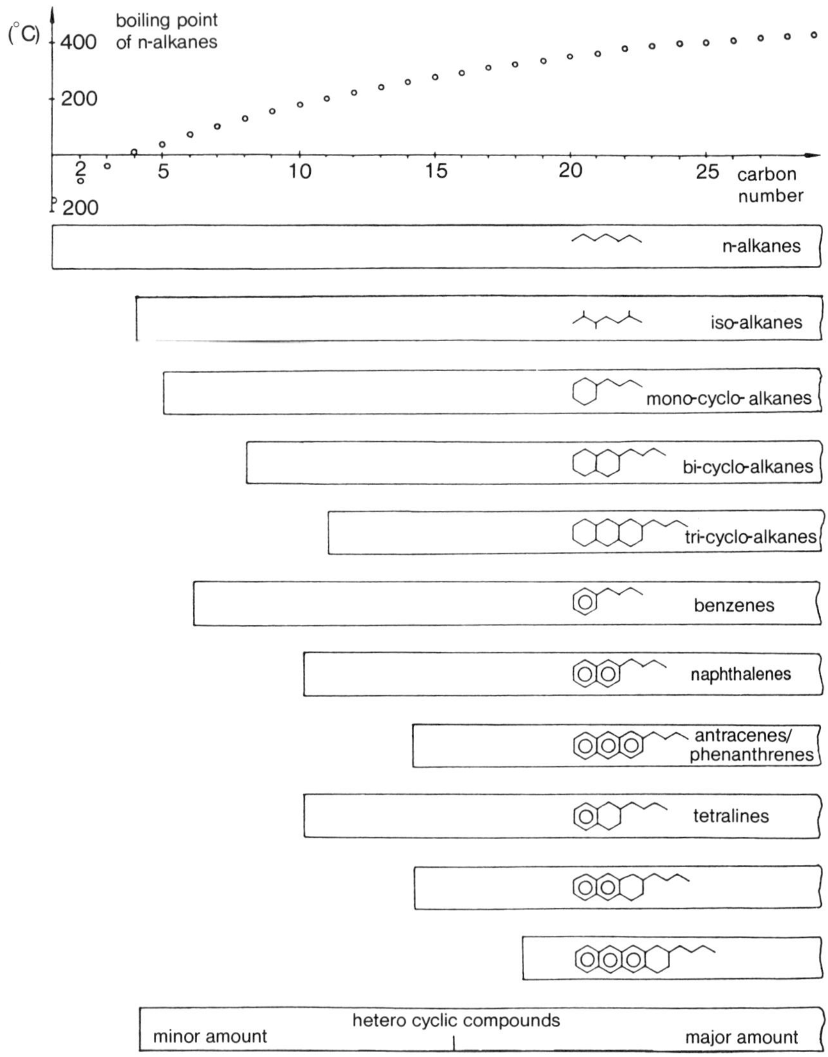

Fig. 4.3 Expected hydrocarbons in a typical crude oil fraction

standing for 5 min, the height (h) and diameter (d) at the base of the droplet were measured. The cosine of the contact angle was calculated from:

$$\cos \theta = \frac{r^2 - h^2}{r^2 + h^2}$$

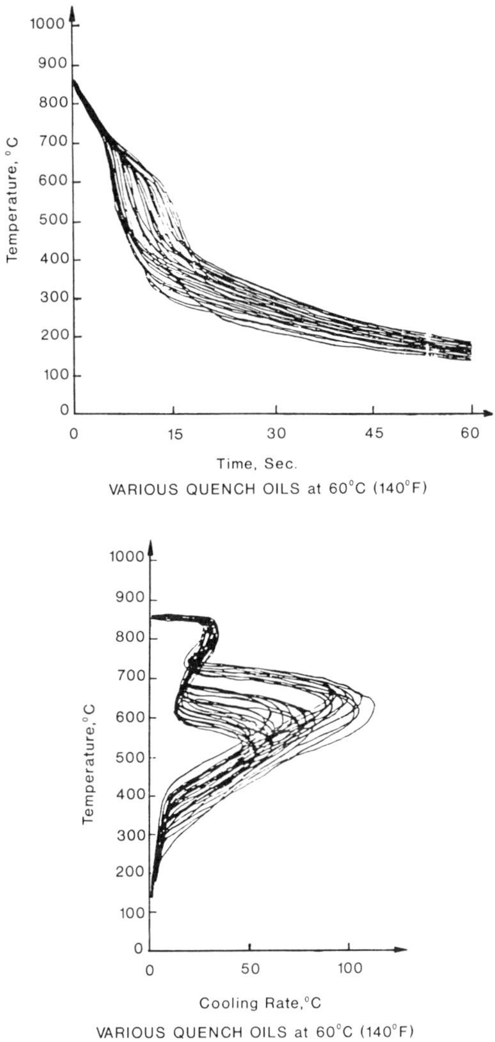

Fig. 4.4 Potential variation of the severity of quench oils

where $r = d/2$. The data showed that increasing contact angles reflect increasing wettability of the oil on the steel surface.

The results of this study also showed that as the viscosity of the quench oil increased, the contact angle decreased and the cooling rate decreased (Fig. 4.9). Although the mean boiling points of the four oils were similar, significant differences in cooling rate were attributable to variations in the transition temperature from A- to B-stage cooling.

Table 4.2 Impact of quench oil composition on physical properties

Property	Oil: MZM-16	I-20A	I-20AR1	I-20AR2
Hydrocarbon composition, %				
Paraffin-Naphthene	73.70	71.48	67.04	62.25
Aromatic groups				
I	11.30	11.96	11.60	16.00
II	6.70	7.80	3.16	10.58
III	3.80	4.00	2.80	4.20
IV	3.60	2.00	3.84	1.40
resins	0.50	2.10	4.00	4.10
losses	0.40	0.66	2.96	1.48
Kinematic viscosity, 50 °C, m^2/s (120 °F, ft^2/s)	17.5 (188.4)	18.3 (197.0)	24.8 (267.0)	25.6 (275.6)
Flash point, °C (°F)				
Open crucible	182 (360)	180 (356)	160 (320)	167 (333)
Closed crucible	188 (370)	186 (367)	180 (356)	188 (370)
Oxidation test, wt% decrease	1.54	1.8	2.2	3.0
Change in flash point in closed crucible, °C (°F)	None	None	+2.2 (+4.0)	+3.0 (+5.4)

Table 4.3 Effect of quench oil composition on viscosity and contact angle

Property	Composition: Paraffin-naphthene	Aromatic hydrocarbons: I	II	III
Kinematic viscosity, 50 °C, m^2/s (120 °F, ft^2/s)	32.7 (352.0)	51.3 (552.2)	66.8 (719.0)	172.1 (1852.5)
Fraction composition, °C (°F)				
Start of boiling	337 (639)	340 (644)	335 (635)	334 (633)
10% boiled off	417 (783)	415 (779)	416 (781)	420 (788)
50% boiled off	458 (856)	456 (853)	454 (849)	462 (864)
90% boiled off	498 (928)	500 (932)	503 (937)	504 (939)
Mean boiling point, °C (°F)	456 (853)	457 (855)	456 (853)	460 (860)
Composition of hydrocarbon, %				
Paraffin	35.6	...	...	...
Naphthene	62.6	25.2	27.1	37.3
Naphthene-aromatic	1.8	74.6	72.7	62.5
Aromatic resin	...	0.2	0.2	0.2
Contact angle of wetting, cos θ	0.908	0.898	0.896	0.889

Quench oils derived from fats (Ref 13) are considerably less common than those derived from petroleum. Table 4.4 summarizes some of the fats that have been used as quenchants. Two mineral oil quenchants are also listed for comparison (Ref 14).

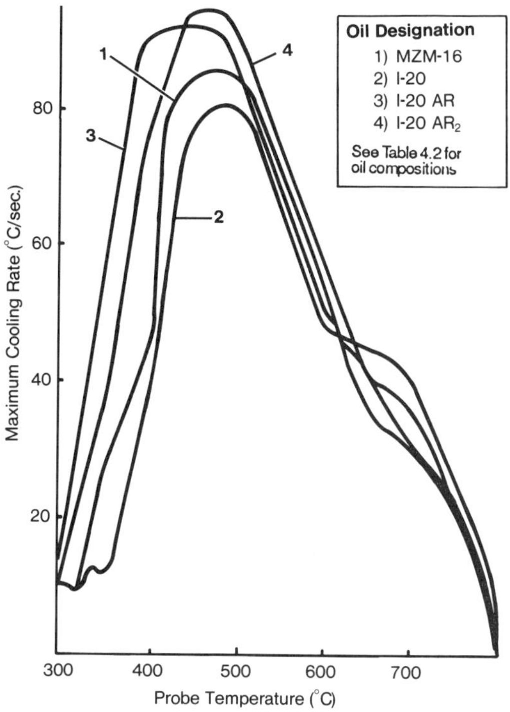

Fig. 4.5 Cooling rate of a copper sphere quenched in different oils

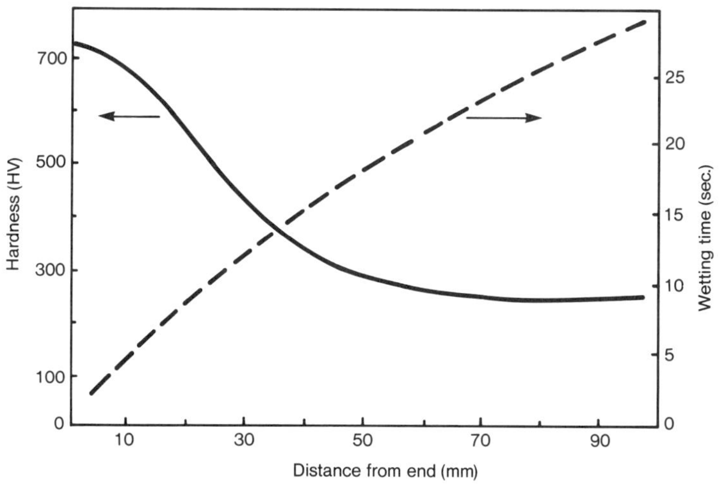

Fig. 4.6 Wetting time and surface hardness of 1045 steel as a function of distance from end of specimen

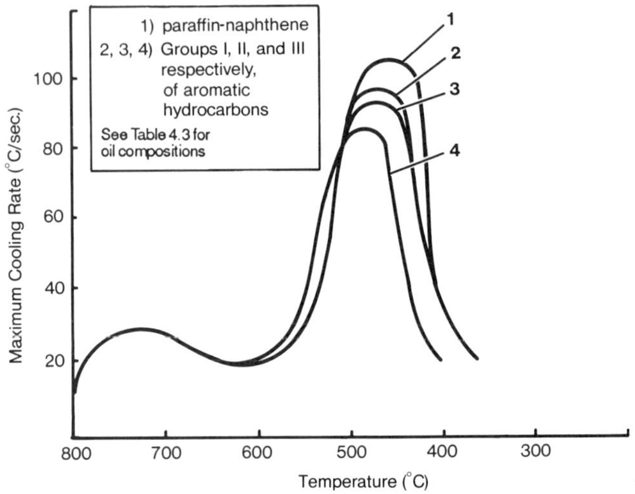

Fig. 4.7 Cooling rates of a silver sphere quenched in different hydrocarbons separated from a single oil

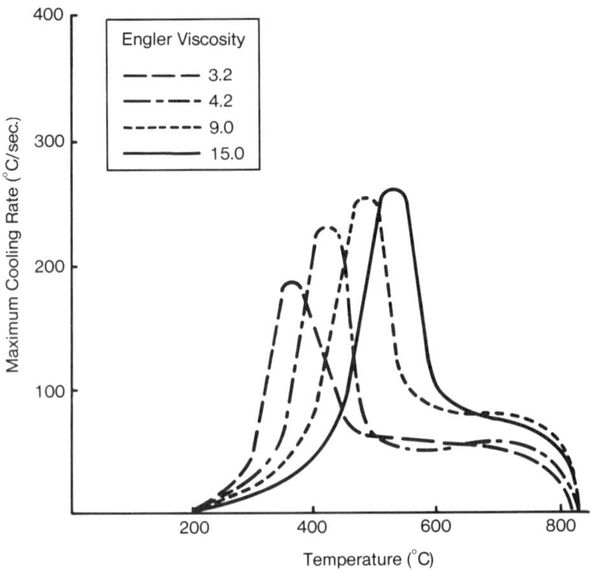

Fig. 4.8 Maximum cooling rate of a 20 mm (0.8 in.) diam silver sphere quenched in mineral oil with various viscosities

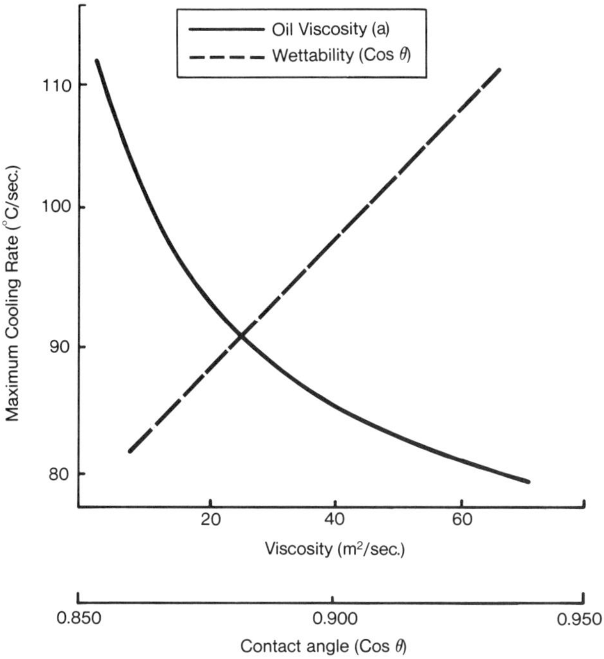

Fig. 4.9 Relationship of the cooling rate of a silver sphere quenched in different oils with varying viscosities

Table 4.4 Quenching properties of various fats compared with two mineral oils(a)

Oil	Acid value	Saponification value	Iodine No.	Redwood viscosity at 40 °C (105 °F)	Cooling time (700-300 °C, or 1290-570 °F), *s*	*H* factor
Soya bean	4.221	195.6	139.8	188	1.42	0.200
Fin whale	1.795	199.0	113.1	153	1.35	0.198
Sperm skin	8.203	133.0	71.6	83	1.45	0.200
Rapeseed	3.2	168.2	104.3	188.1	1.63	0.199
Castor	3.6	169.1	87.6	1229.0	1.80	0.199
Hardened-fish	6.5	138.5	28.3	85	3.30	0.125
Mineral oil A	...	...	...	85.2	3.87	0.142
Mineral oil B	...	...	...	1080	3.34	0.136

(a) Quenching from 800 °C (1470 °F); quench oil temperature of 80 °C (175 °F)

Fats, which are alkyl esters of triglycerides extracted from various fish and animals, exhibit quenching performance that is, in some ways, superior to mineral oils. Fatty oils produce faster cooling at higher temperatures, an important property if the pearlite transformation is to be minimized during steel quenching. The Grossmann quench severity factor, or *H* factor, is also generally higher. (*H* factors increase with increasing quench severity.) One fat, a hardened fish oil, which has a viscosity similar to mineral oil A, exhibited essentially analogous quenching properties.

These data show that quenching oils, whether mineral or fat derived, can be formulated to produce similar quenching properties. Availability, price, stability, and quenching performance currently favor the selection of mineral oils.

Oil Classification and Properties

In order to gain widespread commercial acceptance, a quench oil must possess several important properties (Ref 9):

- Acceptable flash and fire points
- Low sludge formation
- Nonstaining of parts
- Appropriate heat removal properties

Today most quench oils are derived from refined petroleum-base stocks. Higher naphthenic fractions usually result in lower flash points and greater sludge formation. Sludge formation reduces heat-transfer efficiency, which in turn may result in inadequately hardened parts. Increased sludging also reduces oil flow through heat exchangers used to cool the oil during quenching.

Ideally, a quench oil should not stain parts. High-quality paraffinic oils impart a light gray color to quenched parts. Sulfur-containing components may cause unacceptable black stains, but are generally removed during the refining process. Some additives used to increase quench speed may also cause staining. It should be noted that the color of an oil is not necessarily indicative of its staining tendency. Some quench-speed accelerators may produce a dark brown color in the oil but have minimal effect on staining.

A quenching oil obviously must not catch fire during use. The flash point of an oil is used as an indicator of its tendency to ignite. Maximizing the flash point minimizes the fire hazard. Increasing the naphthenic content of an oil generally decreases the flash point.

Quench oils are selected on the basis of their ability to mediate heat transfer during the quench. Oils are usually classified on the basis of quenching speed and temperature of use. The three major quench oil classifications are (Ref 9, 15-19):

- Conventional (nonaccelerated) oils, or cold oils
- Accelerated oils
- Marquenching oils, or hot oils

Although there are other classifications, such as fast, superfast, and water emulsifiable, the above three are the most commonly encountered.

Various characterization criteria for quench oils are given in Table 4.5. Representative cooling curves for oils in each category and for other quench media are compared in Fig. 4.10.

Conventional quenching oils are typically mineral oils, which may contain antioxidants to reduce the rates of oxidative and thermal degradation. Most of these oils have viscosities in the range of 100 to 110 SUS (Saybolt universal seconds) at 40 °C (100 °F), although some have viscosities of up to 200 SUS at 40 °C (100 °F). Conventional quenching oils do not contain additives to increase the cooling rate.

Table 4.5 Characteristics of quenching oils

Type of oil	Bath temperature °C	Bath temperature °F	Flash point °C	Flash point °F	Typical viscosity at 40 °C (100 °F), SUS	GM quenchometer (nickel ball) time, s	Hot-wire test, A
Conventional	< 65	< 150	170	340	105	16.0	30
Accelerated	< 120	< 250	180	355	94	10	39
Marquenching	< 200	< 400	300	570	700	30	30

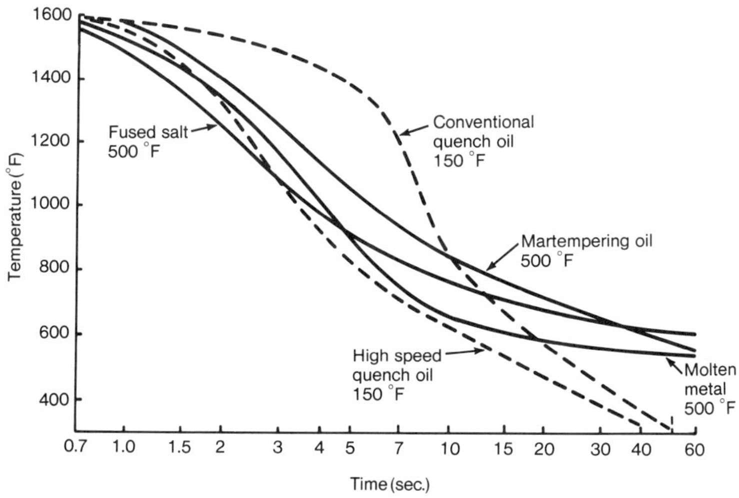

Fig. 4.10 Comparison of cooling curves for various quench media

Accelerated quenching oils are usually formulated from a mineral oil and contain one or more additives to increase cooling rates. Viscosities of these oils may vary from 50 to 100 SUS at 40 °C (100 °F).

Martempering or hot quenching oils are used at temperatures between 95 and 230 °C (200 and 450 °F). They are usually formulated from solvent-refined mineral oils with a very high paraffinic fraction to optimize oxidative and thermal stability. Stability is also enhanced by the addition of antioxidants. Nonaccelerated and accelerated martempering oils are available.

Martempering oils are formulated for use at higher temperatures than conventional oils, thus permitting modified or actual martempering of ferrous metals. Typical temperature ranges for commercially available martempering oils are shown in Table 4.6 (Ref 20). Because martempering oils are used at relatively high temperatures, a protective, nonoxidizing environment is often employed, which allows use of temperatures much closer to the flash point than open-air conditions (Ref 21).

Measurement of Heat-Removal Properties. The cooling rates produced by quench oils are often classified on the basis of the General Motors (GM) quenchometer or nickel ball test (Ref 22, 23). Details of this test are given in ASTM Standard D 3520. The test involves heating a 22 mm (7/8 in.) diam nickel ball to 885 °C (1625 °F) and then dropping it into a wire basket suspended in a beaker containing 200 mL of the quenchant oil at 21 to 27 °C (70 to 80 °F) (Ref 24). A timer is activated as the glowing nickel ball passes a photoelectric sensor. A horseshoe magnet is located outside the beaker as close as

Table 4.6 Typical use temperatures for martempering oils

Viscosity at 40 °C (100 °F), SUS	Minimum flash point		Use temperature: Open air		Use temperature: Protective atmosphere	
	°C	°F	°C	°F	°C	°F
250-550	220	430	95-150	200-300	95-175	200-350
700-1500	250	480	120-175	250-350	120-205	250-400
2000-2800	290	550	150-205	300-400	150-230	300-450

Table 4.7 Classification of quench oils by GM quenchometer times

Classification	GM quenchometer time, s
Fast oil	8-10
Medium oil	11-14
Slow oil	15-20
Martempering oil	18-25
Deionized water	2.0
9% aqueous solution of NaCl	1.5

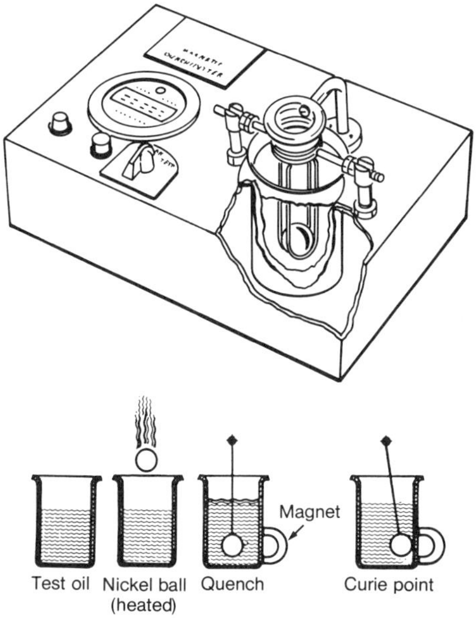

Fig. 4.11 GM quenchometer and principle of operation

possible to the nickel ball. As the ball cools, it passes through its Curie point (354 °C, or 670 °F), the temperature at which it becomes magnetic. When the ball becomes magnetic, it is attracted to the magnet, activating a sensor that stops the timer (Fig. 4.11). The cooling time required to reach the Curie temperature is sometimes referred to as the heat-extraction rate or quenchometer time. Test results thus obtained have been used to classify quench oils as fast, medium, slow, and so on (Table 4.7) (Ref 25, 26).

The hot-wire test is another, although considerably less common, method for describing the heat-removal properties of an oil. The experimental details of this test are available in Ref 27. A correlation between the hot-wire test and GM quenchometer results is illustrated in Fig. 4.12 (Ref 28).

Although the GM quenchometer test is a common method for evaluating quench oils, it is occasionally difficult to obtain reproducible test results. In addition, the quenchometer times, which are based on a relatively limited portion of the total cooling process, often cannot be related to as-quenched hardness or cracking properties. In fact, the test defines the heat-removal rate only over the high-temperature portion of the cooling curve.

Hardening Power. Segerberg (Ref 8) described an empirical approach to quantifying the quench severity, or hardening power (HP), of an oil quenchant using the Wolfson probe. His technique is based on the cooling curve testing procedure submitted for ISO specification approval (Ref 2). The hardening power of a quench oil is calculated by determining three transition temperatures from the cooling rate curve of the quench oil: T_{VP}, CR, and T_{CP} (Fig. 4.13). These parameters are defined as:

- T_{VP}, transition temperature (°C) between the vapor phase (A stage) and nucleate boiling phase (B stage) of the cooling process
- CR, cooling rate (°C/s) over the temperature range of 500 to 600 °C (930 to 1110 °F)
- T_{CP}, transition temperature (°C) between the nucleate boiling (B stage) and the convective cooling (C stage) of the cooling process

The hardening power is then calculated from the regression equation:

$$HP = 91.5 + 1.34\,T_{VP} + 10.88\,CR - 3.85\,T_{CP}$$

This algorithm was successfully used to rate the relative quench severities of a wide range of different types of quench oils, including conventional, accelerated, and martempering oils (Fig. 4.14). Note that this algorithm was developed specifically for oils; a different algorithm must be used for aqueous polymer quenchants.

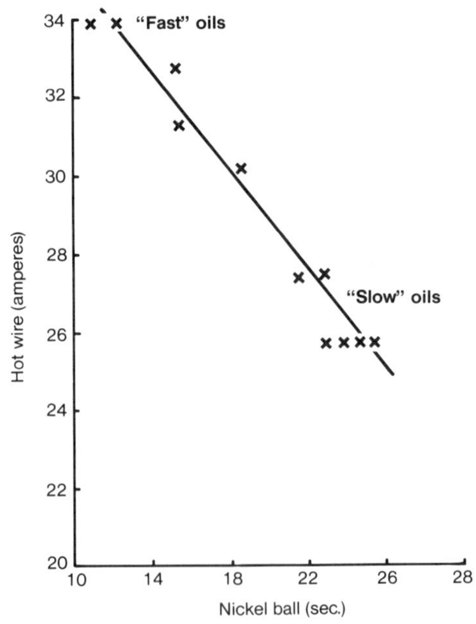

Fig. 4.12 Correlation of hot-wire and GM quenchometer (nickel ball) tests

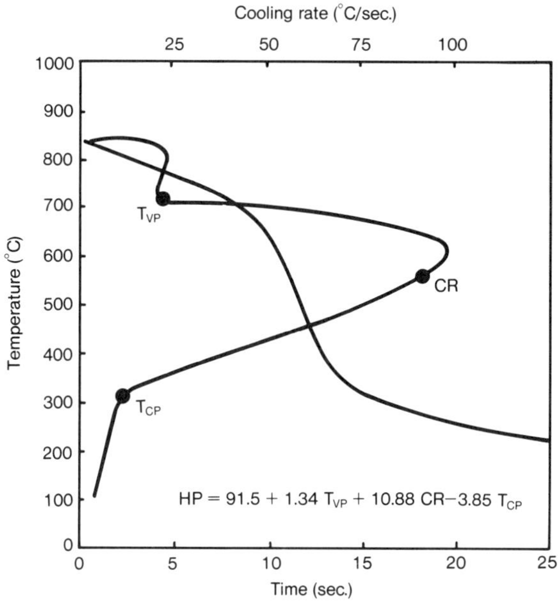

Fig. 4.13 Calculation of hardening power from cooling curves

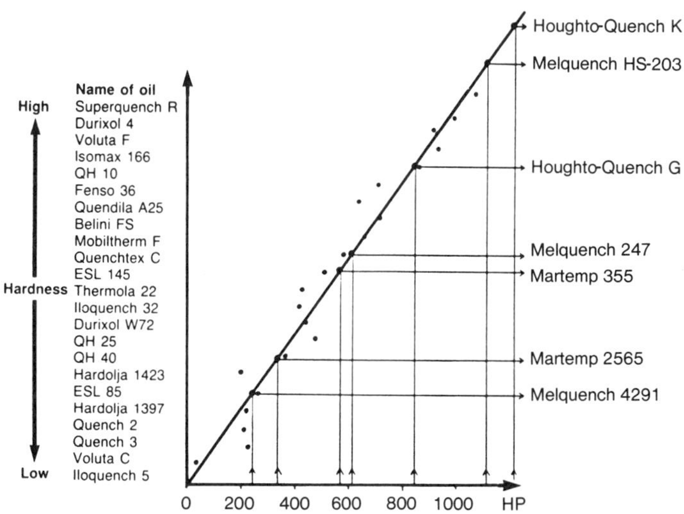

Fig. 4.14 Correlation of calculated hardening power values for various quench oils

Quench Oil Additives

The primary purpose of any quenchant is to produce desirable metallurgical transformations that in turn will dictate the final physical properties of a part, such as toughness, hardness, and strength. In addition, the quenchant must prevent cracking and minimize distortion by controlling cooling rates during quenching.

Figure 4.15 graphically illustrates the metallurgical transformations that may occur in a steel during quenching. The first step in the process is to heat the steel to its austenitizing temperature (Ref 29). The steel is then cooled rapidly to avoid the formation of pearlite; this is necessary to obtain maximum as-quenched hardness. The critical or minimum cooling rate is dictated by the time required for pearlite to form in the particular steel being quenched. As a general rule, a quenchant must produce a cooling rate equivalent to or faster than that indicated at the nose of the pearlite curve in order to achieve maximum as-quenched hardness.

In many heat treating processes, relatively low-hardenability steels may have pearlite transformation curves located considerably farther to the left than the one shown in Fig. 4.15. In such cases, the cooling rates achievable with conventional quench oils are insufficient to avoid the formation of pearlite, leading to unacceptably low as-quenched hardness. Such situations necessitate the use of accelerated quench oils.

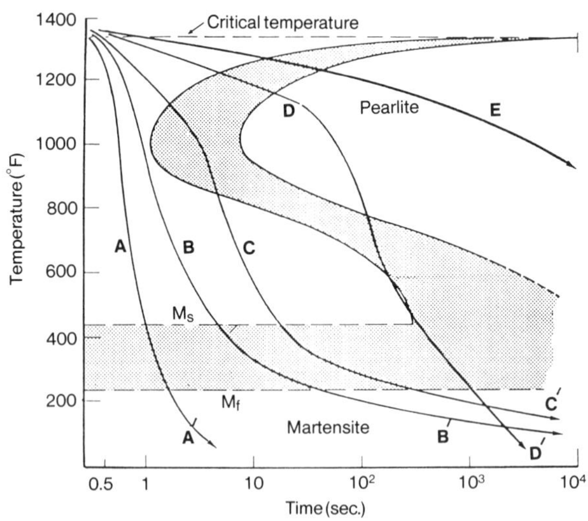

A Cold water or brine
B High speed oil
C Conventional oil
D Hot (180 °F) water
E Air

Fig. 4.15 Transformation diagram of low-alloy steel with cooling curves for various quench media

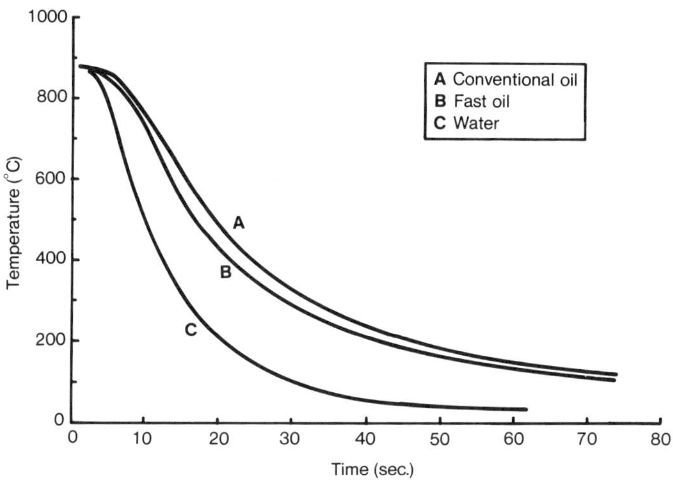

Fig. 4.16 Cooling curves for conventional and accelerated oils versus water

Table 4.8 Cooling characteristics of typical quench oils(a)

	CR_{max}(b)		CR_{232}(c)		*H* factor
Quenchant	**°C/s**	**°F/s**	**°C/s**	**°F/s**	**in.$^{-1}$(d)**
Water(e)	54.3	97.7	20.1	36.2	1.90
Conventional oil(f)	31.6	56.9	6.5	11.7	0.68
Accelerated oil(f)	34.6	62.3	6.9	12.4	0.97

(**a**) Determined using the quenchant testing apparatus and probe described in Ref 30. (**b**) CR_{max}, maximum cooling rate. (**c**) CR_{232}, cooling rate at a centerline probe temperature of 232 °C (450 °F). (**d**) Calculated using the procedure described in Ref 31. (**e**) 32 °C (90 °F) and 30 m/min (100 ft/min). (**f**) 65 °C (150 °F) and 30 m/min (100 ft/min)

Figure 4.16 illustrates the cooling curves produced by a conventional and an accelerated quench oil obtained using a 25 × 100 mm (1 × 4 in.) cylindrical stainless steel instrumented probe (Ref 30). A cooling curve for water is shown for reference. The cooling rates achieved with these quench media are shown in Table 4.8. From these data, it is clear that the accelerated oil produces considerably higher cooling rates than the conventional oil.

Tkachuk *et al.* (Ref 12), Guillemeau (Ref 32), and Tensi and Stitzelberger-Jakob (Ref 11) have shown that quenching rates increase with increasing wettability of an oil on steel (Fig. 4.6 and 4.9). Tkachuk *et al.* (Ref 12) showed that additives that increase the wettability of an oil on steel also produce higher cooling rates (Fig. 4.17). The

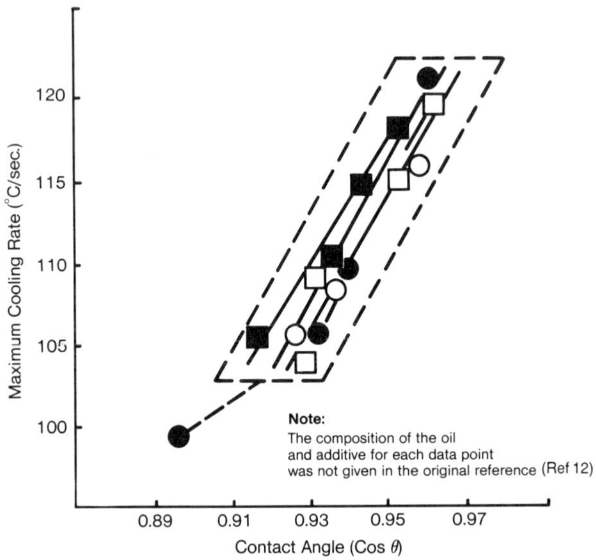

Fig. 4.17 Relationship between cooling rate and wettability of oils with additives

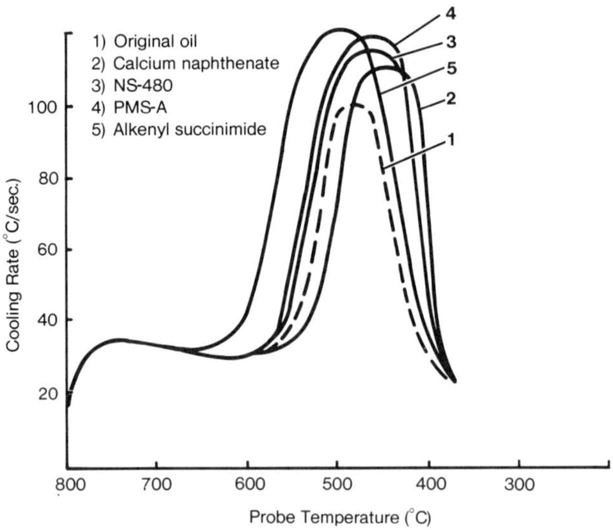

Fig. 4.18 Cooling rates of a silver sphere quenched in oil containing different additives

additives used for this particular study included calcium naphthenate and alkenyl succinimide (Ref 33). The ability of these additives to increase the quench rates is shown by the curves in Fig. 4.18.

Allen *et al.* (Ref 34) examined the effect of sodium sulfonate on the cooling properties of a conventional quench oil (Table 4.9). Sodium sulfonate increased the heat-transfer coefficient, decreased the temperature where the maximum cooling rate occurs, and increased the cooling rate achieved in the quenching process.

Under simulated conditions, Bashford and Mills (Ref 35) studied the effect of various additives on quench oil performance using an Edwin Cooper accelerated aging test (Fig. 4.19). From these studies, the following characteristics of an ideal quenching oil were proposed:

- Accelerated cooling rate to give a maximum hardening response
- Minimum deposition of sludge on steel parts with minimum sludge formation
- Minimum viscosity increase with use to reduce dragout loss of oil on part surfaces
- Minimum acid formation upon aging to prevent staining of bright steel components
- Minimum potential for cracking and distortion of parts

The data of Bashford and Mills (Ref 35) showed that additives may have a dramatic effect on quenchant properties, primarily by accentuating the wettability of the base oil. The magnitude of this effect is determined by the particular additive or additive combination and the concentrations present. Additives may also affect stability properties, including sludging, staining, and so forth. Because the additives operate by increasing the

Table 4.9 Effect of the addition of sodium sulfonate on the properties of a conventional quench oil(a)

Characterization	Base oil	Sodium sulfonate 1.5%	Sodium sulfonate 3.0%
Length of vapor blanket stage, s	36.5	21.5	16.5
Average temperature at end of vapor blanket stage, °C (°F)	600(1110)	650(1200)	650(1200)
Mean cooling rate during nucleate boiling stage, °C/s (°F/s)	10.4(18.7)	14.2(25.6)	20.5(36.9)
Maximum surface heat-transfer coefficient, $W/m^2 \cdot K$ ($Btu/ft^2 \cdot h \cdot °F$)	921(162)	1779(312)	2592(455)
Temperature at maximum surface heat transfer, °C (°F)	476(889)	425(797)	400(752)
Kinematic viscosity, cSt at 40 °C (105 °F)	32	33	34
Surface tension, N/m (dyne/cm)	0.0388(38.8)	0.0334(33.4)	0.0326(32.6)

(a) Still quench properties. Cooling rates were determined using a 120 × 120 × 20 mm (5 × 5 × 0.8 in.) stainless steel plate (0.29Si-1.16Mn-17.43Cr-2.29Mo-10.56Ni)

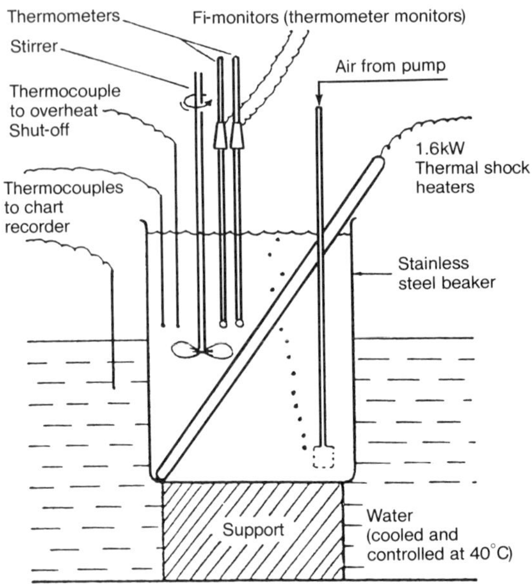

Fig. 4.19 Edwin Cooper accelerated aging test apparatus

wettability of the oil on the steel surface, they are susceptible to selective dragout after the quench. Because they come into intimate contact with the steel, they may also degrade during use. Therefore, quench processes must utilize adequate quality control procedures to monitor changes in the oil during quenching. These monitoring procedures are often provided without charge by the oil supplier.

Quench Oil Selection

The following criteria have been proposed for the selection of a quenching oil (Ref 36):

- An oil with a relatively slow cooling rate (e.g., a conventional oil) should be used for products with thick cross sections.
- Depending on the quenching method and agitation of the oil, the oil temperature should be adjusted to give a viscosity of 5 to 15 cSt.
- The minimum flash point of the oil should be 90 °C (160 °F) above the oil temperature being used.
- If the products are heated in an air furnace, a quenching oil should be selected that gives maximum brightness.

Dragout Properties

The dragout loss of a quenching oil is primarily a function of the viscosity of the oil and the interface temperature when the part is removed from the quench bath. Because higher-viscosity oils are typically used at correspondingly higher temperatures, the actual viscosities of oils during use do not vary greatly. Therefore, the bulk dragout losses of various quench oils are similar. Table 4.10 shows the viscosities of four quench oils as a function of temperature (Ref 20). These oils are typically used at a temperature where the viscosity lies within the range of 40 to 60 SUS (Ref 21).

Cooling Characteristics

It is generally desirable for an oil quenchant to produce relatively high cooling rates during the initial stages of the quenching process to minimize the formation of pearlite transformation products, which reduce as-quenched hardness. However, it is critically important to recognize that quench rates should be minimized as the steel approaches the martensitic transformation temperature, M_S. Typically, although not always, a quenchant that produces relatively high cooling rates during A- and B-stage cooling also exhibits relatively high cooling rates in the M_S transformation region (C-stage cooling).

To illustrate this point, consider the quenching rates produced by water, conventional oil, and accelerated oil under selected quenching conditions. The data in Table 4.8 show that water is the fastest quenchant of the media listed. Conventional oil exhibits the slowest cooling rates and is therefore the least severe quenchant media, as also indicated by the Grossmann *H* factor. The accelerated quenching oil exhibits a severity intermediate between water and conventional oil.

It is interesting to consider the differences in cooling curves produced by these three media. In Fig. 4.16, it is evident that water exhibits the shortest A-stage cooling times

Table 4.10 Viscosities of quenching oils as a function of temperature

Oil temperature		Viscosity, SUS			
				Martempering oils	
°C	°F	Conventional oil	Accelerated oil	No. 1	No. 2
40	100	100	95	720	2630
65	150	55	60	190	555
95	200	40	40	85	180
120	250	35	35	55	90
150	300	30	35	45	60
175	350	...	...	40	45
205	400	...	...	35	40
230	450	...	...	...	35

and conventional oil the longest A-stage time. It would be difficult to achieve optimal hardness in lower-hardenability steels using a conventional oil because A-stage cooling may extend into the pearlite transformation region.

The cooling curve for water in Fig. 4.16 has the steepest slope, indicating that water has the fastest cooling rate in the B-stage cooling region. Conventional oil, on the other hand, has the slowest cooling rate. The cooling curve for water has the greatest curvature in the C-stage region, also indicating that it has the fastest cooling rate. Similarly, the conventional oil has the least curvature or slowest cooling rate in the C-stage region. The accelerated oil has only slightly greater curvature than the conventional oil.

Ideally, a quenching media should have a short A-stage and a high B-stage quench rate. To minimize cracking and distortion, it is generally desirable to minimize quench rates in the C-stage region. However, because most quenchant media do not exhibit ideal behavior for all alloys, masses, and cross sections under all quenching conditions, it is often necessary to use more than one oil or to vary quenching conditions to obtain the desired quench severity.

The quench severity of an oil can be affected by varying bath temperature and agitation. Although it is possible to use oil without agitation, some agitation is almost always used to obtain greater quench uniformity.

To evaluate the impact of variations in bath temperature and agitation on quenchant performance, a three-level statistically designed cooling curve analysis experiment was conducted. The experimental variables and limits of this work are shown in Fig. 4.20. Quenchant bath temperatures of 43, 65, and 88 °C (110, 150, and 190 °F), linear, turbulent flow rates of 0, 15, and 30 m/min (0, 50, and 100 ft/min), probe diameters of 12.5, 25, 37.5, and 50 mm (0.5, 1.0, 1.5, and 2.0 in.), and a typical accelerated oil and a conventional oil were used. The quenchant testing apparatus is shown in Fig. 4.21.

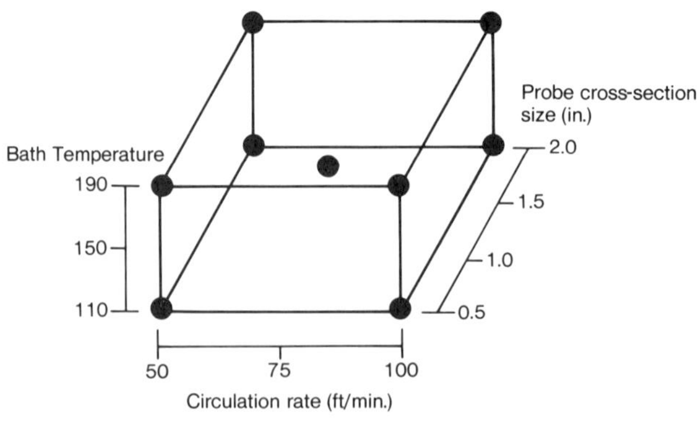

Fig. 4.20 Designed experiment to statistically evaluate the effects of process variables on quench oil performance

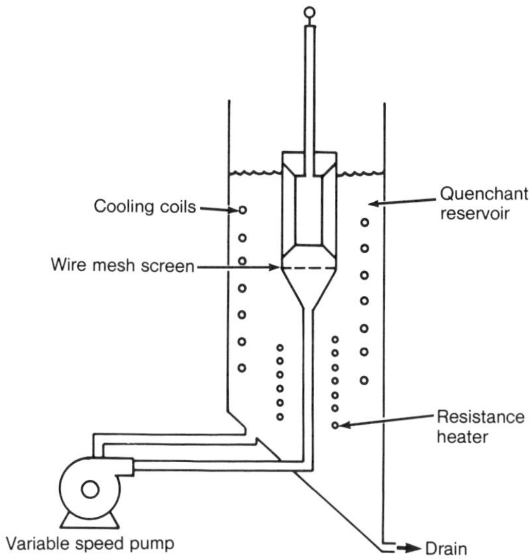

Fig. 4.21 Schematic of cooling curve testing apparatus (Southern Research Institute)

Results of these studies are shown graphically in the contour plots in Fig. 4.22(a) and (b). Several conclusions were drawn:

- The maximum cooling rate (CR_{max}) and cooling rate at 650 °F (CR_{650}) for both oils were essentially independent of bath temperature, but were dependent on agitation rate. The dependence of CR_{max} and CR_{650} on circulation rate was greater for the conventional oil.
- The Grossmann *H* factors for the conventional oil were independent of bath temperature but dependent on agitation rate. However, the *H* factors obtained for the accelerated oil were dependent on both the bath temperature and agitation rates. The *H* factors were calculated according to the procedure described earlier (Ref 31).
- The Rockwell hardness (Rc, or HRC) values for a low-hardenability steel (1045) and a high-hardenability steel (4140) were calculated using the procedure described by Bates (Ref 37). The hardness values obtained for both steel grades were essentially independent of bath temperature for both the accelerated and the conventional quench oils. However, the hardness values for both steel grades were dependent on the agitation rate. Not surprisingly, this dependence was greater for the low-hardenability steel.

These data, coupled with results described previously, show that the quench severity provided by an oil can be significantly affected by additives and by the degree of agitation of the quench oil around the part. However, quench severity is relatively

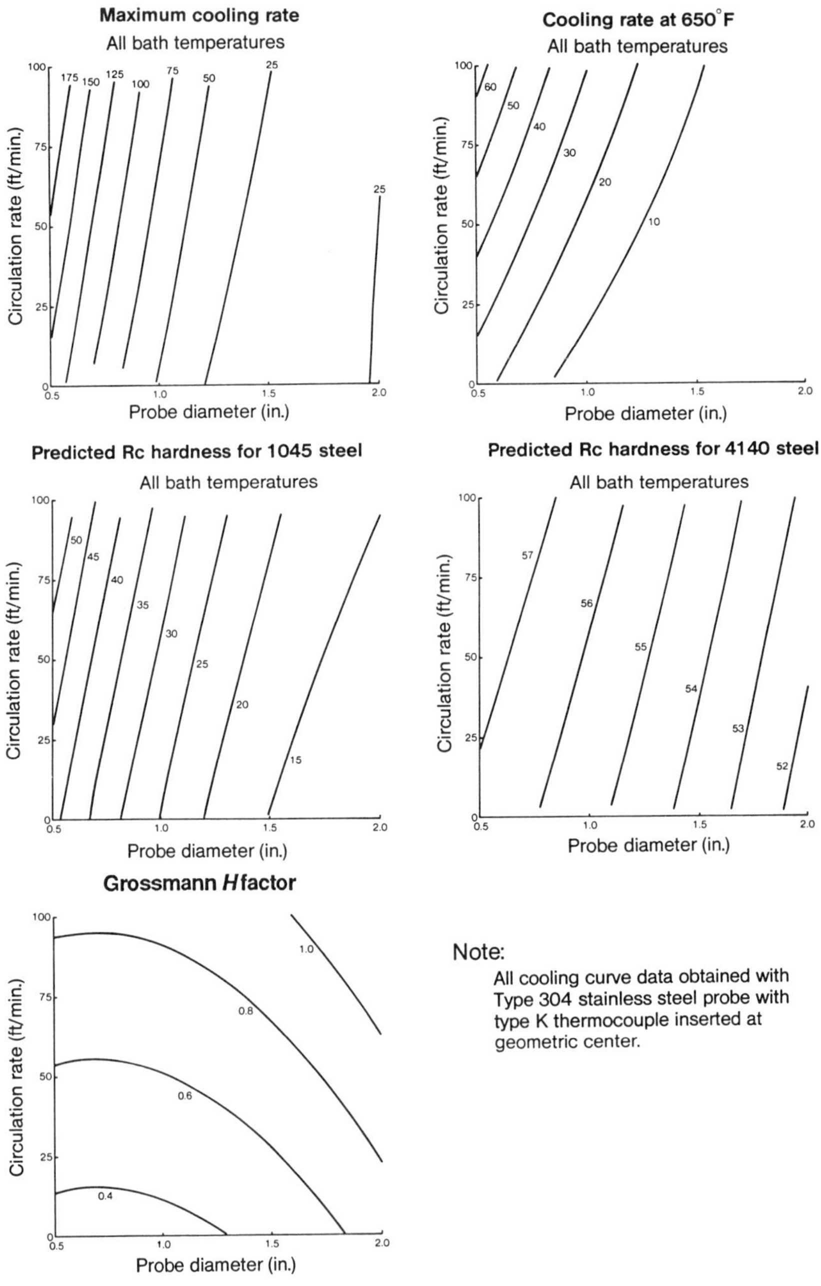

Fig. 4.22(a) Contour plots of quenching data for a conventional mineral oil **(continued)**

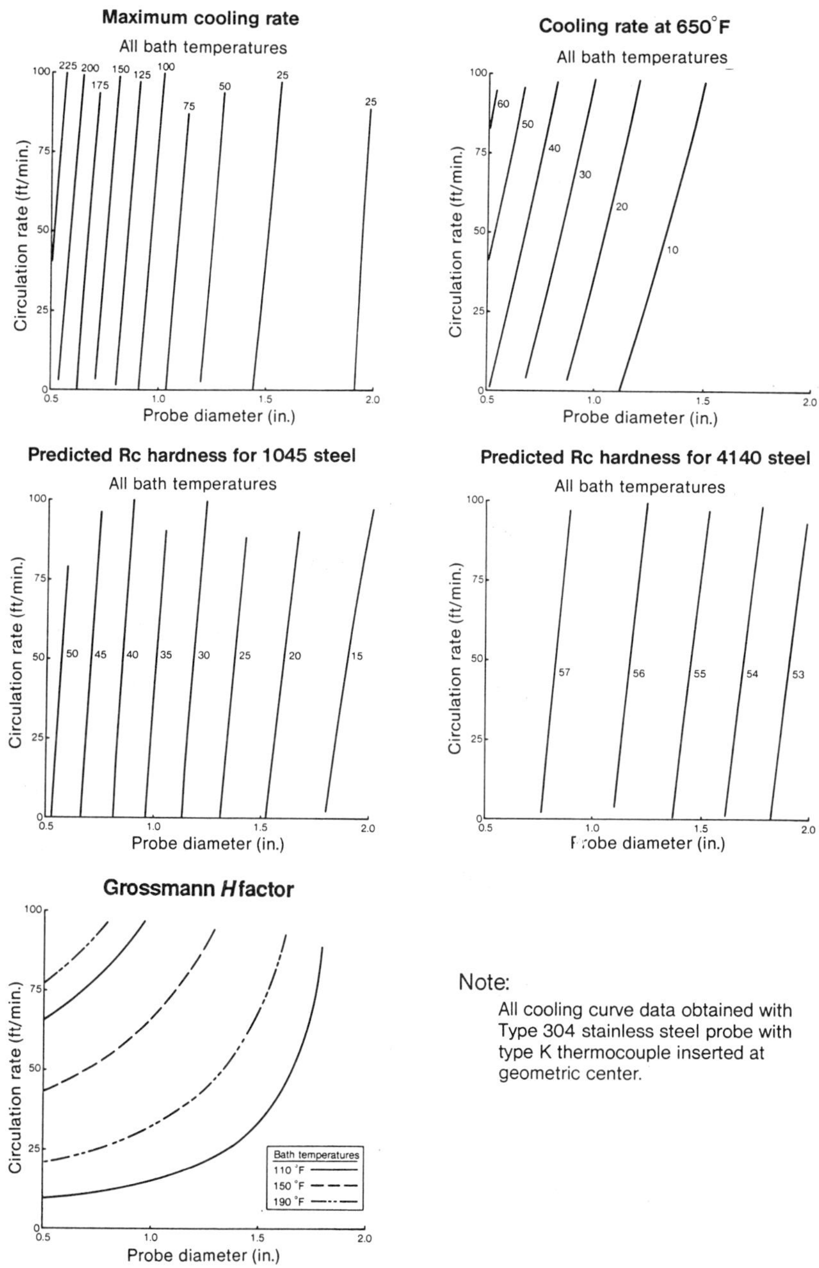

Fig. 4.22(b) Contour plots of quenching data for an accelerated quenching oil

independent of bath temperature. Unfortunately, the effect of quench oil temperature on cracking and distortion was not determined in this study. However, an earlier study showed that the amount of distortion resulting from quenching decreases as the bath temperature increases (Ref 21).

Hampshire (Ref 38) has studied the effect of varying marquenching oil temperature on the distortion of bearings. His results showed that there was a decrease in distortion with increasing oil temperature (Fig. 4.23). The primary cause of this distortion was reported to be internal stresses created by nonuniform martensite formation, which in turn was caused by thermal gradients during the quench.

Oil Degradation

All organic compounds, whether oils, polymers, or additives, are subject to degradation under quenching conditions. Quenchant suppliers often relate degradation to the chemistry of an oil. For example, degradation may be due to the progressive loss of antioxidants caused by selective dragout or to preferential oil component oxidation or volatilization.

The heat treater may be able to visually detect quench oil degradation by observing an increase in viscosity. Sludge content and pour point may also increase. Oil degradation may also be evidenced by an inability to achieve desired as-quenched hardness or by an increase in distortion or cracking of parts.

Quench oil stability can be quantified by using an oxidation test to measure sludging tendency. Quench oil stability will vary with the rate of paraffinic/naphthenic content (Table 4.2).

Similar results have also been obtained by Bashford and Mills (Ref 35) using an accelerated aging test (Fig. 4.19). Cooling curves of two aged oils, a straight mineral oil

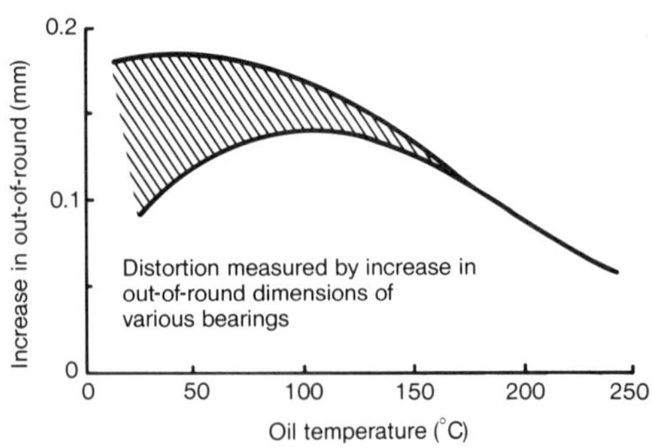

Fig. 4.23 Distortion as a function of oil temperature

and a formulated, high-performance quenching oil (Fig. 4.24), show that the addition of antioxidants can dramatically affect oil stability.

Hampshire (Ref 38) showed that a martempering oil exhibited a dramatic viscosity increase at the end of the bath lifetime (Fig. 4.25). A corresponding increase in cooling rate as measured by the Wolfson test (Ref 2) was observed during this time (Fig. 4.26). These cooling rate increases occurred concurrently with the viscosity increase shown in Fig. 4.24 and 4.25. This would suggest that although the isothermal viscosity at 38 °C (100 °F) increased, the relationship between viscosity and temperature also changed.

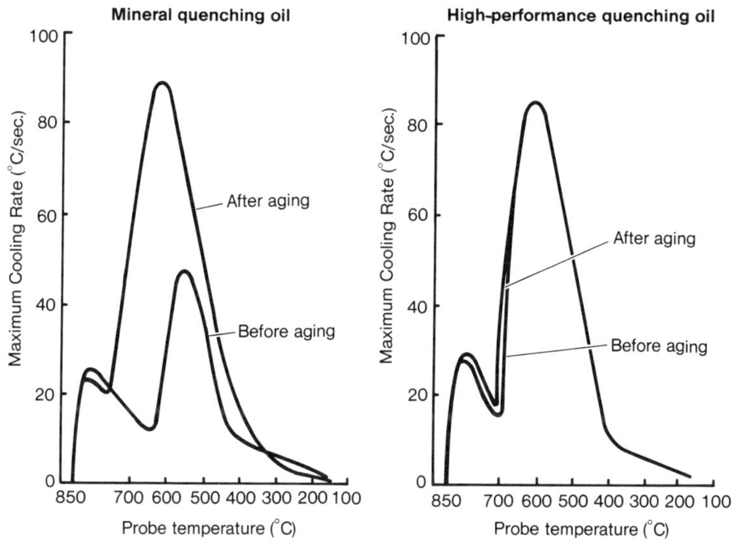

Fig. 4.24 Comparison of cooling rate profiles before and after aging for a straight mineral oil and a high-performance quenching oil

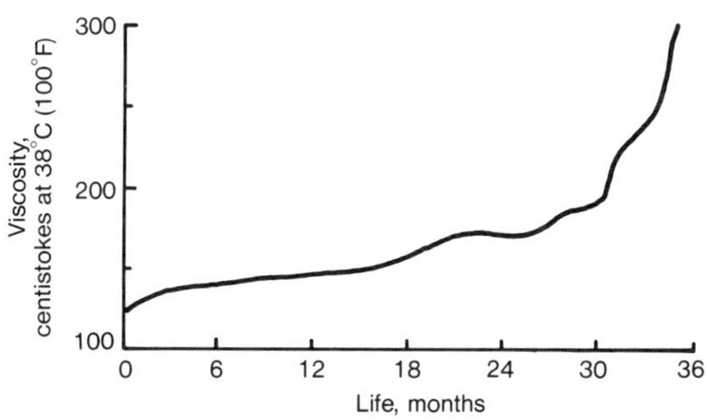

Fig. 4.25 Viscosity of a martempering oil as a function of time in use

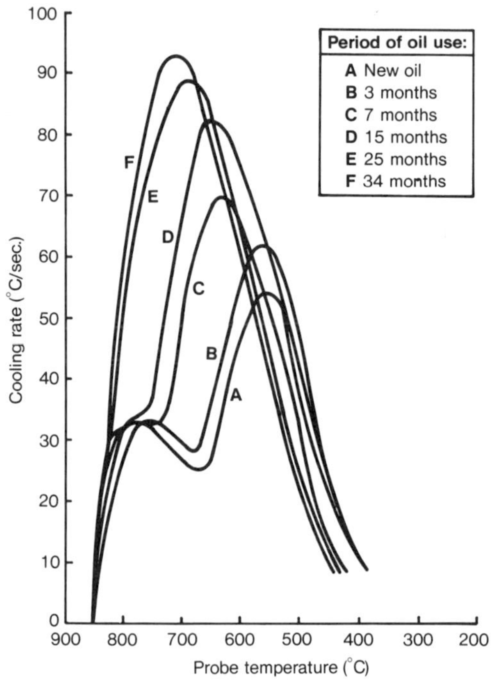

Fig. 4.26 Cooling rate variation of a martempering oil as a function of time in use

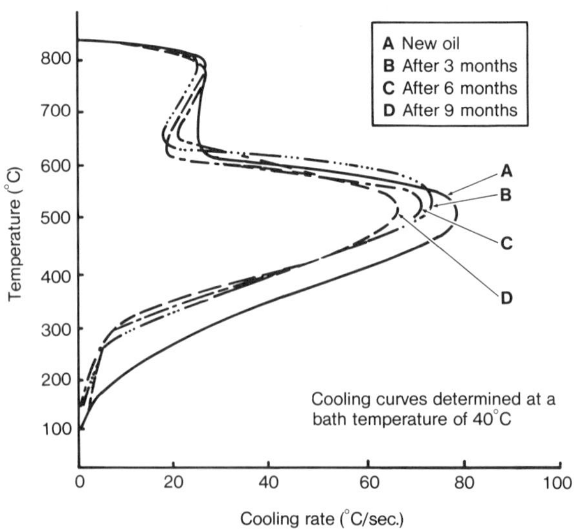

Fig. 4.27 Variation of cooling rate of an accelerated quench oil as a function of time in use

The interfacial temperature between the quench oil and the part being quenched varied over the duration of the quench cycle. Although the low-temperature viscosity increased, the viscosity at higher temperatures decreased. The lower viscosity at the higher temperatures resulted in progressively lower maximum cooling rates.

Hewitt (Ref 39) published a similar paper on the change in cooling curve behavior of an accelerated quenching oil with increasing use (Fig. 4.27). The cooling rate progressively decreased over time, and the results were consistent with the conclusion that the cooling rate accelerator was being removed, presumably by dragout.

Tagaya and Tamura (Ref 14) modeled the quench oil deterioration process with an accelerated test designated the "Indiana test." This test involves blowing dry air at a rate of 10 L/h (2.6 gal/h) into 300 mL (10 oz) of an oil held at 170 °C (340 °F) using the apparatus schematically illustrated in Fig. 4.28. The properties of the quench oil, including kinematic viscosity, sludge formation, and peroxide content, were monitored as a function of time. The stability results for a typical mineral oil are summarized in Fig. 4.29, showing a gradual decrease in quench severity over time. This occurred with relatively little change in viscosity. Although viscosity is a very important parameter, it is not the sole parameter governing quenchant performance.

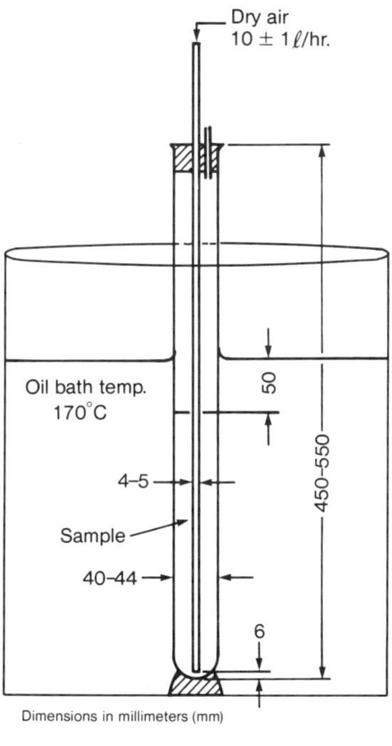

Fig. 4.28 Dimensions of the test cell for the Indiana test

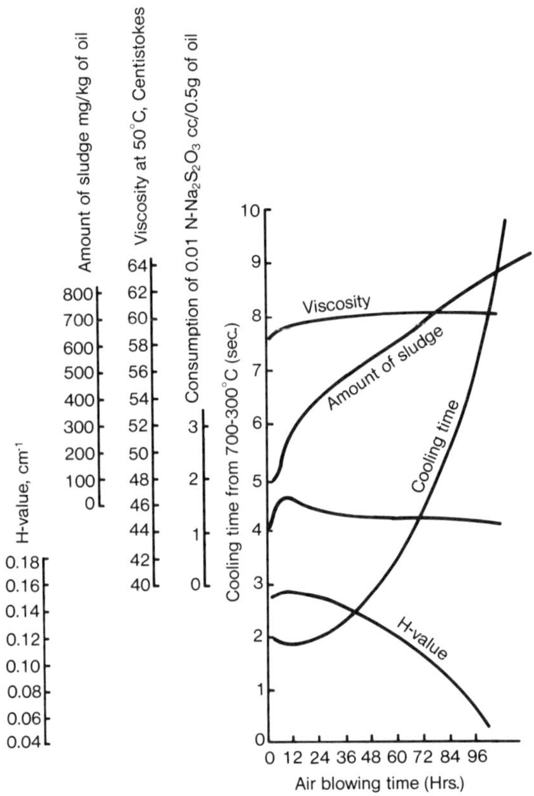

Fig. 4.29 Variations of a commercial mineral quenching oil after the Indiana stability test

These results clearly illustrate the necessity of monitoring the stability of the performance of a quench oil. Various monitoring methods will be reviewed in Chapter 6.

References

1. R.L. Zhou, *Proc. 4th Annual Conf. Heat Treatment*, Institute of Chinese Mechanical Engineering Society, Nanging, 25-31 May 1987
2. "Laboratory Test for Assessing the Cooling Curve Characteristics of Industrial Quenching Media," Wolfson Heat Treatment Centre Engineering Group Specifications, 1982
3. "Industrial Cooling Oils—Determination of Cooling Characteristics—Laboratory Test Method," ISO/DIS 9950, International Organization for Standardization, 1988
4. P.G. Desideri, L. Lepri, D. Heimler, L. Checchini, and S. Giannessi, *J. Chromatogr.*, Vol 322, 1985, p 107-116
5. M.A. Gough and S.J. Rowland, *Nature*, Vol 344, 1990, p 648-650

6. G.D. Hobson and W. Pohl, *Modern Petroleum Technology*, John Wiley & Sons, 1973, chap 6
7. H-J. Neumann, B. Paczynska-Lahme, and D. Severin, *Composition and Properties of Petroleum*, Halstead Press, 1981, p 22
8. S.O. Segerberg, *Heat Treat.*, Dec 1988, p 30-33
9. R-J. Windgassen, paper presented at Metalworking Fluids Today, Society of Tribologists and Lubrication Engineers, Feb 1989
10. P.S. Protsidim, N.Ya. Rudakova, and B.K. Sheremeta, *Metalloved. Term. Obrab. Met.*, Feb 1988, p 5-7
11. H.M. Tensi and P. Stitzelberger-Jakob, *Bedeutungdes H-Wertes Bestimmung Harteverteilung*, Vol 44, 1989, p 99-105
12. T.I. Tkachuk, N.Ya. Rudakova, B.K. Sheremeta, and M.A. Al'tshuler, *Metalloved. Term. Obrab. Met.*, Oct 1986, p 45-47
13. M. Tagaya and I. Tamura, *Technol. Rep. Osaka Univ.*, Vol 4, 1954, p 305-319
14. M. Tagaya and I. Tamura, *Technol. Rep. Osaka Univ.*, Vol 7, 1957, p 403-424
15. G.R. Furman, *Lubrication*, Vol 57, 1971, p 13-24
16. S-W. Han and R. Yul Chev, *Kong Hak Hoe Jie*, Vol 2, 1989, p 59-65
17. T.W. Dicken, *Heat Treat. Met.*, Vol 1, 1986, p 6-8
18. "Quenching Oils," Sun Refining and Marketing Co.
19. S.B. Lasday, *Ind. Heat.*, Oct 1976, p 8-19
20. *Metals Handbook*, 9th ed., Vol 4, American Society for Metals, 1981, p 48
21. H.E. Boyer and P.R. Cary, Ed., *Quenching and Control of Distortion*, ASM International, 1988
22. H-J. Gilliland, *Met. Prog.*, Oct 1960, p 111-114
23. E.A. Bender and H-J. Gilliland, *Steel*, Dec 1957, p 56-59
24. C.A. Barley and J.S. Aarons, Ed., *The Lubrication Engineers Manual*, U.S. Steel Corp., 1971, p 56-57
25. G.R. Furman, *Lubrication*, Vol 57, 1971, p 25-36
26. R.R. Blackwood, Texanol Inc., personal communication, 1987
27. "Quenchability of Oils—Hot Wire Method," Aerospace Material Specification ARP 4206, 28 Sept 1990
28. R.W. Foreman, paper presented at ASM Short Course on Quenching Media, American Society for Metals, Nov 1985
29. "Gulf Super-Quench 70: The Revolutionary Dual-Action Quenching Oil," Gulf Oil Co.
30. R.W. Heins and E.R. Mueller, *Met. Prog.*, Vol 22, 1982, p 33-39
31. M.E. Dakins, C.E. Bates, and G.E. Totten, *Metallurgia*, Vol 56, 1989, p 57-59
32. M. Guillemeau, *Trait. Therm.*, Vol 225, 1989, p 50-60
33. T.I. Tkachuk, N.Ya. Rudakova, B.K. Sheremeta, and R.D. Novoded, *Metalloved. Term. Obrab. Met.*, Oct 1986, p 42-45
34. F.S. Allen, A.J. Fletcher, and A. Mills, *Steel Res.*, Vol 60, 1989, p 522-530
35. A. Bashford and A.J. Mills, *Heat Treat. Met.*, Vol 1, 1984, p 9-14
36. "Heat Treating Oil," DK-Chemical Co. Ltd., Japan
37. C.E. Bates, *J. Heat Treat.*, Vol 6, 1988, p 27-45
38. J.M. Hampshire, *Heat Treat. Met.*, Vol 11, 1984, p 15-20
39. W. Hewitt, *Heat Treat. Met.*, Vol 13, 1986, p 9-14

CHAPTER 5

Polymer Quenchants

Aqueous polymer quenchant solutions are being used increasingly in applications such as the quenching of forgings (Ref 1-4), open-tank quenching of high-hardenability steels (e.g., 4340) (Ref 5), patenting of high-carbon steel wire and rod (Ref 6), induction hardening (Ref 7), quenchants for integral quench furnaces (Ref 8), quenching of railroad rails (Ref 9), and quenching of aluminum forgings and castings (Ref 10). Despite more than three decades of use, aqueous polymer solutions are poorly understood by many heat treaters. For example, it is not generally recognized that the quenching behavior of polymers varies mechanistically with the chemical structure and solution properties of each polymer.

This chapter describes the use of aqueous polymer quenchants from a historical perspective, provides a system for polymer classification with respect to chemical structure and the effect of structure on solution properties, and summarizes polymers used as quenchants. The effect of polymer structure on the quenchant cooling mechanism is also discussed, as are variations in quenching performance as functions of bath agitation, temperature, and polymer concentration. Finally, polymer stability and dragout are reviewed.

Historical Perspective

Aqueous polymer solutions have been used as quenchants for more than 30 years. However, substantial use in the heat treating industry is a relatively recent phenomenon. It has not always been clear how to appropriately use polymer quenchants in view of the numerous polymers available and the varied and sometimes inflated promotional claims of their manufacturers.

Water is the most common quenchant, followed by quenching oils. Water obviously is readily available and nontoxic; however, it is a severe quenchant, often causing cracking soft spots, and distortion in parts. Oils also have several disadvantages, including limited variability in quench rates, fire hazards, smoke emissions, and disposal problems (Ref 10).

Many applications require quench rates intermediate between those achievable with oil and water. For example, parts that are heat treated in an atmosphere, such as during carburizing, are quenched in water to obtain through-hardening, but often distort or crack in the process. The traditional solution to cracking has been to reduce the quench severity by quenching in an oil. The trade-off, however, may be inadequate through-hardening.

Polymer quenchants provide quench rates that fall between those achievable with oil and water (Table 5.1). Under some conditions, quench severities equivalent to those of oil can be attained; by changing the operating conditions, quench severities approaching those of brine are possible.

Perhaps the major driving force for the use of aqueous polymers as quenchants is reduction of the fire risk associated with oil. Water provides substantially greater fire safety and higher quench severities than oil. Warm or hot water may reduce the tendency toward cracking and distortion while maintaining target hardness properties in parts, but the difficulty of controlling hot water often leads to inconsistent results. It is thus desirable to modify, or "thicken," water so that it becomes more "oil-like."

Oil is more viscous and has a higher boiling point than water. Heat-transfer theory suggests that the reduced cracking tendency of an oil is a result of its higher viscosity, which reduces the convective heat transfer and thus reduces the cooling rate in the martensite formation region for many steels. (Heat-transfer rates decrease exponentially with increasing viscosity.) This same logic has been used in the development of aqueous polymer quenchants as replacements for oil quenchants (Ref 12).

An attractive feature of thickened water is the potential cost savings that can be achieved if the amount and cost of the thickening agent are kept low. An early goal in polymer quenchant development was to identify the most effective thickeners for water. It was thought that this would produce the desired quenching properties while minimizing cost by holding the polymer concentration to a low value.

Many polymers are known to be effective water thickeners, including cellulosic derivatives and polyvinyl alcohol (PVA). The thickening efficiency and resistance to bacterial attack of PVA were two factors that contributed to its widespread use as the first polymer quenchant.

Unfortunately, PVA is unstable and the cooling rates produced vary over time. This instability has been shown to be related to a cross-linking reaction (Ref 13) or molecular instability attributed to the high molecular weight of PVA. Another disadvantage of PVA is the hard lacquer finish, commonly called "plastic," which it leaves on the surface of metal parts. This plastic finish was the genesis of one of the common, but incorrect, terms applied to all aqueous polymers. As a consequence, many heat treaters still refer to aqueous polymer quenchants as "plastic" quenchants.

Table 5.1 Typical quench severities achievable with various media

Quenchant	Grossmann *H* factor
Oil	0.25-0.8
Polymer	0.2-1.2
Water	0.9-2.0
Brine	2.0-5.0

Source: Ref 11

Aqueous PVA quenchants have been considerably improved by the addition of 8 to 12% triethanolamine and 0.2 to 0.8% of dinitrobenzoic acid amine, which reduce foaming and promote bacterial stability both during storage and in the bath. Ethylene glycol has been added to reduce the amount of residual film on the workpiece after quenching (Ref 14). Despite these enhancements, PVA maintains a small market share.

Other polymers have been promoted as "oil-like" quenchants, but many still do not model the quenching behavior of oil, although some have actually provided process advantages. Over the past 30 years, polyalkylene oxide, commonly known as polyalkylene glycol (PAG), has captured the largest share of the synthetic quenchant market. Although the popularity of polymers has ebbed and flowed, PAG remains the most often used polymer quenchant in industrialized countries.

Fundamental Descriptive Terms

A *polymer* is defined as a large molecule built up by the repetition of small, "simple" chemical units called *monomers* (Ref 15). The synthesis of a polymer—poly (sodium

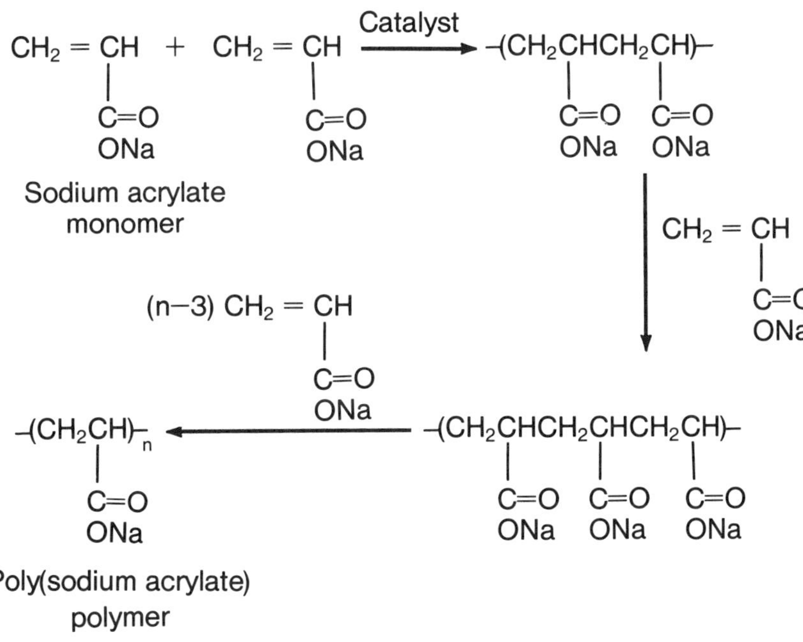

Fig. 5.1 Synthesis of polysodium acrylate from the sodium acrylate monomer

acrylate) (PSA), from the monomer sodium acrylate—is illustrated in Fig. 5.1. Other commercially important polymers used in the heat treating industry are shown in Fig. 5.2.

Polysodium acrylate (Fig. 5.1) is a *homopolymer*, because it is constructed from only one monomer. If a polymer is synthesized from more than one monomer, it is designated as a *copolymer*. Polyalkylene glycol (PAG) (Fig. 5.2) is an example of a copolymer

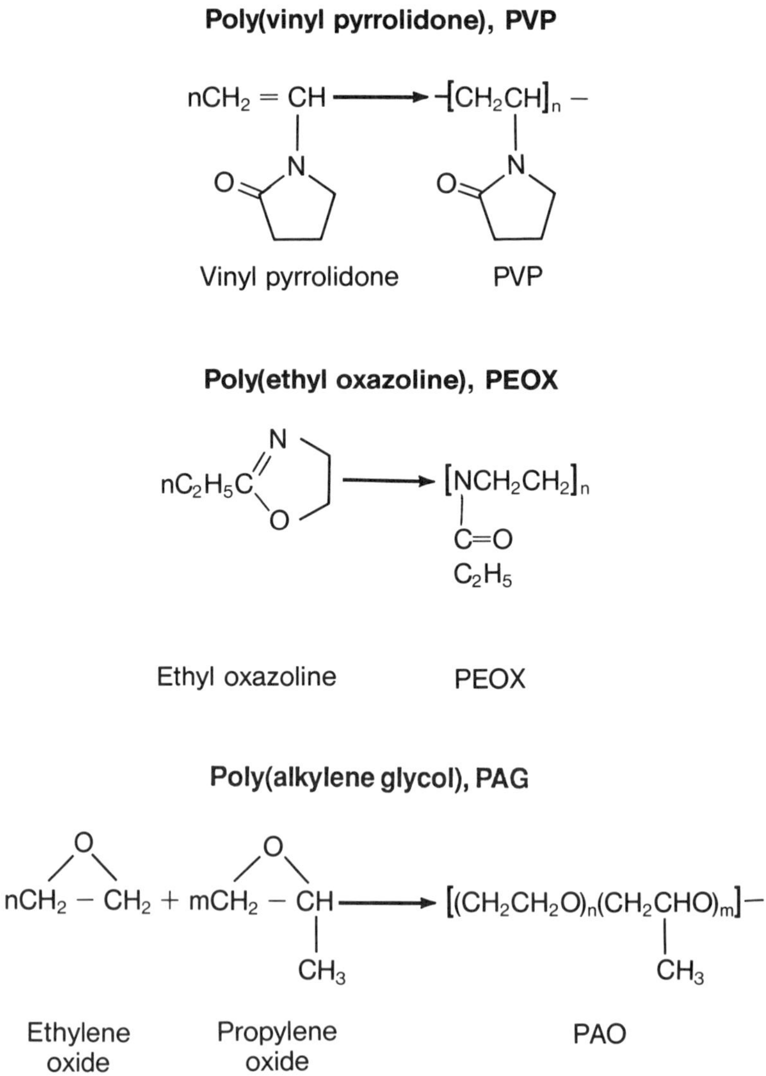

Fig. 5.2 Synthesis of various polymer quenchants

derived from two monomeric units, ethylene oxide and propylene oxide. The term alkylene oxide is a simplification of terminology used to designate that the polymer contains both ethylene and propylene oxide units. In chemical terms, ethylene and propylene are both included in the general class of alkylene.

Inverse Solubility

Polymers are dissolved in water by the solvation of polymer chains through hydrogen bonding interactions (Fig. 5.3). In order to obtain a homogeneous aqueous solution, it is essential that the polymer be capable of hydrogen bond interactions with water.

Some polymers undergo a phase separation as the solution temperature is raised (Fig. 5.4) (Ref 16). At a given temperature, the thermal energy of the system becomes greater than the energy of the hydrogen bonding interactions. When this occurs, the polymer chain loses its solubility in water and coils upon itself. A two-phase system develops at this point, one phase being polymer rich and the other water rich. Both phases usually contain some of the other component. The temperature at which this separation occurs is called the separation temperature or cloud point. If a polymer is not capable of forming a hydrogen bond interaction, it will not be soluble in water.

It should be noted, as shown in Fig. 5.4, that PAG polymers separate as hydrates at temperatures above their cloud points. The water content, which is related to the degree of hydration, is dependent on the ratio of ethylene to propylene oxide in the polymer and to the temperature.

Two polymers in current use that exhibit a separation temperature are polyethyl oxazoline (PEOX) and PAG. The ratio of the monomers in PAG copolymers is selected to control the separation temperature. For example, the separation temperature decreases as the number of the propylene oxide monomer units in the PAG copolymer is increased.

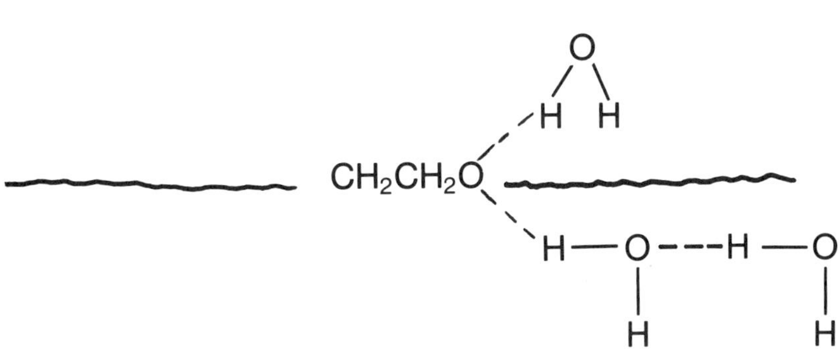

Fig. 5.3 Hydrogen bonding interaction for PAG polymer

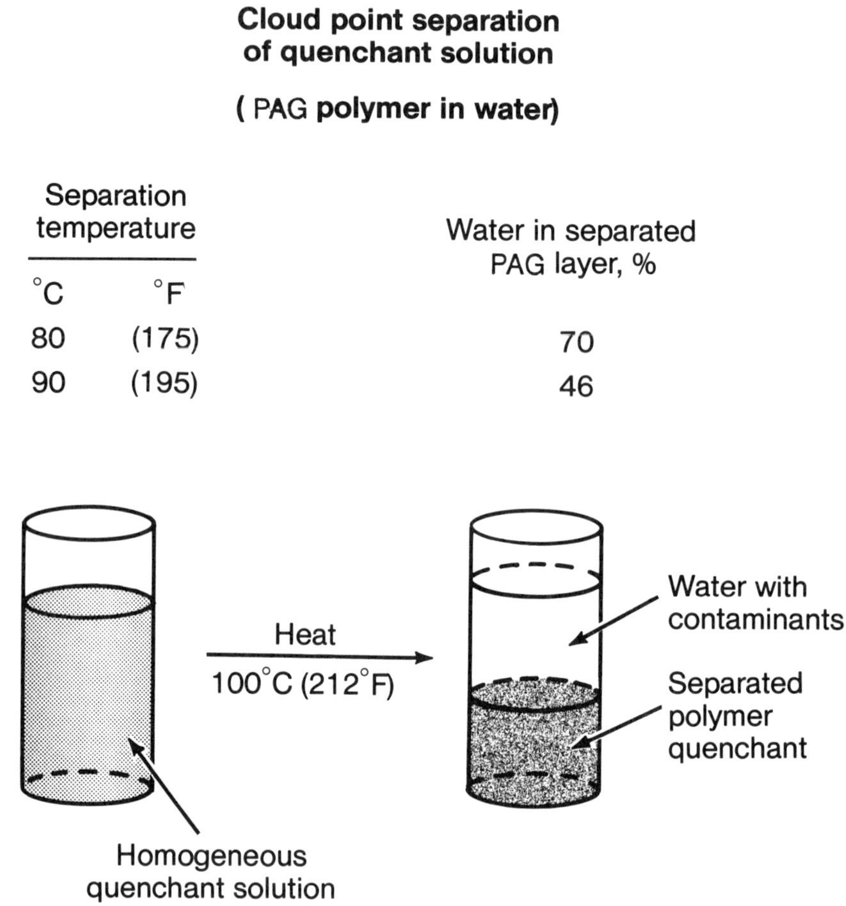

Fig. 5.4 Thermal separation of PAG quenchant

Mikita *et al.* (Ref 17) have shown that the cloud point of PAG copolymers is dependent on the pH of the aqueous solution (Fig. 5.5). Variations in the cloud point affect the viscosity/temperature properties of the solution (Fig. 5.6) and the cooling capacity of the quenchant (Fig. 5.7). Thus, it is important that substantial pH variations be avoided to minimize variations in quench severity.

The cloud point behavior of quenchant solutions is often used advantageously in practice, because it offers an inexpensive method of purification if the quenchant bath is contaminated (e.g., by salts). The temperature of the aqueous solution is raised to effect separation. After separation and layer formation, the aqueous supernatant solution is removed. The purified polymer, which usually forms as the lower layer, is then redissolved with fresh water to the desired concentration.

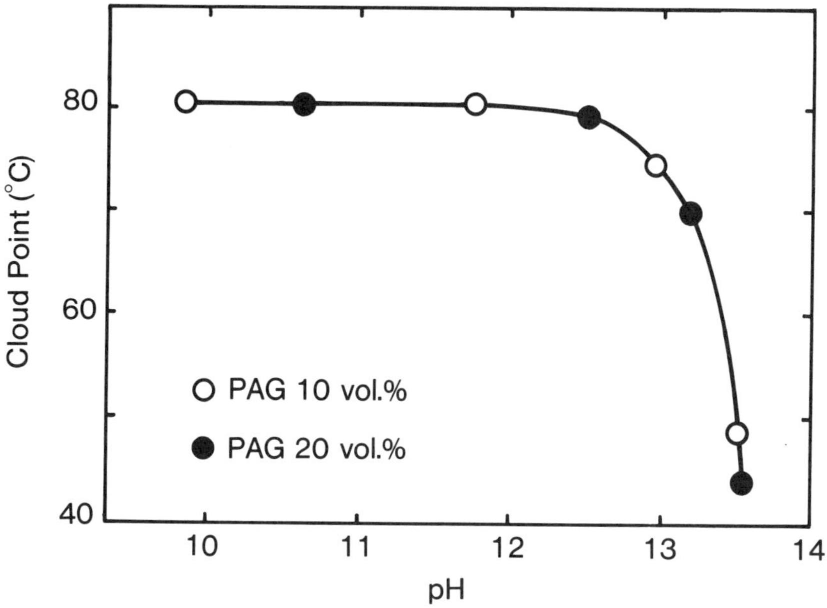

Fig. 5.5 Effect of PAG solution pH on cloud point

Polymer Size

Polymers are large molecules synthesized from one or more smaller simple chemical units. The size of a polymer is designated in terms of molecular weight (Mw) and degree of polymerization (DP).

Molecular weight can be viewed as the total weight of simple polymer units in the polymer and can be calculated as:

$$\text{Mw} = \text{Number of simple repeat units} \times \text{Molecular weight of a repeat unit}$$

It is evident that larger polymers have higher molecular weights.

The DP is calculated as follows:

$$\text{DP} = \frac{\text{Total molecular weight}}{\text{Molecular weight of a repeat unit}}$$

The DP of a polymer is equal to the average number of simple chemical units in an average polymer unit. Therefore, the size of a polymer increases as its DP increases.

Increasing the molecular weight of a polymer results in an exponential increase in polymer viscosity (Fig. 5.8). For a given polymer and concentration, higher molecular

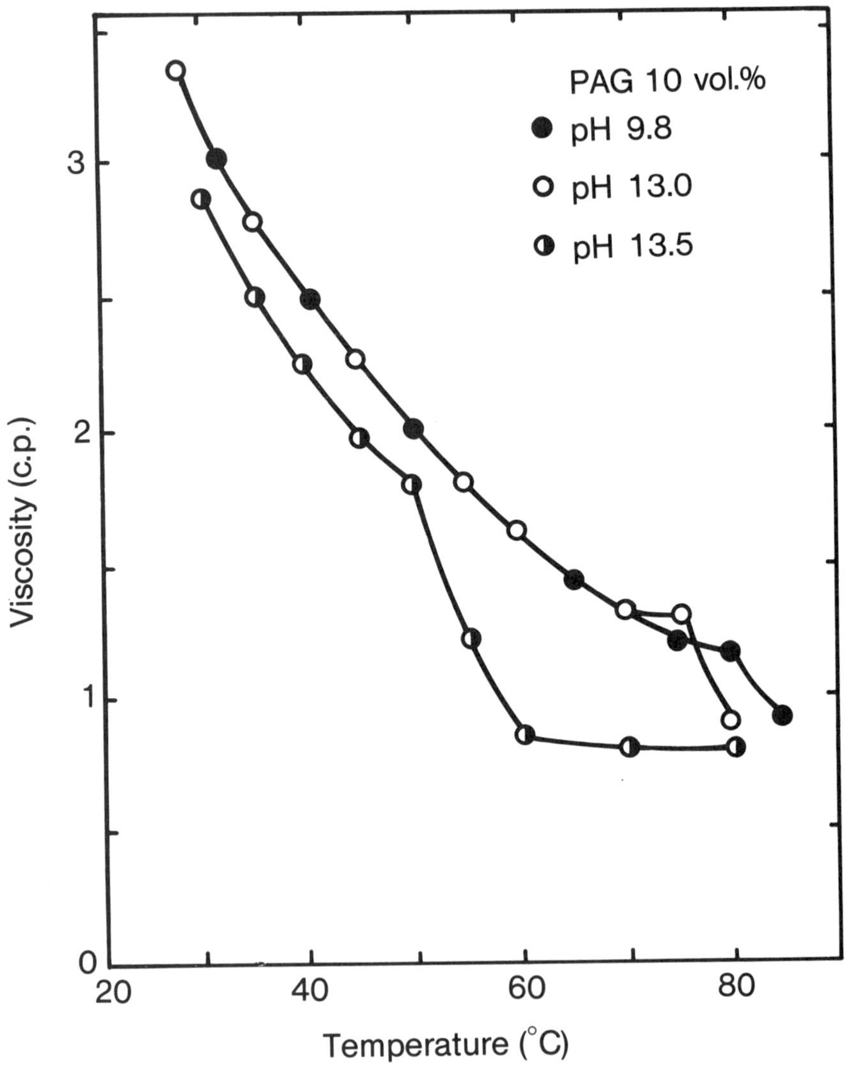

Fig. 5.6 Dependence of PAG solution viscosity on temperature

weights produce aqueous solutions with higher viscosities. Cooling rates usually decrease as the viscosity of a polymer solution increases. However, thermal/oxidative instability and mechanodegradation generally increase with increasing molecular weight. Thus, polymers used for quenchant applications must be optimized with respect to more than one variable.

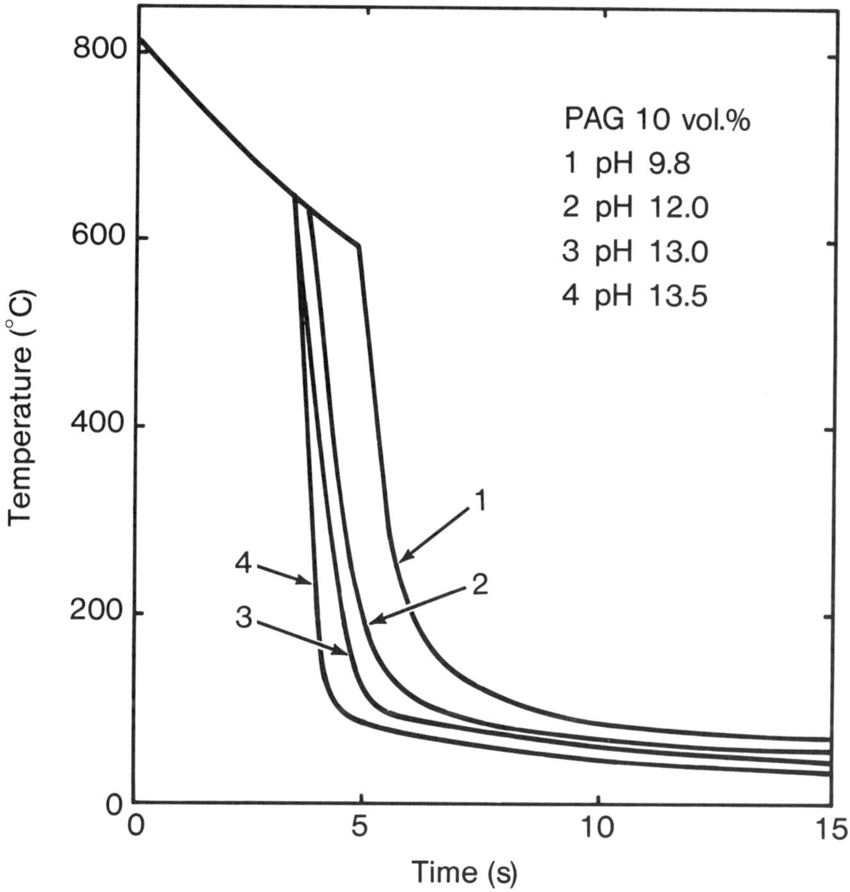

Fig. 5.7 Effect of PAG solution pH on the silver-bar cooling curve

Polymer Classification

Polymers can be categorized according to source, solubility, chain type, and so on. One of the primary classifications is whether a polymer is natural or synthetic. Examples of polymers derived directly from nature include natural rubber and cellulose. Most of the polymers used as quenchants in the heat treating industry are synthetic, or man-made. Examples of synthetic polymers include PAG, PSA, and polyvinyl pyrrolidone (PVP).

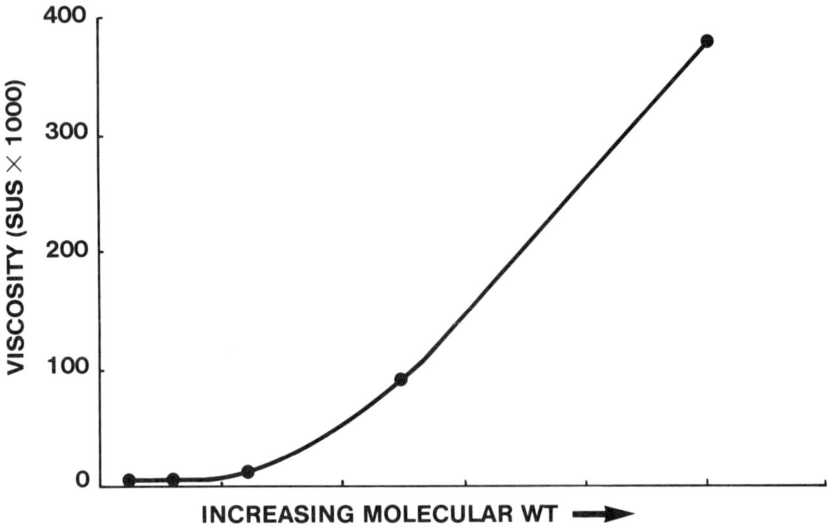

Fig. 5.8 Effect of polymer molecular weight on viscosity

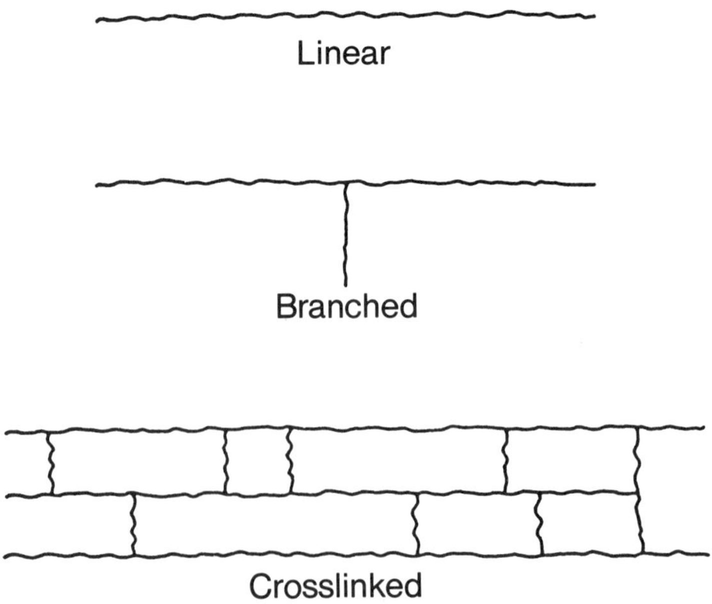

Fig. 5.9 Polymer structural configurations

Polymers can be further classified as water soluble or insoluble. Obviously, polymers used as aqueous quenchants are water soluble.

Polymers can also be classified structurally as linear, branched, or cross-linked (Fig. 5.9). Most of the polymers used as aqueous quenchants are linear. However, some branched polymers are beginning to appear in the marketplace (Ref 18). Cross-linked polymers are generally not used, as they are usually insoluble in any solvent. One exception is a polymer that has slight cross-linking and an unusual viscosity/temperature relationship, where the viscosity increases with temperature (Ref 19).

Finally, polymers can be classified as either ionic or nonionic. Ionic polymers are those that ionize when dissolved in water. A few ionic polymers have been successfully employed in quenchant formulations. The better known include PSA and sodium polyisobutylene maleate (PMI) (Ref 20).

One of the problems associated with the use of ionic polymers is their sensitivity to polyvalent cations in quenching baths. Polyanionic polymers such as PSA and PMI form "polyelectrolyte complexes" with polyvalent cations, which may form a precipitate or a coacervate (an insoluble oil that forms in aqueous solutions as a result of complexation) (Ref 21). Coacervation and precipitation result from the cross-linking action. This reduces hydration and renders the polymer and cation complex insoluble, as shown schematically in Fig. 5.10. Sources of polyvalent metal ions include calcium or magnesium from hard water, or solubilized ferric (Fe^{+3}) ions from iron oxide quenching scale or quench tank corrosion products.

Most of the polymers used in quenchant formulations are nonionic. The more common examples include PAG, PVP, and PEOX. Nonionic polymers are not as susceptible to polyelectrolyte complexation as their ionic counterparts and are more suitable for use in heat treating applications.

In summary, most polymers employed in quenchant formulations have a DP ranging from approximately 100 to greater than 10,000 and are:

- Linear
- Synthetic
- Water soluble
- Ionic or nonionic

"Glycol" Quenchants

The term glycol quenchants is often inappropriately applied to PAG copolymers. This misunderstanding has led to the use of ethylene glycol (a component of automotive antifreeze) as a quenchant, with disastrous results. The term glycol has arisen from the misapplication of the common name, *poly*glycol, used for PAG copolymers.

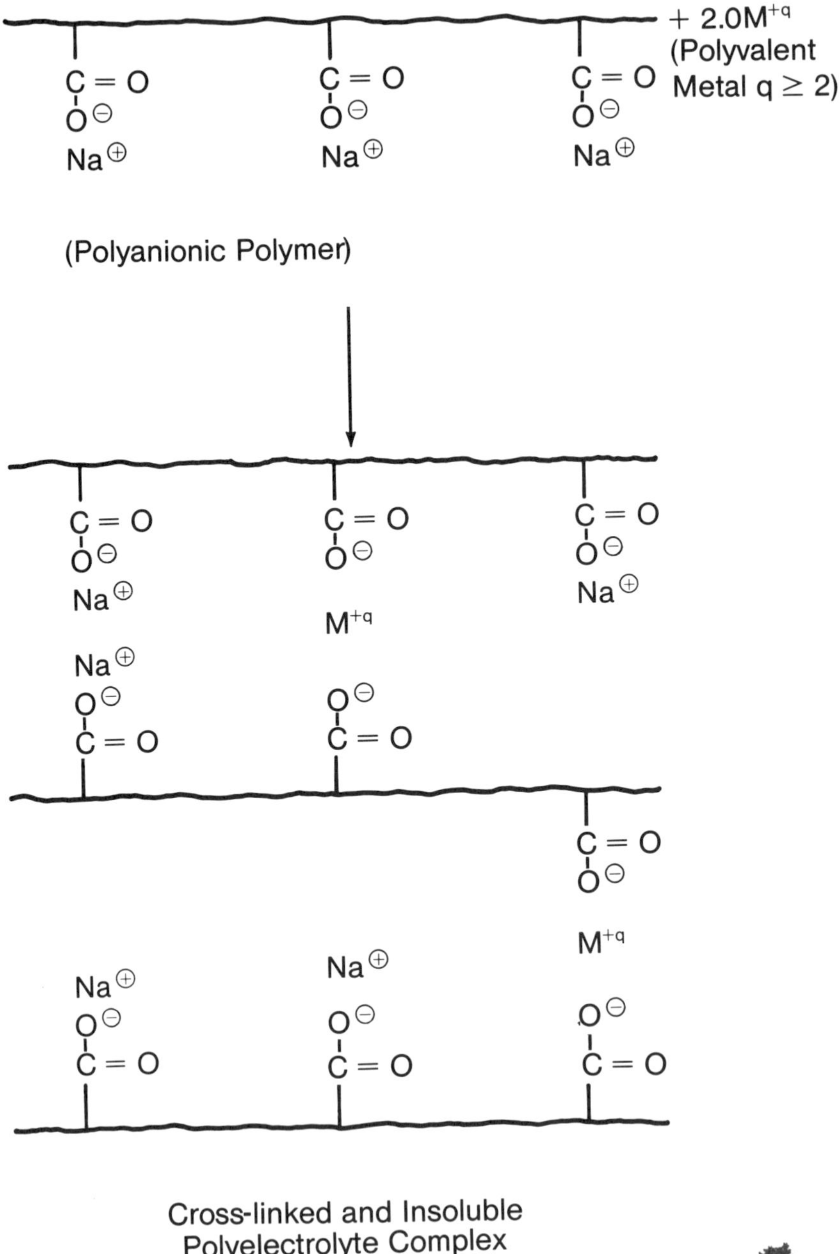

Fig. 5.10 Formation of an insoluble polyelectrolyte complex between a polyanionic polymer and a polyvalent metal

The term polyglycol implies that PAG copolymers are derived from the condensation of monomeric glycols (e.g., ethylene glycol and propylene glycol) to form a polyglycol. Although it is possible to produce polymers by such processes, this approach is not industrially used to prepare PAG copolymers. To aid in communication, specification writing, purchasing, and so forth, it is recommended that the correct and complete terminology, polyalkylene oxide or polyalkylene glycol, be used. Furthermore, when specifying the use of a particular polyalkylene oxide, it is important to include either a trade name or the structure and physical properties of the polymer. These recommendations apply to all polymer quenchants.

Differences in Polymer Properties

Another misconception is that all polymers are essentially equivalent and yield the same quenching results. Consider the differences between the quenching behaviors of PSA and PAG.

PSA is an ionic polymer, whereas PAG is a nonionic copolymer. PAG typically exhibits a cloud point, while PSA does not. Therefore, separation and purification of these two polymers require different approaches. Aqueous solutions of each polymer, as with most polymers of differing structures, are not compatible and should never be mixed.

Similarly, PVP is a nonionic polymer, and PSA is an ionic polymer. Aqueous solutions of these polymers are also mutually incompatible. The heat-transfer properties of each in quenching operations are also quite different. PVP polymers exhibit a broader range of quenchant speeds than equivalent concentrations of PSA.

Even within a single polymer classification, quenching behavior varies. For example, consider the following questions related to the structure of the PAG copolymer shown in Fig. 5.2, where n is the number of oxyethylene units in the polymer and m is the number of oxypropylene units.

- Is the polymer where $n + m = 100$ equivalent to the polymer where $n + m = 1000$?
- Is the polymer where $n/m = 7.5$ equivalent to the polymer where $n/m = 2.4$?

The total number of simple chemical units, or monomers, is equal to the DP, which is proportional to polymer size. Therefore, the polymer whose $n + m$ (DP) = 100 is smaller than the polymer where $n + m = 1000$. In practice, smaller polymers possess a lower viscosity and produce proportionally higher quenching speeds at equivalent total polymer concentrations, agitation rates, and bath temperatures. The quenching performance of a polymer as a function of DP can only be correlated for a given polymer. For example, the performance of PAG, PVP, and PSA polymers with DP values of 1000 are not equivalent, because the different chemical structures of these polymers lead to different solution behaviors at the metal interface during quenching.

PAG copolymers exhibit separation temperatures that are a function of the ratio of ethylene oxide to propylene oxide in the polymer. The aqueous solution cloud point decreases with increasing amounts of propylene oxide. Therefore, the polymer with a ratio of $n/m = 7.5$ would be expected to exhibit a substantially higher separation temperature than the polymer with a n/m ratio of 2.4.

Although a limited number of polymers have been described in this chapter, numerous water-soluble polymers have been reported to be useful as quenchants. A summary of these is provided in Table 5.2.

Additives have been developed in an effort to improve the quenching performance of aqueous polymers. For example, Tensi and Stitzelberger-Jakob (Ref 47) have reported that the addition of alkyl phosphate enhances the ability of solutions to wet the surface of a metal and thus improves the heat-removal properties of the media. One of the major drawbacks to this approach is that alkyl phosphates may be depleted by selective removal, or dragout, which causes variations in quench severity. Furthermore, alkyl phosphate concentrations, or any additive for that matter, may be difficult to monitor under shop conditions.

French workers (Ref 48) have added ionic salts to quenchants to obtain significant quench rate reductions and thus more oil-like performance. Such additions are not likely to gain broad acceptance at the present time, because many of these salts aggravate corrosion.

Effect of Polymer Structure on Quenching Mechanism

If an instrumented metal specimen (probe) is heated to a typical austenitizing temperature (e.g., 850 °C, or 1560 °F) and then immersed in a quenching medium, the temperature/time profile, or cooling curve, of the quenchant can be recorded. Cooling curves reflect the ability of a quenchant to mediate temperature changes in the probe during quenching, which in the case of polymer quenchants can be shown to be dependent on the polymer used to formulate the solution.

Figure 5.11 illustrates a temperature/time profile typical of most quenching processes. As discussed in Chapter 3, a cooling curve generally possesses three regions, traditionally designated as stages A, B, and C.

In the A-stage cooling region, a vapor blanket is formed around the hot metal. Upon further cooling, a transition from stage A to stage B occurs. B-stage cooling is accomplished by nucleate boiling and is the fastest region of cooling during the quenching process. C-stage cooling begins when the temperature of the film adjacent to the probe falls below the boiling point of the quenchant. At this point, boiling stops and heat transfer occurs by convection and conduction.

Table 5.2 Polymers reported as quenchants

Polymer	Comment	Ref
Polyalkylene oxide	High molecular weight, branched polymers analogous to those described in Ref 22	18
	Linear, with Mw of 7-10,000	22
Polyacrylamide	Soda ash used as a corrosion inhibitor	23, 24
Cellulosic derivatives	Quenching performance compared with water and oil	25
	Carboxy methyl cellulose, methyl cellulose, hydroxyethyl cellulose, etc.	26-28
Polyethylene oxide	Mw ranges from 500,000-5,000,000	29, 30
	Considerably lower Mw than those cited in Ref 29, 30	31
Polyethyl oxazoline	Mw ranges from 50,000-500,000	32
Capped polyalkylene oxide	Hydrophobe linear, branched derivatives of PAG	33
Polysodium acrylate	Viscosities of 20% solutions may vary from 5000-100,000 cps at 25 °C (77 °F)	34
Vinyl copolymer	Polyvinyl alcohol, polyvinyl acetate, polyacrylonitrile, polyvinyl chloride, etc.	35
Polyethyl oxazoline blends	Blended with various vinyl heterocyclic polymers (e.g., PVP) and copolymers (e.g., PVP-maleic anhydride)	36
Glycol blends with PAG	Ethylene glycol, propylene glycol, and hexamethylene glycol	37
Polyvinyl pyrrolidone	Mw ranges from 5000-400,000	38
Poly-*n*, *n*, *n*, *n*-dialkyldimethyl ammonium chloride		39
Partial iron (III) salt of PSA (designated PK-2)	Polyvalent cations, both inorganic and organic, will form a polyelectrolyte complex with a polyanionic polymer such as PSA. These complexes may be a liquid (coacervate) or a solid. It is not stated, but perhaps this is a liquid, thus simulating a soluble oil quenchant	40, 41
Polyamide-polyalkylene glycol copolymer	Alkoxylated copolymer of a polyamide derived from ethylenediamine	42
Polyisobutylenemaleic acid	Anionic copolymer of isobutylene and maleic acid similar to PSA; Mw approximately 60,000	43, 44
Latex polymers ("Narit" latex)	Emulsified terpolymers of chloroprene, methyl methacrylate, and methylacrylic acid	45
Polyurethane	Prepared by reacting a low molecular weight PAG with a diisocyanate (e.g., isophorone diisocyanate)	46

The quenchant, whether oil, water, or an aqueous polymer, alters the cooling mechanism by virtue of its effect on the liquid and vapor film characteristics at the hot metal interface. To illustrate this point, photographs of the metal-solution interface for

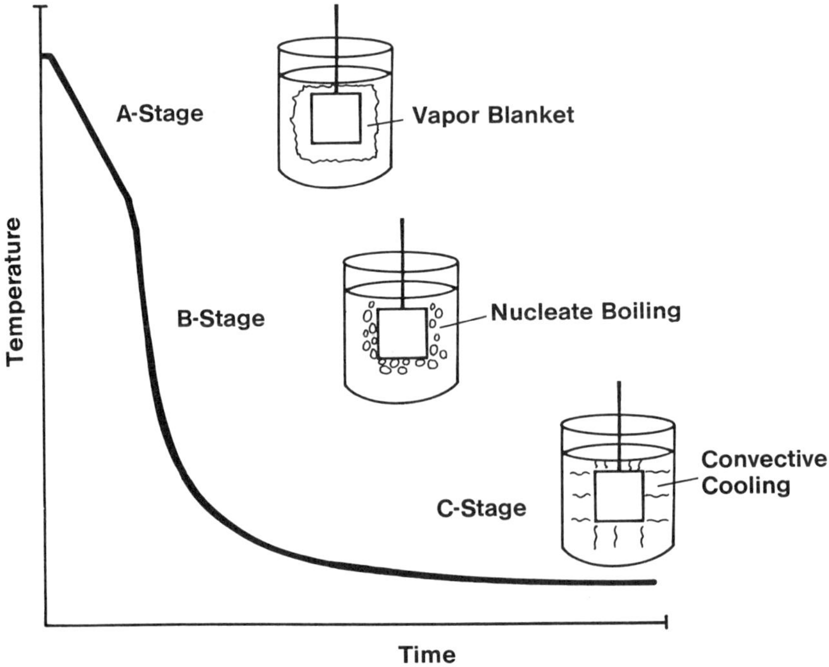

Fig. 5.11 Three stages of a cooling curve

Inconel 600 were taken during the A, B, and C cooling stages for four major types of commercial polymer quenchants: PAG, PVP, PEOX, and PSA. These photographs are superimposed on actual cooling curves in Fig. 5.12(a) to (d), respectively.

All of the cooling curves in Fig. 5.12 were produced by still-quenching with a 20% solution of the aqueous quenchant concentrate as sold by the manufacturer. The results of this study should not be overinterpreted, as a broad range of quenching behavior can be obtained with each quenchant by varying the bath temperature, concentration, and agitation. The point is simply to illustrate that the film-forming properties of polymers are different at the hot metal interface.

Each of the polymer solutions in Fig. 5.12 forms a vapor film blanket around the metal piece during A-stage cooling. Clearly, the nature of the film (e.g., strength, thickness, viscosity, etc.) controls cooling in this region.

The film is broken by a rupture process during cooling. This is not nucleate boiling, as traditionally described, because the characteristic film strength affects the boiling process. For example, the PSA and PVP quenchants appear to cool primarily by film rupture. PSA appears to undergo film rupture with much smaller film lesions than PVP, at least under the nonagitated conditions of this test.

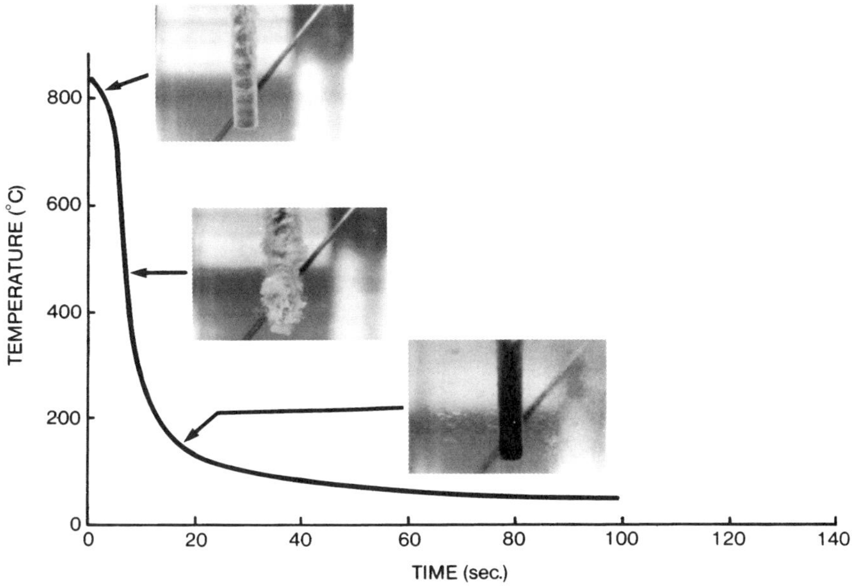

(a)

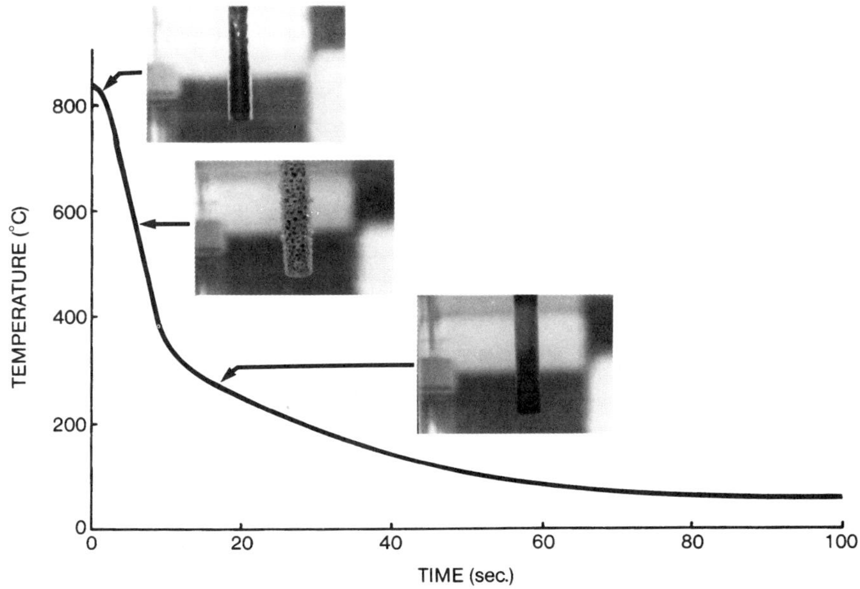

(b)

Fig. 5.12 Cooling of Inconel 600 in various still-quench conditions. Bath temperature, 40 °C (105 °F). **(a)** PAG. **(b)** PVP **(continued)**

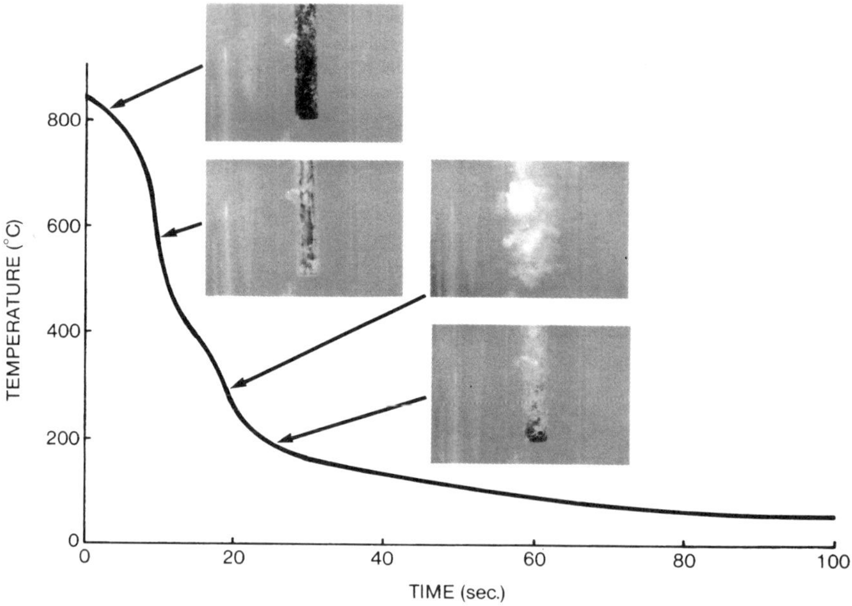

(c)

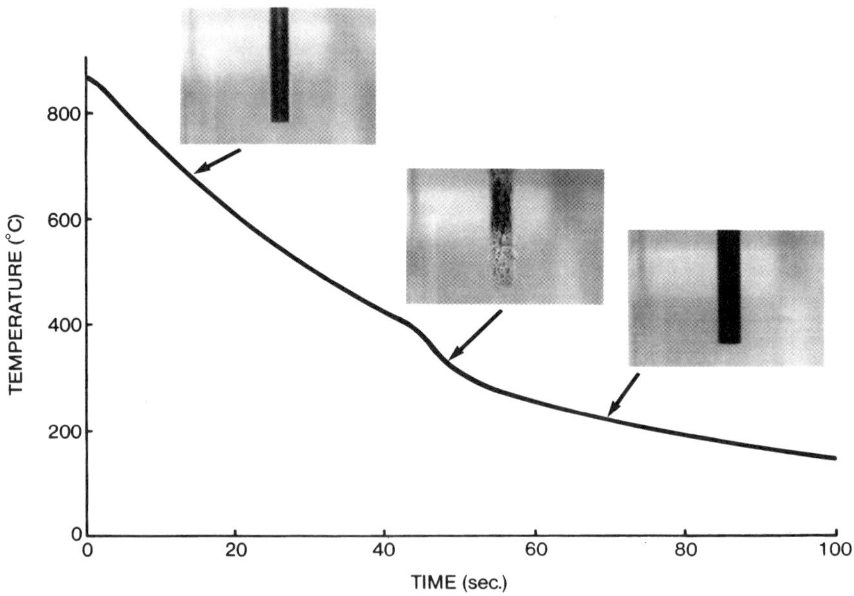

(d)

Fig. 5.12 Cooling of Inconel 600 in various still-quench conditions. Bath temperature, 40 °C (105 °F). **(c)** PEOX. **(d)** PSA

The PEOX quenchant also undergoes a film rupture process, but the rupture appears to be accompanied by a partial release of polymer to the solution. The haziness of the solution is caused by polymer insolubility, because the film temperature is above the polymer cloud point.

The PAG quenchant does not appear to form a continuous film, nor does it appear to undergo film rupture. The difference in behavior compared with PVP, PSA, and PEOX is probably due to the lower molecular weight and lower polymer film strength of PAG. Again, the haziness of the solution is the result of the temperature of the region being above the cloud point of the copolymer.

On continued cooling, the films of PVP and PSA collapse and remain on the probe. Partial dissolution of PEOX is observed as the interfacial temperature decreases below the separation temperature. The PAG polymer appears to be nearly completely redissolved in the bath as the interfacial temperature decreases below the cloud point.

The ability of a polymer to conduct heat by either convection or conduction through the film remaining on the workpiece during C-stage cooling is one of the primary factors controlling heat transfer in this region. The viscosity of aqueous polymer solutions is a factor; the effectiveness of heat transfer would be expected to decrease as the solution viscosity increases (Ref 12). Of course, excessive solution viscosities, typically greater than 10 to 12 cSt, results in substantial polymer dragout as parts are removed from the bath at the conclusion of the quenching process.

The nature of the various polymer films and their response to the thermal conditions during cooling are very different from one another. Figure 5.12 indicates that the film strength of the polymers may be strong (as in the case of PVP and PSA), weak (PAG), or intermediate (PEOX). In addition, the character of the film varies with the distance from the metal interface to the bulk solution. It follows that heat transfer also varies under these conditions.

One parameter that usually has a dramatic effect on the heat-transfer properties of polymer quenchants is solution agitation, which accelerates the removal of polymers at the metal interface. This produces faster cooling, as heat transfer is more efficient because of the lower polymer film thickness at the interface. The effects of agitation on heat-transfer properties must be considered during quench system design.

Quenchant Viscosity Effects

The chemical structure of a polymer influences the cooling mechanisms involved in quenching. The molecular weight of a polymer affects film strength and thus its dissolution rate during quenching, which in turn affects heat transfer.

Other parameters that affect the performance of polymer quenchants include bath temperature, agitation, and concentration of the polymer (Ref 49-51). The performances of representative PVP and PAG quenchants under varying conditions are compared in Fig. 5.13. The results were obtained with the cooling curve apparatus described by Hines

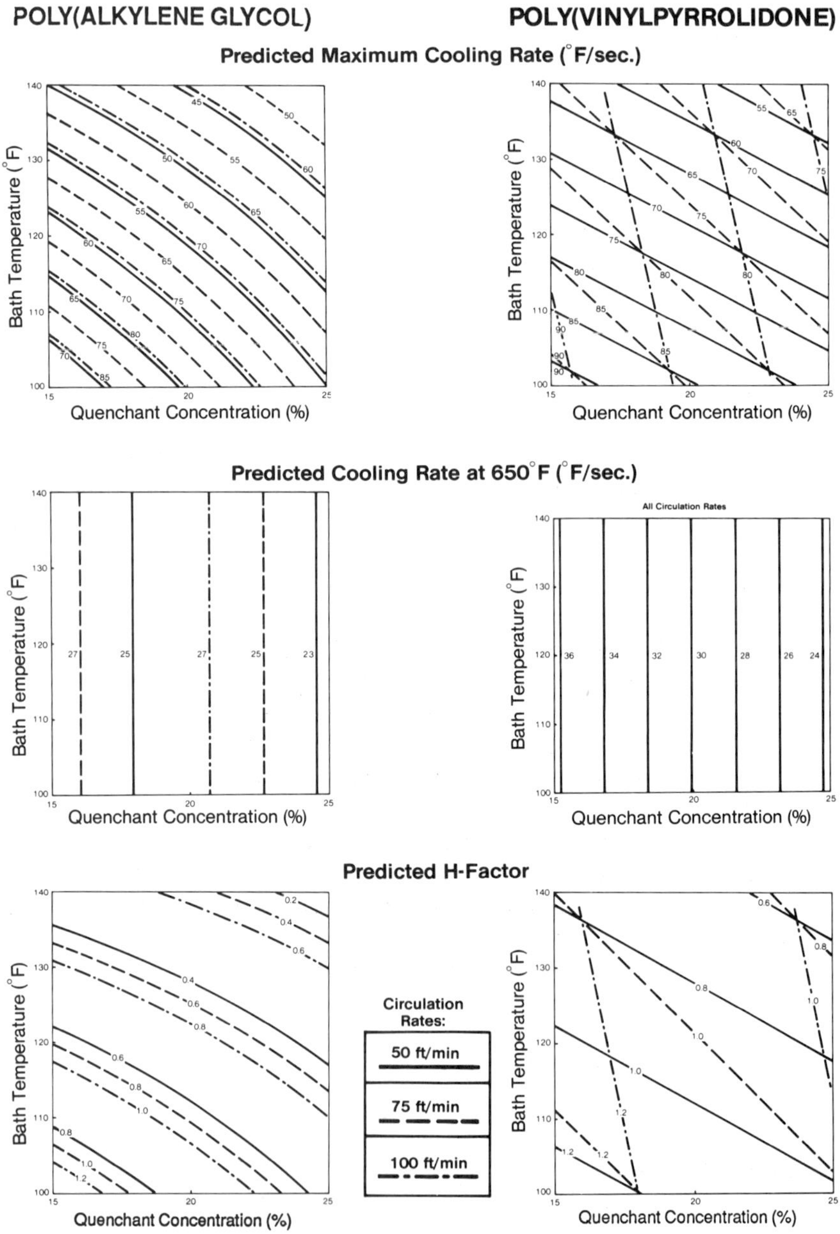

Fig. 5.13 Comparison of the quenching performance of PAG and PVP polymers under varying conditions

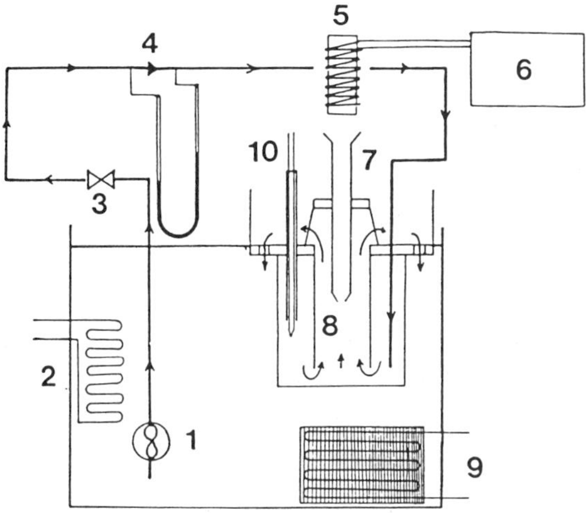

Fig. 5.14 Schematic of Hines and Mueller cooling curve apparatus

and Mueller (Ref 52) (see Fig. 5.14) using a 25 × 100 mm (1 × 4 in.) cylindrical stainless steel probe with a spring-loaded type K thermocouple inserted at the geometric center. Although various parameters can be used to quantitatively characterize cooling curve behavior, only three are used here for illustrative purposes: the maximum cooling rate (CR_{max}), the cooling rate at 650 °F (CR_{650}) (which is within the martensitic transformation temperature region of many steels), and the Grossmann hardenability factor (H). CR_{max} is generally assumed to occur within the B-stage region of the cooling curve, and CR_{650} typically occurs either in the C-stage region or within the transition from stage B to stage C.

These data show that:

- Increasing bath temperature decreases quench severity.
- Increasing agitation generally increases quench severity. The effect of agitation on quench severity depends on the direction of flow impingement on the quenched surface (e.g., horizontal, linear, etc.) and on the degree of solution turbulence.
- Increasing polymer concentration decreases quench severity; the extent of the effect varies with the particular polymer quenchant.
- PAG and PVP quenchants exhibit significantly different responses to these quenching variables with respect to CR_{max}, CR_{650}, and the H factor. These differences are caused by differences in interfacial cooling between the hot metal and the quenchant solution. Similar differences occur with other polymers and even with different PAG structures.

Factors Affecting Polymer Stability

Quenchant stability can be affected by several factors, including mechanodegradation, thermal/oxidative stability, dragout, and hydrolytic stability. The magnitude of these effects depends on the polymer structure and on bath and quenching conditions.

Mechanodegradation

Polymers do not exist in solution in their chain-extended form (Fig. 5.9). If they did, they probably would not be soluble. Instead, the polymer chains coil around each other, providing a relatively long-range order that increases viscosity (thickening effect) with concentration (Fig. 5.15). The degree of polymer chain entanglement increases with increasing molecular weight (chain length). Therefore, at the same concentration, the viscosity of a polymer in solution increases with increasing molecular weight.

When mechanical energy (shear) is applied to a polymer solution—for example, by pump or impeller agitation—a shear-induced thinning of the solution occurs. If sufficient

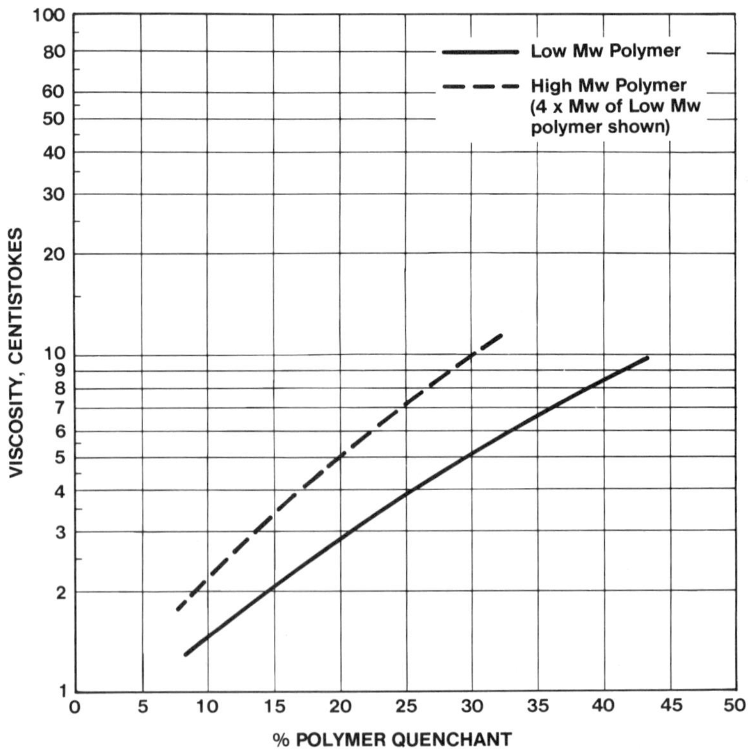

Fig. 5.15 Variation of aqueous solution viscosity with polymer molecular weight

energy is applied, it may take less energy to break covalent C–C bonds in entangled polymer chains than to pull them apart. When C–C bond scission occurs, the average molecular weight of the polymer is reduced. This is called mechanodegradation, or shear degradation (Ref 53). If there is a reduction in average molecular weight, there will be a corresponding reduction in solution viscosity. Because heat-transfer rates increase with decreasing viscosity, shear degradation can be expected to result in proportional increases in cooling rate.

The amount of shear stress applied to a quenchant in use varies widely. For example, the shear stress produced by a slowly turning impeller stirrer in the corner of a large tank is considerably less than that produced by spraying the quenchant at 275 kPa (40 psi) through a 6.4 mm (¼ in.) orifice or by pumping it at a rate of 2270 L/min (6000 gal/min) through a 150 mm (6 in.) pipe. It is difficult to develop a procedure to exactly describe the shear stability of a polymer in a quenching system. However, tests that reflect a worst-case scenario do exist.

One of the more rigorous tests is the ASTM D 3519 Waring blender test. In this test, the viscosity is evaluated after the solution is stirred in a blender at 9000 rev/min for a specified time. Table 5.3 presents results obtained when two polyacrylamide polymers, representing two degrees of hydrolysis (10 and 90%) (see Fig. 5.16) were tested. Similar polymers have been patented as polymer quenchants (Ref 23). The results of a PAG copolymer subjected to this same test are also shown for comparison in Table 5.3. The data show that both polyacrylamide polymers were shear unstable, whereas the PAG copolymer was shear stable. The thickening efficiency and shear instability increased with the degree of hydrolysis, further proof that various polymers behave differently.

Most quenchant solutions are not subjected to shear rates comparable to those produced in the Waring blender test. To determine how this test compares with actual quenching conditions, the shear stability of 10% hydrolyzed polyacrylamide polymer was evaluated by continuously pumping it through a wire screen in a 25 mm (1 in.) pipe at a linear flow rate of 6 m/min (20 ft/min). The data in Table 5.4 reveal a substantial and progressive reduction in viscosity.

Variations in solution viscosity are indicative of polymer stability and reflect potential variations in quench severity; therefore, a polymer must be stable to the shear field imparted by the process. Polymers may exhibit different shear stabilities when subjected

Table 5.3 Waring blender shear stability test of polyacrylamides and a PAG copolymer

Polymer	Polymer concentration, %	Viscosity at 40 °C (100 °F), cSt — Shear time, min: 1	2	3	Change
Polyacrylamide					
20% hydrolyzed	2.5	6.3	6.2	6.0	−0.3
90% hydrolyzed	0.25	6.3	5.9	5.6	−0.7
PAG copolymer	7.6	6.0	6.0	6.1	+0.1

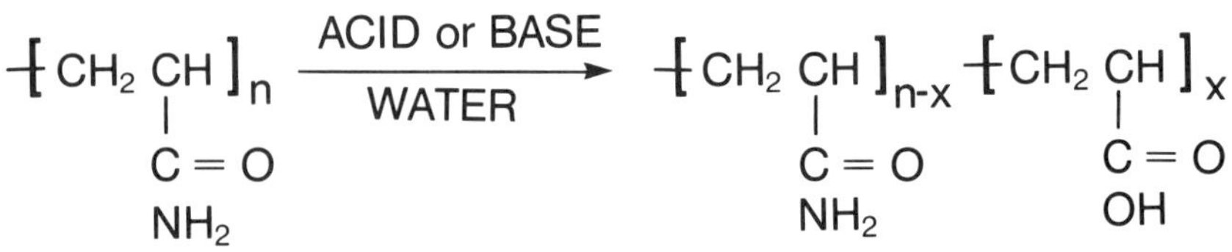

X	% Hydrolysis
0.1	10
0.9	90

Fig. 5.16 Hydrolysis of polyacrylamide polymers

Table 5.4 Pump shear stability of 10% hydrolyzed polyacrylamide (2.5% solution)

Start time, h	Viscosity at 40 °C (100 °F), cSt
0	7.9
3	6.7
5	6.5
8	6.5
11	6.3
16	6.0
20	5.6
24	4.8

to the same shear field, and a variety of results can be obtained with different polymers. The heat treater should request data on the relative shear stability of a polymer quenchant before putting the quenchant to use.

Thermal/Oxidative Stability

All organic compounds, including oil and polymer quenchants, are subject to thermal/oxidative instability when exposed to quenching conditions. Although instability under quenching conditions is known to occur, there is little published information on the thermal/oxidative stabilities of polymers or oils—perhaps because of the difficulty of quantitatively separating effects that cause changes in bath performance, such as hydrolysis, thermal instability, shear instability, metal contamination, and selective dragout.

Thermal/oxidative stability may substantially affect the cooling characteristics of a quenchant. For example, Kopietz (Ref 54) studied the effects of polymer service life on cooling curve properties (Table 5.5). The time and polymer service conditions were not described, but samples were taken from production quenchant tanks. Tank use conditions undoubtedly vary from shop to shop and perhaps from tank to tank. As a consequence, stability rankings of quenchants cannot be made from these data. However, variations in quench performance relative to new quenchants are apparent. Table 5.5 shows that bath aging affects quenching properties, as evidenced by the general acceleration in cooling rates over the B- and C-stage regions of cooling curves.

Polymer Dragout

Polymer dragout on a metal workpiece after quenching can be influenced by several parameters, including degree of redissolution of the polymer after quenching, shape of the workpiece, interaction between the polymer and the workpiece, degree of wetting of the polymer and the workpiece, concentration of the polymer in the solution quenchant, agitation, and bath temperature. It is thus difficult to obtain absolute measures of polymer dragout for comparison. Nevertheless, it is clear that if the workpiece is wetted by the quenchant (as it must be if the quenchant is effective), there also must be some dragout as the part is removed from the quenchant. If it is assumed that the polymer concentrate at the metal interface redissolves after the quench cycle is complete and that the interaction of the polymer with the workpiece is not great (not always a valid assumption), then the amount of dragout must be at least equal to the amount of solution remaining on the wet workpiece multiplied by the polymer concentration. Even under ideal conditions, all polymer quenchants experience some dragout, as do oil quenchants.

Ideal dragout conditions seldom exist. Differences in polymer-workpiece interactions are caused by variations in polymer structure. Also, depending on the system engineering (e.g., agitation and bath temperature), the polymer may not completely redissolve at the conclusion of the quench. Therefore, the previous analysis provides only an estimate of the minimum dragout that may occur. Furthermore, because various polymers are actually used at different concentrations, the dragout rates should be reported as relative and not absolute values.

Table 5.5 Comparison of the cooling characteristics of fresh and used PAG quenchants(a)

Temperature range			Average cooling rate			
			Fresh solution		Used solution	
°C	°F	Cooling stage	°C/s	°F/s	°C/s	°F/s
870-480	1600-900	A	14.0	25.3	15.5	27.9
480-200	900-400	B	21.2	38.2	34.7	62.5
200-95	400-200	C	4.1	7.4	8.2	14.7

(a) The used quenchant was taken from a commercial quench tank. Both the used and fresh quenchant solutions contained 15.7 wt% PAG polymer.

Hilder (Ref 50) designed a test workpiece with an intricate shape (Fig. 5.17) intended to model many part-irregularities. This test piece was heated and quenched in aqueous PAG, PVP, and PSA solutions under static and agitated conditions. The data showed substantial differences in polymer dragout. Unfortunately, the relative dragout rates and resulting impact on cooling rates were not reported. Nevertheless, Hilder concluded that the average dragout of a polymer was considerably lower than the average dragout of a quench oil at 40 °C (100 °F). Segerberg (Ref 55) obtained somewhat different results in a similar study.

Hilder (Ref 50) also showed that the magnitude of dragout was a linear function of solution viscosity (Fig. 5.18). Consequently, as the concentration of a polymer in a quenchant is increased, a corresponding increase in polymer dragout must be expected.

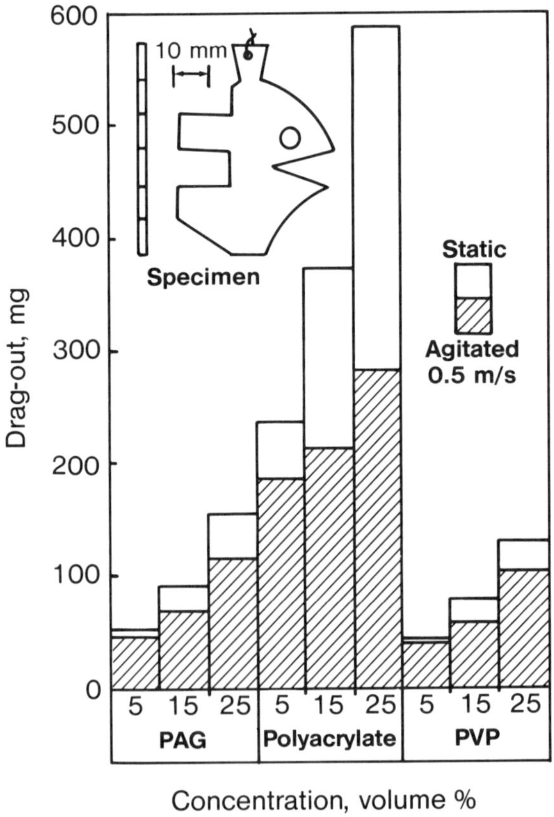

Fig. 5.17 Dragout specimen shape and results for polymer quenchants of various concentrations in both the static and agitated conditions

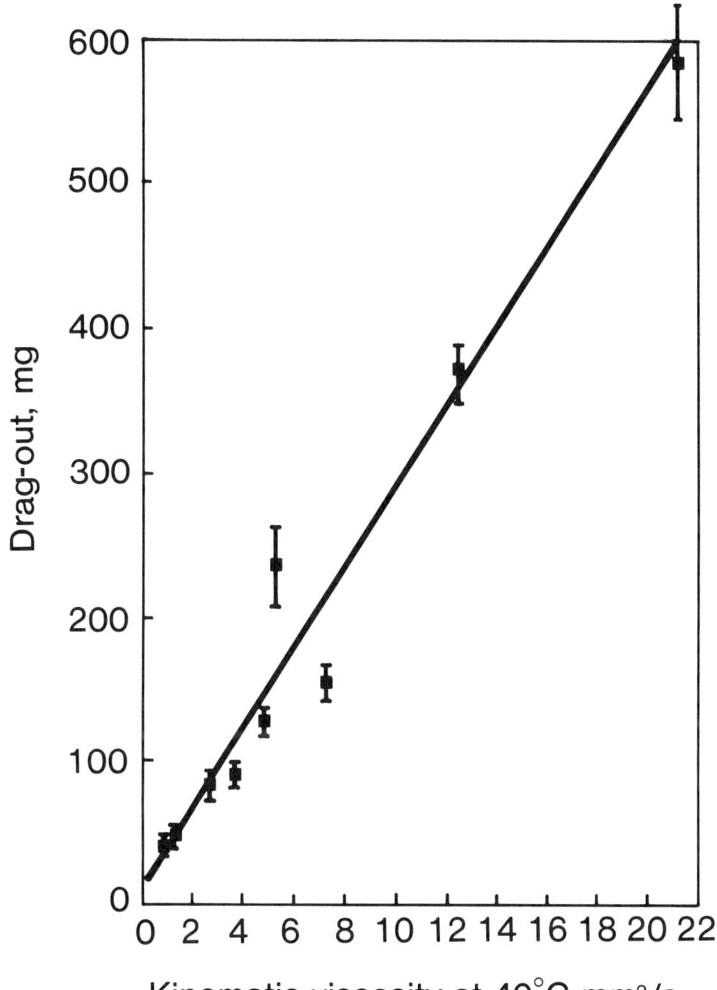

Fig. 5.18 Effect of polymer solution viscosity on the level of dragout in static tests

Chemical Stability

If variations in quench severity are to be minimized, the quenchant must be chemically stable. As shown in Fig. 5.12, polymer quenchants work by mediating heat transfer by film formation. The nature of the film is dependent on the chemical structure of the polymer. Therefore, if a quenchant polymer is not chemically stable, the film-forming and film-rupture properties will vary during use, creating a corresponding variation in heat-transfer properties.

Because quenchant polymers are used as aqueous solutions, it is essential that the polymer selected be hydrolytically stable over time. Therefore, it is preferable that a polymer for quenching applications not possess a chemical functional group susceptible to hydrolytic degradation under quench bath conditions (e.g., elevated temperature and basic pH).

This is one of the factors that has restricted the widespread use of polyacrylamide (Fig. 5.13). Esters and amides are known to be susceptible to hydrolytic degradation. Because these groups include PVA, polyacrylamide, and PEOX, quenchants formulated from these polymers may be susceptible to hydrolytic degradation under some conditions. Possible hydrolytic degradation reactions are illustrated in Fig. 5.19 and are based on well-known organic chemistry reactions (Ref 56). Although amides generally hydrolyze at lower rates than esters, substantial variations in film-forming properties still occur over the lifetime of the quenchant bath.

Ester **Amide**

Some quench oils and quench-rate accelerators (e.g., alkyl phosphates) are also subject to hydrolysis and other chemical instabilities. It is essential that the chemical

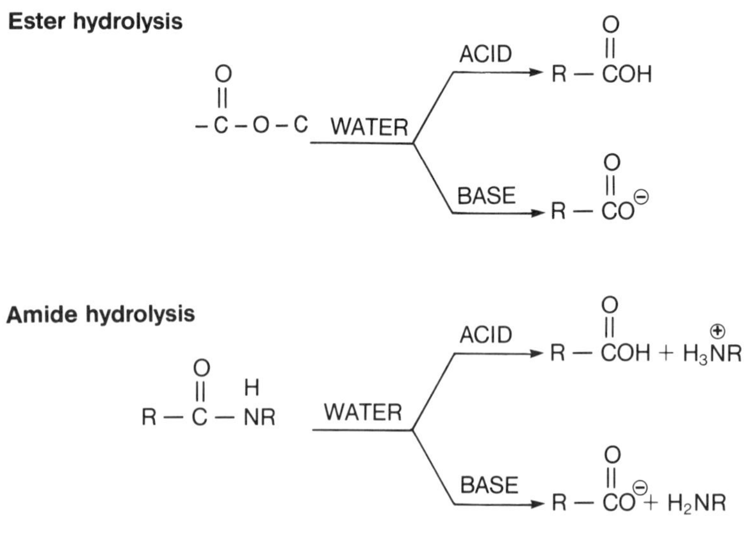

Fig. 5.19 Hydrolysis of esters and amides

stability of any potential quenchant medium or additive be considered in the selection process. A major goal of quenchant development is to minimize degradation reactions and to improve long-term quenchant stability.

References

1. W.G. Patton, *Iron Age*, Vol 212 (No. 9), 1973, p 46
2. H. Yamamoto, Y. Odagiri, M. Miyashita, and F. Nishiyama, *Netsu Shori*, Vol 29 (No. 1), 1989, p 14-18
3. E. Sprow, *Tool. Prod.*, Vol 49 (No. 9), 1983, p 22
4. R.S. Krysiak, *Mech. Work. Steel Proc. Proc.*, 1988, p 255-259
5. R. Creal, *Heat Treat.*, 1982, p 21-23
6. K.J. Mason and T. Griffin, *Heat Treat. Met.*, Vol 3, 1982, p 72
7. O. Sparks, *Heat Treat.*, 1978, p 18-22
8. A.J. Beck, *Heat Treat.*, May 1977, p 55-60
9. E.L. Kolosova, L.P. Shcherbakova, Yu.G. Ersmondt, and A.V. Kharov, *Steel USSR*, Vol 18, 1988, p 557-559
10. R.K. Singh and C.R. Chakravorty, *Tool Alloy Steels*, Vol 21 (No. 5), 1987, p 145-147
11. M.E. Dakins, C.E. Bates, and G.E. Totten, *Metallurgia*, Vol 56, 1989, p S7-S9. Note: The algorithm to calculate the Grossmann H factor should read: $H = AX^C \exp(BX^D)$.
12. G.T. Bozhko, O.A. Bannykh, M.N. Tropkina, P.I. Mznikhin, and A.V. Popov, *Metalloved. Term. Orab. Met.*, Vol 11, 1983, p 12-14
13. E. Burgdorf, *Z. Wirtsch. Fertigung*, Vol 74 (No. 9), 1979, p 431-436
14. V.M. Chudnova, T.P. Dumenskaya, and N.V. Golubeva, *Plast. Massy.*, Vol 6, 1989, p 80-84
15. G.E. Totten, *Adv. Mater. Proc.*, Vol 137 (No. 3), 1990, p 51-53
16. L.M. Jarvis, R.R. Blackwood, and G.E. Totten, *Ind. Heat.*, Nov 1989, p 23-24
17. Y. Mikita, I. Nakabayashi, N. Ohga, and K. Ohsaka, *Nippon Kikai Gakkai Ronbunshu (A-hen)*, Vol 53 (No. 496), 1987, p 2211-2215
18. E. Schoch, F. Reyrol, G. Beck, and F. Moreaux, British Patent 2,132,638A, 1984
19. K. Haas, E. Klahr, T. Proll, G. Welzel, and H. Tensi, German Patent, 3,429,611A1, 1986
20. T. Narumi, *Kogyo Kanetsu*, Vol 26 (No. 12), p 19-26
21. L. Holiday, *Ionic Polymers*, John Wiley & Sons, 1975, chap 6
22. R.R. Blackwood and W.D. Cheesman, U.S. Patent 3,220,893, 1965
23. N.F. Terekhova, O.N. Tanicheva, N.M. Tiunova, and N.M. Lobanova, Soviet Patent 1,250,586, 1984
24. A.A. Zakamaldin, Yu.I. Kochurkina, and G.I. Tikhonov, *Met. Sci. Heat Treat.*, Vol 255 (No. 1-2), 1983, p 13-18
25. J. Wyszkowski and St. Sobol, *Proc. 5th Int. Cong. Heat Treat. Mater.*, Budapest, Vol 3, 1986, p 1836-1843
26. M. Gordon, U.S. Patent 2,770,564, 1956
27. M. Trusculescu, D. Trusculescu, D. Serban, and A. Raduta, *Cercet. Metal.*, Vol 27, 1986, p 223-228
28. V.M. Ezhor, S.V. Semakin, V.V. Svetlichnyi, A.M. Mazaer, and Yu.V. Soboler, *Energomashinostroenie*, Vol 1, 1988, p 35-38
29. A.S. Bedarev, G. Ye, G. Il'Yushko, G.I. Beloborodov, G.P. Konyukhov, and E.V. Al'tergot, Soviet Patent 600,190, 1978

30. F.L. Chase, D. Ewing, and C.W. Ewing, U.S. Patent 3,022,205, 1962
31. T. Hibi, *Netsu Shori*, Vol 25 (No. 1), 1985, p 8-13
32. J.F. Warchol, U.S. Patent 4,486,246, 1984
33. J.F. Warchol, U.S. Patent 4,381,205, 1983
34. K-H. Kopietz and F.S. Munjat, , U.S. Patent 4,087,290, 1978
35. E.R. Cornell, U.S. Patent 2,600,290, 1952
36. J.F. Warchol, U.S. Patent 4,528,044, 1985
37. M. Lewis and P.J. Welsh, U.S. Patent 3,475,232, 1969
38. A.G. Meszaros, U.S. Patent 3,902,929, 1975
39. M.J. Koraleva, A.A. Kats, M.M. Rozin, N.Y. Kudryartseva, E.Y. Markus, V.V. Kryuchkov, L.A. Amburg, and Y.S. Parkhamovich, Soviet Patent 1,257,102, 1984
40. V.V. Goryushin, V.F. Arifmetchikov, A.K. Tsvetkov, and S.N. Sinetskii, *Met. Sci. Heat Treat.*, 1986, p 709-714
41. O.D. Khutoranko and E.I. Marchuk, *Teknol. Organ. Proizvod*, Vol 4, 1987, p 40
42. R.L. Zhou, *Proc. 4th Ann. Conf. Heat Treat. Inst. Chinese Mech. Eng. Soc.*, Nanjing, Vol I, 25-31 May 1987
43. E. Nakamura, *Junkatsu*, Vol 32 (No. 8), 1987, p 570-574
44. Y. Murakami, *Netsu Shori*, Vol 25 (No. 1), 1985, p 14-15
45. A.S. Paronyan, N.A. Arakelyan, Sh.S. Shushanyan, and A.V. Gevorkyan, *Metalloved. Term. Obrab. Met.*, Vol 6, 1989, p 13-15
46. R.J. Knopf, U.S. Patent 4,481,367, 1984
47. H.M. Tensi and P. Stitzelberger-Jakob, *Härt.-Tech. Mitt.*, Vol 44, 1989, p 99-105
48. F. Moreaux, J-M. Naud, and G. Beck, British Patent 2,133,047A, 1984
49. G.E. Totten, M.E. Dakins, and R.W. Herms, *J. Heat Treat.*, Vol 6 (No. 2), 1988, p 87-95
50. N.A. Hilder, Ph.D. thesis, University of Birmingham (Aston), 1988
51. G.E. Totten, M.E. Dakins, and L.M. Jarvis, *Heat Treat.*, Dec 1989, p 28-29
52. R.W. Hines and E.R. Mueller, *Met. Prog.*, Vol 122 (No. 4), 1982, p 33-39
53. L.S. Zarkhin, S.V. Sheberstov, N.V. Panfilovich, and L.I. Manevich, *Russ. Chem. Rev.*, Vol 58 (No. 4), 1989, p 381-392
54. K-H. Kopietz, *Ind. Heat.*, Vol 46 (No. 6), 1979, p 30-39
55. S. Segerberg, paper presented at 4th Int. Cong. Heat Treat. Mater., Berlin, June 1985
56. J. March, *Advanced Organic Chemistry—Reactions, Mechanisms and Structure*, 2nd ed., McGraw-Hill, 1977, p 349-356

CHAPTER 6

Quench Bath Maintenance

The purpose of a quenchant is to mediate heat transfer from the hot metal workpiece. Most factors that affect heat-transfer rates during quenching also affect metallurgical properties. Oil and polymer quenchants demand periodic maintenance to ensure continued process control.

The performance of quench oils depends on chemical composition, viscosity, and additive concentration. The various chemical structures (Fig. 4.3) that constitute a quench oil undergo oxidative and thermal degradation at different rates to provide different by-products during use. These degradation reactions produce compositional variations that affect the viscosity of the oil and thus the heat transfer. Normal quench oil use may also result in selective additive dragout, which affects cooling rates. Other factors that affect quench oil performance include sludge formation and contamination by liquids such as water and hydraulic oils.

The more common polymer quenchants mediate heat transfer by the reversible formation of a polymer-rich film around the metal part during the cooling process. Polymer degradation, by oxidation or mechanodegradation (Ref 1), reduces the viscosity of the polymer film, which increases heat transfer. The presence of contaminants (e.g., hydraulic oil, forging lubricants, salts, etc.) also affects the viscosity and continuity (uniformity) of the film, which may lead to cracking, soft spots, or distortion.

To avoid such problems, it is essential that periodic quality control tests be performed on quenchants. This chapter will discuss several tests that are used to maintain the consistency of oil and polymer quenchant baths. The tests used to monitor oils and aqueous polymer quenchants are different; therefore, bath maintenance procedures for each will be discussed separately.

Other quenching media, such as brine, molten salt, and even water quenchants, should also be periodically monitored. However, with the exception of water, these media are encountered less often and, with the exception of cooling curve analysis, special testing procedures may be required. These procedures will not be discussed in this chapter.

Quench Oil Bath Maintenance

Oil Degradation Processes

The immersion of hot parts into quench oils will cause chemical changes analogous to those that occur in engine oils during use (Ref 2). When the hot oil rises to the surface during quenching and contacts air, the oil may become oxidized and subsequently polymerized. The hot metal surface itself may also produce chemical changes in the oil. Degradation reactions produce volatile and nonvolatile by-products (Ref 3). The specific degradation reactions that occur depend on the interfacial temperature and contact time during the quench, the amount of oxygen present, the composition of the oil, and the concentration of the additives used to enhance cooling rates and oxidative stability.

Hydrocarbons are the major components of most oil quenchants. Representative reactions of a typical hydrocarbon will be used to illustrate the oxidation reactions that may occur during quenching.

Oxidation of a saturated hydrocarbon will produce alcohols, esters, and acids, as illustrated by Eq 6.1 to 6.3, respectively. The relative amounts of these by-products are quantified by various analytical techniques, such as acid number and saponification number, and will be discussed later.

$$\underset{\text{(alkane)}}{R-CH_2CH_2-R} + O_2 \rightarrow \underset{\text{(alcohol)}}{R-CH_2\underset{\displaystyle OH}{\underset{|}{C}}H-R} \quad \text{(Eq 6.1)}$$

$$R-CH_2\underset{\displaystyle OH}{\underset{|}{C}}H-R + O_2 \rightarrow \underset{\text{(alkene)}}{R-CH=CH-R} \quad \text{(Eq 6.2)}$$

$$R-CH=CH-R+O_2 \rightarrow \underset{\text{(ester)}}{R-\overset{\displaystyle O}{\overset{\|}{C}}OCH_2R} + \underset{\text{(acid)}}{R\overset{\displaystyle O}{\overset{\|}{C}}OH} + HOCH_2-R \quad \text{(Eq 6.3)}$$

The initial high temperature of the metal surface may also result in chain scission (molecular weight reduction) of the hydrocarbon, increasing the amount of undesirable volatile by-products. Such increases are reflected by a decrease of the flash point of the oil. One potential volatile-forming reaction is shown by Eq 6.4. Substantial flash point decreases may result in unacceptable fire hazards.

$$R-CH=CH-R \rightarrow R-CH=CH\cdot + H-R \quad \text{(Eq 6.4)}$$

Another degradation process that may result in high molecular weight by-products (sludge) is the coupling, or cross-linking, of reactive intermediates, as shown in Eq 6.5

to 6.8. One of the properties of a cross-linked material is that it is usually insoluble, or only partially soluble, in solvents, including the oil itself. Therefore, such cross-linking reactions may lead to the formation of insoluble sludge.

Scheme 1

$$R-\overset{\overset{O}{\|}}{C}OH + R-CH_2OH \rightarrow R-\overset{\overset{O}{\|}}{C}OCH_2-R \qquad \text{(Eq 6.5)}$$

or

$$R'H + O_2 \rightarrow R'\cdot + HOO\cdot \qquad \text{(Eq 6.6)}$$

$$R'\cdot + R-CH=CH\cdot \rightarrow R'-CH=CH-R \qquad \text{(Eq 6.7)}$$

$$R-CH=CH-R + RH \rightarrow R-C\cdot H-CH_2-R + R\cdot \qquad \text{(Eq 6.8)}$$

All of these processes, and many others not shown, result in the production of compounds that may enhance the wetting characteristics of the oil (Ref 4) and increase the cooling rates of the quenching process. Usually a "fresh," unaccelerated oil has a lower quench severity than a used oil (Ref 5-7). Increasing oxidation may increase the cooling rate—a phenomenon at least partially due to a decrease in viscosity of the oil and the buildup of water that often accompanies oil oxidation (Ref 8).

Test Procedures

Two types of tests are usually performed on oil quenchants: (1) physical and chemical characterizations and (2) quench severity measurements. Physical and chemical characterizations include viscosity, flash point, water content, neutralization number, carbon residue, ash, and sludge tests (Ref 4, 9, 10). These tests are analogous to those used for the characterization of industrial engine and machine oils and have been

Table 6.1 ASTM test methods for oil quenchants

Property	ASTM method
Viscosity	D 445
Flash point	D 92
	D 93
	D 1310
Fire point	D 92
Water content	D 95
	D 1744
	D 4007
Neutralization number	D 974
Saponification number	D 94
Ash	D 482
Conradson carbon residue	D 189
Precipitation number or sludge	D 91
	D 2773
Specific gravity	D 287

Source: Ref 9

incorporated into various industrial standards (Table 6.1). The results of these analytical tests provide insight into the condition of an oil and its suitability for continued and safe use. Only the most common physical and chemical characterization tests used to monitor the quality of a quench oil will be reviewed here (Ref 9).

Viscosity. The performance of a quench oil is strongly dependent on viscosity, which varies with temperature and use. Increased oil viscosities typically decrease heat-transfer rates (Ref 11) and result in higher dragout rates, thus increasing operating costs (Ref 12). Oil viscosity variation with temperature affects heat-transfer rates throughout the quenching process.

The "pumpability" of a quench oil depends on both viscosity and temperature. This is especially critical for marquenching oils, which are used at higher temperatures where oxidation may produce dispersed by-products that are often colloidal or "gel-like" at room temperature. Although the viscosity of a marquenching oil may not fluctuate substantially at elevated temperature, the oil may become almost solid upon cooling. Thus, the viscosity-temperature relationship of a quench oil may be critically important from the dual standpoints of quench severity and pumpability.

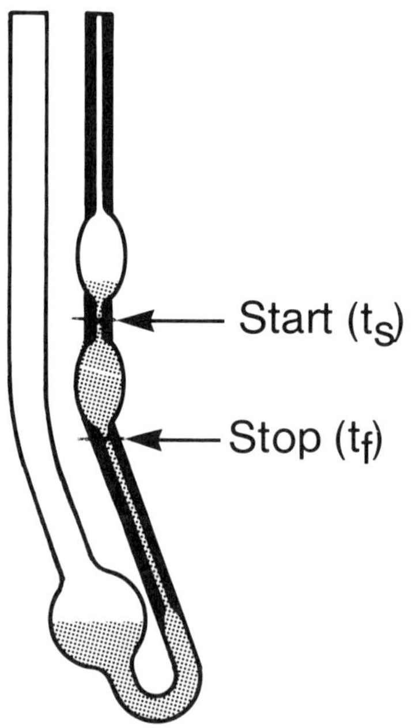

$$\text{Viscosity (cS)} = \text{Constant} \times (t_f - t_S)\ \text{sec.}$$

Fig. 6.1 Viscometer used to measure the kinematic viscosity of a quench oil

A calibrated glass capillary viscometer (Fig. 6.1) and a constant temperature bath are used to conduct the ASTM D 445 test method for kinematic viscosity. Kinematic viscosity is determined by measuring the time (seconds) for a fixed volume of fluid to flow through a capillary tube of the calibrated viscometer held at a constant temperature. Kinematic viscosity (η) is measured in units of centistokes (cS) and is determined by multiplying the measured flow time (t) by the viscometer constant (C):

$$\eta = C \times t$$

Another occasionally encountered unit of viscosity measurement is centipoise (cP). Centistoke viscosity units are related to centipoise by the density (ρ) of the fluid at the same temperature:

$$cP = cS \times \rho$$

Flash and Fire Points. The flash point of a liquid organic compound is the temperature at which the liquid, in equilibrium with its vapor, produces a gas-phase mixture that is ignitable but does not continue to burn when exposed to a spark or flame source. There are two types of flash points: closed-cup and open-cup. In the closed-cup measurement, the liquid and vapor are heated in a closed system. Traces of flammable, low-boiling-point components accumulate in the vapor phase, producing a relatively low flash point. Volatile by-products are lost while conducting the open-cup flash point procedure; therefore, traces of residual flammable components have less impact on the flash point. Closed-cup flash points are lower than the corresponding open-cup values, because the volatiles are left in the test system.

The fire point is the temperature at which the air-vapor mixture over the fluid ignites *and* continues to burn after exposure to an ignition source. The fire point is always higher than the flash point. However, flash points are more commonly used as equalizing control measures than fire points. As a quench bath is used, the flash and fire points decrease as a result of the chain-scission degradative processes described earlier.

In the case of a partially submerged load, the quench oil adjacent to the hot metal may become heated above its fire point, causing the hot metal to act as an ignition source and resulting in a fire. ASTM D 92 (the "Cleveland open-cup" method) can be used to determine an open-cup flash point. This procedure uses an apparatus with a small cup to hold a reproducible amount of the liquid to be tested. The cup is fitted with a thermometer and is heated rapidly to a temperature about 50 °C (90 °F) below the expected flash point. The heating rate is then reduced to 5 to 6 °C/min (9 to 11 °F/min). A flame is passed over the liquid at 2 °C (4 °F) intervals. The flash point is the temperature at which a flash appears on the surface of the oil.

The ASTM D 93 method for closed-cup flash points is similar, except that the apparatus has a cover fitted with a shutter that is opened only for the time required to test the cup for a flash.

General guidelines have been developed for the use-temperature of a bath relative to its flash point. One author suggests that a bath temperature of less than 28 °C (50 °F) below the oil flash point should be considered hazardous (Ref 13).

Water Content. The presence of water in an oil quenchant, which may arise from a degradation process or by external contamination (such as a leaking cooling coil), presents a potentially serious problem. Water concentrations as low as 0.1% may cause the bath to foam during quenching, greatly increasing the risk of fire. Overflowing oil from the foaming bath may result in a more serious fire than if the flames were contained by the bath, as the oil may contact nearby furnaces or other ignition sources. If a sufficient amount of water accumulates in a hot bath, an explosion caused by steam generation may also result (Ref 4, 14).

The presence of water in a quench oil may produce variable cooling rates, depending on the nature and amounts of rate-accelerating additives present in the oil (Ref 5, 6, 15, 16). For example, increasing water concentrations in a normal-speed quenching oil produce corresponding increases in cooling rates (Fig. 6.2a) (Ref 15). Conversely, increasing water concentrations in an accelerated quench oil have been reported to produce lower cooling rates (Fig. 6.2b) (Ref 15). Although the magnitude and direction of these effects depend on the particular quench oil, it is clear that water contamination may dramatically affect quench severity.

Water contamination may also result in staining (Ref 4), uneven hardness, and soft spots on the workpiece (Ref 9). Experience has shown that more than half the problems encountered with oil quenchants are associated with water contamination (Ref 17).

A common qualitative field test for water contamination is the so-called crackle test (Ref 17), which is conducted by heating a sample of the oil and listening for an audible crackling sound. If the oil is contaminated with water, a crackling sound will be detected before the oil has reached its smokepoint.

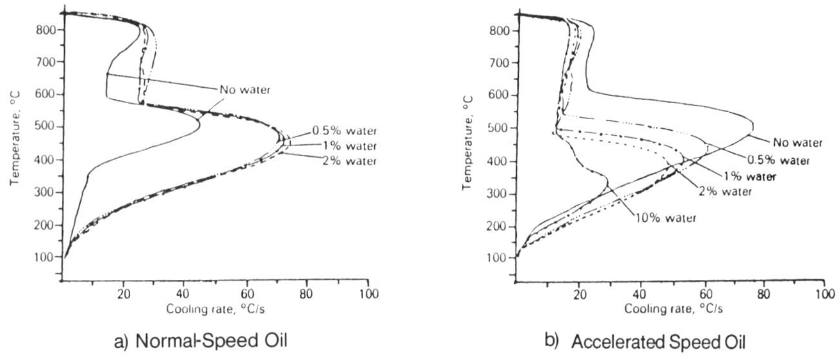

Fig. 6.2 Effect of water contamination on quench oil cooling rates

A quantitative test for water contamination involves titration of the oil with Karl Fisher chemical reagent to a electrometric endpoint (ASTM D 1744). This test is recommended for water levels of 50 to 1000 ppm (less than 0.1%). Because this titration requires Fisher reagent and a special apparatus, it is usually performed by quench oil suppliers or analytical laboratories, rather than by heat treaters. Higher levels of water contamination can be quantified by centrifugation of a sample and measurement of the volume of separated water according to ASTM D 4007.

Water in a quench oil can be removed by allowing the water to settle to the bottom of the tank and then draining. Alternatively, agitated oil can be carefully heated to 120 °C (250 °F). When the bath no longer foams or bubbles, the water has been completely removed (Ref 14).

Automated detectors are available to monitor oils continuously for water contamination. These systems are instrumented with an alarm set to detect critical levels of water concentration. Such systems are usually limited to sensitivities of ±0.5% water; however, several recent models have been reported to be sensitive to 0.1 to 0.3% water (Ref 18).

Neutralization Number. Quench oil oxidation results in the formation of carboxylic acids and esters. Similar compounds are also used as cooling rate accelerators. These acids and esters significantly affect the viscosity and viscosity-temperature relationship of the oil, which in turn affect quench severity. Carboxylic acids may also act as wetting agents and increase the quench rate by increasing the wettability of the quench oil on the metal surface (see Chapter 4).

Oxidation of an oil can be monitored by tracking changes in the neutralization number. Because the fresh oil may be either alkaline or acidic, depending on the additives present, the absolute value of the neutralization number itself is not indicative of quality. However, changes in the neutralization number from the initial condition may be used to indicate the degree of oxidation.

In some cases, the carboxylic acids produced immediately adjacent to the hot metal part are volatilized during quenching, while polymeric components remain in the oil. If the oxidation by-products are vaporized, oxidation may not be indicated directly by the neutralization number. However, increasing neutralization numbers generally signify that the amount of sludge is increasing.

Neutralization numbers should be used in conjunction with other measurements, such as viscosity and water content, to determine the chemical changes occurring in the quench oil. These analyses may be used to construct charts similar to that shown in Fig. 6.3 (Ref 5, 6). Such charts are an essential component in any quality control program.

The neutralization number or acid number is determined by titrating the acidity of a sample of known size with a known amount of standard base. The test is performed by dissolving a sample in a mixture of isopropanol and toluene in a burette and titrating it with a standard solution of potassium hydroxide (KOH) in alcohol. The end point can be determined colorimetrically with a pH-sensitive indicator. The acid number is determined in units of milligrams of KOH per gram of sample (mg/g). A standard sample

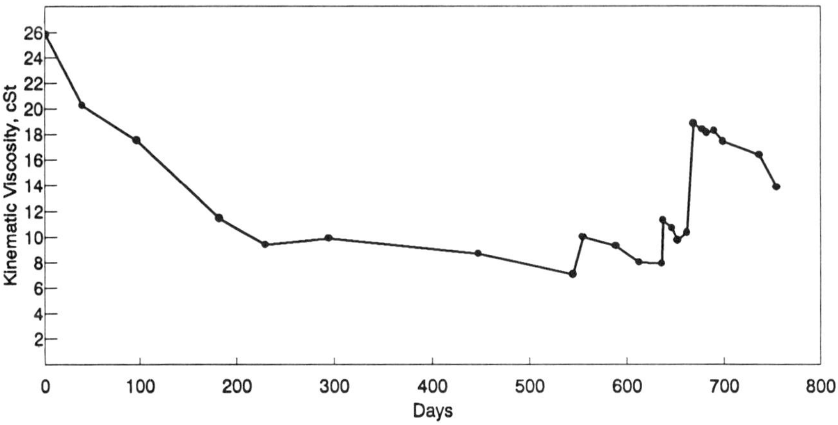

Fig. 6.3 Viscosity control chart

size of the solvent is titrated to give a "blank" value B. This allows correction for any acid present in the used quenchant that is not due to degradation. The used oil is dissolved in a standard amount of the solvent and titrated immediately. The acid number of the quench oil is calculated as:

$$\text{Acid number (mg of KOH/g of sample)} = \frac{(A - B)\, N \times 56.12}{W}$$

where A is the KOH solution required for titration of the sample (mL), B is the KOH solution required for titration of the blank (mL), N is the normality (meq. of base/g of solution) of the KOH solution, and W is the weight of the sample (g).

Saponification Number. Oil degradation may produce both acidic and ester degradation by-products. The neutralization number quantifies the amount of acidic degradation products in the oil, whereas the saponification number is a measure of the presence of esters or fatty materials in the oil. The saponification number of an oil is determined by heating a sample of the oil with a known amount of basic reagent and measuring the amount of reagent consumed. Because some quench oils are formulated with components that also have saponification numbers, it is necessary to monitor trends over time rather than to rely on an absolute value (Ref 9).

An increase in the saponification value indicates an increased propensity to sludge formation. It has been suggested that if the results of other tests of the oil are satisfactory, then saponification numbers below 3 mg KOH/g oil may be acceptable (Ref 17).

Determination of saponification number involves heating a known weight of the oil with a stoichiometric excess of alcoholic KOH in a solvent (methyl ethyl ketone). The technique requires an Erlenmeyer flask, condenser, hot plate, standard solutions of acid and base, methyl ethyl ketone, and a burette. The saponification reaction consumes some of the KOH, and the excess is titrated. The difference between the amount originally added and the amount titrated is the saponification number:

$$\text{Saponification number} = \frac{56.1N(V - V_1)}{W}$$

where N is the normality of the hydrochloric acid, V_1 is the volume of acid used in titrating the blank (mL), V is the volume of acid used in titrating the sample (mL), W is the weight of the sample (g), and 56.1 is the molecular weight of KOH. Saponification numbers should be considered in conjunction with the neutralization number to quantify both ester and acidic by-products.

Advances in the Analysis of Used Oil. Although analytical methods such as neutralization and saponification number continue to be important in determining the condition of a used oil, more advanced instrumental methods such as infrared (IR) spectroscopy and gel permeation chromatography (GPC) are being increasingly applied. IR spectroscopy is an excellent method for determining the chemical identity of functional groups (e.g., alcohols, acids, esters, unsaturation, etc.) that may be present in the quench oil as degradation by-products, contaminants, or additives. Figure 6.4 illustrates typical IR

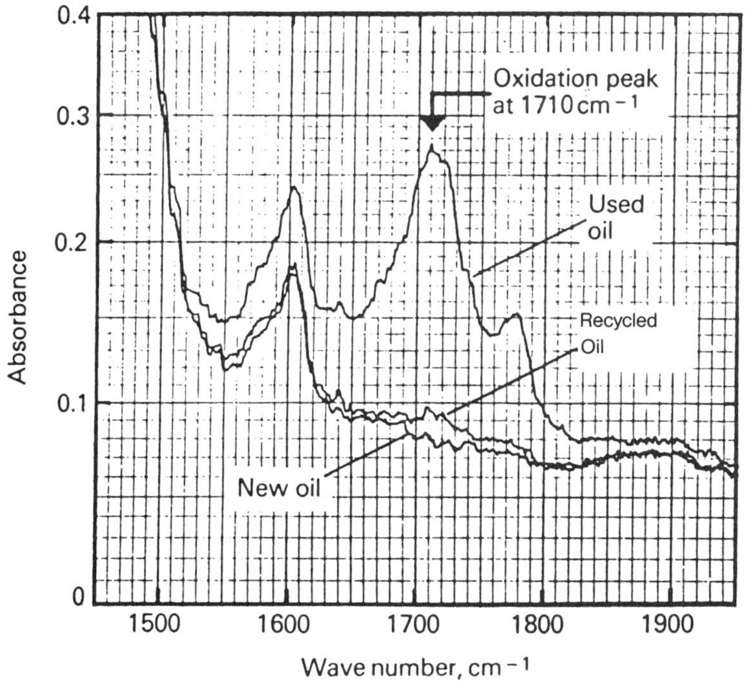

Fig. 6.4 Oil analysis by IR spectroscopy

spectra for an oil that has undergone oxidation and for the initial unoxidized oil. A regenerated oil is also shown; oil regeneration will be discussed later (Ref 19).

Gel permeation chromatography is another analytical method now finding broad application for quench oil analysis (Ref 20). GPC separation is based on molecular size (molecular weight). This technique uses a column packed with a microporous, cross-linked polymer, through which the sample is passed. The smaller molecules enter the pores of the cross-linked polymer packing and are retained in the column for a longer period of time than the larger molecules, which do not enter the pores and hence move with the liquid (solvent) flowing through the column. The size-exclusion property of the pores is the basis of the separation. The molecules are detected as they exit the column, and the resulting response of the detector with respect to elution time yields (with proper calibration) the molecular weight and distribution of the components of the sample. This analysis requires special instruments, but the comparison of a new and a used oil (or polymer) provides unambiguous information on degradation.

Equation 6.8 shows that oil oxidation produces larger molecular species by chain extension and cross-linking, which is why oil viscosity increases with use. Thus, a used oil would be expected to possess a higher average molecular weight than a fresh oil. Research has shown that oil oxidation by-products with molecular weights greater than approximately 50 to 1000 are generally insoluble in oil (Ref 20). This is one reason why sludge, which typically has a molecular weight substantially in excess of 1000, precipitates from a quench oil during use. Figure 6.5 shows a GPC trace obtained on a typical hydraulic oil. Similar traces would be obtained with a quench oil (Ref 20).

Ash content is determined by heating the quench oil under conditions that burn off the organic components but leave metallic species such as metal oxides or hydroxides (ASTM D 482). Although straight mineral oils have very low ash values, many of the formulated oils contain metallic components that contribute to ash. Again, ash contents are only useful when monitored over time or if the ash content of fresh quench oil is known. If the ash content of a bath filled with a formulated oil is decreasing, it is likely that an ash-containing additive is being removed by dragout or some other process. (Some quench oil additives are organometallic.) If the ash value is increasing, the additive is either accumulating in the bath or metallic contamination is increasing, perhaps in the form of scale accumulation.

Ash content is determined by igniting and burning a sample in a weighed vessel and then heating in a muffle furnace at 775 °C (1425 °F). The ash content of the sample is calculated as:

$$\text{Ash\%} = \frac{w}{W} \times 100$$

where w is the weight of the ash (g) and W is the weight of the sample (g).

Conradson Carbon Residue. One of the greatest problems encountered with the use of a quench oil is the accumulation of sludge. Although the various analyses described previously may indicate that a quench oil is adequate for continued use, the amount of sludge buildup in the tank may demand that the system be drained and cleaned. Cleaning and sludge disposal are growing problems for the heat treating industry. Therefore, determination of the sludge-forming potential of a quench oil prior to use is becoming increasingly important.

One method for quantifying sludge-forming potential involves determination of the Conradson carbon residue value (Ref 9)—a measure of the amount of polymeric material remaining in an oil after heating to elevated temperatures in the absence of sufficient oxygen to burn all of the organic compounds present. The value of this test for predicting the bath condition has been significantly reduced by the development of specially formulated quench oils. Many additives currently used give an elevated value

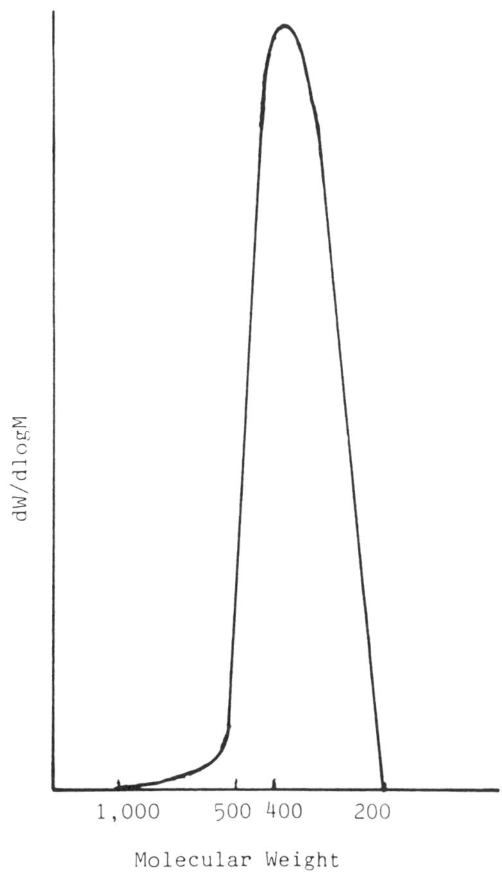

Fig. 6.5 Oil analyses by gel permeation chromatography

in the Conradson carbon test (due to the formation of ash), but these same additives also reduce the formation of deposits in use (Ref 9).

The Conradson carbon residue is determined by placing a weighed sample of the oil in an iron crucible (ASTM D 189). The crucible is heated with a Meeker-type gas burner to a sufficiently high temperature to evaporate and burn the oil. The sample is further heated until the bottom and sides of the crucible are cherry red and is held at this temperature for 30 min. The crucible is then cooled and weighed. The amount of tar remaining in the crucible and the amount of original sample define the Conradson carbon residue value:

$$\text{Carbon residue} = \frac{A \times 100}{W}$$

where A is the weight of the carbon residue (g) and W is the weight of the sample (g). Alternative tests, such as the panel coker test, yield similar information (Ref 21).

Precipitation Number or Sludge. Sludge formation in a quench oil is caused by oxidation of various components, leading to polymerization and cross-linking reactions. These cross-linked and polymerized by-products are sufficiently high in molecular weight to cause them to be insoluble in the oil. Other sources that contribute to sludge are dirt and carbon contamination.

Sludge can plug filters and foul heat-exchange surfaces. The loss of heat-exchanger efficiency may result in overheating, excessive foaming, and possibly a fire (Ref 6). Increasing sludge formation often indicates increasing oxidation of the oil. In addition, sludge can adsorb on a part, causing decreased and nonuniform film heat transfer during the quench. This causes thermal gradients sufficiently high to cause cracking or distortion. Sludge has been identified as a leading cause of problems in oil quench baths, second only to water (Ref 9).

If an oil is not seriously degraded, the sludge can be removed by filtration performed by a batch operation or by continuous centrifugal separation (Fig. 6.6). It is important to maintain particulate contamination to ≤1 μm to optimize quench performance (Ref 8).

Sludge levels of 0.2% or greater may produce staining of normally bright metal surfaces (Ref 9). The formation of sludge from cross-linking reactions is only one possible degradation reaction. Sludge formation is typically accompanied by chain-scission reactions, which also produce volatile by-products. Volatile by-product formation reduces both the open and closed flash points and thus causes a simultaneous increase in fire potential.

Sludge formation is undesirable not only from a process and quenching perspective, but is also an indication of significant chemical changes occurring in the oil. The continual reduction of flash points leads to increasingly unsafe conditions. The viscosity of a quench bath also changes with the formation of sludge, affecting both heat transfer and quench severity. The amount of sludge can be quantified by adding naphtha solvent

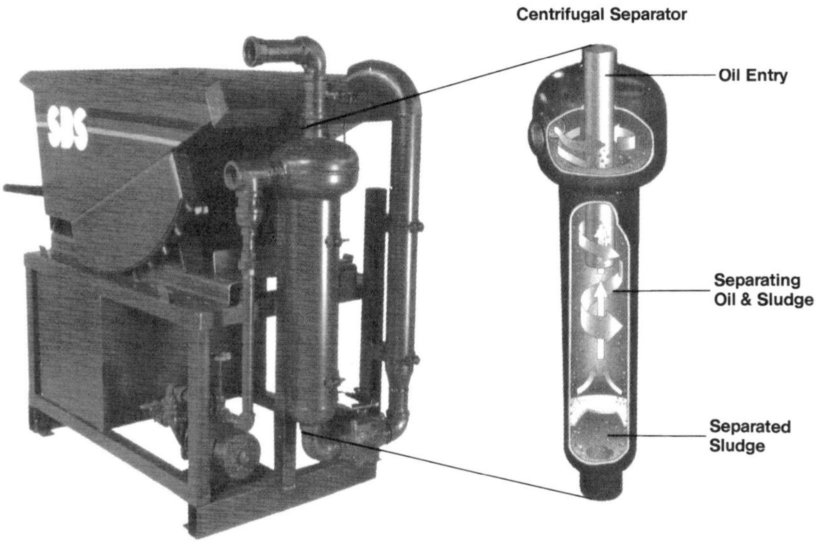

Fig. 6.6 Continuous centrifugal oil sludge separator. Courtesy of SBS Corporation, Inc.

to the oil sample and determining the volume of precipitate (sludge) after centrifuging (ASTM D 91 and D 2773).

Specific gravity depends on the chemical composition of the base stock used to formulate a quenchant. As described in Chapter 4, the chemical composition of an oil (e.g., the paraffinic/naphthenic ratio) affects its oxidative stability. Therefore, the specific gravity of a quench oil is an indirect measure of its potential oxidative stability. However, specific gravity has only limited utility in monitoring the quality of a used quench oil.

The ASTM D 287 test for specific gravity requires the use of either a hydrometer and an accurate thermometer or a thermohydrometer. Because specific gravity is temperature dependent, the temperature of the oil at the time of measurement must be determined precisely.

Quench Severity Measurement

All of the techniques described thus far are conducted to determine relative changes in the physical and chemical properties of an oil quenchant. Variations in physical properties are caused by oxidative degradation of the oil and/or by contamination by external sources (e.g., water).

Traditionally, physical property characterization has been extremely useful for identifying the changes that occur in a quench oil during use. However, these techniques do not directly indicate the quench severity variations caused by these fluid chemistry changes. Therefore, various laboratory tests have been developed for use in conjunction with physical property characterization. These tests include Jominy hardness (Ref 22), cross-sectional hardness survey (H-bar) (Ref 3), and the GM quenchometer (Ref 7). Of these, the GM quenchometer has traditionally been the most common method used by quenchant suppliers.

Today, cooling curve analysis (see Chapter 3) is increasingly being used to determine quench severity because of its substantial advantages over methods such as the GM quenchometer (Ref 23, 55). One of the principal advantages of cooling curve analysis is that it provides the total time-temperature cooling profile. This is illustrated by Fig. 6.7, which shows three cooling curves (A, B, and C) that exhibit the same GM quenchometer times but different cooling profiles. This may be especially significant if an alloy with relatively low hardenability is being quenched; it may be more difficult to harden the alloy in the quenchant represented by cooling curve C than in the quenchant represented by cooling curve A.

Recently, von Bergen (Ref 5, 6) described the use of cooling curve analysis to monitor the effect of oil oxidation during use on cooling rates (Fig. 6.8) using a standard analysis procedure (Ref 25). Similar results have been obtained with different oils by

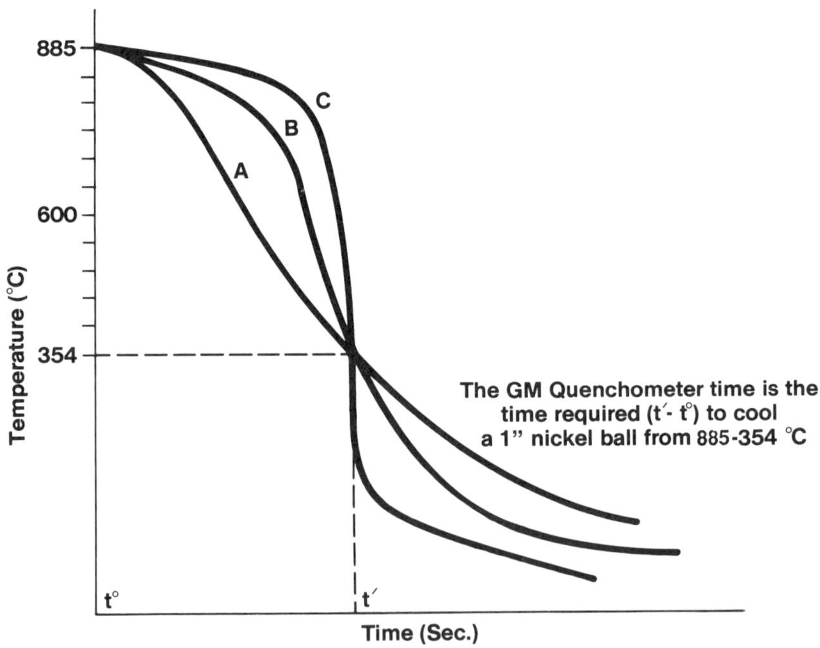

Fig. 6.7 Comparison of the GM quenchometer test with cooling curve analysis

Bashford and Mills (Fig. 4.24) (Ref 26), Hampshire (Fig. 4.26) (Ref 27), Hewitt (Fig. 4.27) (Ref 15), and Tagaya and Tamura (Fig. 4.29) (Ref 11).

Tagaya and Tamura's work (Fig. 4.29) (Ref 11) was particularly interesting. An attempt was made to correlate cooling rate variations with the extent of oxidation as determined by conventional analytical techniques. Although statistical correlation equations were developed, they are specific to the laboratory apparatus used to perform the accelerated degradation analysis. However, similar correlations could be developed for a particular commercial quench oil system.

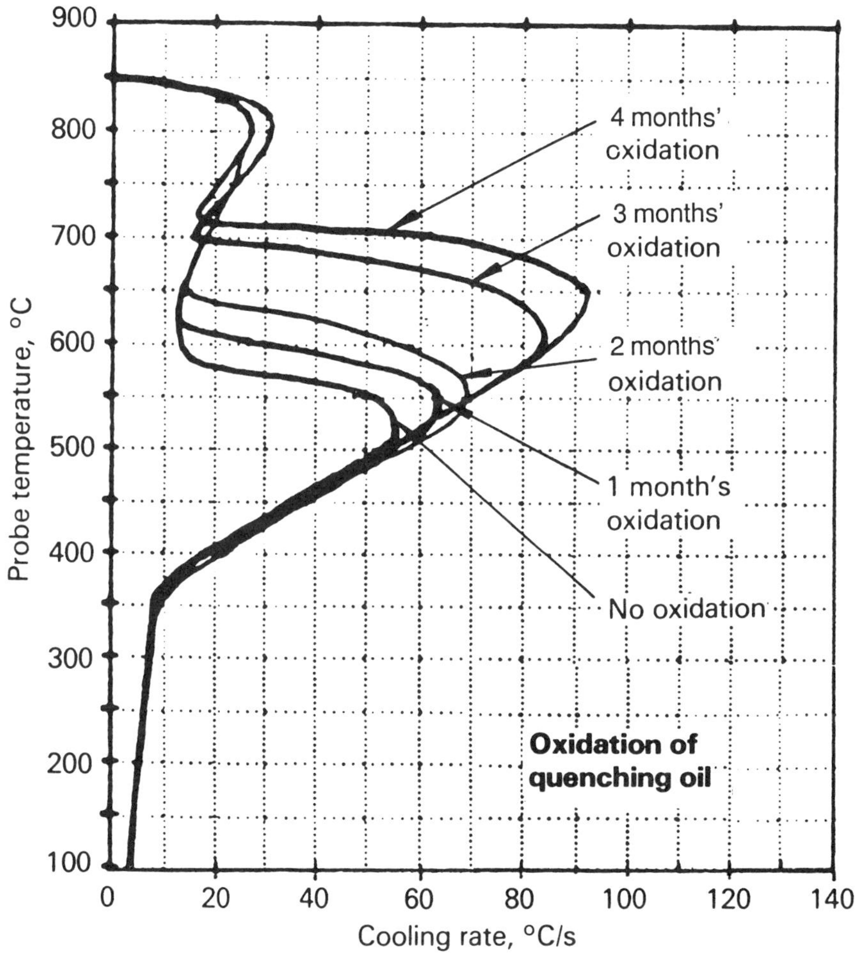

Fig. 6.8 Effect of quench oil oxidation on cooling rates

Hewitt (Fig. 6.3) and von Bergen (Fig. 6.9) used cooling curve analysis to evaluate the effect of varying levels of water contamination on different quench oils. It was shown that increasing concentrations of water produced corresponding cooling rate increases for a conventional (normal-speed) quench oil. However, Hewitt (Ref 15) obtained the opposite effect for an accelerated quench oil: a cooling rate decrease with increasing water concentration.

Hasson (Ref 28) showed that cooling curve analysis is a vital part of a maintenance program to monitor variations in quench severity of an oil quench bath during use.

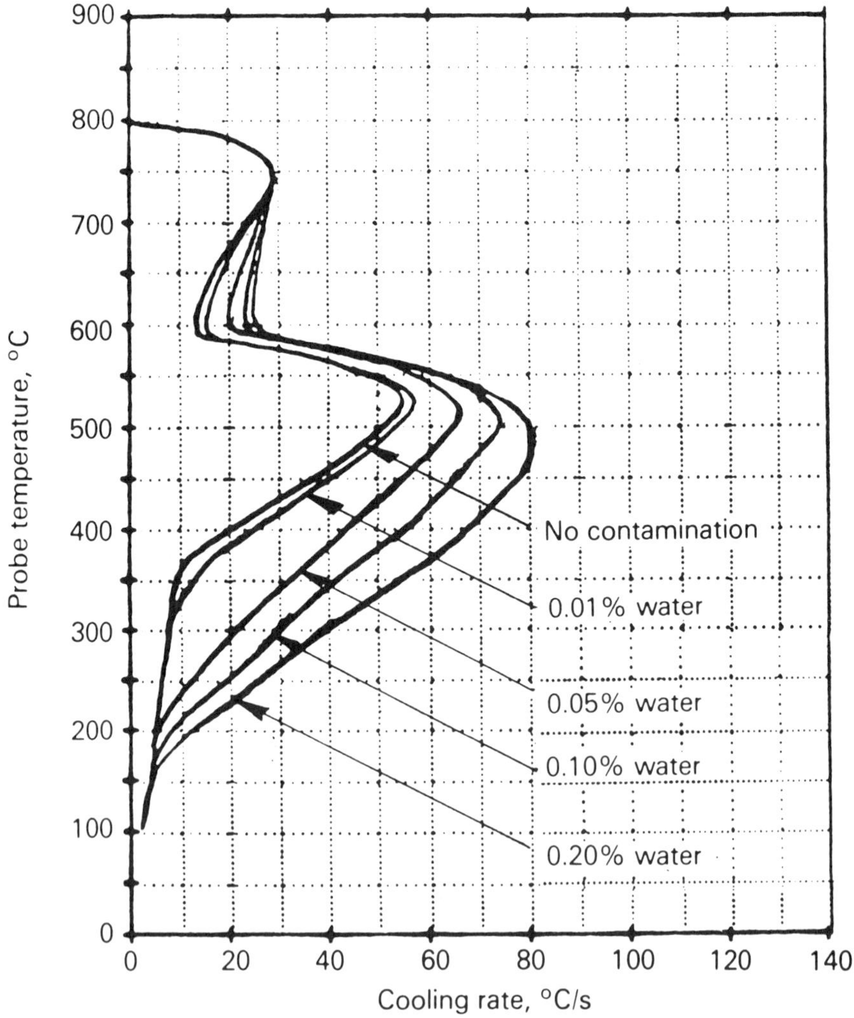

Fig. 6.9 Effect of water contamination on a normal-speed quench oil

Cooling rates were used to classify the relative quench severity of an oil and were also correlated with more traditional GM quenchometer test data (Table 6.2).

It has been shown that oil chemistry variations cause changes in the physical properties of an oil. Physical property variations can be measured by a variety of tests, such as viscosity, flash point, neutralization number, and so on, and their impact on quench severity determined by cooling curve analysis. However, although the effect of physical property variations on quench severity can be quantified by a standard laboratory cooling curve test, other variables that may affect quench severity cannot be readily monitored by such a test.

For example, the only unambiguous method of determining the effects of agitation, bath temperature, and so forth in the quench tank during use is to perform a cooling curve test utilizing the quenching conditions that exist within the tank itself. This is a complex task, however, and should only be done with appropriate control experiments. Brennan (Ref 16) has studied the effect of agitation variation under commercial quenching conditions by cooling curve analysis. Using a 25 mm (1 in.) Inconel 600 probe (Fig. 3.12) and a fluid velocity probe, it was possible to correlate cooling rate variation with agitation and hardness of various steels. Similar studies have been conducted by Segerberg (Ref 30).

Fluid flow in a quench tank can be measured by a number of methods with varying success. One of the classic techniques utilizes a pitot tube (Ref 31). The pitot tube is relatively easy to use; however, its presence actually disrupts the flow pattern (Ref 32).

The Mead velocimeter (Ref 33) is another low-cost, simple instrument that is often used by heat treaters to measure flow. When properly calibrated, it is capable of providing reasonably accurate flow-rate information.

Quench Oil Contamination

The various tests described above are used to monitor variations in physical and chemical properties of quench oils with respect to time. It is also important to monitor for potential bath contamination. Contaminants, such as water and hydraulic oils, may either increase or decrease cooling rates. Soaps, fats, greases, and salts usually increase

Table 6.2 Cooling rate classification of quench oils

Oil classification	Maximum cooling rate(a) °C/s	Maximum cooling rate(a) °F/s	GM quenchometer time(b), s
Fast	140-165	250-300(c)	7-9
Medium	110-140	200-250	10-12
Slow	90-110	165-200	13-20

(a) Cooling rates were obtained with a 10 mm (0.4 in.) austenitic stainless steel probe with a laboratory servodyne propeller agitation device at 750 rev/min. The oil temperature was 65 °C (150 °F) (Ref 29). **(b)** This test was conducted according to ASTM D 3520 using a nickel ball and 27 °C (80 °F) oil. **(c)** This would correspond to a cooling rate of 75 to 90 °C (170 to 190 °F) using the proposed Wolfson standard method with a 25 mm (1 in.) Inconel 600 probe described in Ref 25 (Ref 29).

the quench severity of conventional (unaccelerated) oils and decrease the quench severity of accelerated oils (Ref 4). These same contaminants may also produce sooty or flaky deposits on quenched parts and promote foaming. Volatile contaminants may also cause foaming and enhance the vapor blanket phase during the quench, retarding the initial cooling rate and thus decreasing overall quench severity. The furnace atmosphere may interact with the accelerating additives in a quench oil, resulting in a slower quench (Ref 15).

Quench Oil Recovery

Ecological constraints have been increasingly placed on the use of quench oils in the heat treating industry (Ref 34, 35). The reasons for these constraints are varied, involving such air, water, and human toxicological concerns as:

- Smoke production (both within the plant and ventilated to the outside)
- Fire safety
- Use and disposal of cleaning solvents
- Water pollution from plant effluents

These concerns have driven the industry to search for alternative quench media and to develop procedures for recycling quench oils (Ref 19, 36, 37). Recycling procedures include trapping and reuse, additive readdition, and oil re-refining with subsequent additive addition.

Trapping and Reuse. One of the simplest methods of recycling a quench oil is to trap or skim it from a rinse tank and then put it back in the quench oil reservoir. An example of this process was described by Kopke (Ref 37), who showed that waste was reduced from 2300 to 230 L (600 to 60 gal) per week, producing a cost savings of $127,000 per year. The installation cost of the recycling system was $15,000.

Additive Addition. One method of extending the lifetime of a quench oil is by the readdition of additives to stabilize the oil, reduce ash formation, and accelerate cooling rates (Ref 36). For example, it has been shown that additive dragout typically results in a reduction in the desired cooling rate (Fig. 4.27) for an accelerated quench oil.

The use of additive readdition to salvage a used accelerated quench oil is shown in Fig. 6.10 (Ref 29). In this example, the customer had experienced a substantial cooling rate reduction from the original fast oil. Additive readdition to the used oil provided a quench oil with improved cooling rates that were intermediate between a medium oil and the original fast oil. Interestingly, simple additions did not produce a fast oil identical to the original. The base oil also probably underwent at least some oxidation, which would have resulted in a different viscosity-temperature profile compared with the original base oil.

Refining of Used Quench Oils. When the base quench oil has undergone substantial oxidation, it must be re-refined before it can be reused (Ref 19, 36). However, the oil may exhibit a different cooling rate profile after refining; this is partially due to selective

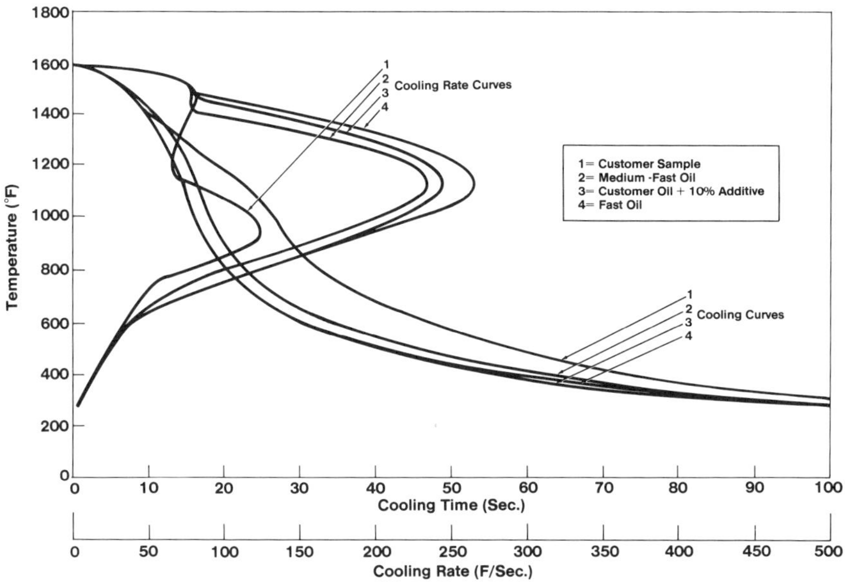

Fig. 6.10 Regeneration of used quench oil by additive addition. Courtesy of E.F. Houghton & Company, Inc.

oxidation of the many molecular components of the base stock (Fig. 4.3). Because the viscosity-temperature profile is affected by the chemical constituents of the oil, it is not surprising that the cooling curves of the original oil and re-refined oil would be different. Fortunately, current additive technology has permitted the formulation of quench oils that exhibit cooling profiles essentially equivalent to those obtained with the re-refined oils (Fig. 6.11) (Ref 19).

Polymer Bath Maintenance

Aqueous polymer quenchants are also susceptible to degradation reactions. Some of these processes have been discussed in Chapter 5 and include both oxidative and mechanical shear degradation (Ref 1). Prior work by von Bergen (Ref 5, 6), Segerberg (Ref 38), Hilder (Ref 39), Hibi (Ref 40), and Kopietz (Ref 41) has detailed the effects of degradative processes on quenching performance.

The effects of degradative reactions on the properties of polymer quenchants are manifested differently from those described for oil quenchants. For example, polymer degradation has relatively little, if any, impact on flash point. Sludge formation with aqueous polymer quenchants by oxidative degradation rarely occurs. However, aqueous polymer quenchants are more susceptible to biodegradation reactions (Ref 40), especially by anaerobic processes (Ref 42). Therefore, many of the characterization procedures used for aqueous polymer quenchants are different from those used for oils.

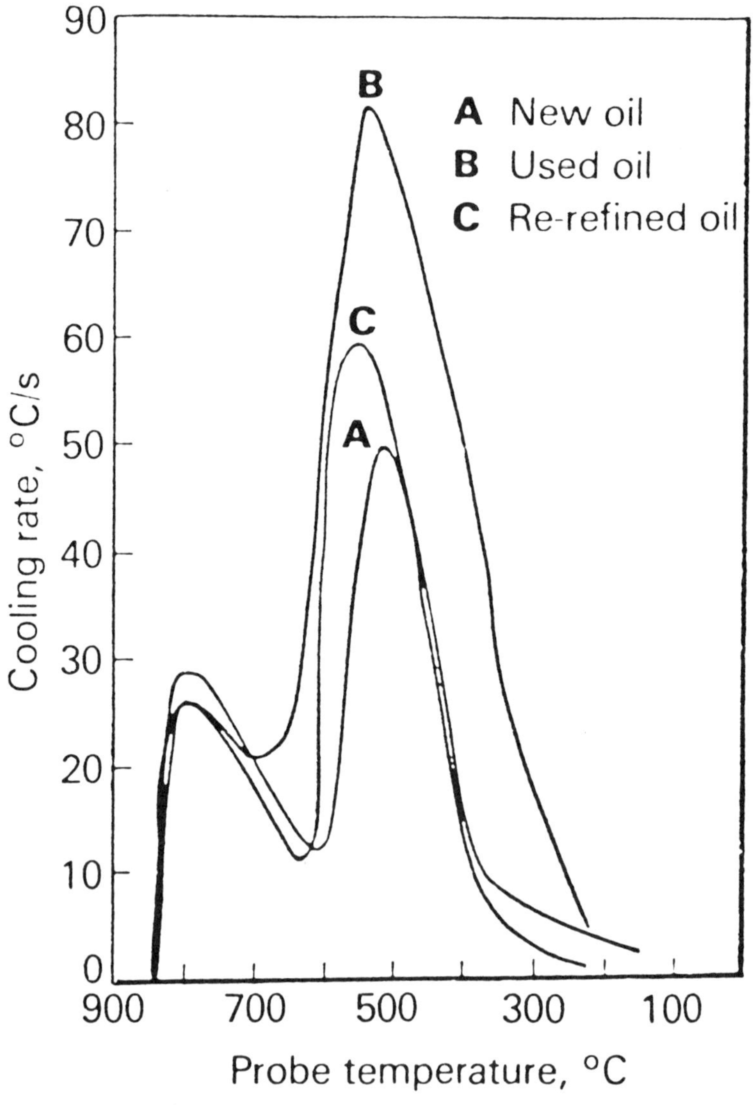

Fig. 6.11 Cooling curve properties of new, used, and re-refined oil

Procedures that are generally applicable to polymer systems (Ref 24, 43, 44) are shown in Table 6.3. Most suppliers will perform the tests most appropriate to their particular polymer quenchant as a free service. Only the most common testing procedures will be described here.

Table 6.3 Methods for polymer quenchant analysis

Test	Procedure
Concentration viscosity	ASTM D 445
Refractive index	ASTM D 1749, D 1218
Salt content	Conductance
Corrosion protection	pH additive analysis
Biological activity	"Dipstick" test

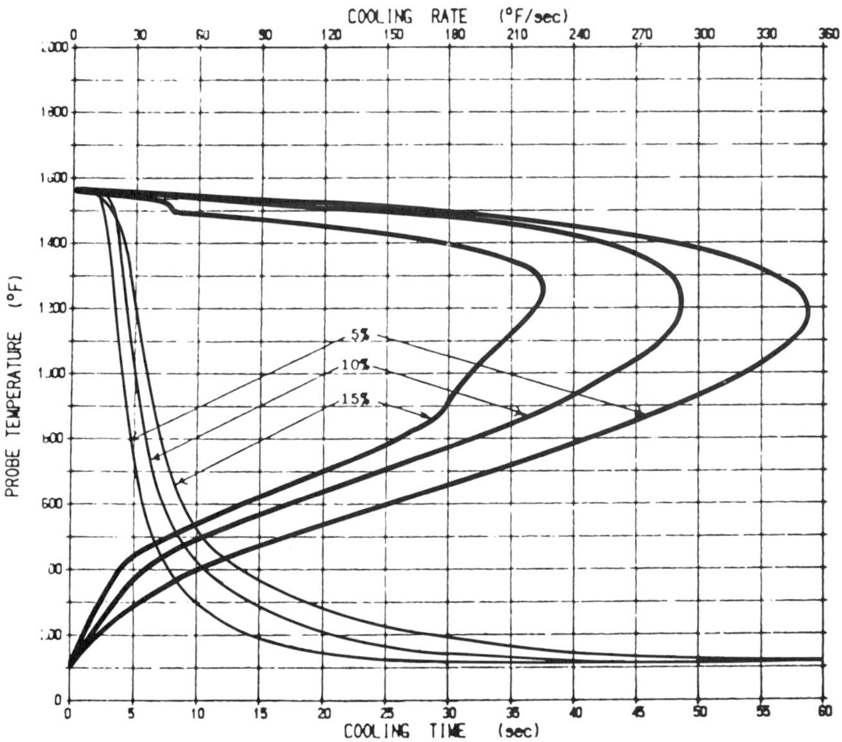

Fig. 6.12 Effect of polymer concentration on cooling rates

Polymer Concentration

The cooling rate produced by an aqueous polymer quenchant is critically dependent on the concentration (Fig. 6.12) and the condition (molecular weight and oxidation) (Fig. 6.13) of the polymer. Polymer loss occurs primarily through dragout and, to a lesser extent, polymer degradation (Ref 38, 40, 43).

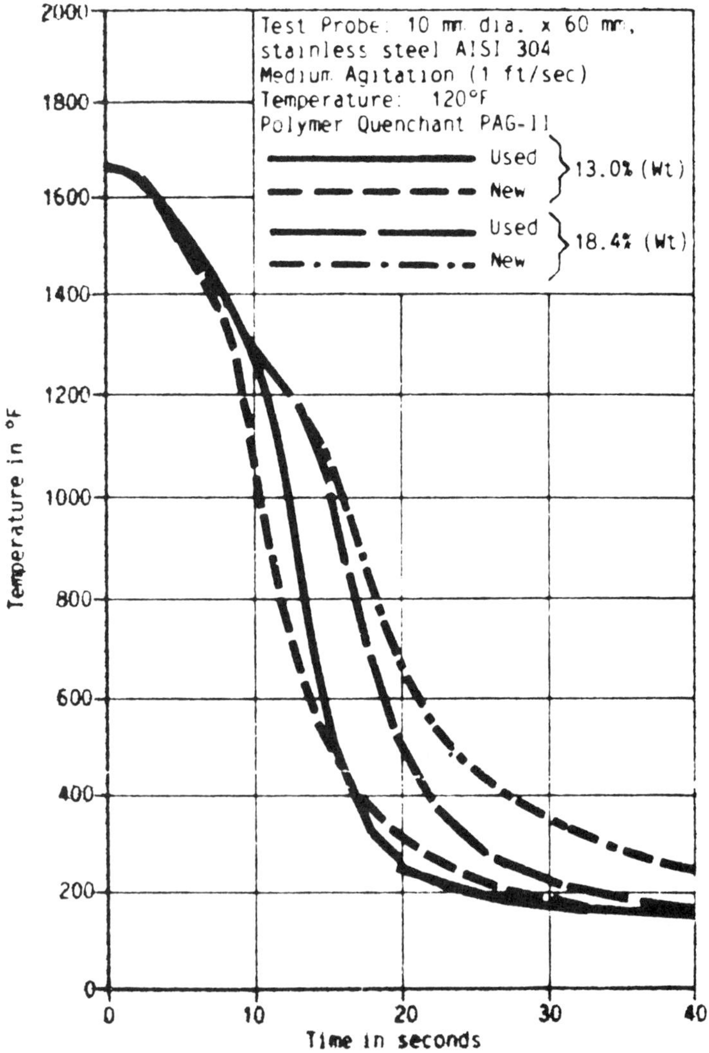

Fig. 6.13 Comparison of fresh and used PAG quenchant

Refractive Index. The most common method of measuring polymer concentration in the heat treating shop is by refractive index (n_D) (Ref 43, 44), which is linearly related to polymer concentration (Fig. 6.14). A detailed procedure for refractive index determination is provided in ASTM Standards D 1747 and D 1218. In the heat treating shop, this test is easily performed using a hand-held refractometer (Fig. 6.15). One drop of the aqueous quenchant solution from the bath is placed on the lens of the instrument, which is then held up to the light to read the refractive index from the optical scale. This value is converted to a concentration using a calibration plot or an appropriate tabulation of concentration versus refractive index, which is different for each quenchant. It is important to note that a different calibration must be used for each individual quenchant.

A determination of the quenchant concentration by refractive index is primarily limited to quenchants derived from polyalkylene glycol (PAG). Other polymers used as quenchants, such as polyvinyl pyrrolidone (PVP) and polysodium acrylate (PSA), are usually too dilute to provide acceptable sensitivity for polymer concentration measurements by refractometry.

Refractive index measurements are also influenced by factors other than polymer concentration (Ref 38, 44). Degradation of the polymer and contamination of the quench bath by salts or organic materials will result in refractive index variations that do not necessarily reflect actual changes in polymer concentration. This is illustrated in Fig. 6.16,

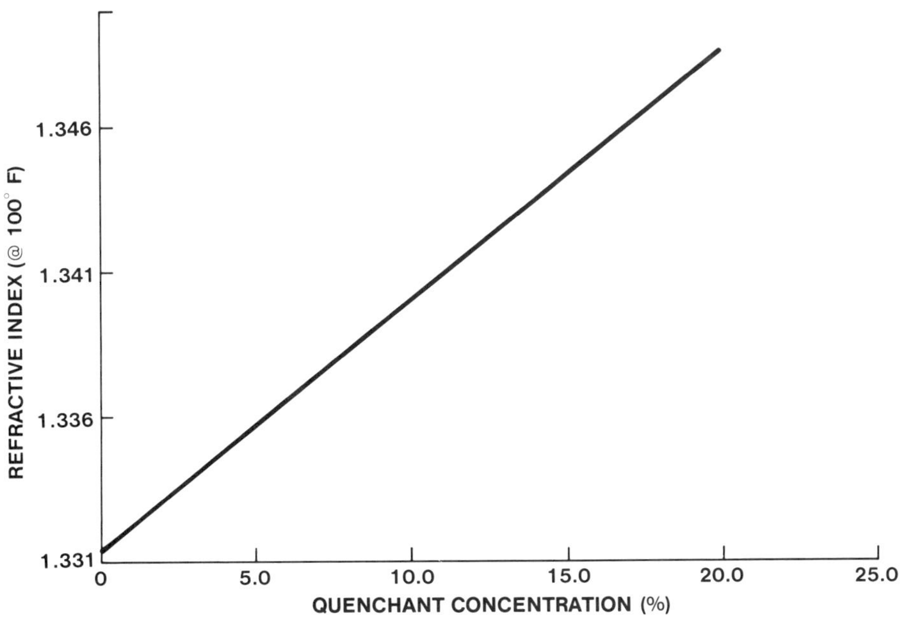

Fig. 6.14 Refractive index versus quenchant concentration

Fig. 6.15 Use of a hand-held refractometer. Courtesy of Leica Corporation

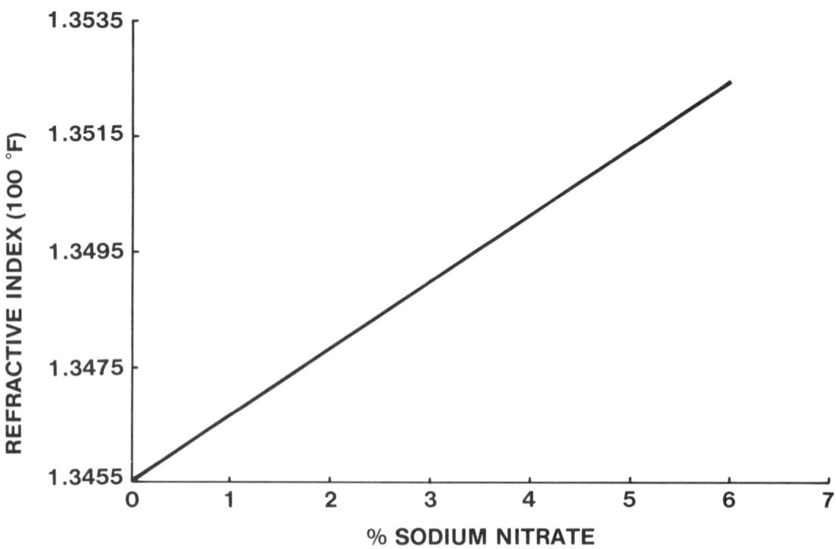

Fig. 6.16 Sodium nitrate concentration versus refractive index of a polymer quenchant at 40 °C (100 °F)

where the refractive index is shown to increase with increasing salt concentration, although the polymer quenchant concentration remains constant. Therefore, although the refractive index provides a simple and rapid determination of quenchant concentration, it is essential that periodic validation of the quenchant concentration be obtained by another procedure, such as viscosity.

The viscosity of a polymer depends on molecular size (molecular weight), concentration, and temperature (Fig. 6.17). Therefore, it is essential that the viscosities of fresh and used polymer quenchants be compared at the same temperature. Viscosity is less affected by salt concentration (Fig. 6.18) when compared with the refractive index (Fig. 6.16).

The viscosity of an aqueous polymer can be determined using the procedure described in ASTM D 445. This method is identical to that discussed earlier for oil.

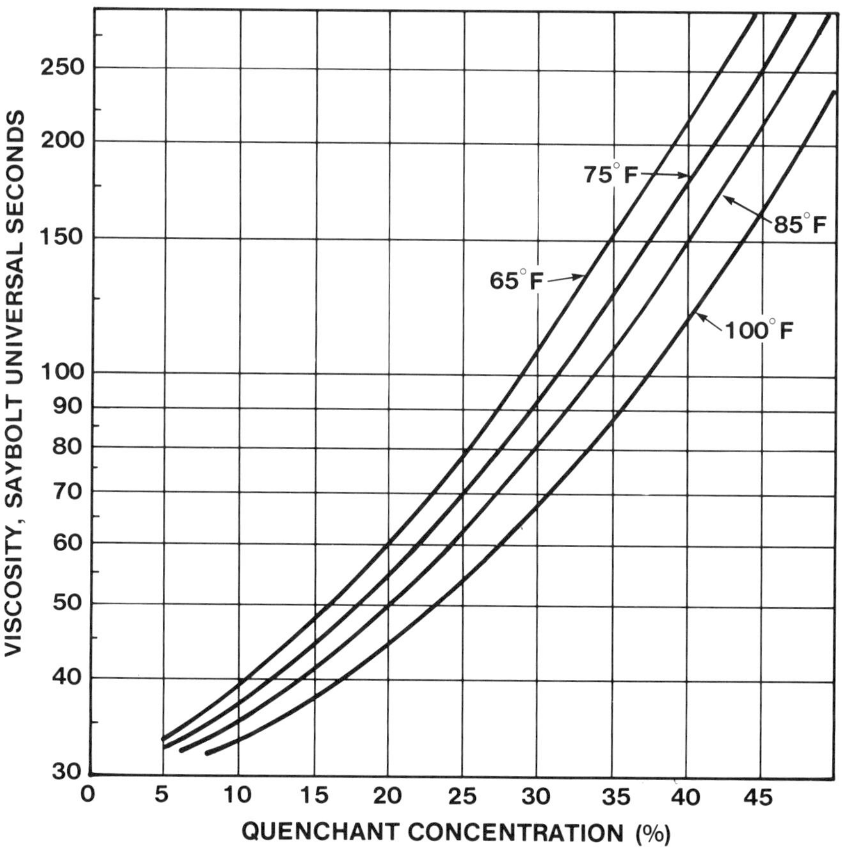

Fig. 6.17 Viscosity/concentration relationship for PAG polymer quenchant

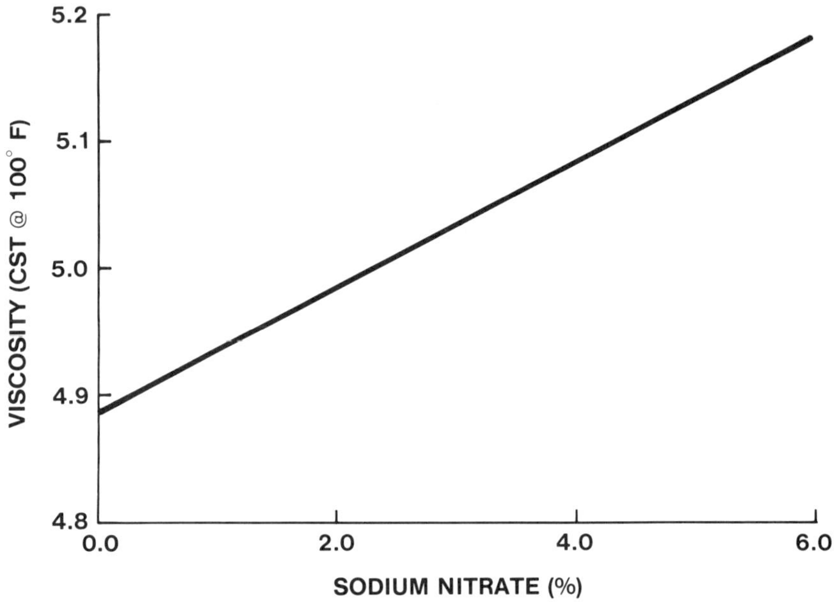

Fig. 6.18 Viscosity versus sodium nitrate concentration

Table 6.4 Effect of salt contamination on cooling rates of an aqueous polymer quenchant(a)

Sodium nitrite concentration, %	Maximum cooling rate °C/s	Maximum cooling rate °F/s	Cooling rate at 343 °C (650 °F) °C/s	Cooling rate at 343 °C (650 °F) °F/s
0.0	39.0	70.2	25.3	45.5
3.0	54.0	97.2	31.8	57.2
6.0	65.8	118.4	35.0	63.0
Water(b)	61.8	111.2	33.0	59.4

(**a**) Data was obtained using a 25 × 50 mm (1 × 2 in.) cylindrical type 304 stainless steel probe instrumented with a type K thermocouple inserted at the geometric center. Agitation was provided by horizontal flow of the probe surface at 23 L/min (6 gal/min) at 40 °C (100 °F). (**b**) Distilled water containing no polymer

Viscosity and refractive index measurements of quenchant concentration are complementary procedures. Refractive index is more sensitive to salt contamination effects, and viscosity is primarily affected by polymer degradation. Taken together, these measurements provide an excellent insight into the condition of a polymer quenchant (Ref 45).

Generally, refractive index is not suitable for measuring the polymer concentration in aqueous quenchants formulated with low concentrations of relatively high-molecular-weight polymers such as PSA and PVP. Viscosity is the preferred method of concentration measurement for these polymers.

The salt content of a polymer quenchant bath can increase during use. Salts can be introduced either from hard water or by contamination from salt baths. Salt contamination can significantly affect quenchant cooling rates (Table 6.4). To minimize contamination

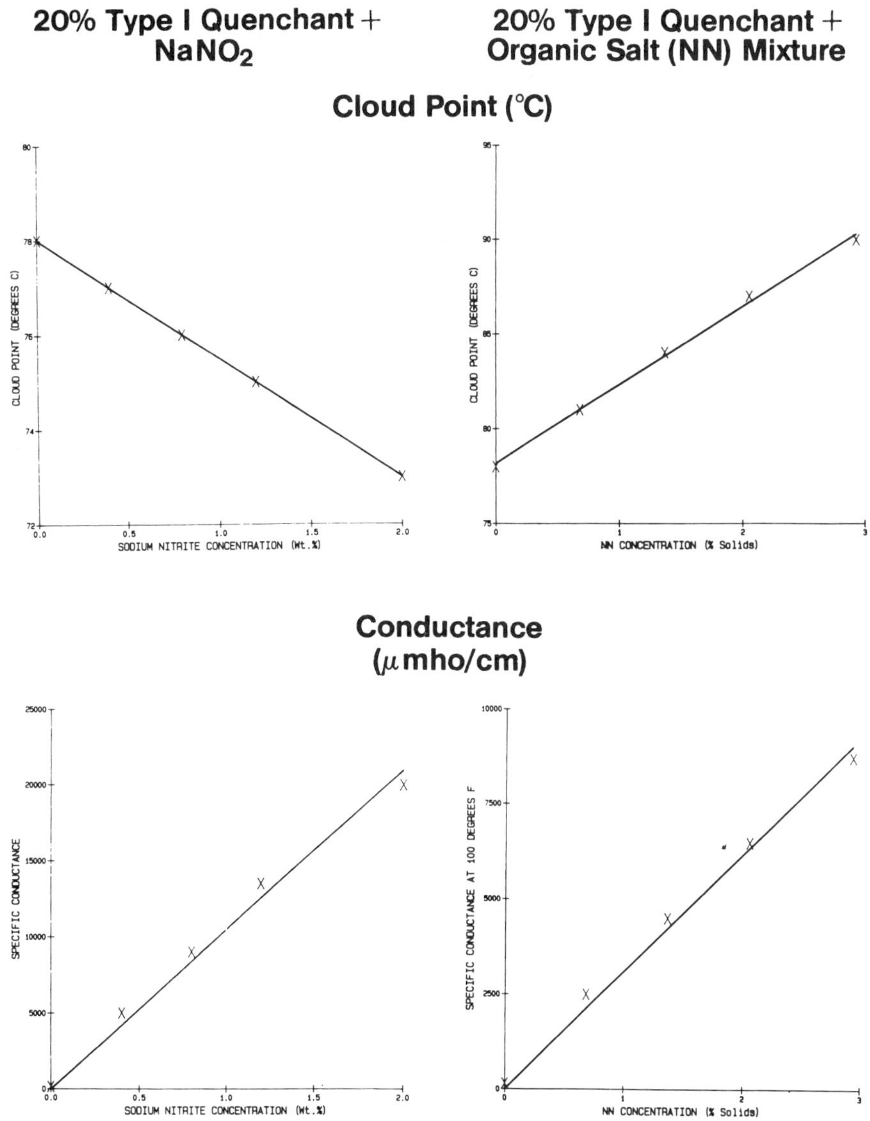

Fig. 6.19 Comparison of the physical and cooling properties of a 20% solution of a type I quenchant spiked with sodium nitrite and an organic salt non-nitrite (NN) blend **(continued)**

from hard water salts (e.g., iron, calcium, magnesium, etc.) (Ref 54), the makeup water for the quench bath should be distilled or at least deionized. Ion exchange and reverse osmosis are common deionization techniques.

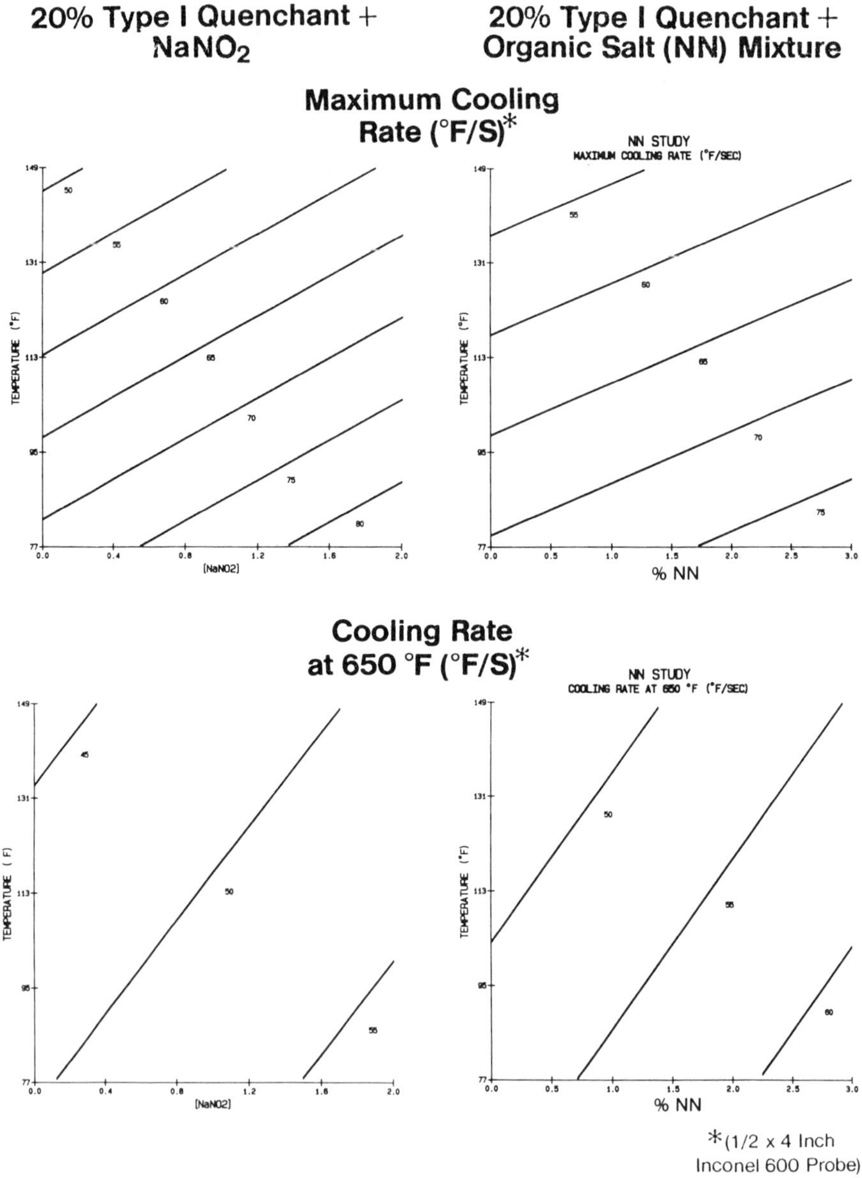

Fig. 6.19 Comparison of the physical and cooling properties of a 20% solution of a type I quenchant spiked with sodium nitrite and an organic salt non-nitrite (NN) blend

Effects on both the physical and the cooling properties of a quenchant will vary with a particular salt and its concentration. Figure 6.19 compares the effects of two types of salts commonly used as corrosion inhibitors: sodium nitrite and an organic salt blend. (Various organic amine/carboxylic acid salt combinations are used as alternatives to sodium nitrite in non-nitrite-containing polymer quenchants.) It is apparent that increasing concentrations of sodium nitrite increase the temperature required for thermal separation of the solution (cloud point), whereas the addition of increasing concentrations of the organic salt blend has the opposite effect. As expected, both salts increase the conductance of the solution and cooling rates.

Two other inorganic salts, sodium nitrate and sodium chloride, are compared in Table 6.5. The data show that increasing concentrations of both salts decrease the cloud point, although the magnitude of this effect depends on the particular salt. As expected, increasing salt concentrations increase the conductance of the solution. One notable difference between the two salts is that sodium nitrate increases cooling rates, whereas sodium chloride has the opposite effect.

The general effects of sodium nitrite, sodium nitrate, and potassium nitrite on a type I quenchant (Ref 46) are compared in Fig. 6.20. The effects of class I and class II salts

Table 6.5 Comparative effects of sodium nitrate and sodium chloride on physical properties and cooling rates of a 20% solution on type I(a) quenchant

	$NaNO_3$, %			NaCl, %				
Property	**0.0(b)**	**3.0**	**6.0**	**0.0(b)**	**0.5**	**1.0**	**3.0**	**6.0**
Physical properties								
Viscosity, cSt at 40 °C (100 °F)	4.9	5.0	5.2	4.9	4.9	5.0	5.8	5.0
Cloud point, °C (°F)	76.0 (169.0)	69.0 (156.0)	64.0 (147.0)	76.0 (169.0)	68.0 (154.0)	66.0 (151.0)	56.0 (133.0)	49.0 (120.0)
Conductance, µmho at 25%	5025	>20,000	>20,000	5025	12,000	16,000	$>10^5$	$>10^5$
Refractive index at 40 °C (100 °F)	1.3455	1.3490	1.3525	1.3455	1.3470	1.3480	1.3500	1.3525
Cooling curve properties(c,d)								
Maximum cooling rate, °C/s (°F/s)	39.0 (70.2)	54.0 (97.2)	65.8 (118.4)	35.0 (63.0)	30.2 (54.4)	29.5 (53.1)	24.4 (43.9)	20.8 (37.4)
Cooling rate at 343 °C (650 °F) °C/s (°F/s)	25.3 (45.5)	31.8 (57.2)	35.0 (63.0)	24.6 (44.3)	26.3 (47.3)	26.5 (47.7)	20.9 (37.6)	19.5 (35.1)

(**a**) Type I quenchants are defined in Ref 46. (**b**) Each of these runs was made independently. (**c**) Cooling curve properties were determined at a bath temperature of 40 °C (100 °F), 6 m/min (20 ft/min) using a 12.5 × 100 mm (1/2 × 4 in.) Inconel 600 probe preheated to 844 °C (1550 °F). (**d**) Distilled water at 40 °C (100 °F) exhibits a maximum cooling rate of 61.8 °C/s (111.2 °F/s) and cooling rate of 33 °C/s (59.4 °F/s) at 343 °C (650 °F).

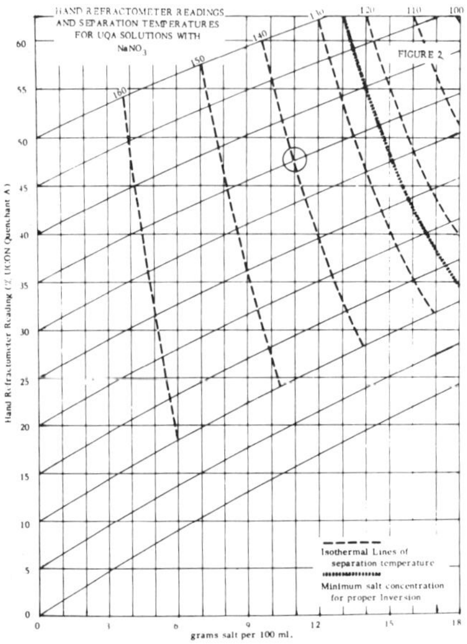

Sodium Nitrite

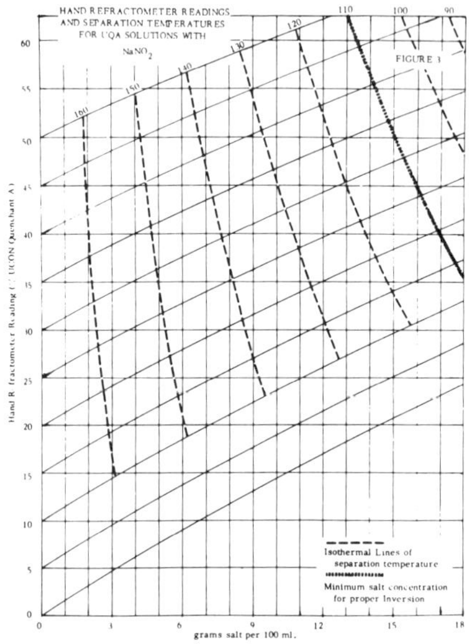

Sodium Nitrate

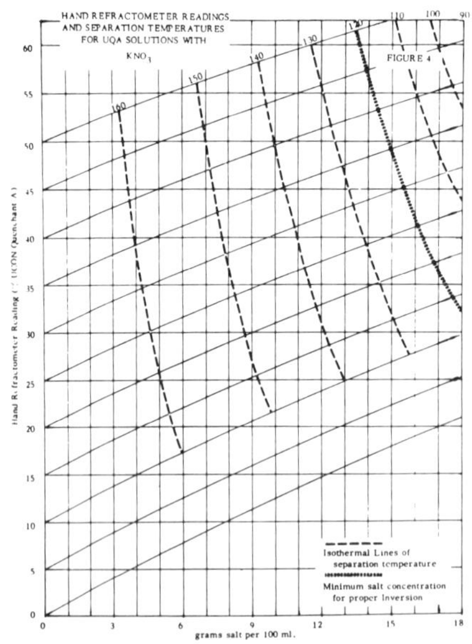

Potassium Nitrite

Isothermal Lines of separation temperature

Minimum salt concentration for proper Inversion

Fig. 6.20 Effects of various salts on refractometer readings and separation temperature for a type I quenchant **(continued)**

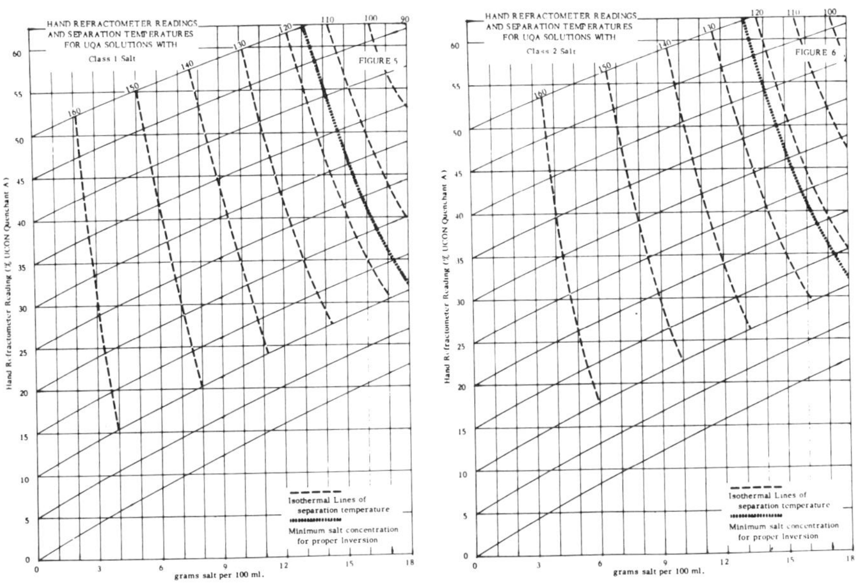

Class 1 Salt Class 2 Salt

Fig. 6.20 Effects of various salts on refractometer readings and separation temperature for a type I quenchant

Table 6.6 Definition of class I and class II salts(a)

	Salt class	
	I	II
	Working temperature range, °C (°F)	
Salt	**160-595 (325-1100)**	**290-595 (550-1100)**
$NaNO_2$,%	37-50	...
KNO_3,%	50-60	45-55
$NaNO_3$,%	0-10	45-55

(a) Defined according to MIL-S-10699A

(see Table 6.6) are also shown. It is apparent that different salts affect the refractive index of a quenchant differently and that quenchant concentration must be periodically validated. Most suppliers provide this support service free of charge.

The simplest and most convenient method of determining salt concentration is to measure the electrical conductivity (Ref 44). The conductance of an aqueous solution is the sum of the conductivity of each ionic component. Conductance increases with increasing concentrations of the different ionic species. Figure 6.21 shows that the

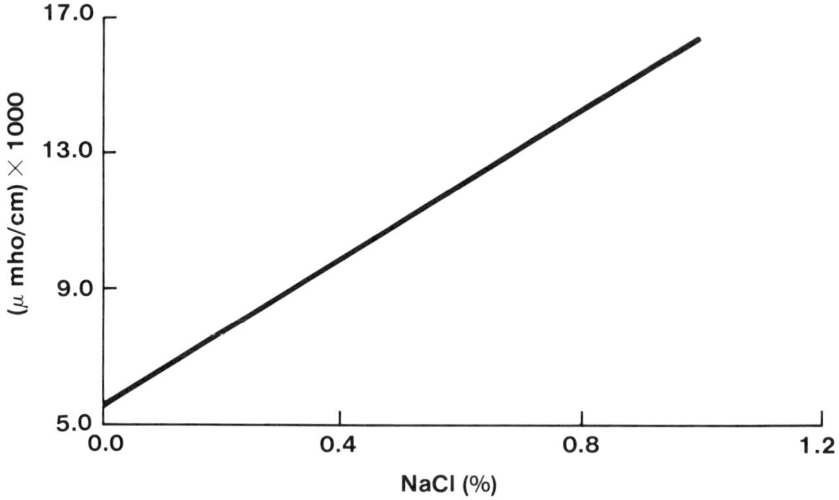

Fig. 6.21 Conductance versus salt (NaCl) concentration in 20% polymer quenchant solution

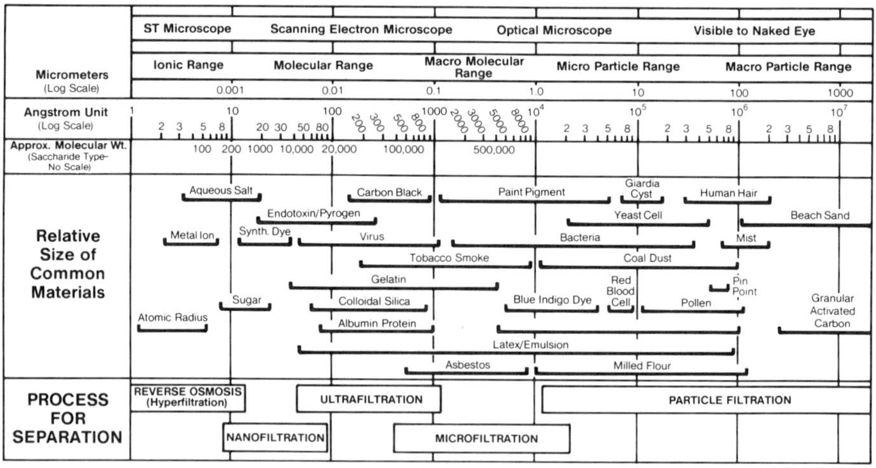

Fig. 6.22 Microporous membrane separation. Courtesy of Osmonics, Inc.

conductivity of an aqueous solution increases linearly with increasing concentration of a single salt species. However, if contamination with a mixture of varying salt compositions occurs, the conductance of the solution will increase, but will be proportional to the concentration of each salt component. The results are nonlinear, because the conductivity of each salt component is different. Therefore, this procedure will detect overall contamination, but not chemical identity.

An aqueous quenchant solution will exhibit a characteristic conductance for a particular concentration range. Continuous monitoring for any significant variation in conductance can be a valuable quality control procedure for some heat treating products. Abnormal increases in conductance, or even small but consistent increases, suggest that salt contamination may be occurring; this in turn may increase the cooling rate. Conversely, decreases in conductance generally indicate depletion of the corrosion inhibitor, which is usually ionic.

Membrane Separation

There is increasing interest in microporous membranes for reverse osmosis (RO) and ultrafiltration (UF) to remove salts and produce subsequent quenchant concentration variations (Ref 56-58). Figure 6.22 shows that RO and UF separation processes are based on membrane pore size. Smaller membrane porosity is associated with the relatively high-pressure RO process. Larger membrane porosity requires a lower pressure drop and is associated with the UF process. Both processes can be viewed as molecular filtration.

The principles of a membrane separation process such as UF or RO are shown schematically in Fig. 6.23 (Ref 47). The quenchant, usually composed of a polymer additive package and water, is pumped through the inner tube of the membrane cartridge. If the porosity of the membrane is sufficiently small, only "pure" water will pass through the membrane, and the polymer and additives will concentrate in the inner core of the membrane cartridge. If the porosity of the membrane is large, the additives, or at least a

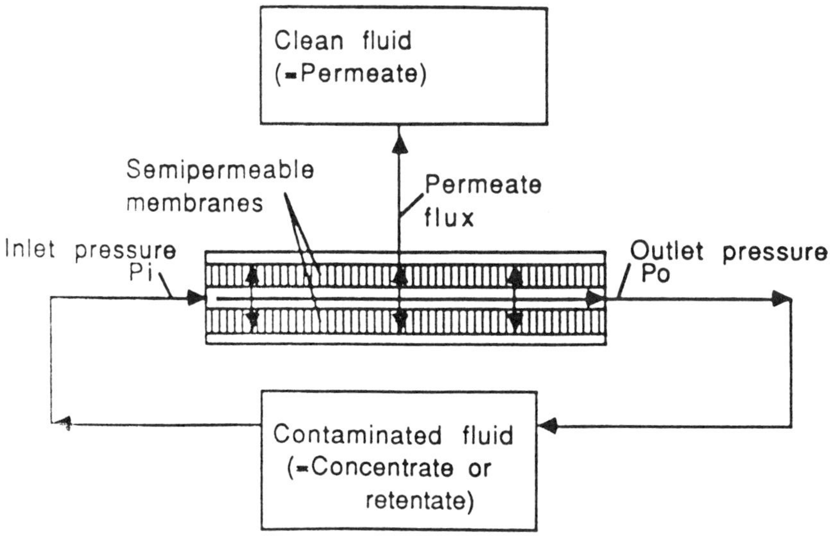

Fig. 6.23 Principles of membrane filtration with a cross-flow design

portion of them, will pass through the membrane with the water. The solution concentrating in the inner core of the membrane will contain the polymer and a fraction of the additive system. When the polymer solution has concentrated to the desired level, it is then pumped to a storage tank (or the quench tank) to be diluted to the desired concentration.

This process can be used to change the concentration of a polymer quenchant, to recycle it (e.g., from a rinse tank), or to separate the organic and inorganic residues from a wastewater system for subsequent disposal. If the quenchant is to be reused, it is essential that appropriate analyses be performed to ensure that the corrosion inhibitors are present at the required level. Such analyses are usually performed as a free service by the quenchant supplier.

The advantage of membrane separation is that it is a low-energy process that can be engineered as a continuous system. The system expands the applicability of polymer quenchants and may also help to solve waste disposal problems (Ref 47). Similar systems can be used for quench oil separation and regeneration. More details on membrane separation procedures are provided in Chapter 9.

Thermal (Cloud Point) Separation

The cloud point exhibited by a PAG copolymer is dependent on the type of salt present (Tables 6.4 and 6.5) and on the particular chemical composition of the polymer (Table 6.7). The cloud points of a PAG polymer characteristically increase as the degree of polymer oxidation increases. This usually increases both cooling rate and quench severity. Therefore, periodic cloud point determinations should be part of any quality control program.

With polymers that exhibit a cloud point, such as PAG and poly (ethyl oxazoline) (PEOX), it is possible to remove salts that have accumulated in the system by a thermal

Table 6.7 Variation of cloud point with PAG composition(a)

$$\mathrm{H\,[(OCH_2CH_2)_m\,(OCH_2\overset{\displaystyle CH_3}{\overset{|}{C}}H)_n\,]\,OH}$$

Ratio m/n	Approximate cloud point °C	Approximate cloud point °F
	> 100	> 212
7.47	85	185
3.95	75	165
2.45	65	150
1.32	50	120

(a) 20% aqueous solutions of the quenchant concentrate

(cloud point) separation (Fig. 6.24) (Ref 48). Heating a homogeneous aqueous solution to a critical temperature (the cloud point), which is dependent on the composition of the polymer, results in the formation of two phases, one polymer rich and the other water rich. The efficiency of this fractionation procedure depends on the separation temperature (Table 6.8). The salts, being ionic, concentrate in the aqueous phase and can be separated. Care must be exercised, as higher salt contamination levels may result in the reversal of the separated layers shown in Fig. 6.24.

pH Measurement

The pH range of most polymer quenchants usually varies from 7 to 10. The pH of the fresh quenchant depends primarily on the polymer and additives. Excessively high

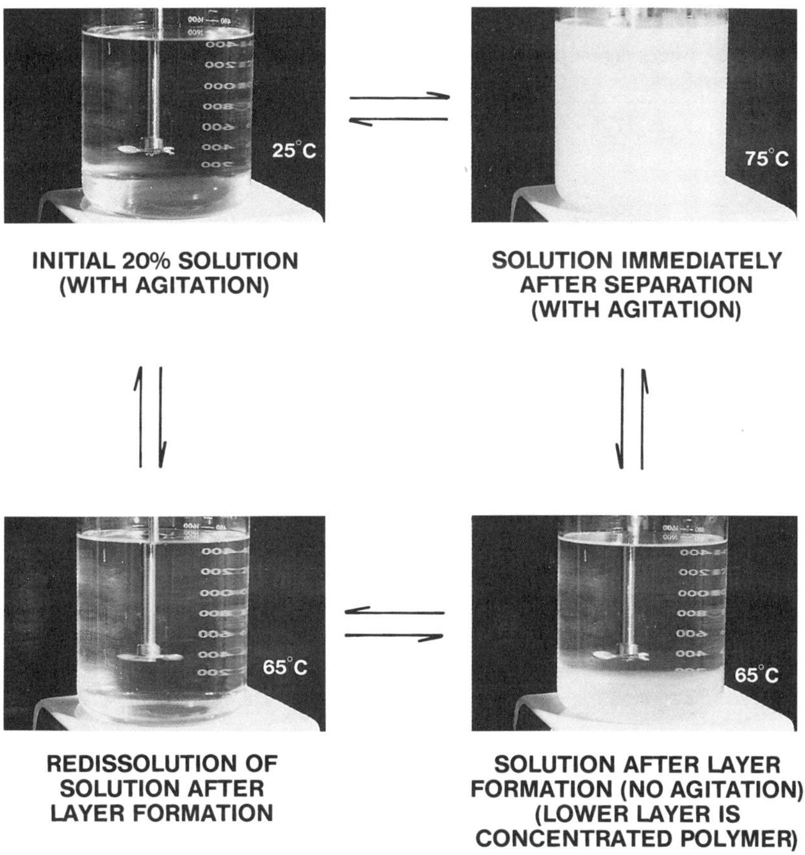

Fig. 6.24 Thermal separation process

pH values for nonferrous applications should be avoided, because the propensity for caustic corrosion increases when the pH exceeds 9.5 to 10.0.

The pH value varies with the dilution of the quenchant (Table 6.9). In fact, pH has little meaning in nonaqueous or partially aqueous media, since pH is defined as:

$$pH = -\log [H^+]$$

where $[H^+]$ is the concentration of the hydrogen ion, which is dependent on the degree of ionization:

$$H_3O^+ \rightarrow H^+ + H_2O$$

Clearly, the ionization constant of this process varies with the water concentration. In view of the potential for anomalous results, it is suggested that the caustic (OH^-) content of the desired partially aqueous solution be obtained directly by titration.

Table 6.8 Effect of solution temperature on the efficiency of polymer fractionation of a PAG quenchant

Separation temperature °C	Separation temperature °F	Water in the separated PAG layer, %
80	175	70
90	195	46

Table 6.9 Quenchant pH as a function of dilution

Concentration, %	pH
2.5	9.45
5.0	9.90
10.0	10.45
20.0	10.90
30.0	11.40
40.0	11.50

Table 6.10 Correlation of pH with actual base content

Experimental quenchant No.	Average undiluted base content, mEq/g	pH of 10% aqueous solution: Calculated(a)	Obtained	Difference
A	0.0195 ± 0.0014	11.29	9.90	1.39
B	0.0298 ± 0.0032	11.47	10.00	1.47
C	0.0396 ± 0.0037	11.60	10.45	1.15

(a) The aqueous solution pH values were calculated from the average base content (mEq/g) after conversion to mEq/mL by multiplying the mEq/g by the density (g/mL). Also, because 10% aqueous solutions were used, only 10% of the total base content is indicated by the titration.

By definition, pH is related to the caustic (OH^-) concentration by:

$$pOH = -\log [OH^-]$$

$$pH + pOH = 14$$

$$pOH = 14 - pH$$

A comparison of caustic content as measured by pH and titration is shown in Table 6.10. It is evident that pH measurement by a glass electrode is too susceptible to media effects to be useful as a precise quality control test. Quantification of total caustic content by titration is preferable.

Corrosion Protection

Polymer quenchant baths, fixtures, and so on are susceptible to both liquid and vapor phase corrosion caused by the water used to dissolve the polymer and additive system. Corrosion problems can be avoided by the addition of corrosion inhibitors (Ref 45), which protect the quench tank and fixtures in both the liquid and the vapor space above the tank. Some, though not all, of these additives may impart corrosion protection to the parts being quenched after their withdrawal from the tank and (less commonly) after tempering.

Corrosion inhibitors can be classified as either inorganic or organic. The most common inorganic corrosion inhibitors are sodium nitrite and potassium nitrite. These salts provide corrosion protection by surface passivation at the points of oxidative attack. Organic corrosion inhibitors are usually formulated from fatty acid and amine combinations and function primarily by the formation of a protective film.

Aqueous quenchant solutions are known to develop both liquid and vapor phase corrosion problems if the corrosion inhibitors are depleted. The most common modes of inhibitor depletion are by reaction, such as surface passivation, and by dragout, which will occur concurrently with protective film formation. Therefore, if corrosion problems are to be avoided, it is essential that minimum inhibitor concentrations be maintained. Should inhibitor depletion occur, supplemental additives must be added to the bath.

Corrosion inhibitor maintenance is relatively complex. Monitoring and addition recommendations usually are provided as a service by quenchant suppliers.

Biological Activity

The polymers used for quenching are, in general, relatively resistant to microbiological degradation. However, growth of microorganisms in a bath is often enhanced by oil contamination, which acts as a nutrient. Problems caused by microbial growth include foul odors, slime, depletion of nitrite corrosion inhibitors, and staining of parts. Occasionally, microorganisms develop that can degrade the polymer. When this happens, the performance of the quenchant can be affected. The simple dipstick test for detecting bacteria and fungi is illustrated in Fig. 6.25.

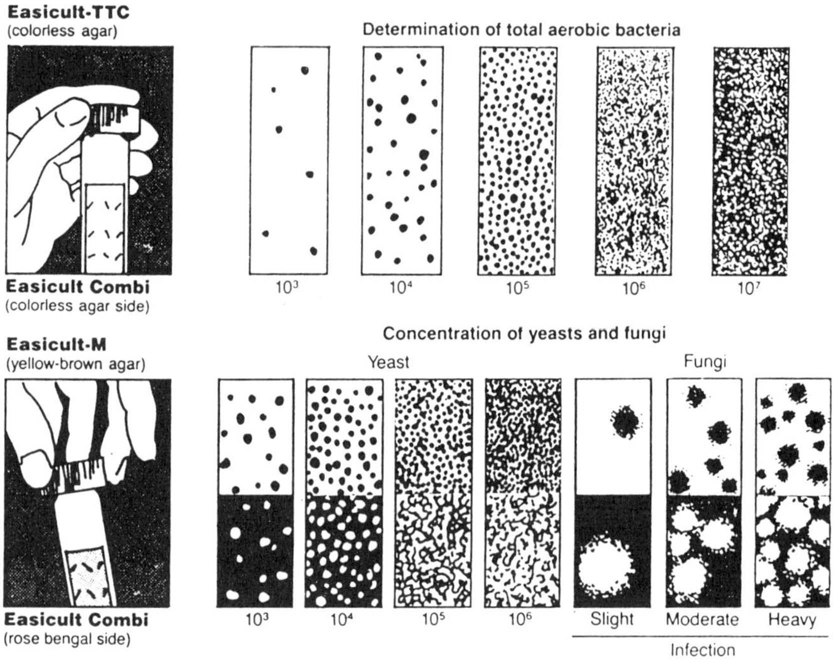

Fig. 6.25 Dipstick test for bacteria and fungi detection

Problems related to biological activity can usually be controlled by the addition of an appropriate biocide. Users faced with such problems should consult with a biocide supplier or with the polymer supplier.

Care must be taken to ensure that the organic moieties (e.g., polymer and additive systems) are chemically stable (compatible) in the presence of the biocide that is selected. For example, many effective biocides contain active chlorine compounds that will spontaneously react with PAG polymers, producing vinyl and chloroethyl ether by-products. In addition, several common biocides are amine based and can react with nitrite corrosion inhibitors to produce nitrosamines. Some nitrosamines have been shown to cause cancer in laboratory animals (Ref 49, 50). Although there is no direct evidence firmly linking human cancer to nitrosamines, it seems unlikely that humans should be uniquely resistant to their action (Ref 51). Therefore, users should work with suppliers to avoid potentially costly or hazardous selection errors.

The following example illustrates the successful use of a biocide to provide microbial control in a quenchant system.

The system in question had an in-ground stainless steel quench tank located in an open area outside the plant. The tank contained approximately 22,000 L (5800 gal) of a water-base quenchant containing PAG polymer plus sodium nitrite and was used to process precast and machined aluminum parts. The system experienced periodic episodes of microbiological activity, resulting in foul odors and staining of parts. When this occurred, the heat treater would thermally separate the quenchant and recover a portion of the glycol. The glycol would then be diluted with fresh water and sodium nitrite would be added. This process normally required 2.5 days to complete.

Only 3 weeks after the system had undergone a thermal separation, it again experienced problems with biological activity. Instead of performing another thermal separation, the heat treater decided to treat the system on a regular basis with a glutaraldehyde-base biocide. Laboratory tests indicated that this biocide had excellent compatibility with the quenchant in use. Because fouled systems typically show an initial demand for biocide, the system was first treated with a 300 ppm dose and thereafter with a weekly 100 ppm dose. Testing showed that this treatment regimen maintained a residual concentration of biocide in the system and as a result kept it free of microorganisms. No problems of microbiological activity have been encountered since this treatment program was initiated, and the biocide has had no other effects on the properties of the quenchant.

If a bath remains unagitated for a long period of time at room temperature, anaerobic bacterial growth may develop. Active anaerobic bacteria can contribute to corrosion in the bath and often produce foul odors. This condition can be prevented by the use of an appropriate biocide (not all biocides are effective against anaerobic bacteria) and/or by agitating the bath daily or bubbling air into the system during periods of inactivity.

The biological activity of even a contaminated bath is also sensitive to pH. If the pH is above 9.6, then bacterial and fungal activity may be reduced (Ref 40).

Contamination of Polymer Quench Baths

Quenchant contamination has many possible sources. For example, salt contamination from salt baths and hard water can dramatically affect both the solution properties and quench rates. Both oil and aqueous polymer quenchants are susceptible to oil contamination by other quenching oils, corrosion inhibitors, forging lubricants, and hydraulic fluids (Fig. 6.26) (Ref 52). In some cases, the addition of water-glycol hydraulic fluid to an aqueous polymer results in extended A-stage cooling, as shown in Fig. 6.27 (Ref 59). In another example, a PAG quenchant contaminated with a hydraulic oil exhibited substantial reduction in the maximum cooling rate (Fig. 6.28) (Ref 29).

A persistent foam may form in a contaminated, highly agitated quench bath, necessitating the addition of an antifoam composition. Antifoam compositions used for aqueous polymer quenchants are generally classified as either "cloud point" or "silica supported." Silica-supported antifoams are composed of a silicone either reacted to or absorbed on a silica substrate. These types of antifoams are the most potent; however, they do not form homogeneous solutions and are readily removed by adsorption on parts and by the system filters. Cloud point antifoams, the type more commonly used for

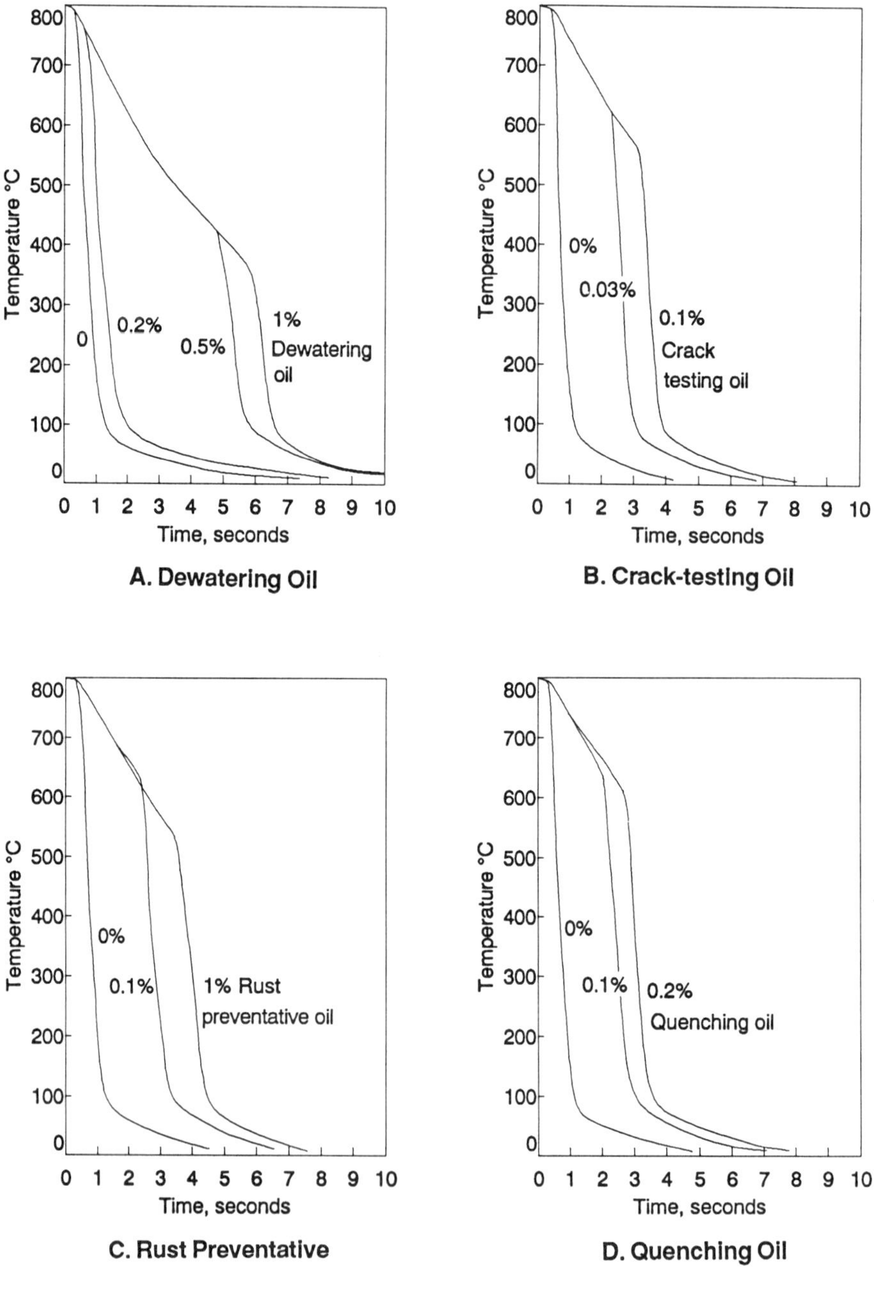

Fig. 6.26 Polymer quenchant contamination by various types of oils

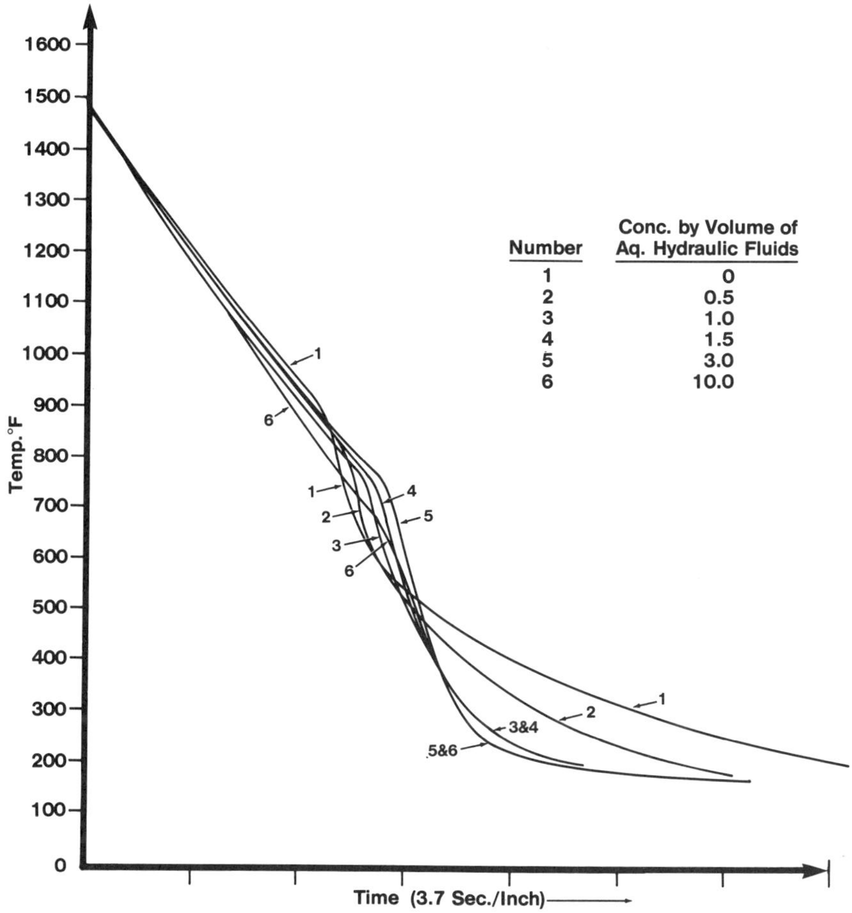

Fig. 6.27 Effect of water-glycol hydraulic fluid contamination of an aqueous PVP quenchant solution

aqueous polymer quenchants, are composed of one or more low-molecular-weight (≤3000 to 5000) PAG copolymers with a cloud point of ≤30 °C (85 °F).

One type of antifoam that has been used in the water coolant industry is oil/water emulsion. However, because of their instability, oil/water emulsions are generally not the antifoam composition of choice. The emulsion may break, leading to an oil-contaminated quenchant and spotty hardness or cracking. This phenomenon is caused by the formation of a heterogeneous film around the hot metal during the quench, which results in thermal gradients across the surface during the cooling process. Interestingly, this problem may not be readily apparent from cooling curve analysis. Figure 6.29 illustrates

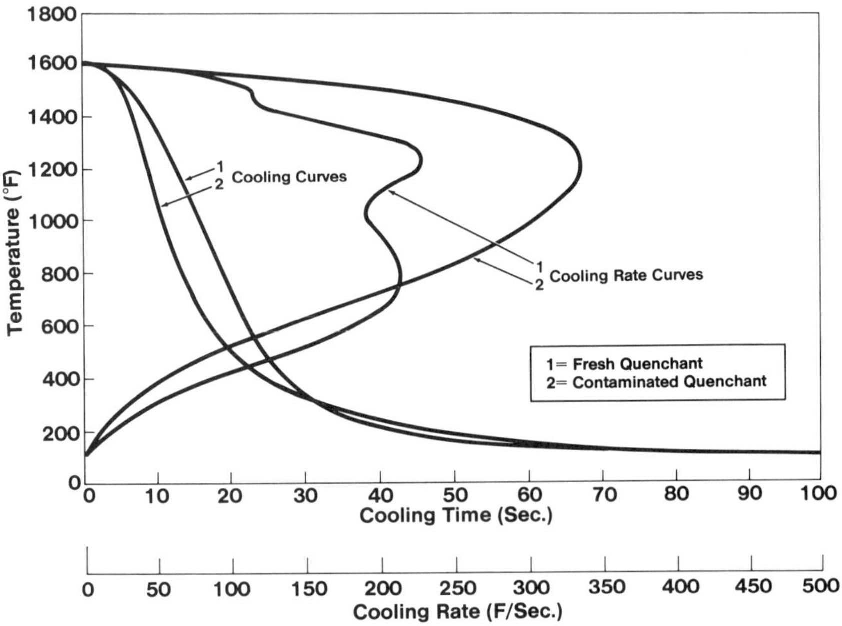

Fig. 6.28 Effect of hydraulic oil contamination of an aqueous PAG quenchant. Courtesy of E.F. Houghton & Company, Inc.

the cooling curve performance of a quenchant contaminated by an oil/water antifoam versus a fresh, uncontaminated solution. Cooling curve analysis shows only a very slight difference between the contaminated and uncontaminated quenchants. (A slight decrease in C-stage cooling rates was observed.) Several cooling rate values along with the physical properties of the quenchants are summarized in Table 6.11. GPC analysis (Fig. 6.29) revealed that no unusual degradation of the polymer occurred. In this case, simple visual observation of a floating oil contaminant provided the critical information needed to successfully troubleshoot the system.

As discussed in the section on biological activity, the presence of oil can lead to bacterial growth (Ref 40). Use of an oil stripper should help deal with both issues.

Changes in a Polymer Quenchant With Use

Most commercial polymer quenchants are generally stable, but some degradation during use is inevitable (Ref 53). Such degradation is usually evidenced by a loss of viscosity even at the same apparent polymer concentration.

The use of refractive index to determine polymer concentration is relatively insensitive to polymer degradation. Therefore, measurement of viscosity, which is more

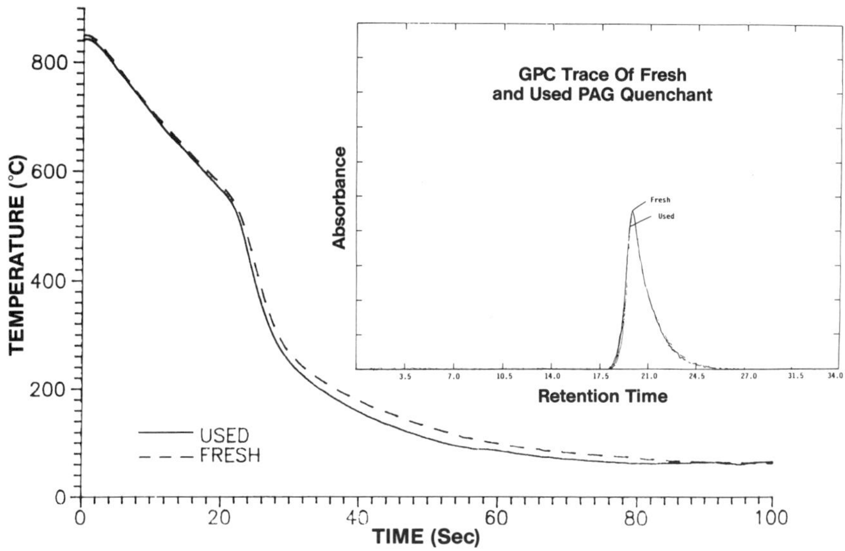

Fig. 6.29 Effect of oil/water emulsion contamination of an aqueous PAG polymer quenchant

Table 6.11 Physical and cooling curve properties of a polymer quenchant contaminated by an oil/water emulsion

	Fresh quenchant	Contaminated quenchant
Physical properties		
Concentration, %	24.7	24.4
Viscosity, cSt at 40 °C (100 °F)	9.90	9.77
Conductance, μmho/μm	4600	4400
Cloud point, °C (°F)	76.0 (169.0)	76.0 (169.0)
GPC area shift, %	...	2.0
Cooling curve results (bath temperature of 40 °C, or 100 °F)		
Time to cool from 732-260 °C (1350-500 °F), s	16.0	15.4
Maximum cooling rate, °C/s (°F/s)	59.6 (107.3)	64.8 (116.6)
Temperature at maximum cooling rate, °C (°F)	469 (876)	482 (900)
Cooling rate, °C/s (°F/s), at:		
704 °C (1300 °F)	17.3 (31.1)	16.6 (29.9)
343 °C (650 °F)	38.9 (70.0)	44.3 (79.7)
232 °C (450 °F)	12.0 (21.6)	16.1 (29.0)

sensitive to the molecular weight of the polymer, should be used to periodically validate polymer concentration. The difference in apparent polymer concentration as determined by refractive index and by viscosity will increase with time if polymer degradation is occurring (Ref 49).

To ensure long-term process control, it is essential that both the physical properties and the severity of the quench bath be monitored periodically. A specific example involves a quenchant that, after many years of use, had undergone severe degradation, resulting in quenched parts with unacceptably high reject rates.

Table 6.12 compares the physical properties of the "used" quenchant and a "fresh" solution made up to the same total polymer concentration. The cooling curves for these quenchant solutions are shown in Fig. 6.30. Comparison of the viscosities of the fresh (5.65 cSt) and used (1.93 cSt) quenchants shows that substantial polymer degradation has occurred. This observation is substantiated by the GPC area shift of 40%.

The GPC peak of a used polymer quenchant is usually broader than that obtained for a fresh quenchant normalized to the same total area. Therefore, if the GPC peak for the used polymer is superimposed on a peak for a fresh polymer, an area shift will be observed. It is not unusual for a polymer quenchant in use to exhibit a steady-state GPC area shift of 8 to 10%. The amount of polymer degradation indicated in Fig. 6.30 is clearly much greater than an acceptable level.

Further evidence of substantial polymer degradation lies in the loss of the characteristic cloud point (>100 °C, or 212 °F, versus 88 °C, or 190 °F) for the used polymer quenchant. In general, used polymer quenchants do not exhibit greater than a 1 to 2 °C (2 to 4 °F) cloud point elevation during normal use. Therefore, this quenchant was in particularly bad condition.

The used quenchant had also undergone salt contamination, which is evident from the higher conductance in comparison with the fresh quenchant (Table 6.12). Although high conductance can arise from overdosing of the corrosion inhibitor (in this case, sodium nitrite), it is more likely that the high conductance reflects hard water contami-

Table 6.12 Characterization of a severely degraded and contaminated polymer quenchant

	Fresh quenchant	Contaminated quenchant	Water
Physical properties			
Concentration, %	20.0	20.0	...
Viscosity, cSt at 40 °C (100 °F)	5.65	1.93	...
Conductance, μmho/μm	5000	>20,000	...
Cloud point, °C (°F)	88.0 (190.0)	>100 (212)	...
GPC area shift, %	...	– 40	...
Cooling curve results (bath temperature of 40 °C, or 100 °F)			
Time to cool from 732-260 °C (1350-500 °F), s	15.4	10.8	10.4
Maximum cooling rate, °C/s (°F/s)	39.2 (70.6)	69.3 (124.7)	61.8 (111.2)
Temperature at maximum cooling rate, °C (°F)	534 (993)	650 (1200)	610 (1130)
Cooling rate, °C/s (°F/s), at:			
704 °C (1300 °F)	20.6 (37.1)	65.0 (117.0)	56.8 (102.2)
343 °C (650 °F)	29.6 (53.3)	29.6 (53.3)	33.0 (59.4)

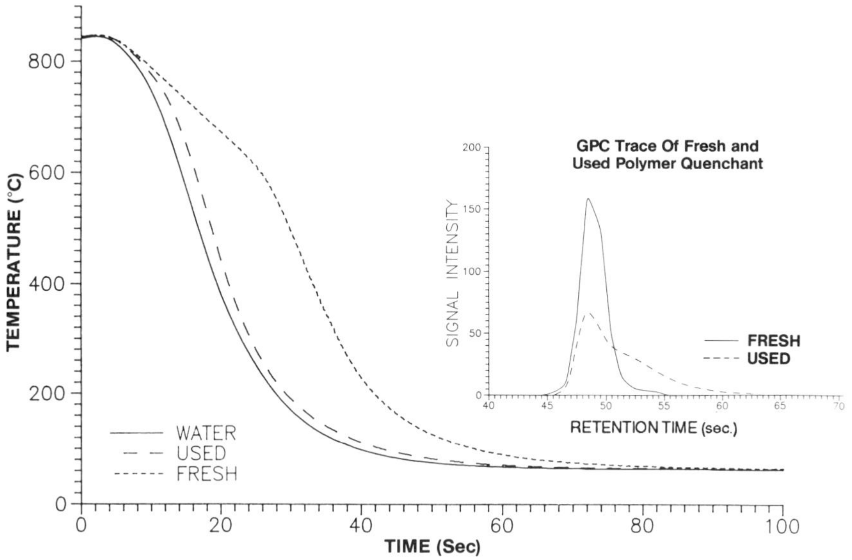

Fig. 6.30 Cooling curves for an excessively degraded aqueous PAG quenchant and a fresh quenchant

nation. If deionized water is not used in a polymer quenchant, hard metal ions will build up in the quenchant due to the successive water additions required to compensate for evaporation. Therefore, to obtain optimal lifetimes for a polymer quenchant, it is generally recommended that some method of water deionization, such as ion exchange, reverse osmosis, or distillation, be used.

Cooling curve analysis illustrates the combined effects of polymer degradation and salt contamination (Table 6.12). In this case, the quench severity of the used quenchant was much closer to water than the fresh quenchant. The degradation level of a polymer quenchant seldom reaches this point. In fact, if the physical properties and the quench severity of the bath had been periodically monitored, appropriate corrective action probably would have been taken.

Gaseous Contamination

Quenchants can also be contaminated by gases, such as ammonia, which may occur during nitriding or carburizing. Increasing ammonia contamination will result in corresponding increases in A-stage cooling times (Fig. 6.31) (Ref 52).

Similar results would be expected for quenchants contaminated by other gases, such as carbon dioxide or carbon monoxide. The extended A-stage cooling times caused by the gaseous contaminants generally result in insufficient hardening (Ref 52).

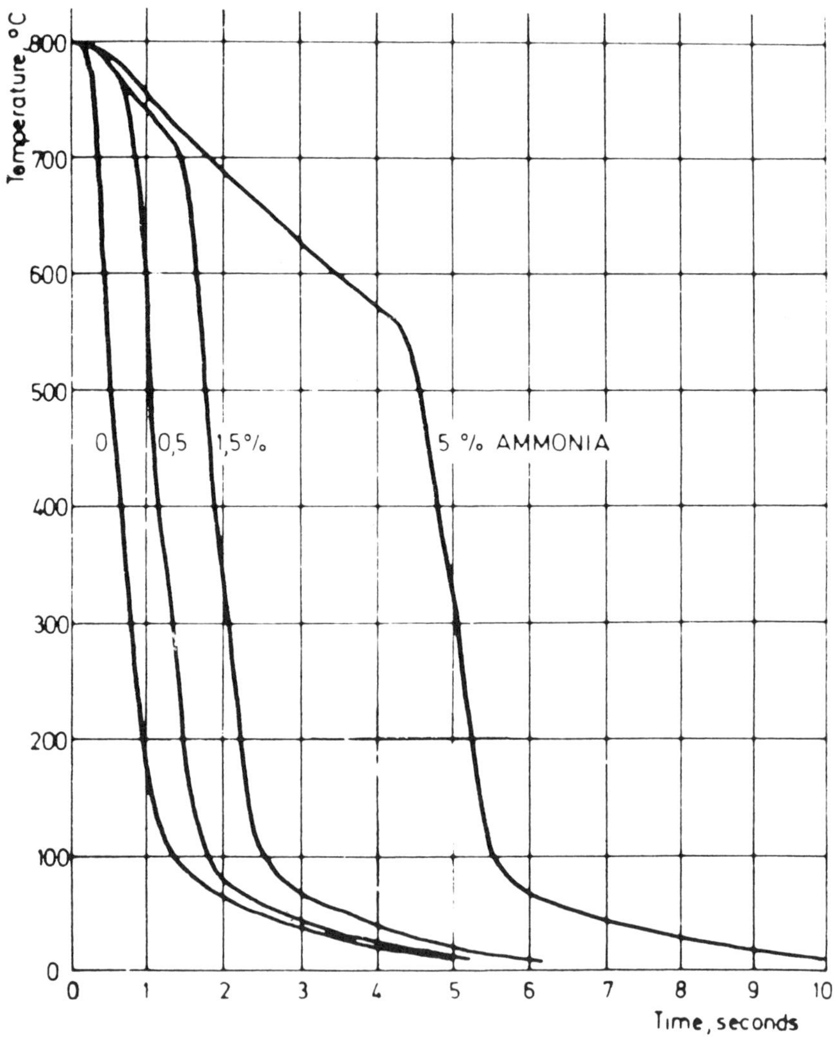

Fig. 6.31 Effect of ammonia contamination of an aqueous polymer quenchant

References

1. L.S. Zarkhin, S.V. Sheberstov, N.V. Panfilovich, and L.I. Manevich, *Russ. Chem. Rev.*, Vol 58 (No. 4), 1989, p 381-392
2. J. Igarashi, *Jpn. J. Tribol.*, Vol 35, 1990, p 1095-1105
3. S. Kodama, *Netsu Shori*, Vol 25, 1985, p 23-26
4. G.R. Furman, *Lubrication*, Vol 57, 1971, p 25-36

5. R.T. von Bergen, *Heat Treat. Met.*, Vol 2, 1991, p 37-42
6. R.T. von Bergen, *Proc. Conf. Heat Treat. Steel*, Glasgow, 5 Sept 1989, Scottish Association for Metals
7. H-J. Gilliland, *Met. Prog.*, Oct 1960, p 111-114
8. V. Srimongkolkul, *Heat Treat.*, Dec 1990, p 27-28
9. J.A. Hasson, *Ind. Heat.*, Sept 1981, p 21-23
10. *Lubrication*, Vol 41, 1955, p 1-12
11. M. Tagaya and I. Tamura, *Technol. Rep. Osaka Univ.*, Vol 7, 1957, p 403-424
12. "Houghton on Quenching—Quenching Speed to Fit Your Need," product brochure, E.F. Houghton & Co., Inc.
13. H.E. Boyer and P.R. Cary, Ed., *Quenching and Control of Distortion*, ASM International, 1988, p 169
14. H.E. Boyer and P.R. Cary, Ed., *Quenching and Control of Distortion*, ASM International, 1988, p 33-35
15. W. Hewitt, *Heat Treat. Met.*, Vol 1, 1986, p 9-14
16. R. Brennan, *Ind. Heat.*, Jan 1991, p 25-28
17. H.E. Boyer and P.R. Cary, Ed., *Quenching and Control of Distortion*, ASM International, 1988, p 44-45
18. These detectors are available from various manufacturers; an example is the "Quench Guard" from Lindberg, a unit of General Signal.
19. B.R. Horton and R. Weetman, *Heat Treat. Met.*, Vol 2, 1984, p 49-51
20. A. Sasaki, T. Tobisu, S. Uchiyama, and M. Kawasaki, *Lubr. Eng.*, Vol 47, 1991, p 525-577
21. W.D. Stofey, "Recycling Quench Oil: Filtration, Reclamation, or Disposal," presented at ASM International National Meeting, Cincinnati, Oct 1991
22. The experimental details of a Jominy test are described in a number of specifications, including: (1) "End-Quench Test for Hardenability of Steel," ASTM A 255-67 and (2) "Hardenability Test by End-Quenching," DIN 50,159, Sept 1987
23. T. Croucher, *Heat Treat.*, Dec 1990, p 17-19
24. G.E. Totten, K.B. Orszak, L.M. Jarvis, and R.R. Blackwood, *Ind. Heat.*, Oct 1991, p 37-41
25. "Laboratory Test for Assessing the Cooling Curve Characteristics of Industrial Quenching Media," Wolfson Heat Treatment Centre Engineering Groups Specifications, 1982
26. A. Bashford and A.J. Mills, *Heat Treat. Met.*, Vol 1, 1984, p 9-14
27. G.E. Hampshire, *Heat Treat. Met.*, 1984, p 15-20
28. J. Hasson, *Heat Treat.*, Dec 1990, p 29-31
29. R.J. Brennan, E.F. Houghton & Co., Inc., personal communication, 1991
30. S.O. Segerberg and J.L. Bodin, *Heat Treat.*, May 1988, p 26-28
31. C.E. Bates, G.E. Totten, and R.L. Brennan, Quenching of Steel, *ASM Handbook, Heat Treating*, ASM International, Vol 4, 1991, p 114-115
32. J.Y. Oldshue, *Fluid Mixing Technology*, McGraw-Hill, 1983, p 163
33. The Mead velocimeter, in particular the HP-302 open-stream velocity probe, is used most often for the measurement of flow rates in water streams for pollution control. It is available from Mead Instruments Corp., P.O. Box 367, Sussex, NJ 07461.
34. V.G. Kaufman, M.B. Gutman, R.G. Gol'tseya, and O.I. Makarychev, *Metalloved. Term. Obrab. Met.*, Jan 1989, p 30-33
35. E.H. Burgdorf, *Proc. 3rd Int. Sem. Quenching Carburizing*, International Federation for Heat Treatment and Surface Engineering, Sept 1991, p 66-77

36. P.S. Protsidim, N.Ya Rudakova, and B.K. Sheremeta, *Metalloved. Term. Obrab. Met.*, Feb 1988, p 5-7
37. W. Kopke, *Heat Treat.*, Dec 1990, p 25-26
38. S. Segerberg, *4th Int. Cong. Heat Treat. Mater.*, Berlin, 3-7 June 1985, Vol II, p 1252-1265
39. N.A. Hilder, Ph.D. thesis, University of Birmingham (Aston), 1988
40. T. Hibi, *Netsu Shori*, 1985, Vol 25, No. 1, p 46-50
41. K-H. Kopietz, *Heat Treat.*, Sept 1984, p 20-26
42. R.R. Blackwood, M.M. Lorbach, and E.R. Mueller, *Heat Treat.*, July 1985, p 40-42
43. N. Kobayashi, *Netsu Shori*, Vol 25, 1985, p 51-54
44. G.E. Totten, R.R. Blackwood, and L.M. Jarvis, *Heat Treat.*, March 1991, p 16-18
45. E.R. Mueller, *Heat Treat.*, Sept 1986, p 24-26
46. "Polyalkylene Glycol Heat Treat Quenchant," AMS 3025, Society of Automotive Engineers, 15 April 1980
47. S.M. Mahdi and R.O. Skold, *Lubr. Eng.*, Vol 47, 1991, p 686-690
48. L.M. Jarvis, R.R. Blackwood, and G.E. Totten, *Ind. Heat.*, 1989, p 23-24
49. P. Issenberg, *Fed. Proc.*, Vol 35, 1976, p 1322-1326
50. P.N. Magee and J.M. Barnes, *Adv. Can. Res.*, Vol 10, 1967, p 163-240
51. H.M. Fribush, "The Effect of TSCA on the Metalworking Fluid Industry: Increased Awareness of the Presence of Nitrosamines in Metalworking Fluids," presented at the American Chemical Society National Meeting, Las Vegas, 1 April 1982
52. E.H. Burgdorf, *Ind. Heat.*, Oct 1981, p 18-25
53. *Heat Treat.*, Oct 1983, p 40-41
54. R. Wang, *Jin-Shu Re Chu-Li*, Vol 34, 1987, p 43-46
55. C.E. Bates and G.E. Totten, *Adv. Mater. Proc.*, Vol 3, 1991, p 25-28
56. R. Foreman, U.S. Patent 4,251,292, 1981
57. K. Anderson, T. Croucher, and D. Butler, *Met. Prog.*, May 1983, p 37-40
58. T. Croucher, *Heat Treat.*, Sept 1984, p 27-30
59. R. Foreman, "Polymer Quenchant Short Course" offered by ASM International, Chicago, Sept 1988

CHAPTER 7

Spray Quenching

The term spray quenching refers to a wide variety of quenching processes that involve heat removal facilitated by the impingement of a quenchant medium on a hot metal surface. Some of these processes have obvious differences, while others are similar and differ only in degree. Examples include the addition of droplets of water (or other volatile liquids) to a gas quenching stream in fog quenching (Ref 1); quenching with water or water/air streams (Ref 2); sprays of volatile liquid quenchants other than water (Ref 3); and high-pressure jets of an oil (Ref 4), water (Ref 5-7), or aqueous polymer solution (Ref 8) under the liquid level in a bath.

Spray quenching is used to optimize heat transfer from a part to allow the capture of a metastable structure with the required physical properties, while simultaneously developing the desired distribution and level of stress (Ref 9). Therefore, the time-temperature history of the part must be accurately controlled during the quench. The heat-transfer coefficient is a measure of the heat flow from the hot metal surface. With spray quenching, the heat-transfer coefficient from the part to the quenching medium is directly related to the flow rate, turbulence, and impingement pressures of the quenchant at the hot surface. It is possible to adjust these parameters during the quench to yield a cooling profile achievable in no other way (Ref 2, 5, 10-12). Computerized adjustment of these variables during the quench has been extensively explored (Ref 2, 5, 10, 13-18). This ability to control and vary the heat-transfer coefficient during the quench may produce properties that would otherwise require more expensive, higher-alloy materials (Ref 9, 19-22).

Spray quenching usually requires more expensive equipment than that needed for immersion quenching (Ref 5). Maintenance of a spray quench system is also demanding (Ref 12). However, these disadvantages are counterbalanced by the fact that spray quenching may allow the user to achieve optimal as-quenched properties for a given alloy and, in many cases, to achieve properties superior to those obtained by other quench methods (Ref 23, 24). Another potential benefit of spray quenching is that it is sometimes possible to achieve the needed properties while using a lower-cost, lower-alloy-content steel.

This chapter will examine the fundamentals of spray quenching, the concept and mathematics of heat transfer from a hot metal surface to droplets of volatile liquids, how heat transfer is controlled by spray variables, results achieved by spray quenching of steel and aluminum parts, the use of spray quenching in induction hardening; and some less common variants of the spray quenching process.

Concepts of Spray Quenching

The quenching of a hot metal part in a water bath will follow the cooling mechanisms illustrated in Fig. 7.1 (Ref 13, 25, 26), which shows the heat transfer from the part as a function of its surface temperature. Figure 7.1 also relates the surface temperature and heat-transfer coefficient to the mechanism of heat removal. Upon immersion, the part will first be surrounded by a vapor blanket, which will collapse as the part cools. The heat transfer through this vapor blanket is poor, and the part will cool slowly in this region.

The second region of the cooling curve is called the nucleate boiling region and corresponds to rapid heat transfer caused by direct contact of the part with the water. In

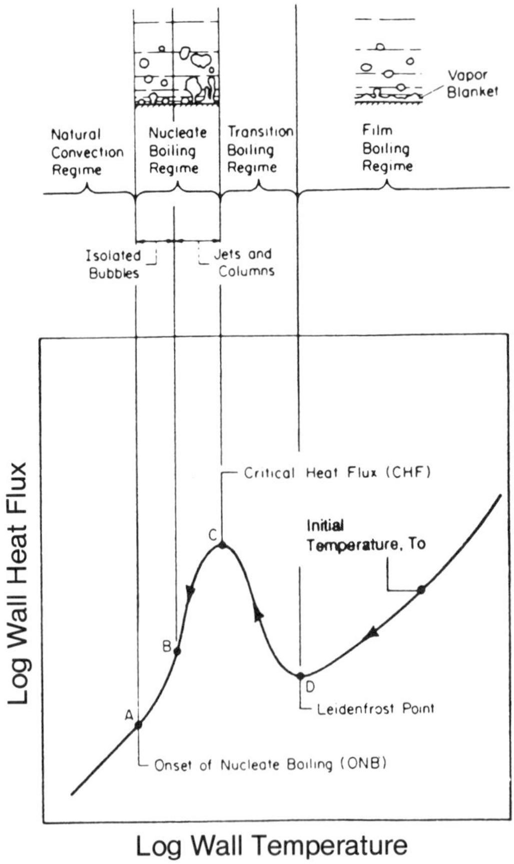

Fig. 7.1 Heat transfer versus surface temperature

this region, the part is still very hot and the water will boil vigorously. The high heat of water vaporization accounts for the very rapid heat transfer. In the third, or convective, cooling region, the surface of the part has cooled to a temperature below the boiling point of water. Only convective heat transfer occurs in this region.

The rate of heat removal from a quenched part can be increased by agitation, which reduces the stability of the vapor blanket that surrounds the part during the early stage of the quench. The effect of agitation on the cooling mechanism of a silver probe quenched by immersion in a 60 °C (140 °F) water bath is shown in Fig. 7.2 (Ref 6). A stream of water with a velocity of V_i was injected below the surface of the bath and directed at the probe. The greater the flow rate (agitation), the higher the temperature at which the vapor blanket mechanism was displaced by the much more effective nucleate boiling mechanism.

Cooling by the impact of a stream of droplets or liquid in spray quenching is also represented by Fig. 7.1. Fluid flow is used to accelerate the quench by rupturing the vapor blanket. The high agitation rates inherent in a spray quench also accelerate cooling in the nucleate boiling and convective cooling portions of the quench. For more

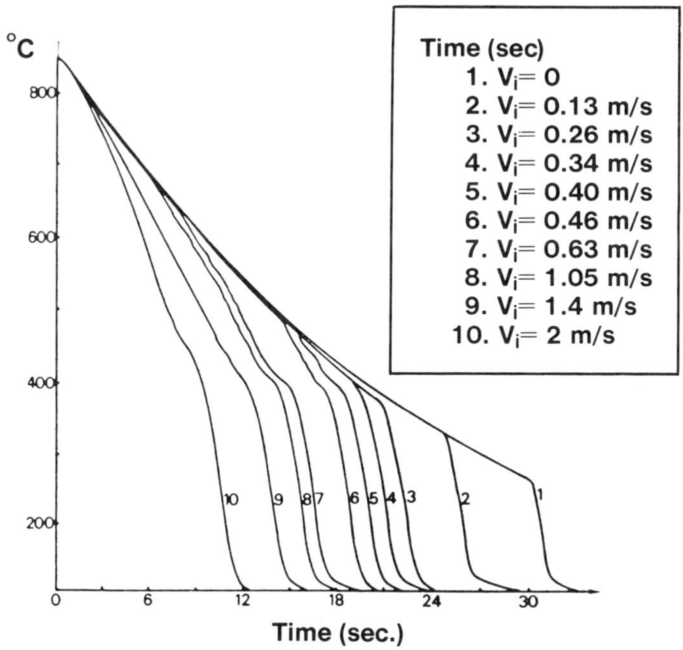

Fig. 7.2 Effect of agitation on the cooling curve

information on the influence of agitation on the cooling mechanism stages, see Chapter 10. Based on the mathematics needed to describe the spray systems, one author has stated that "spray cooling is also a sort of turbulent flow cooling" in all three regions of the cooling curve (Ref 27). In fact, it is useful to use the same terms to describe the heat-transfer mechanisms in both spray and immersion quenching.

Before continuing, some commonly used terms will be defined. The *saturation temperature* is the temperature at which liquid water is at equilibrium with its vapor at the system pressure. *Superheat* refers to components of the quench system that have temperatures higher than this equilibrium temperature, while *subcooled* refers to components or regions that are cooler than this reference temperature.

Consider a series of water droplets moving toward the surface of a hot metal plate. The first droplet will contact the plate, and the surface temperature of the droplet at the point of contact will instantly rise. If the plate is sufficiently hot that the liquid in the contact zone is raised above its limiting superheat temperature (260 °C, or 500 °F, for water), then a portion of the liquid will evaporate in approximately 0.001 s (Ref 28). This evaporation will occur before the original spherical shape of the droplet has been significantly deformed. Because the contact region represents a small fraction of the mass of the drop, no significant heat removal from the plate will have occurred. The vapor film that forms will push the remaining portion of the droplet away from the plate. This vapor blanket provides a partial insulating barrier and inhibits the impact of additional droplets. Only those droplets with sufficient kinetic energy (momentum) will be able to penetrate the vapor film and strike the plate. The result will be a film of liquid separated from the plate by a layer of vapor. This is called the *film boiling* or *nonwetting* phase.

The heat-transfer process for this state can be pictured as shown in Fig. 7.3 (Ref 2, 29). The heat-transfer rate through the vapor film is relatively slow; therefore, there is a relatively slow decrease in the temperature of the plate. As the temperature falls, the thickness of the vapor film will decrease (Ref 5). At a characteristic temperature, which is dependent on the quench conditions, the droplets with higher-than-average kinetic energy will penetrate the vapor film, contact the plate, and begin to spread out on the surface. The temperature at which this occurs is called the *Leidenfrost point*. At this temperature, regions of the plate will be observed to be "wetted" by the quenchant.

The increased contact of the droplets with the metal surface will result in a corresponding increase in the heat transfer from the plate, causing the plate temperature to decrease more rapidly. A larger number of droplets will penetrate the film, and heat transfer will increase until the entire plate is wet with boiling liquid. At this point, the heat-transfer process is characterized by nucleate boiling. Further reduction of the plate temperature leads to an end of boiling, and the heat-transfer mechanism becomes dependent on convective cooling. In practice, the Leidenfrost temperature is found to be a function of the surface physical properties (Ref 6, 11, 30, 31) as well as its temperature (Ref 32, 33).

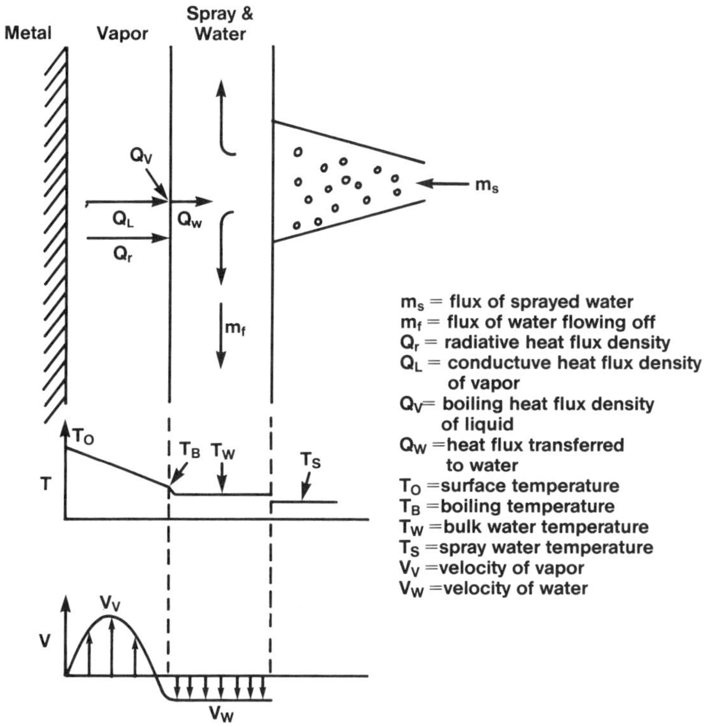

Fig. 7.3 Heat transfer in the vapor blanket portion of a spray quench

Quantification of Heat Transfer in Spray Quenching

When the temperature of a metal plate is well above the Leidenfrost temperature, cooling by spray quenching depends on the total water flow, the distribution of the water spray on the surface, and the size distribution of the droplets in the spray. Total water flow can be easily measured by using turbine flow meters or rotometers, or by measuring the amount of water delivered in a given amount of time. Air flow rates are usually measured with rotometers. The water flux of an air-water spray nozzle is determined by both the water and air pressure used, as illustrated in Fig. 7.4. The amount of water delivered through the nozzle will be a complex function of these variables (Ref 34).

Droplet size is a critical factor in the efficiency of the spray quench system. Both the kinetic energy and the surface-to-volume ratio of the droplet depend on droplet size, which in turn depends on air and water pressures, the inlet orifice designs for air and

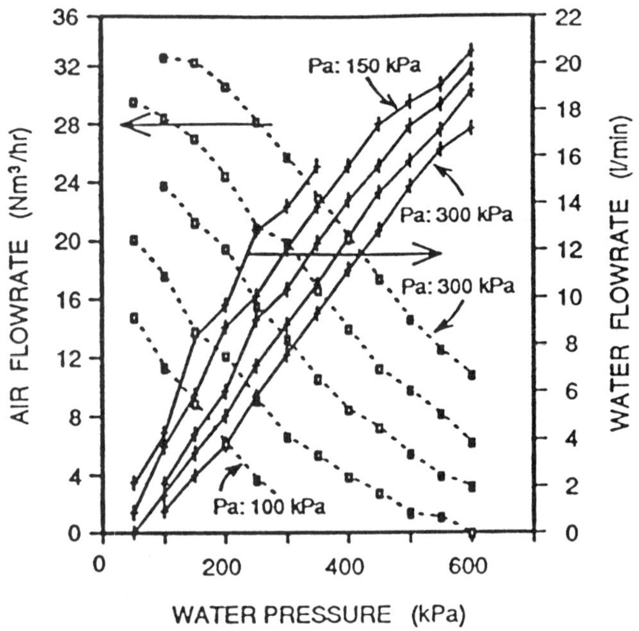

NOTE: The dashed lines represent air flowrate while the solid lines denote water flowrate

Fig. 7.4 Water flux versus water pressure and air flow

water, and the geometry of the mixing chamber and nozzle tip (Ref 34). The initial velocity of a droplet is a complex function of water and air pressures, the tube diameter of the nozzle, water flow rate, density of the water, and head loss through the tube, as indicated by Eq 7.1 (Ref 35):

$$v_0 = \sqrt{\frac{\dfrac{2(P - P_0)^{2/\rho}}{\rho} + \left(\dfrac{W}{15\pi D^2}\right)^2}{1 + \xi}} \qquad \text{(Eq 7.1)}$$

where v_0 is initial droplet viscosity, P is spray pressure, P_0 is air pressure, D is tube diameter, W is water flux, ρ is water density, and ξ is the head loss in the tube. The quenchant velocity at the surface of the plate is a function of the initial velocity, water and air density, droplet size, and the spray distance, as shown by Eq 7.2 (Ref 35):

$$V_t = V_0 \exp\left[-0.33\left(\frac{\rho_a}{\rho}\right)\frac{s}{d}\right] \qquad \text{(Eq 7.2)}$$

where V_t is the velocity of the droplet at the plate, V_0 is the initial drop velocity, ρ_a is air density, ρ is water density, s is distance traveled, and d is droplet size.

Consider a plate with a temperature well above the Leidenfrost point. The cooling of the plate by a spray quench will depend on the flow rate of water striking the plate. This in turn depends on the water flux, which is the amount of water striking a given area in a given amount of time. A given water flux can be delivered by a spray composed of a large number of small droplets or by a smaller number of large droplets. The velocity and kinetic energy of the droplet as it strikes the plate are a function of the droplet size.

Equation 7.2 indicates that the smaller drops will have a larger loss in velocity as they move toward the plate. The smaller drops will also have a smaller kinetic energy (momentum) because of their smaller mass. A droplet with a low velocity and a low momentum will flatten on the vapor film and run off. The larger droplets will have a low surface-to-volume ratio and will penetrate the vapor film more easily; however, the resulting water surface in contact with the hot metal plate is relatively small (Ref 34). The highest heat flux is obtained by droplets that have sufficient momentum to penetrate the film and reach the plate, but still have relatively small surface-to-volume ratios. This produces the maximum water surface area that comes in contact with the plate.

Optimum droplet diameter for heat transfer is reported to be on the order of 0.5 to 0.8 mm (0.02 to 0.03 in.) (Ref 28). With air-mist nozzles, the optimum range for droplet diameter is reported to be 0.1 to 0.2 mm (0.004 to 0.008 in.) (Ref 34). Hydraulic nozzles produce acceptable droplet diameters of 0.08 to 4.0 mm (0.003 to 0.16 in.) when the air flow to water flow is in the ratio of 1:10 to 1:100. If the cost of compressed air is balanced against cooling efficiency, then the optimum droplet diameter is reported to be 0.2 to 0.3 mm (0.008 to 0.012 in.). This is produced with an air flow to water flow ratio of 1:20 to 1:30 (Ref 34).

Experimental determination of droplet size distribution can be made using a Malvern particle sizer. This technique employs the diffraction of a laser beam by the spray droplets. Each droplet generates a cone of diffracted light that is dependent on the size of the droplet but not on its position in the beam. Smaller droplets generate a greater angle of diffraction. A series of detectors measure the intensity of light as a function of the diffraction angle. If it is assumed that the droplets are spherical, the angular dependency of intensity gives the distribution of droplet size. In order to allow a mathematical treatment of its distribution, a spray is often represented as being made up of droplets of only one mean diameter. A number of these mean or average diameters have been developed and have been summarized by Mugele and Evans (Ref 36). Two, which will be used in the subsequent discussion, are the Sauter mean diameter and the mass medium diameter, both of which can be used to mathematically model mass transfer.

The velocity distribution of spray droplets can be determined by using a two-dimensional gray-scale optical array imaging probe. This technique uses a linear array of photodiodes illuminated by a laser. As a droplet moves through the laser beam it casts a shadow, which can be tracked across the array. If it is assumed that the droplet is spherical, then the size of the shadow and the rate at which it moves across the array gives both the size and the velocity of the droplet. This technique can be used to make a measurement at any point within the spray field. One study, which reported the distribution of the particle sizes as mean diameters based on the weight frequency distribution (Ref 34), resulted in the data presented in Fig. 7.5, which shows changes across the spray pattern as a function of distance from the nozzle.

Deiters and Mudawar (Ref 15) have measured the distribution of the spray flux, particle size as represented by the Sauter mean diameter, and the mean drop velocity for many types of flat, full-cone, and hollow-cone sprays. To allow a mathematical treatment of the results, they developed an exponential function of values for spray flux, droplet size, and droplet velocity on an X and Y grid:

$$\varphi\,(X,Y) = A_0 e^{(A_1X + A_2Y + A_3X^2 + A_4XY + A_5Y^2)} \quad \text{(Eq 7.3)}$$

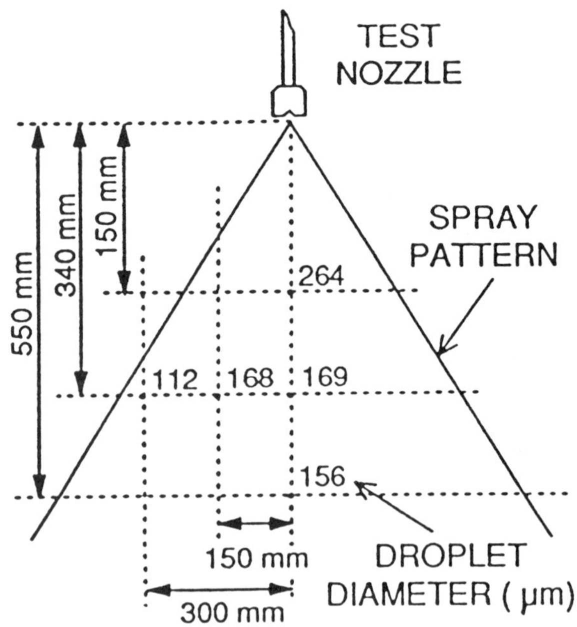

Fig. 7.5 Droplet size distribution for an air-mist nozzle

A least-squares fit of the experimental data gives values for A_1 through A_5. Figure 7.6 represents the distribution of the calculated values for these terms across the spray field.

Spray patterns are typically classified as either circular or elliptical (Ref 34). Circular spray heads are designed to deliver a spray that is symmetrical around the center point of the spray. Elliptical sprays, on the other hand, have very different distributions along the major (long) and minor (short) axes of the spray pattern. The water distribution along the major axis can be measured using an inclined tray divided into a series of channels (see Fig. 7.7) (Ref 34). These channels are of equal size and are arranged so that the water in each channel drains into a tube.

By measuring the volume collected in each channel and knowing the time duration of the spray, it is possible to calculate the average water flux distribution of the spray along this axis. The water flux is simply the volume of water delivered in a given amount of time to a given unit of area:

$$WF = \frac{W}{A} \qquad \text{(Eq 7.4)}$$

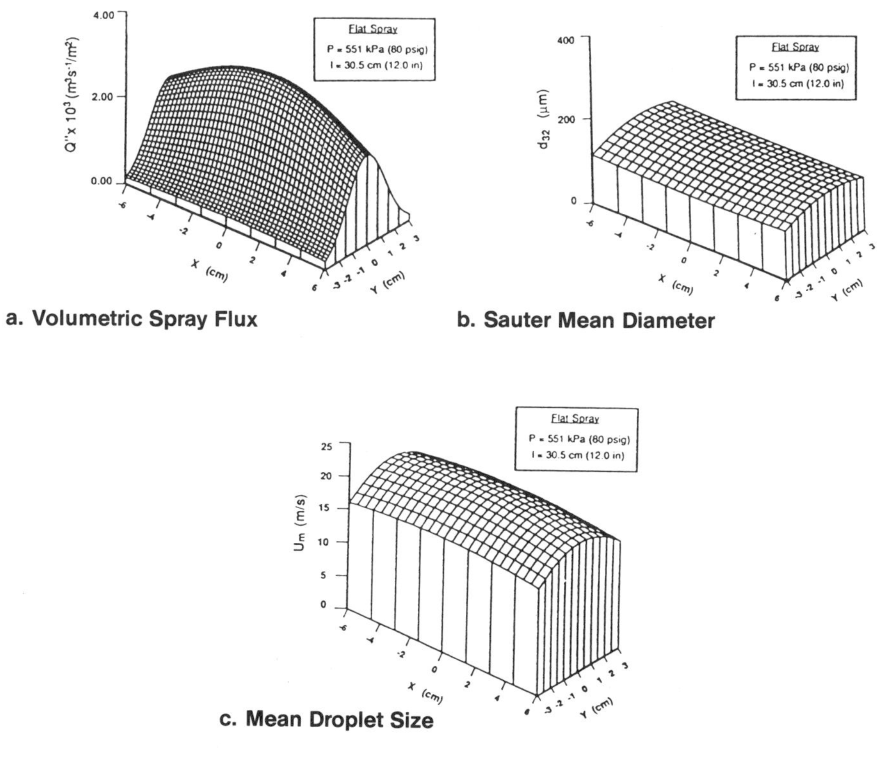

Fig. 7.6 Distribution of fundamental spray properties

where WF is water flux, W is water flow, and A is area. To measure the water flux along the minor axis, a series of transparent tubes were mounted in a rack (Fig. 7.7) and exposed to the spray for a given length of time. Using Eq 7.4, the flux was measured by dividing the volume of water in each tube by the diameter of the tube and the time the system was exposed to the spray.

The distribution of the water flux along the major and minor axes for an elliptical spray nozzle is shown in Fig. 7.8. The distribution is shown for a constant air pressure as a function of water pressure. The geometric dispersion of the spray is clearly dependent on the water pressure at the nozzle. Figure 7.9 shows the distribution of water flux for a Casterjet 3/8 CJL-5-120 cone nozzle as a function of water pressure (Ref 37).

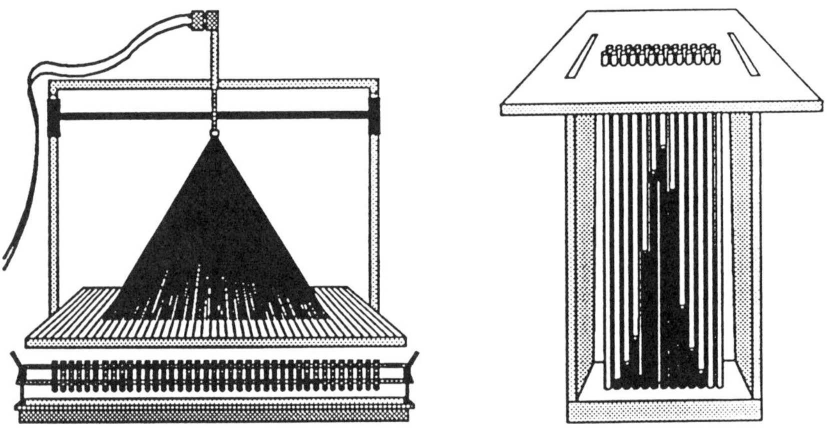

Spray flux on the major axis **Spray flux on the minor axis**

Fig. 7.7 Measuring devices for elliptical sprays

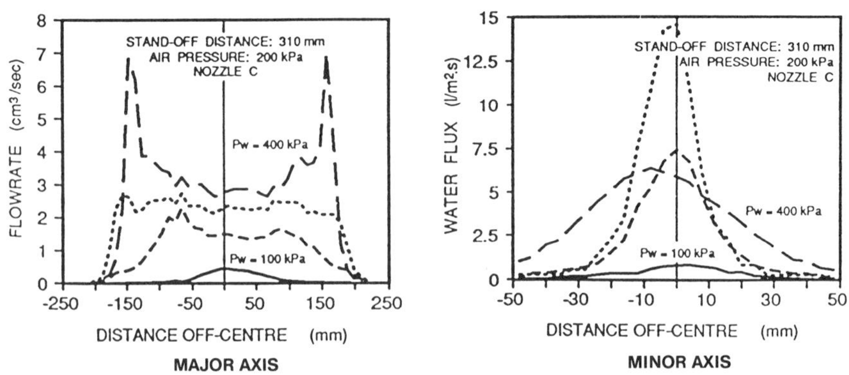

Fig. 7.8 Water flux versus water pressure

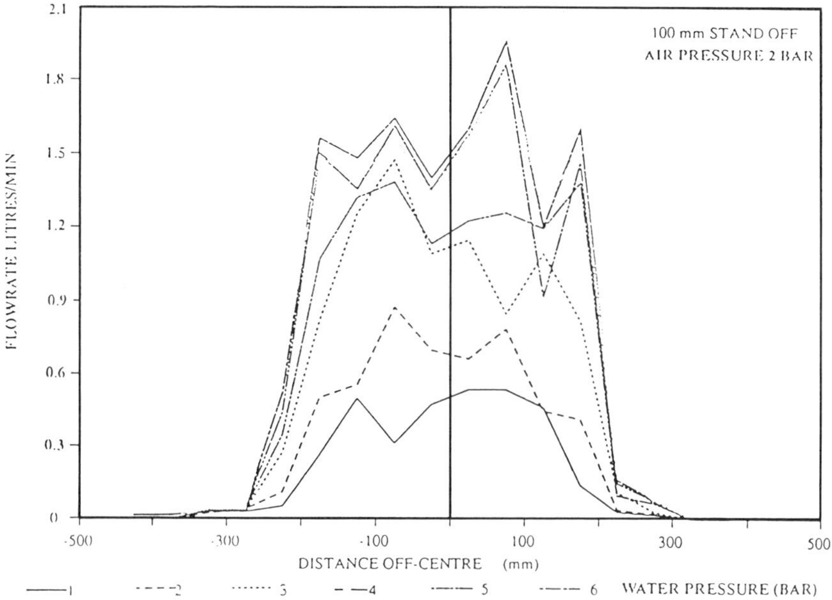

Fig. 7.9 Water flux distribution versus water pressure for a typical nozzle

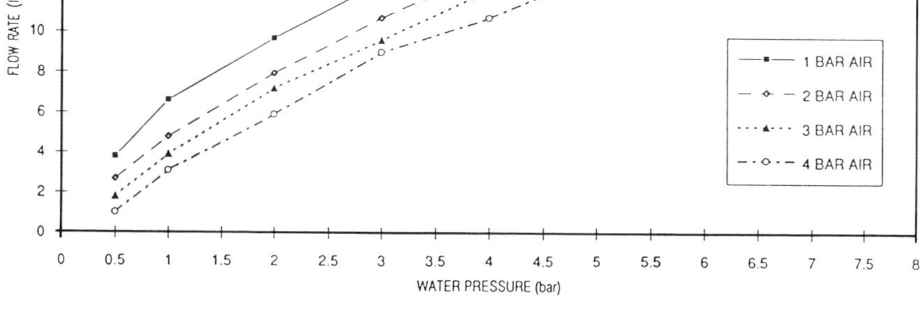

Fig. 7.10 Water flux for different water and air pressures

Examination of Fig. 7.8 and 7.9 shows that both the sensitivity of the water flux to the nozzle used and the distribution of the water flux over the sprayed surface are a function of water pressure. For a flat plate, the ideal nozzle would deliver a uniform spray flux across the entire spray pattern (Ref 38), and, if necessary, a series of nozzles would be arranged to give a uniform water flux over the entire surface. This spray pattern would produce the most uniform cooling and minimum stress in the plate. However, such spray uniformity is not possible in practice. In many cases, cooling uniformity is achieved by combining a good distribution of water over the surface with motion of the part. Moving the part through an uneven spray pattern may result in an averaging out of the asymmetries of the spray pattern and even cooling of the part.

The water flow from a nozzle is dependent on both water and air pressure. To fully characterize a nozzle, this dependence must be defined. Figure 7.10 shows reported results (Ref 37) for one nozzle design. The spray pattern was maintained over the full range of the pressures used. To allow mathematical modeling, the water flow from this nozzle was empirically fit to Eq 7.5:

$$W = 4.59\left(\frac{AP^{0.65}}{WP^{0.15}}\right) - 1.0 \qquad \text{(Eq 7.5)}$$

where *W* is water flow (L/min), *AP* is air pressure, and *WP* is water pressure. The data, presented in Fig. 7.10, and the model of those data, given by Eq 7.5, show the sensitivity of water flux to changes in water and air pressure. From these results, it is apparent that careful control of these pressures is necessary for reproducible cooling rates. These data also illustrate how the water flux, and thus the cooling rate, can be adjusted by changing these variables. Mathematical models of this type are useful for developing computer programs to control cooling rates by adjusting air and water pressures, which in turn change the water flux.

If the hot metal plate is moved farther from the nozzle, the droplet velocity will be reduced, because the droplet is progressively slowed by air resistance (hence the term for distance traveled in the denominator in Eq 7.2). As a result of this loss in kinetic energy, the droplet will be less likely to penetrate the vapor film and will therefore provide less cooling. The spray pattern will also spread out to cover a wider area, reducing the water flux at any given point and thus reducing heat removal. These two effects can be controlled by changes in water and air pressure, design of the nozzle, and the use of multiple nozzles. However, it is apparent that the optimum design of a spray quench system requires knowledge of the properties of the nozzle to be used and the amount and rate of heat removal required, as well as careful location of the nozzles.

Segerberg (Ref 5) has reported the cooling rates for a standard test piece as a function of the water flow rate in a spray. These data, presented in Fig. 7.11, were gathered by using a test probe containing a thermocouple. As expected, at low flow rates the point of maximum cooling is about 300 °C (570 °F), but the temperature of maximum cooling increases dramatically as the water flow rate increases. Figure 7.2 illustrates how increasing levels of agitation in a bath raises the temperature at which the vapor blanket is broken, thus allowing much more effective heat transfer due to nucleate boiling. In a similar manner, Fig. 7.11 demonstrates that higher water flow rates in a spray produce a

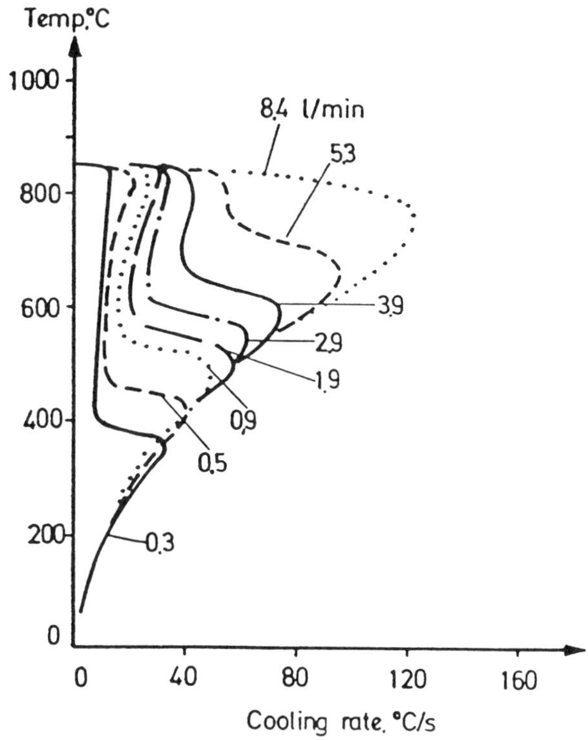

Fig. 7.11 Influence of water flux on the cooling curve with constant air flow

more severe quench. This increase in quench severity can be attributed to the mechanical energy of the water/air stream, which causes the vapor blanket to break down at higher and higher temperatures. The higher the temperature at which the vapor blanket mechanism is replaced by the much more effective nucleate boiling mechanism, the higher the overall heat-transfer coefficient and the more severe the quench.

A number of experimental procedures have been applied to the problem of determining the heat-transfer coefficient from a hot surface during spray quenching. A procedure called the steady-state technique uses a hot metal plate and measures the amount of energy needed to keep the surface at a constant temperature while the plate is exposed to a spray. The energy needed to maintain the surface temperature is then equal to the energy removed by the spray.

One steady-state study (Ref 34) used a plate with two thermocouples in line and centrally located at different depths in the plate. The plate was heated from below with a gas burner [other studies (Ref 13, 15) have used electrical heaters] to about 1000 °C

(1830 °F) and a constant spray was applied to the top surface of the plate. The burner was adjusted to give the desired temperature near the top surface and the heat input held steady until the thermocouples each gave a constant temperature reading. The assumption was made that the heat flow was one-dimensional, that is, only perpendicular from the heated surface to the cooled surface. Equation 7.6 was used to calculate the heat-transfer coefficient:

$$H_{tot}(T_s - T_w) = \frac{k(T)(T_2 - T_1)}{\Delta x} = \frac{Q}{A} = \text{heat flux} \quad \text{(Eq 7.6)}$$

where H_{tot} is the total heat-transfer coefficient, $k(T)$ is thermal conductivity, T_s is surface temperature, T_w is water temperature, T_1 is the temperature of thermocouple 1, T_2 is the temperature of thermocouple 2, Δx is thermocouple separation, Q is heat, and A is area. The temperature range was limited to 950 to 1200 °C (1740 to 2190 °F). Above this range the test plate underwent excessive oxidation (scaling), and below this range the steady-state conditions could not be maintained because of the breakup of the stable film boiling state.

Studies of aluminum quenching by Mudawar and coworkers (Ref 13, 15) have used an electrically heated plate similar in principle to the one described above. Because of the relatively long time spent at high temperatures, the aluminum oxidized severely. The oxidized surface influenced the transition temperature between the various regions of the cooling curve, resulting in poor reproducibility. After demonstrating the near equivalence of heat transfer with aluminum and copper test pieces, the work was completed using copper as the test material. In these studies, after the steady-state condition was reached and the data gathered, the power to the heater was increased slightly to produce a new steady-state condition. Thus, a series of data points could be generated in one set of experiments.

An alternate technique involves heating a thermocouple-equipped probe to a high temperature, followed by quenching. The temperature-time curve (cooling curve) is recorded and used to calculate the heat-transfer coefficient from the surface. Because the temperature is allowed to change, this is called an unsteady-state determination.

In one reported case of an unsteady-state determination (Ref 34), a steel block with a number of thermocouples was used as the test piece. The depth of the thermocouples was known, and successive approximations could be used to solve the heat-transfer coefficient at the surface if the temperature dependence of density, specific heat, and thermal conductivity of the metal were also known. Assuming the specific heat and density to be constant, a finite difference method was used to solve Eq 7.7 (Ref 34):

$$\rho C \frac{\partial T}{\partial t} = \frac{\partial}{\partial x}\left[k(t)\frac{\partial T}{\partial x}\right] \quad \text{(Eq 7.7)}$$

where ρ is density, C is specific heat, T is temperature, t is time, x is distance in the x direction, and k is thermal conductivity. Steady-state and unsteady-state experiments have been compared (Ref 34); their advantages and disadvantages are summarized in Table 7.1.

Mathematical modeling of the temperature-time profile of quenched parts is critical to the understanding and proper design of spray quenching systems. In a study of the secondary cooling of continuous castings using a pulverization spray head (water and no air), the heat-transfer coefficient at the hot surface was found to be (Ref 39, 40):

$$H_c = 333(WF)^{0.55} \quad \text{(Eq 7.8)}$$

where H_c is the heat-transfer coefficient and WF is water flux.

More recent work (Ref 40) has examined the heat-transfer coefficient from the upper surface of a horizontal steel plate over a temperature range of about 600 to 900 °C (1110 to 1650 °F) using a variety of pulverizing fan nozzles. In this study, the heat-transfer coefficient was found to be best represented by an equation of the same form as Eq 7.8 but with slightly different coefficients:

$$H_c = 423(WF)^{0.556} \quad \text{(Eq 7.9)}$$

Table 7.1 Advantages and disadvantages of unsteady-state and steady-state determinations

Unsteady-state test	Steady-state test
Advantages	
More convenient for determining heat transfer as a function of temperature (especially below the Leidenfrost temperature)	More convenient for determining heat transfer as a function of nozzle operating conditions
Use of three or more thermocouples allows cross check of results	3 to 4 min per determination
	Accuracy not dependent on response time of data acquisition
	Test piece does not move during test, leaving the thermocouple undamaged
Disadvantages	
Inaccurate at high heat flux because of the limited response time of thermocouples and the data acquisition system	Careful design of test piece required to obtain accurate heat transfer at high heat-extraction rates
Computationally difficult data analysis	Steady-state condition difficult to obtain below Leidenfrost temperature
Test piece is moved, which can damage thermocouples	Requires careful alignment of spray and burner to achieve one-dimensional heat flow
45 to 60 min per determination	Complex test piece

where H_c is the heat-transfer coefficient when sprayed from above.

Equations 7.8 and 7.9 were devised to describe the average heat-transfer coefficient over a temperature range. Further work has demonstrated that the rate of cooling is dependent on surface temperature as well as on the water flux (see Fig. 7.2). Mitsutsuka (Ref 27, 30) has used data from his own work, as well as from a number of reports from the literature, to define the empirical relationship between the heat-transfer coefficient and both the surface temperature and water flux:

$$H_c = 2.292 \times 10^8 \left(\frac{(WF)^{0.616}}{(\theta_s)^{2.445}} \right) \qquad \text{(Eq 7.10)}$$

where θ_s is surface temperature (for θ_s = 400 to 800 °C, or 750 to 1470 °F, WF = 10 to 2000 L/m^2 · min, or 0.25-49 gal/ft^2).

At the lower end of the applicable temperature range (400 to 500 °C, or 750 to 930 °F) the exponent for water flux in equations of the form of Eq 7.8 and 7.9 is found to be dependent on the water flux. A larger value is found for water flow rates of less than 200 L/m^2 · min and a smaller value for higher water flows (Ref 30). Other studies have shown the heat-transfer coefficient to be less sensitive to surface temperature as the temperature exceeds 600 °C (1110 °F) (Ref 27, 30). It is also known (Ref 27) that the heat-transfer coefficient goes through a maximum in the region of 200 to 300 °C (390 to 570 °F), as shown by Fig. 7.2. Equation 7.10 has been modified to account for both the high and low temperature deviations. The experimental work used a 100 × 50 × 16 mm (4 × 2 × 0.6 in.) stainless steel block cooled from an initial temperature of between 700 and 900 °C (1290 and 1650 °F). This block was moved through the spray header at accurately measured times ranging between 0.5 and 2 s. Heat loss was determined from the entire block and the results corrected for the radiative heat loss. The refined version of Eq 7.10 is shown in Eq 7.11 (Ref 41):

$$H_c = 3.15 \times 10^6 WF^{0.616} \left[700 + \frac{T_s - 700}{\exp\left(T_s - \frac{700}{10}\right) + 1} \right]^{-2.455}$$

$$\left[1 - \frac{1}{\exp\left[\frac{(T_s - 250)}{40}\right] + 1} \right] \qquad \text{(Eq 7.11)}$$

Equation 7.11 is reported to be valid for surface temperatures from 150 °C (300 °F) to at least 900 °C (1650 °F) and for water fluxes from 0.16 L/m^2 · s to at least 62 L/m^2 · s (4×10^{-3} to 1.5 gal/ft^2). Figure 7.12 shows the excellent agreement between the experimentally determined and the calculated results. The results of this work also demonstrate another

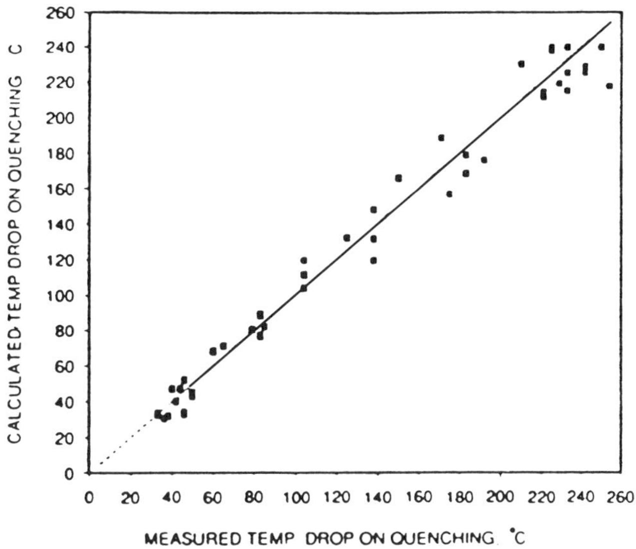

Fig. 7.12 Calculated and measured temperatures for Eq 7.11

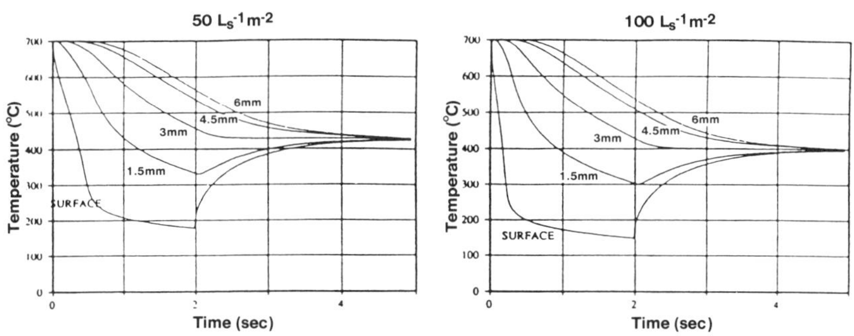

Fig. 7.13 Temperature distribution in a block with different water flow rates

important property of spray (and other) quench systems. Heat is removed from the surface so that, under conditions of high heat flux, the surface will cool rapidly, but the cooling of the interior depends on the thermal conductivity, specific heat, and density of the metal and not exclusively on the rate of removal of energy from the surface (Ref 19, 42). Figure 7.13 illustrates this principle. The cooling of a block is examined at two water flow rates. Doubling the water flow rate drops the surface temperature of the block much faster than is found with the standard water flow rate. However, after the surface reaches about 200 °C (390 °F), the rate of change in surface temperature is quite slow and not very dependent on water flow rate. The interior temperatures are not very sensitive to the higher water flux, and at the end of the quench cycle the blocks differ only marginally in total heat content.

The heat flux during spray quenching would be expected to vary with the metal being quenched (see Fig. 7.14) (Ref 43, 44). Heat transfer from a surface will depend on the rate at which heat is transferred to the surface from the interior of the metal part. As has been shown, this transport of heat to the surface is a function of the thermal conductivity as well as the density and specific heat of the metal. Surface effects can also be important, as some metals (for example, aluminum) will oxidize quickly at the temperatures of interest, while others (copper or stainless steel) will be more resistant (Ref 13, 15). In the case of steel, scale (which arises from oxidation) has a thermal conductivity of only about 10% of that of a clean, unoxidized steel surface. This surface oxidation can significantly affect the measured heat-transfer coefficient (Ref 45). The heat-transfer coefficient for spray cooling of a number of metals using different spray heads, various drop sizes, and velocities is given in Eq 7.12 (Ref 43):

$$H_c = 0.69 \log\left(\frac{V_m}{0.0006}\right)\left(1.4\sqrt{\lambda\rho C}\ ^{[0.32(\theta_s - \theta_c)/(\theta_b - \theta_c)]} + \alpha_v\right) + \alpha_{rad} \qquad \text{(Eq 7.12)}$$

where V_m is water flux density, λ is thermal conductivity, θ_s is the surface temperature of the quenchant, θ_c is the boiling point of the quenchant, θ_b is the quenchant temperature,

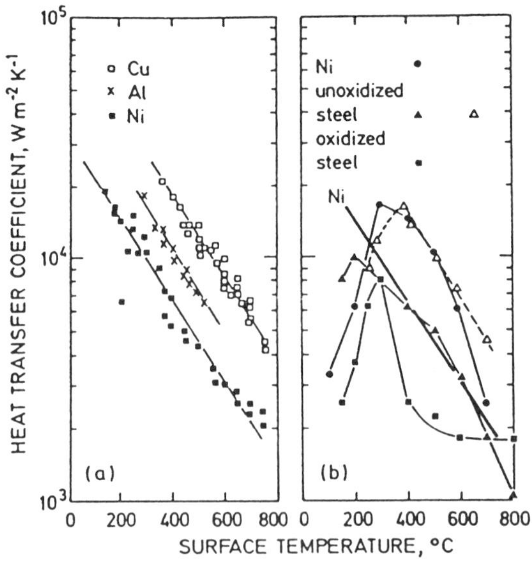

Fig. 7.14 Heat-transfer coefficient versus surface temperature for various metals

α_v is the film boiling heat-transfer coefficient, and α_{rad} is the radiative heat-transfer coefficient.

Special considerations are necessary when modeling the spray quenching used with continuous casting units and hot rolling mills. The hot plate moves in a horizontal plane and is subject to spray from above and below. Figure 7.3 shows that the two surfaces will experience different cooling environments. For a horizontal hot plate, the spray from above will produce a vapor blanket and water will spread over the flat vapor surface. The spray from below will also produce a vapor blanket, but excess water will fall away from the surface (see Fig. 7.15) (Ref 31). Therefore, even with identical water fluxes impinging on the top and bottom surfaces, the bottom surface will cool at a slower rate than the top surface. Under the same conditions used to develop Eq 7.9, but with the spray from below, the heat-transfer coefficient over a temperature range of about 700 to 1000 °C (1290 to 1830 °F) was determined to be:

$$H_c = 360(WF)^{0.556} \quad \text{(Eq 7.13)}$$

These results indicate that the cooling of the bottom surface of the plate is 15% less effective than the cooling of its top surface. With the same spray intensity, the difference

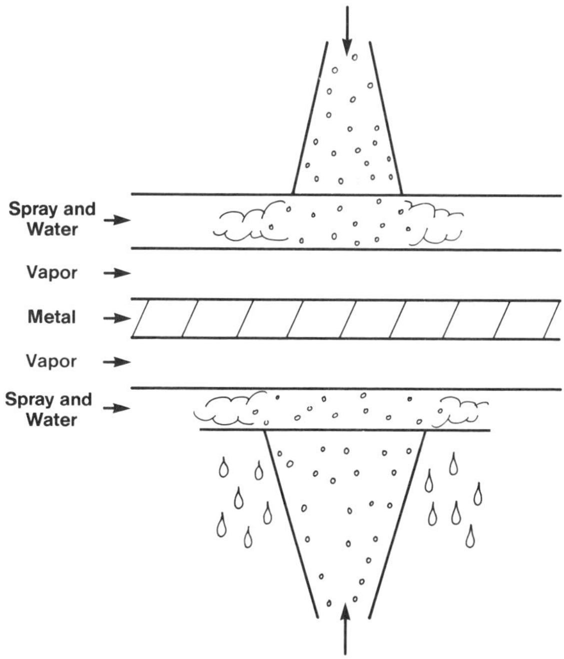

Fig. 7.15 Heat transfer in the vapor blanket portion of a spray quench with spray from above and below

in cooling rates of the top and bottom surfaces of a plate may need to be compensated for to avoid distortion of the plate.

Motion of the plate relative to the spray direction also affects cooling behavior. Plate motion results in a deflection of the falling droplets relative to the plate. To simulate this in the laboratory, a vertical water spray stream was deflected using a fan to induce a transverse air velocity over the plate (Ref 40). Relative to the quiescent system, the heat-transfer coefficient at higher water flow rates increased with higher transverse air flows.

Cooling for runout tables is generally one of two types. A number of headers and circular spray nozzles can be used. The disadvantage of this system is that the spray nozzlcs arc pronc to blockagc and thcrefore require substantial maintenance (Ref 12). An alternative is to use a laminar water curtain. This setup uses a long header that is mounted transverse to the motion of the plate. The header has a slit nozzle, which supplies a low-turbulence vertical sheet of water that impinges on the plate.

A study (Ref 12) of the cooling efficiency of a laminar water curtain as a function of plate velocity revealed a discontinuity in the cooling behavior. At a fixed spray rate, faster motion of the plate must result in a lower spray flux at any given point on the plate. A decrease in the water flux results in a decrease in the water film thickness. Slower motion of the plate produces a thick water film and efficient cooling. The film boiling mechanism provides sufficient cooling to quickly bring the temperature low enough for vapor film collapse, initiating the onset of nucleate boiling.

Table 7.2 Heat-transfer coefficients for spray-cooled systems

Equation	Correlation	Ref
A	$H_c = \dfrac{2.6612 \times 10^5\, W^{0.616}}{T_s^{\,2.445}}$	30
B	$H_c = 708W^{0.75}\, T_s^{\,-1.2} + 0.116$	47
C	$H_c = 0.581W^{0.451}\,(1 - 0.0075T_w)$	48
D	$H_c = 1.57W^{0.55}\,(1 - 0.0075T_w)\left(\dfrac{1}{\alpha}\right)$	49
E	$H_c = 0.423W^{0.556}$	50
F	$H_c = 1.571W^{0.55}\,(1 - 0.0075T_w)$	51
G	$H_c = 0.360W^{0.556}$	52
H	$H_c = 0.77\mathrm{b}W$	53
I	$H_c = 0.1W$	53
J	$H_c = 1.1611 \times 10^{-3}$ antilog $(0.663 \log W - 0.0014T_s)$	54
K	$H_c = 1.611 \times 10^{-3}$ antilog $(2.030 + 0.793 \log W - 0.00154T_s)$	54

Nucleate boiling increases the heat transfer, and the plate is more efficiently cooled. However, as the velocity of the plate increases, the water flux and thus the film thickness decrease, and the cooling of the plate becomes less efficient. At some plate velocity, the water flux and film thickness have decreased to a point where the plate surface does not remain below the Leidenfrost temperature. The residual heat in the interior of the plate is such that, after passing through the spray, the surface temperature increases. If the film thickness is thin, which is the case for fast plate motion, then the surface will reheat to above the Leidenfrost temperature. Therefore, above a critical velocity, some of the heat transfer occurs in the less efficient film boiling region of the cooling curve (see Fig. 7.2), while below this critical velocity the much more efficient nucleate boiling mechanism will define the heat-transfer coefficient.

Thus, there is a discontinuity in the relationship between heat transfer and plate velocity. Under the conditions of one study (Ref 12), the critical velocity was 1.2 m/min (4 ft/min). This critical plate velocity would be expected to be a strong function of water flux and plate thickness. The retention of heat in the core of a part and the resulting heating of the surface are necessary for self-tempering, which will be discussed later in this chapter.

Many studies have examined the cooling of steel plates using empirical models developed by methods similar to those just described. Although many of these studies generate equations that give an excellent fit to the systems for which they were developed, they do not seem to be generally applicable to other systems. A comparative analysis of a number of these models was conducted to determine how well they could predict the cooling behavior of steel rods in the precooling portion of a Stelmor machine (Ref 46). Eleven different models, some synthesized using the results from a number of different studies, were compared with the experimentally obtained results. The models used are given in Table 7.2. Model I was found to give the best fit to the data.

This work clearly shows the limitation of using empirical models. A model derived for one system may give an excellent fit to data for that system but a poor fit to data for another, even closely related system. Factors that constrain the generalization of empirical models include:

- Type and arrangement of nozzles
- Overlap of spray patterns
- Distance between nozzle and surface
- Water pressure and flow
- Air pressure and flow
- Droplet size distribution of spray
- Flow of sprayed quenchant over part
- Quenchant composition and temperature
- Part size, shape, surface roughness, and alloy
- Ratio of cooled surface to total surface and to volume

- Quench temperature range examined
- Duration of quench
- Temperature distribution in part
- Rate of data acquisition
- Presence and amount of scale or surface oxidation
- Orientation of spray
- Orientation of part relative to spray (above or below)
- Motion of part relative to spray

Spray quenching is used to cool steel from temperatures well above the Leidenfrost temperature; therefore, studies of steel cooling rates tend to focus on the film boiling region of the cooling curve (see Fig. 7.1). For aluminum, the focus is on cooling from about 525 °C (975 °F). Depending on the spray, this will be near, or perhaps below, the onset of nucleate boiling (Ref 15). To completely characterize the thermal history of aluminum during spray quenching, it is necessary to fully describe the complex shape of the cooling curve below the Leidenfrost temperature. Mudawar and Valentine (Ref 13) attempted to derive general relationships for the spray quenching of aluminum by using localized measurements of fundamental spray parameters at the heat-transfer surface. They developed equations for heat flux using volumetric spray flux, mean drop diameter, and mean drop velocity. This use of fundamental properties of the spray gives results independent of the type and arrangement of nozzles, nozzle pressure or exit velocity, and distance of the nozzle from the surface. This approach was developed to avoid errors caused by lack of control of some of the variables listed above.

In the region of the cooling curve in which only convective cooling occurs (see Fig. 7.1), it was possible to correlate the cooling rate using terms that take the mathematical form of the Reynolds, Prandtl, Stanton, and Nusselt numbers. In the case of heat transfer from a hot surface to a turbulently flowing liquid, the dimensional relationship between these terms is given by:

$$N_{\mathrm{N}} = N_{\mathrm{St}}\, N_{\mathrm{Re}}\, N_{\mathrm{Pr}} \qquad \text{(Eq 7.14)}$$

where N_{N} is the Nusselt number, N_{St} is the Stanton number, N_{Re} is the Reynolds number, and N_{Pr} is the Prandtl number. In the case of a flowing liquid, the Reynolds number is related to the turbulence of flow, the Prandtl number refers to the relationship of the momentum and thermal diffusivities of the flowing liquid, and the Nusselt number refers to the relationship between the temperature gradient at the wall and in the liquid. The Stanton number is the slope of the relationship defined by Eq 7.14 (Ref 55). For spray quenching, these terms were modified: the flow velocity was converted to the velocity of the droplets at the surface, and the term for mass of flow was converted to a term based on spray flux and the mean droplet diameters. The relationship of these terms can be described by:

$$\mathrm{N}_{32} = 2.512(\mathrm{Re}_{32})^{0.76}\mathrm{Pr}_{\mathrm{f}}^{0.56} \qquad \text{(Eq 7.15)}$$

$$N_{0.5} = 2.569(\mathrm{Re}_{0.5})^{0.78}\mathrm{Pr}_f^{0.56} \quad \text{(Eq 7.16)}$$

In the case of heat transfer in laminar flow, the exponent for the Reynolds number is 0.5, while that for the Prandtl number is 0.33. In Eq 7.15 and 7.16, these terms are based on the volumetric spray flux and the droplet diameter. The subscript 32 refers to the values derived when the mean droplet diameter is expressed as the Sauter mean diameter; the subscript 0.5 refers to the use of the mass medium diameter.

The temperature of the metal surface in the incipient boiling region was related to the Reynolds and Prandtl numbers, the thermal conductivity of the liquid, and the temperature of the water:

$$T_{sur} = 13.43(\mathrm{Re}_{32})^{0.167}\mathrm{Pr}_f^{0.123}\left[\frac{K_f}{d_{32}}\right]^{0.220} + T_{eq} \quad \text{(Eq 7.17)}$$

$$T_{sur} = 13.50(\mathrm{Re}_{0.5})^{0.172}\mathrm{Pr}_f^{0.123}\left[\frac{K_f}{d_{0.5}}\right]^{0.220} + T_{eq} \quad \text{(Eq 7.18)}$$

where T_{sur} is the temperature at the metal surface, K_f is the thermal conductivity of the water, $d_{0.32}$ is the Sauter mean diameter of the droplet, $d_{0.5}$ is the mass medium diameter of the droplet, and T_{eq} is the temperature of the water. The hydrodynamic properties (water flux, droplet velocity, and size) of the spray did not seem to impact on the heat flux in the nucleate boiling portion of the cooling curve. The only factors found to impact on the heat flux were the temperatures of the metal surface and of the water:

$$q = 1.87 \times 10^{-5}(T_{sur} - T_f)^{5.55} \quad \text{(Eq 7.19)}$$

where q is heat flux and T_f is the temperature of the water. This was examined over the narrow water temperature range of 22.5 to 23.5 °C (72.5 to 74.3 °F). Although the fit of the equations to the experimental data is excellent, there must be a lower level of spray flux that will impact on the heat flux during the nucleate boiling portion of the cooling curve.

The point where the greatest heat flux occurs during the cooling process is referred to as the critical heat flux. The surface temperature at this point is defined by the densities of the water vapor and the liquid; by the specific heat, latent heat of vaporization, and surface tension of the water; by volumetric spray flux; and by drop diameter, as shown in Eq 7.20 to 7.23:

$$\frac{q_{max}}{\rho_g h_{fg} Q} = 122.4\left[1 + 0.0118\left(\frac{\rho_g}{\rho_f}\right)^{0.25}\left(\frac{\rho_f c_{pf}\Delta T_s}{\rho_g h_{fg}}\right)\left(\frac{\sigma}{\rho_f Q^2 d_{32}}\right)\right]^{0.198} \quad \text{(Eq 7.20)}$$

$$T_s = 18\left((\rho_g h_{fg} Q)\left[\frac{\sigma}{\rho_f Q^2 d_{32}}\right]^{0.198}\right)^{\frac{1}{5.55}} + T_f \quad \text{(Eq 7.21)}$$

$$\frac{q_{max}}{\rho_g h_{fg} Q} = 134.3\left[1 + 0.0118\left(\frac{\rho_g}{\rho_f}\right)^{0.25}\left(\frac{\rho_f c_{pf} \Delta T_s}{\rho_g h_{fg}}\right)\left(\frac{\sigma}{\rho_f Q^2 d_{0.5}}\right)\right]^{0.192} \quad \text{(Eq 7.22)}$$

$$T_s = 18.3\left((\rho_g h_{fg} Q)\left[\frac{\sigma}{\rho_f Q^2 d_{0.5}}\right]^{0.192}\right)^{\frac{1}{5.55}} + T_f \quad \text{(Eq 7.23)}$$

where q_{max} is critical heat flux, ρ_g is vapor density, ρ_f is water density, C_{pf} is the specific heat of the water, ΔT_s is saturation temperature less water temperature, σ is the surface tension of the water, h_{fg} is the latent heat of vaporization of the water, and Q is volumetric spray flux. The form of the equations for heat flux were derived from earlier work on saturated flow boiling systems with a term to account for the experimentally determined subcooling.

In the transition boiling region, the ratio of the heat flux to the maximum heat flux was correlated with the mean drop velocity, the volumetric spray flux, the temperatures of the surface and of the water, and the maximum surface temperature:

$$\log_{10}\left[\frac{q}{q_{max}}\right] = 4.78 \times 10^5\left[\frac{u_m}{Q}\right]^{-1.255}\left[\log_{10}\left(\frac{T_s - T_f}{T_m - T_f}\right)\right]^3 - 1.90 \times 10^4\left[\frac{u_m}{Q}\right]^{-0.903}\left[\log_{10}\left(\frac{T_s - T_f}{T_m - T_f}\right)\right]^2 \quad \text{(Eq 7.24)}$$

based on the use of d_{32}, and

$$\log_{10}\left[\frac{q}{q_{max}}\right] = 1.90 \times 10^5\left[\frac{u_m}{Q}\right]^{-1.144}\left[\log_{10}\left(\frac{T_s - T_f}{T_m - T_f}\right)\right]^3 - 1.06 \times 10^4\left[\frac{u_m}{Q}\right]^{-0.834}\left[\log_{10}\left(\frac{T_s - T_f}{T_m - T_f}\right)\right]^2 \quad \text{(Eq 7.25)}$$

based on the use of $d_{0.5}$, where u_m is mean drop velocity, Q is volumetric heat capacity, and T_{max} is surface temperature at critical heat flux.

The heat flux at the Leidenfrost temperature was fit to an equation that relates vapor density, latent heat of vaporization of water, volumetric spray flux, and mean drop velocity:

$$\frac{q_{\mathrm{Leid}}}{\rho_g h_{fg} Q} = 0.145\left(\frac{u_m}{Q}\right)^{0.834} \quad \text{(Eq 7.26)}$$

where q_{Leid} is heat flux at the Leidenfrost temperature and Q is volumetric spray flux.

This study illustrates the detailed information that is necessary to describe, in a fundamental way, the full range of heat-transfer coefficients of a spray cooling curve. Empirical correlations can be developed more easily, but they will be limited in terms of generality by the impact of some or all of the variables listed earlier.

Results Obtained With Spray Quenching

One advantage of spray quenching, relative to other quench methods, is that a large and adjustable range of cooling rates is achievable by simple changes in flow rates and pressures. The high rates of heat extraction possible with sprays are critical for attaining a good depth of hardness. Segerberg (Ref 5) has examined this and illustrated the results with the cooling curves shown in Fig. 7.16. Immersion quenching was performed using an oil and a polymer solution. The water flux for the spray quench was chosen to give a cooling curve with a vapor film of the same duration as that for the oil immersion quench. The quantitative impact of the higher rate of cooling found with the spray system was determined using 35 × 100 mm (1.4 × 4 in.) cylindrical samples of SAE 52100 ball bearing steel. These cylinders were heated to 850 °C (1560 °F) and quenched in a fast oil at 50 °C (120 °F), or a 15% aqueous solution of a polyalkylene glycol (PAG) polymer at 25 °C (77 °F), or by the water spray quenching system. Hardness distributions along the diameter of the cylinder are shown in Fig. 7.17. The higher cooling rates achieved by the immersion quench in the PAG solution and by the water spray quench account for the greater depth of hardness shown in Fig. 7.17.

Depth of hardness is limited by part size and shape for all types of quenching (Ref 19). Temperature profiles as a function of depth for air, oil, water, and spray quenching have been reported (Ref 19). It was shown that heat removal from the core of the part can become limited by the rate of heat transfer from the interior to the surface. Thus, for a 50 mm (2 in.) carbon steel bar, the temperature 10 mm (0.4 in.) below the surface was found to be dependent on the water pressure up to a point beyond which a further increase in the pressure had a negligible impact on the rate of cooling of the interior (as in Fig. 7.12). The high rate of surface cooling relative to that of the interior resulted in a high compressive residual stress in the surface, which improved the fatigue strength of the part.

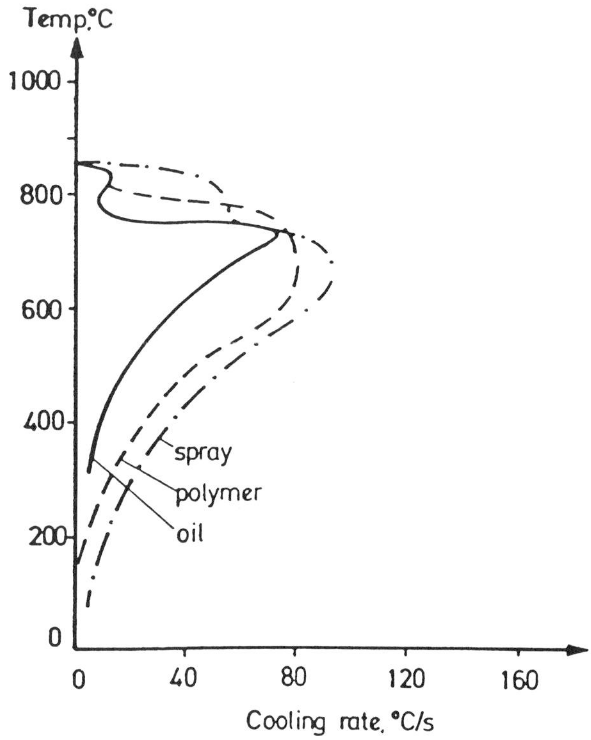

Fig. 7.16 Cooling rates for spray, polymer, and oil immersion quenching

The hardness profiles and surface residual stresses for six grades of spray-quenched commercial carbon steel shafts containing 0.33 to 0.54% C were reported. The hardness profiles are shown in Fig. 7.18 (Ref 19). The surface hardness, core hardness, and depth of hardening all increase with carbon content. In one study of a high-pressure water spray, a correlation between depth of hardness and hardenability was developed (Fig. 7.19) (Ref 19). The fatigue strength of the parts (shafts) was superior to that found with parts heated by induction and then quenched with a low-pressure water spray (Fig. 7.20). The high-pressure spray gave greater heat-transfer rates and higher surface residual stresses. The use of spray quenching allowed the use of carbon steel instead of a more expensive alloy steel (Ref 19).

Fast surface cooling can lead to stress cracking and distortion. It is reported that for steels with a carbon content of between 0.5 and 1.0%, quench cracks can be avoided if there is uniform cooling of all surfaces. This uniform cooling reduces the thermal stress between different regions of the parts (Ref 56). Self-tempering is also reported to be

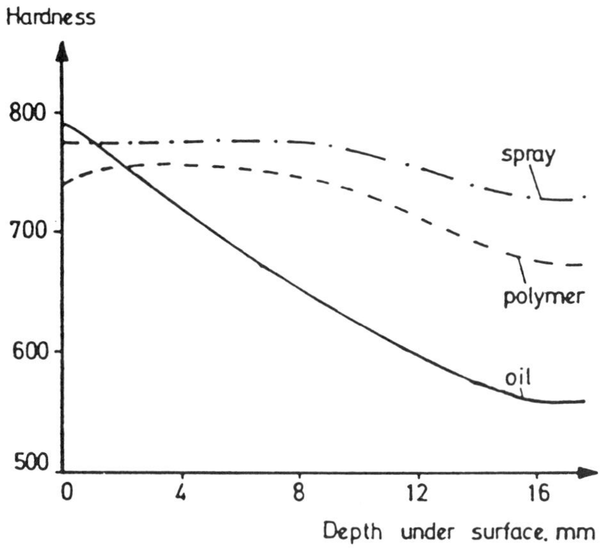

Fig. 7.17 Hardness distributions for spray, polymer, and oil immersion quenching

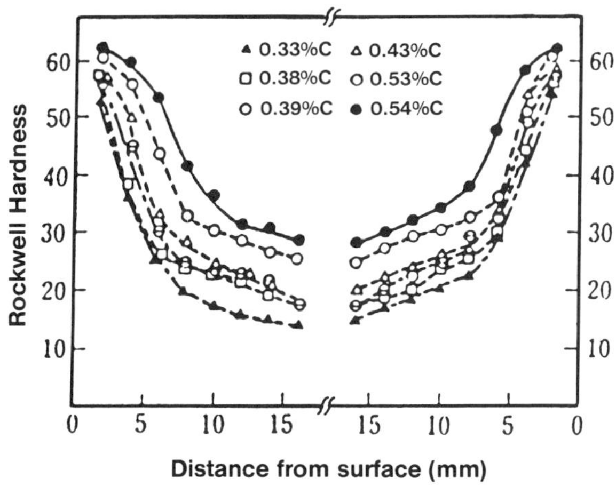

Fig. 7.18 Hardness profiles for high-pressure water-quenched steel shafts

valuable in the reduction of stress cracking for through-heated parts. Optimal self-tempering is reported to occur at 150 to 200 °C (300 to 390 °F), but in some cases it may be necessary to furnace temper even after self-tempering (Ref 56).

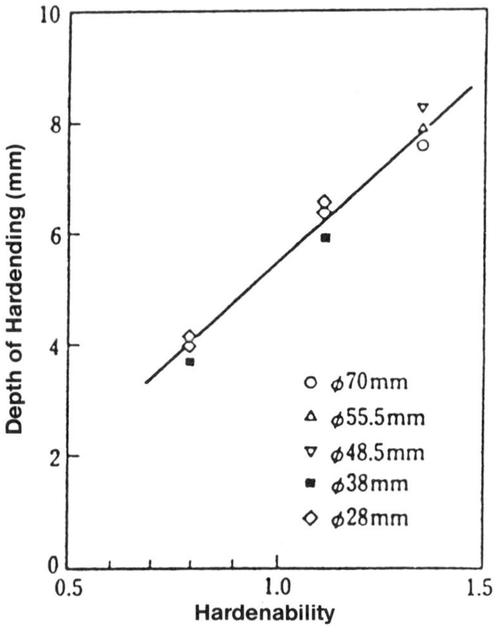

Fig. 7.19 Comparison of hardenability and hardened depth

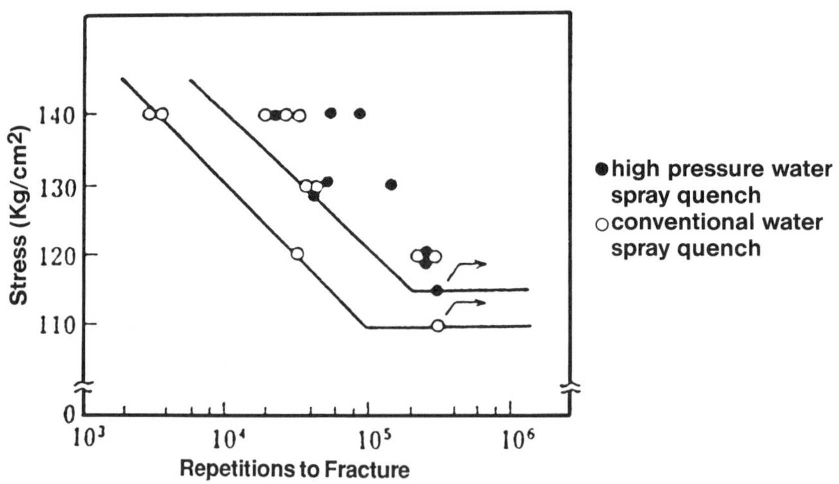

Fig. 7.20 *S-N* chart for hollow shafts

Deformation of parts that have been only surface heated and spray quenched is limited and is no greater than that of oil-quenched parts (Ref 56). Parts that have been through heated experience greater deformation but, by appropriate design of the spray quench assembly, a uniform increase in the dimensions of the part can be achieved. This uniform increase is realized by ensuring similar cooling rates on all surfaces to minimize distortion. Uneven cooling caused by clamps or supports that hold or position the part should be avoided. In some cases, it is possible to calculate and compensate for dimensional changes that occur on quenching (Ref 56).

Surface-heated and spray-quenched parts will have a surface hardness that is much greater than the hardness of the interior (Fig. 7.21). An exception to this rule occurs for thin parts in which the cooling through the entire part is fast enough to leave the part through-hardened. More rapid cooling of the surface than the core results in a surface compressive stress that improves fatigue resistance. Rail liner pads that have been spray quenched are reported to have ductility values similar to those for water-quenched parts, but their fatigue limit is increased by a factor of two (Ref 56).

Surface cooling must be uniform to avoid quench cracking. Nozzle arrangement, size of spray openings, and spray patterns that do not excessively overlap are all critical (Ref 57). If the cooling is not uniform, then the more rapidly cooled regions will be the first to form martensite. If this region is small (e.g., the point of impact of a stream from the spray), then there will be large local surface stresses, which may be sufficient to initiate cracks. This is reported to be the main cause of crack formation (Ref 57) during spray quenching.

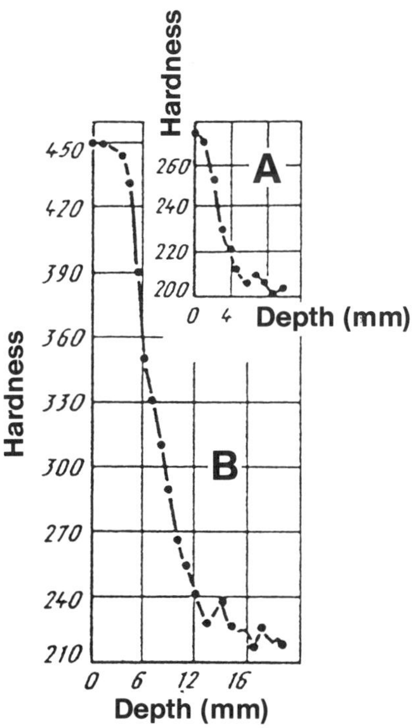

Fig. 7.21 Hardness profile after spray quenching of steel

Steel produced in a rolling mill undergoes a plastic deformation, resulting in a more uniform distribution of alloy elements in the austenite grain and dissolution of deleterious impurities in the region of the grain boundary (Ref 58). This produces finer austenite grains, which give carbon and low-alloy steels high strength, high fatigue limits, and improved low-temperature toughness (Ref 59). Improvements have been claimed in the production of reinforcement rods, low-carbon wire rods, large-diameter pipe, rolled mill balls for the mining industry, railroad wheels and rails, and structural steel from continuous rolling mills (Ref 58).

The Kobe Steel controlled rolling and accelerated cooling process for steel plate has been described and is shown in Fig. 7.22 (Ref 60). Chiefly used in the production of steel plate for shipbuilding, the top of the plate is cooled using slit-jet cooling in combination with laminar cooling at the high-temperature zone. The cooling of the bottom of the plate uses spray cooling, also in combination with slit-jet cooling. This results in a high and uniform hardness across the plate and can be used for cooling ranging from accelerated cooling to direct quench. The need for top restraint rollers is avoided, and the resulting plate has uniform mechanical properties and minimal distortion (Ref 60).

Thermomechanical-controlled processing (TMCP), also called thermomechanical treatment (TMT), has been described along with the use of a laboratory spray cooling unit (Fig. 7.23) to simulate the in-line heat treatment of steel (Ref 37). These methods

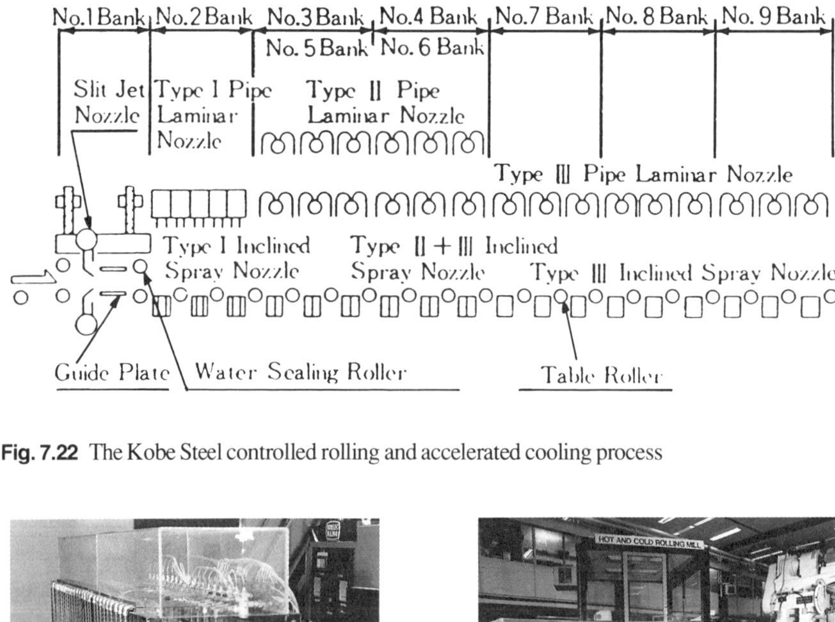

Fig. 7.22 The Kobe Steel controlled rolling and accelerated cooling process

Fig. 7.23 Laboratory spray cooling unit. Courtesy of BHP Research, Melbourne, Australia

depend on the control of microstructure both by quenching and by deformation (Ref 59). Applications include recrystallization-controlled forging (RCF) followed by direct quenching and self-tempering (QST) and the processes to produce high-strength, high-elongation hot-rolled strip steel (Ref 37). The system uses an experimental rolling mill and a series of spray nozzles. The microstructure shows a tempered martensite layer on the surface, with bainite-martensite, bainite-pearlite, and finally ferrite-pearlite at progressively greater depths (Fig. 7.24) (Ref 37).

The use of water to cause the descaling of thin steel plates has been described (Ref 61). Cooling of the surface oxide coating sets up a stress in the oxide layer that causes it to flake off the steel surface. The stress in the oxide is related to the modulus of elasticity, the thickness, and the coefficient of thermal expansion of both the oxide and metal as well as the temperature change produced by the evaporation of the impinging droplet. The mathematical model used to define the stress assumes the complete evaporation of the droplet and uses a two-dimensional finite-difference method to determine the temperature distribution of the moving plate. Unfortunately, the required physical properties of the oxide layer (the coefficient of thermal expansion and the modulus of elasticity) are not available. The authors concluded that droplet diameter and quenching speed were important variables and that a significant portion of the stress in the oxide layer is attributable to convective heat transfer.

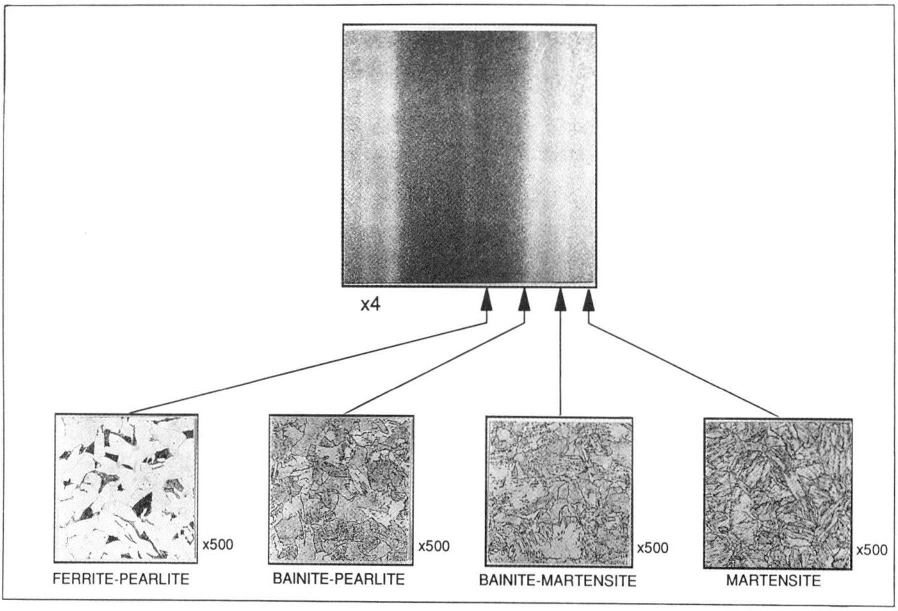

Fig. 7.24 Microstructure produced by thermomechanical treatment and spray quenching. Courtesy of BHP Research, Melbourne, Australia

Spray Quenching and Induction Hardening

Spray quenching is often associated with induction hardening. Induction heating can be used with other quenching methods, but it is most often used with either a spray or a simple agitated immersion quench; in many cases, the immersion quench is simply an open or submerged spray quench arranged so that the flow is sufficient to keep the part completely covered with quenchant. This section will examine the fundamentals of induction heat treatment, the design of induction quench systems, and the stress and hardness profiles for induction-surface-hardened steel and will provide examples of applications.

Induction heating is performed by using a fluxional magnetic field to heat a conductive material. When a rapidly alternating electric current passes through a conductor, it generates a rapidly alternating magnetic field. This magnetic field will, in turn, induce a current in an adjacent electrical conductor. The induced current will alternate at the same rate as the original current. The induced current flow generates heat because of the electrical resistance of the material.

Electrical power is a function of the amperage and the resistance as shown by:

$$P = \frac{W}{t} = (I \cdot E) \cos \phi \qquad \text{(Eq 7.27)}$$

where P is electrical power, W is watts, t is time (s), I is amps, E is volts, and ϕ is the phase angle. If the material to be heated is magnetic, then there is also a heating effect caused by the hysteresis of the alternating magnetic field that is induced. In the case of steel, this hysteresis effect will contribute to heating until the Curie point of the steel is reached. The Curie point is the temperature above which a material is no longer ferromagnetic. Above this temperature, only the electrical resistance contributes to heating. The Curie point for steel is about 730 °C (1365 °F) (Ref 62).

The depth to which the part is heated by induction is a function of the frequency of the current, the power applied, the relative positions of the part and the coil, and the time the sample is exposed to the heating cycle. If all other factors are held constant, a higher-frequency current will heat the part to a more shallow depth. This is an important consideration, because it is possible to treat the surface with a time-temperature profile that is significantly different from that applied to the bulk of the part. This control of the depth of heating allows hardening of the surface while retaining desirable properties in the core. Table 7.3 (Ref 63) gives common frequencies used for different hardened depths (after quenching), and Table 7.4 (Ref 63) gives values for through hardening of various size parts.

A significant advantage of induction in either surface or through-heating applications is the rapid rate of energy transfer, which can be important in reducing surface oxidation (Ref 63) and decarburization (Ref 22, 62). Protective atmospheres can be provided in cases of high sensitivity to oxidation (Ref 63). For example, a 50 mm (2 in.) diameter steel shaft that weighs 1.54 kg (3.4 lb) can be through heated to 850 °C (1560 °F) in 20 s (Ref 62). The high rate of power input that is possible with induction heating can lead to a high level of thermal stress. Just as quenching can lead to cracks because of thermal stress, it is possible to develop sufficient temperature gradients within a part being heated to cause it to crack (Ref 62). It is necessary to determine the source of any cracking that occurs with induction heating, rather than assuming that stress cracks must be due to the quench.

Induction heating units are classified as single shot (static) (Ref 64) or progressive (scanning) (Ref 65). In single shot heating, the entire region to be heated is exposed to the heating coil at once. This usually requires relatively small parts. In progressive induction heating, either the part or the induction coil is moved relative to the other. Thus, only a portion of the part is heated at any given time. With either technique, it is possible to apply heat to only portions of the part or to apply different amounts of power and different depths of heating to different portions of the part.

Table 7.3 Induction hardening frequencies for ferrous metals

Hardened depth		
mm	in.	Frequency, Hz
<1.59	< 0.0625	450,000
1.59-2.38	0.0625-0.094	100,000
2.38-3.18	0.094-0.125	10,000
3.18-6.35	0.125-0.25	3000
6.35-9.53	0.25-0.375	1000

Table 7.4 Induction through-hardening frequencies

Cross section		Frequency, Hz		
mm	in.	Magnetic	Nonmagnetic	Nonferrous
> 175	>7	60	60	60
100-175	4-7	60	200	60
75-100	3-4	200	600	200
50-75	2-3	600	1000	600
25-50	1-2	1000	3000	1000
13-25	0.5-1	3000	10,000	3000

Table 7.5 Control parameters for heat treating

Process	Heating	Cooling	Properties
Hardening	Temperature	Critical	Strength/hardness
Tempering	Temperature	Noncritical	Toughness
Normalizing	Temperature	Noncritical	Grain refinement
Annealing	Temperature	Critical (slow)	Softness
Stress relieving	Temperature	Noncritical	Straightness
Solution heat treating	Time at temperature	Critical (some)	

Control of the heating cycle as well as the cooling cycle is critical in induction heating and quenching. With induction heating, the rate at which energy is transferred to the part is very high. Thus, unlike furnace heating, it is possible to heat carbon steel to austenitizing temperatures in several seconds (Ref 66). Achieving optimum properties from different heat treating processes may require the use of accurate controls (Ref 67). The various heat treatment processes and the required control accuracy for each are shown in Table 7.5 (Ref 65). The types of controls and examples of their use have been discussed (Ref 66, 67).

A number of factors must be considered in the design of a quench system for induction heaters (Ref 8, 68). The size and geometry of the part determine the amount of heat that must be removed; the ratio of surface area to volume is important in determining how fast the part will be cooled. Parts containing abrupt changes in geometry due to flanges or keyways require care to avoid uneven cooling, which can result in stress cracking. The depth of hardening required is also important. A part that is to be surface hardened will require less heat removed than a similar part that is to be through hardened.

Fig. 7.25 Combined inductor and spray head

The structure and composition of the metal to be heat treated are also influential factors. Obtaining acceptable properties of the quenched part over the full range of the variations of composition allowed by the specifications for the alloy used is critical. The previous thermal and mechanical history of the part may be as important as the composition of the alloy. Characteristically, induction heating involves fast heating followed quickly by quenching. The distribution and size of the carbides in the part will be important because the smaller particles will more readily go into solution. Therefore, rolled, forged, annealed, or quenched and tempered parts will each be affected differently by the rapid heating and cooling cycles.

In many spray-quenched induction systems, the quench is closely coupled in both time and space to the heating elements; in fact, they are often part of the same assembly. An example of a combined inductor and spray head is shown in Fig. 7.25 (Ref 22). Many of the units designed for scanning a part are used to surface harden cylindrical shapes. To achieve even heating and quenching, the part is generally rotated during the heating/cooling cycle. If only the surface is to be hardened, then the part may not need to be supported, because the cooler core will be rigid and maintain the shape. For deeper heating, it may be necessary to provide supports to help maintain the shape during the heating cycle. It is important that the system be designed so that the part is centered; otherwise the surface temperature, depth of hardening, and rate of cooling will be uneven.

Figure 7.26 illustrates how the spray pattern will impact on a rotating shaft that is not mounted in the center of the assembly. If the shaft is scanned by an inductor as it rotates (the shaft has a motion relative to the inductor as shown), then a "barber pole" pattern of spiral hardness lines may result (Ref 8, 68). The uneven cooling of rotating shafts can also be a problem when the shaft deviates from the ideal cylindrical geometry. Splines can cause the quenchant to be thrown off (Ref 8), thus producing uneven cooling and an increase in stress. It may even be necessary to stop the rotation when the spline section is being quenched (Ref 8). The presence of splines and keyways is important, because the deviation from cylindrical geometry can also cause uneven heating due to distortions of the current density at the sharp corners (Fig. 7.27) (Ref 62).

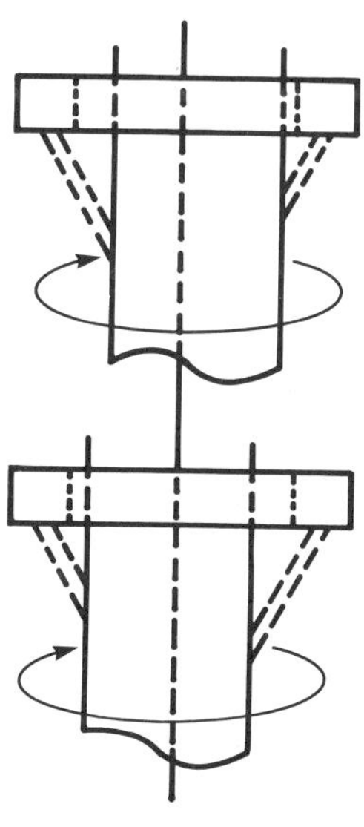

Fig. 7.26 "Barber pole" effect produced by an off-center shaft

The system for supplying the quenchant must be sized to allow accurate control of flow and temperature. It has been suggested that the size of the system be such that it would take 5 min at normal flow rates to completely deplete the system (Ref 8). The spray system requires temperature control; consistency is perhaps best achieved by a closed recirculating system (Ref 8). The system must be kept clean. Dirt, hard water, metal chips, and scale must be filtered so that the small holes in the spray heads do not become plugged. Contamination by cutting oils and hydraulic fluids can change the cooling rate or lead to foam problems.

The holes in the spray system must have the correct placement and size. Table 7.6 (Ref 69) lists typical orifice sizes for single shot systems. Larger holes can be used in scanning induction, because the rotation of the part and the motion of the spray head relative to the part help to supply a more even spray to the surface (Ref 69). The ratio of the surface area of the spray head to the total orifice area is typically 10 for systems with narrow coils and

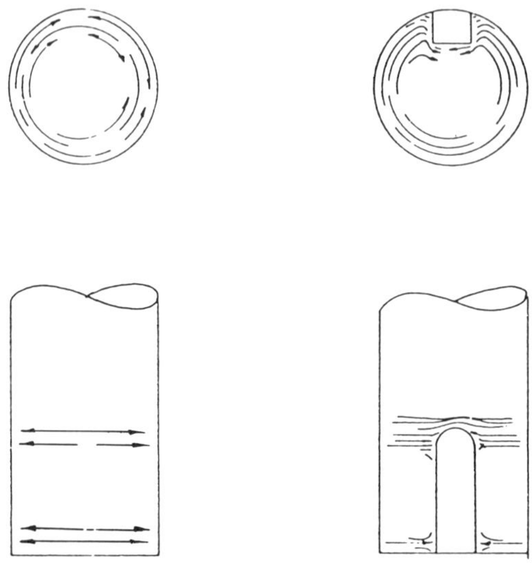

Fig. 7.27 Distortions of current density caused by sharp corners

Table 7.6 Typical orifice sizes for single shot induction systems

Orifice size		Part diameter	
mm	in.	mm	in.
1.6	0.06	6.4-12.7	0.25-0.50
3.2	0.125	12.7-38	0.50-1.5
6.4	0.25	>38	>1.5

20 for systems with wide coils (Ref 69). The spray for single shot systems is typically aligned perpendicular to the part; for scanning induction systems, the spray orifices are optimally aligned at a 30° angle to the axis of the part. This angle has been selected as a compromise between the need for most effective heat removal and the need to keep the quenchant from flowing toward the inductors and thus causing uneven heating.

The very close distances between the part and the spray head can prevent the quenchant from flowing evenly over the part (Ref 8). In some cases, problems of uneven flow can be corrected by the use of baffles. One type of baffle takes the form of a ring between the inlet and outlet with holes two to four times the diameter of the inlet orifices (Ref 69). Baffles are more likely to be necessary if the inlet and outlet orifices are parallel, if the inlet pressure is low and the inlet diameter is large, or if the inlet orifice is large (Ref 69).

Alternatively, uneven flow may be overcome by changes in the spray head. In one case (Ref 8), the flow of quenchant was so high that low clearance between the part and the spray head led to severely restricted flow. Blocking some of the outer holes in the spray ring allowed the quenchant to flow more uniformly over the part, correcting a soft spot problem. Blocking the holes allowed the quenchant flowing from the center of the spray ring to escape, and the formation of steam pockets was avoided.

To achieve "autotempering" or "self-tempering," sufficient heat content must be present in the part to bring the surface back to the desired temperature. This requires precise control of the quench duration. As quenched, a part with martensite will be hard but may be too brittle for use. The brittle character can be attributed to defect dislocations in the atomic structure of the martensite and to the presence of nonequilibrium concentrations of carbon in solution (Ref 69). To relieve both the brittle character and stress, the part is quenched sufficiently to cool the surface but still leave enough residual heat in the core to allow reheating of the part to a temperature of generally less than 675 °C (1250 °F), which is well below that required for austenitizing. This process allows some of the as-quenched martensite to transform to ferrite and carbide (Ref 70), which will cause a slight decrease in hardness but a significant increase in toughness.

Quenching of noncylindrical parts may require specially designed spray heads to achieve uniform cooling. For spray quenching of gears, the spray head should be configured so that the spray impacts on the roots of the gear teeth. To achieve this with a large gear, a quench ring has been designed to be congruent with the gear and to have the spray orifices arranged so that the flow is into the region between the teeth, as shown in Fig. 7.28 (Ref 69). The flow is such that a good depth of hardening is achieved at the roots, the sides, and the tips of the gear teeth (Ref 69). In the absence of this special spray ring, the flow of liquid at the teeth roots will be limited, perhaps leading to the formation of steam pockets, and cooling will be insufficient.

Fig. 7.28 Device for quenching inner-diameter gear teeth

Induction surface hardening results in residual stresses in the part. Consideration of the stress distribution is simplified by using a cylindrical example. Quenching a cylinder that has a hot surface but a cool core will result in the contraction of the surface layer, inducing a compressive residual stress on the core. The very rapid cooling of the surface and the slower cooling of the region between the cool core and the surface result in a compressive stress at the surface and increasing tension stress with increasing depth from the surface of the cylinder. This stress profile is shown in Fig. 7.29 (Ref 69).

The formation of martensite in the surface region of a part contributes to the surface compressive stress (Ref 69). This stress increases as the hardened depth is increased (Fig. 7.30) (Ref 69). The depth of this compressive stress is directly correlated to the depth of hardening (Fig. 7.31) (Ref 69).

Tempering and machining of the part after quenching will influence the final residual stress. Tempering will reduce the stress; the higher the temperature used in the tempering process, the lower the stress (Ref 69). Grinding the surface can reduce the compressive stress and can even lead to a tensile stress. The heat generated during the operation, as well as the deformation caused by intense grinding, produces this effect (Fig. 7.32) (Ref 69).

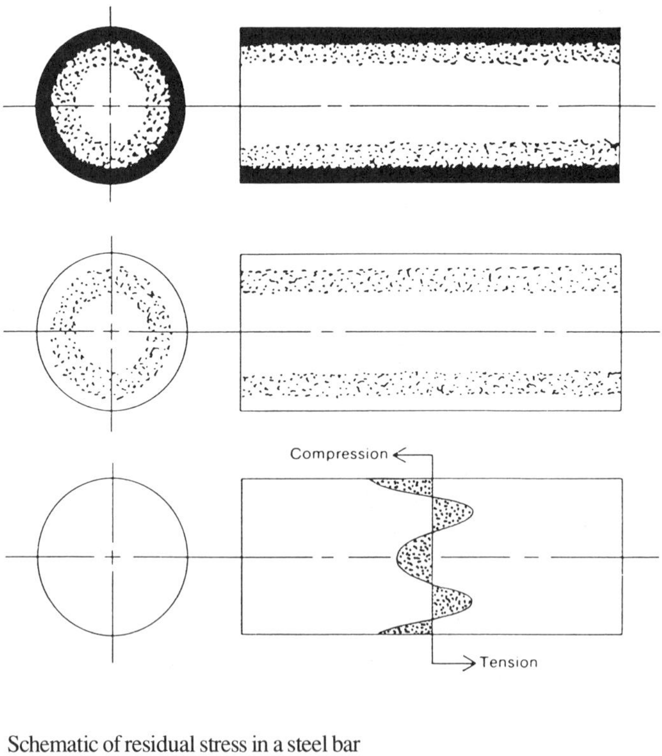

Fig. 7.29 Schematic of residual stress in a steel bar

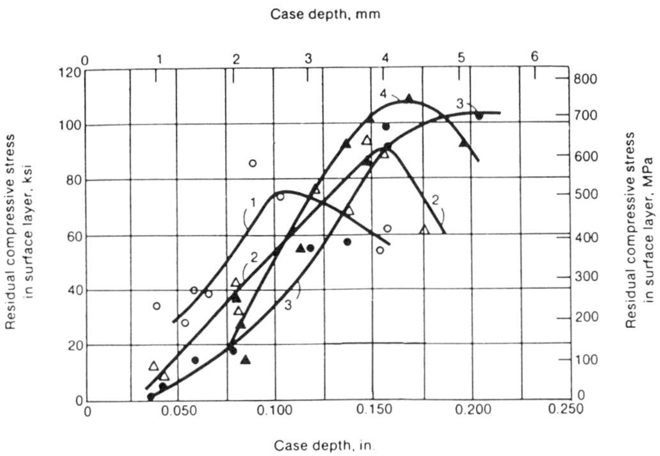

STEEL COMPOSITIONS
1- 0.44C-0.24Si-0.73Mn
2- 0.12C-0.20Si-0.45Mn-1.3Cr-4.45Ni-0.85W
3- 0.39C-0.26Si-0.65Mn-0.68Cr-1.58Ni-0.16Mo
4- 0.38C-0.28Si-0.99Mn-1.33Cr-0.36Mo

Fig. 7.30 Residual stress versus case depth in steel bars surface hardened by induction

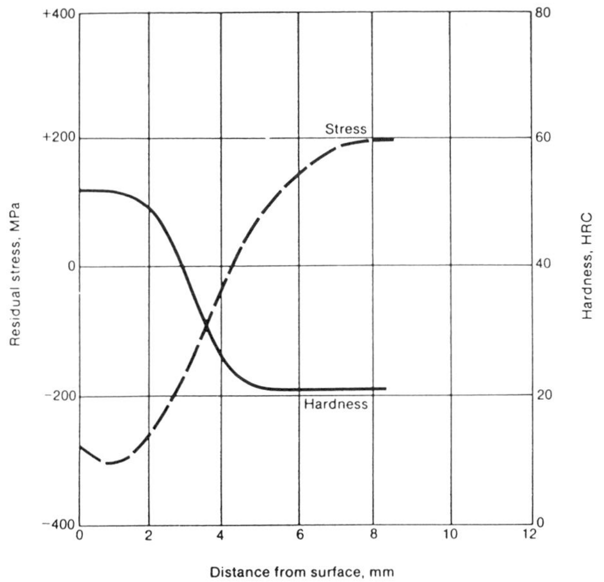

Fig. 7.31 Variation of hardness and residual stress with depth for steel bars surface hardened by induction

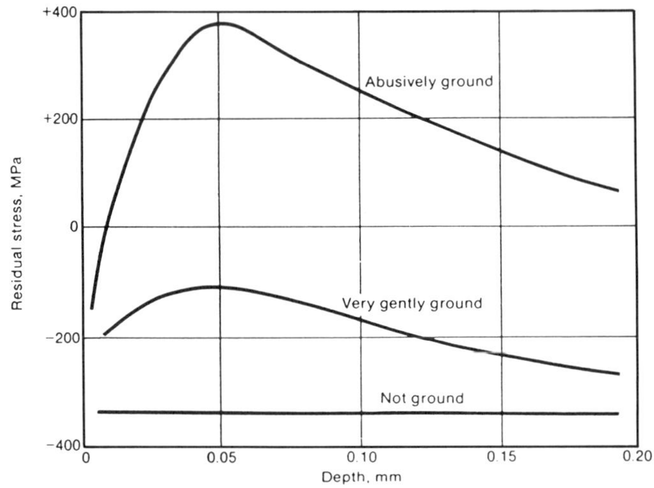

Fig. 7.32 Effect of grinding severity on surface residual stresses in induction quenched and tempered 1046 steel bars

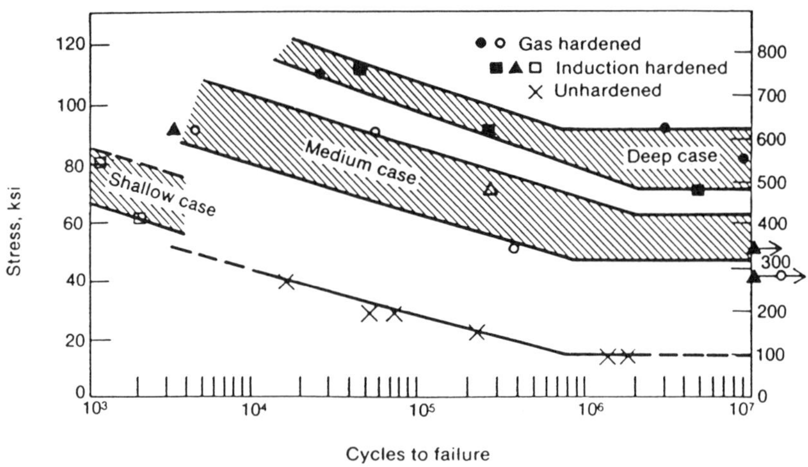

Fig. 7.33 Fatigue *S-N* curves of unhardened and surface-hardened steels of different case depths

Greater compressive stress and greater depth of compressive stress occur with greater case depth (see Fig. 7.30 and 7.31). Higher compressive surface stress and a greater depth of compressive stress will lead to a higher fatigue life in the part. Thus, Fig. 7.33 shows a large increase in fatigue resistance as the depth of hardening is increased.

The stress along the axis of an induction heated and quenched cylinder has also been analyzed. Figure 7.34 (Ref 71) shows the stress pattern and hardness profile produced in a statically heated steel bar that was quenched and tempered at 190 °C (375 °F). The development of the stress pattern has been explained by using a model (see Fig. 7.35) of a heated zone that, on quenching, develops a rigid surface that does not deform; the interior of the heated zone cools more slowly and, during this process, contracts. The inner zone is then under a tensile stress, while the surface and core are under compressive stress. This matches the picture of stress for the ideal cylinder. The heated and quenched surface is held by the adjacent surface regions that are not heated. If the surface-heated region forms martensite on quenching, expansion will occur. The residual stress at the edges of the heated zone will be the sum of the stresses caused by the contraction due to cooling and the expansion due to the formation of martensite. The sum of these stresses will depend on the shapes of the inner and outer heated zones and on the stress produced by each (Fig. 7.35) (Ref 71).

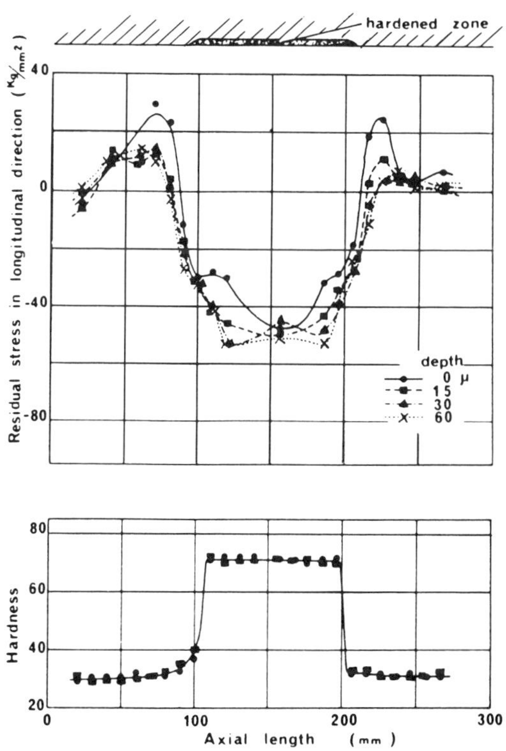

Fig. 7.34 Distribution of residual stress and hardness for a static induction heated and quenched surface

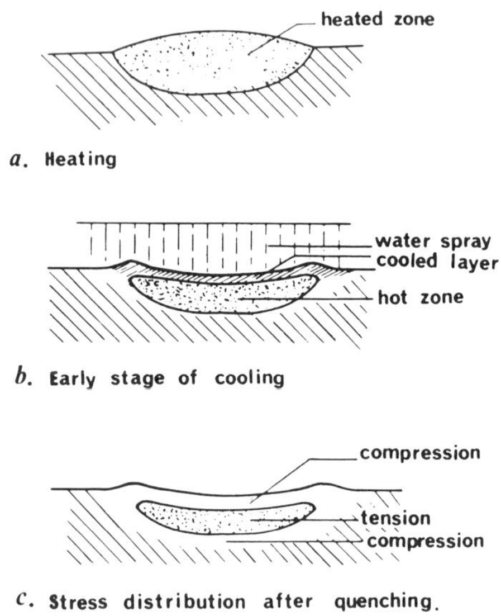

Fig. 7.35 Model of the origin of residual stress in static quenching

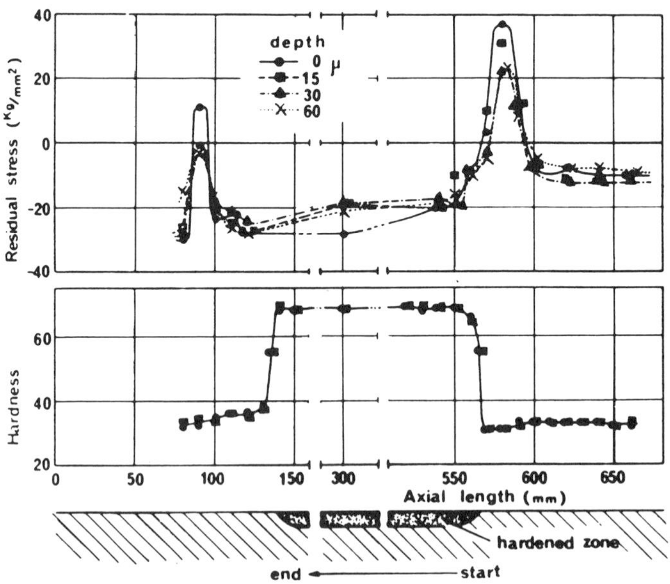

Fig. 7.36 Distribution of residual stress and hardness for a progressive induction heated and quenched surface

Progressive or scanning induction hardening produces an asymmetric stress and hardness profile. An example is illustrated in Fig. 7.36 (Ref 71). The temperature profile during the hardening is shown in Fig. 7.37 (Ref 71). The asymmetric temperature profile results in the hot zone being more constrained on one side than the other. The compressive stress is less than with static heating and quenching (Ref 71) and reaches a maximum at the ends because of the rigid character of the nonheated zone. Tensile stress zones are found at the beginning and end points and are thought to be caused by the uneven cooling pattern produced by the quench. The size of the stress will be influenced by the size of the part, induction frequency, current density, and size and shape of the inductor, as well as by the rate of cooling of the quench. Excessive tensile stresses may decrease the service life of the part.

Results Obtained With Induction Heating

The use of induction heating and quenching to produce surface hardness is important in the production of rolls used for the shaping of metal. The surface must be very hard, while the core and journals must be able to hold up under the application of large cyclic

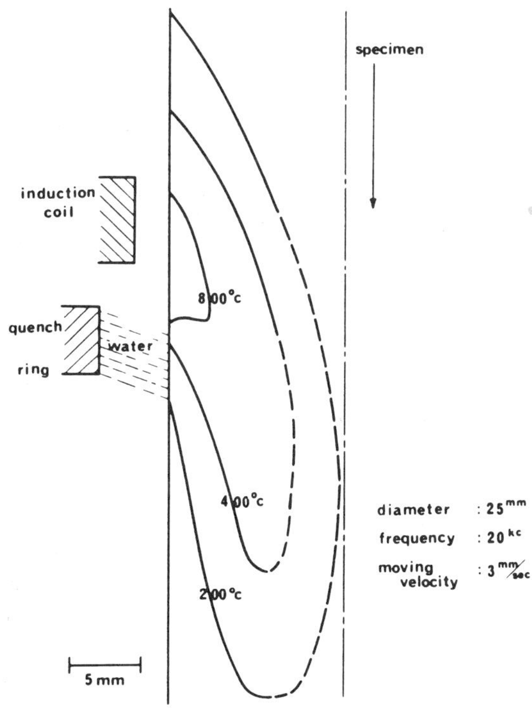

Fig. 7.37 Temperature distribution for induction heating with spray quenching

loads (Ref 72). One process (Ref 73) to achieve this uses a dual-frequency progressive induction heating procedure followed by a high-pressure spray, which is then followed by immersion in a water tank. The first induction coil operates at 50 Hz and the second at 250 Hz. As can be seen in Table 7.3, the first coil will heat to a significant depth, while the second will focus more of the heat on the surface. This process produces a hardened depth significantly greater than when the same roll is through heated and quenched (Fig. 7.38) (Ref 73).

The use of induction heating to produce surface hardness with minimal distortion has been reviewed (Ref 63). Wear resistance and fatigue strength are enhanced by induction heating and spray quenching (Ref 74). In one example, a 12 m (39 ft) railroad rail was prebent by bowing the ends down 0.9 m (3 ft). The inductor and quench ring were designed to follow the curve defined by the bend, and the rail was allowed to distort to give a straight product with a hard crown and tough interior (see Fig. 7.39) (Ref 63).

In the case of strut rods used in shock absorbers, it was necessary to surface harden the part with very little distortion. The struts were 300 mm (12 in.) long, had varying diameters, and had been machined to the final tolerances. A high-power inductor heated the surface to 955 °C (1750 °F) in 0.4 s, followed by quenching and then by tempering with lower-power heating. The case depth was 0.5 mm (0.02 in.) (Ref 63).

Aqueous polymer solutions are often used for induction spray quenching (Ref 75). It is possible to vary the cooling rate by changing the polymer concentration. This avoids the problem of cross-contamination that would occur if a change between oil and water was made. Polymers of different concentrations can be stored and used to change the solution in the quench reservoir, with system draining being the only cleaning necessary. Changes in the temperature of the quenchant can be used to broaden the range of obtainable cooling rates, as discussed in Chapter 5. This sensitivity of cooling rate to polymer solution temperature allows the use of one concentration of quenchant to heat treat steels that require different cooling rates.

The mechanisms for heat transfer and their influence on cooling rate that were presented in Chapter 5 for immersion quenching with aqueous polymers also apply to spray and flood quenching. In general, polymer solutions are used to reduce the fast cooling rate of water. One example (Ref 8) involves the production of large AISI 15B28 steel tractor axle shafts. The parts were surfaced hardened using induction heating and water quenching. Although the depth of hardness was good, the surface hardness was lower than that found at a depth of 1.0 mm (0.04 in.). Examination of the surface layer showed it to be tempered martensite that formed when the surface was rapidly cooled but while the interior still held enough heat to temper the surface. Switching to a 5% PAG polymer solution reduced the cooling rate of the surface sufficiently to produce more uniform cooling, which resulted in the desired surface hardness and case depth. The same tractor axle shaft made from AISI 1541 steel was induction hardened with a 1 KHz power source, followed by quenching with a 10% PAG polymer solution at 32 °C (90 °F) (Ref 8).

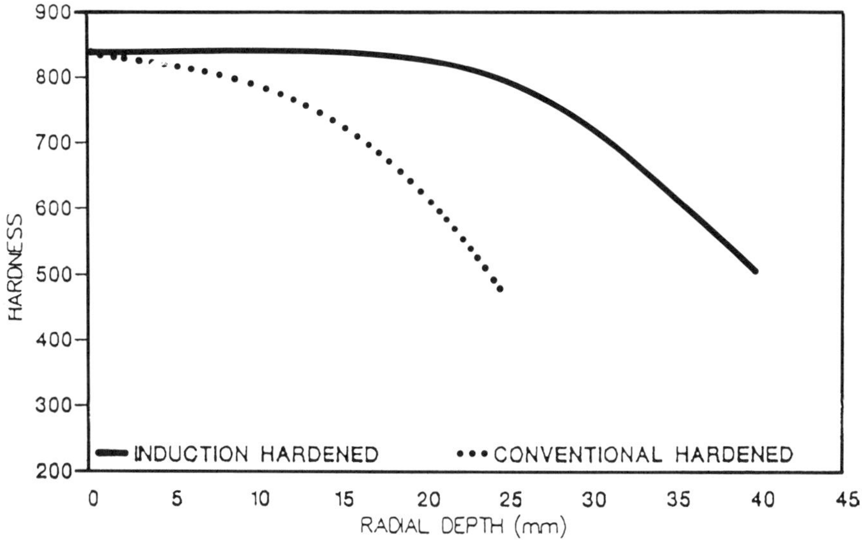

Fig. 7.38 Comparison of conventional and induction hardening profiles for steel rolls

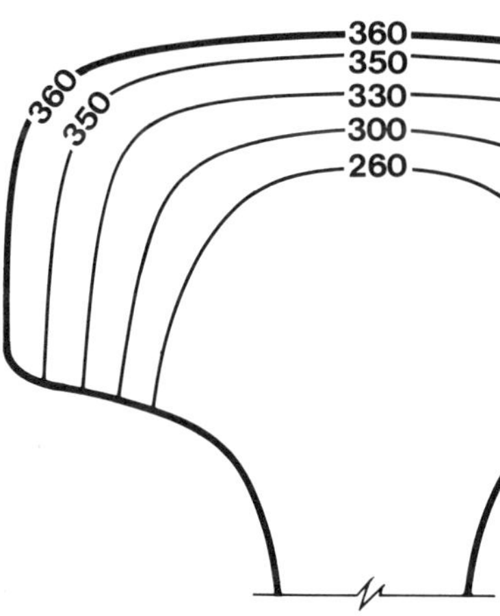

Fig. 7.39 Hardness profile of a railroad rail

Spray Quenchant Variants

The heat-transfer media used in spray quenching are usually water or aqueous polymer solutions. However, it is also possible to use oil in a spray quench application (Ref 3). As with water, an oil spray suppresses the vapor film portion of the cooling curve and raises the temperature at which the surface is wetted by the heat-transfer agent. Therefore, heat transfer with an oil spray is much more effective than heat transfer obtained by immersion in an oil bath.

In one study (Ref 3), a strong oil spray was found to have a heat-transfer coefficient in the pearlite transformation range that was seven or eight times larger than that obtained with an oil immersion quench. In the martensitic transformation region, the heat-transfer coefficient was two to three times larger than with an oil immersion quench. The hot metal surface resulted in immediate ignition of the oil, but the flame was soon extinguished by the spray as the surface of the part cooled below the ignition point. The flame was prevented from propagating back to the spray head by the presence of sufficient carbon dioxide or nitrogen to provide an inert atmosphere. Figure 7.40 compares the heat-transfer coefficients produced by immersion oil quenching and oil spray quenching using two spray velocities. Figure 7.41 provides a comparison of water and oil sprays, still water, and still oil as heat-transfer media.

An unusual approach to using mechanical energy to control the formation of a vapor blanket involves immersing the part in a pressurized water quench bath (Ref 76). The

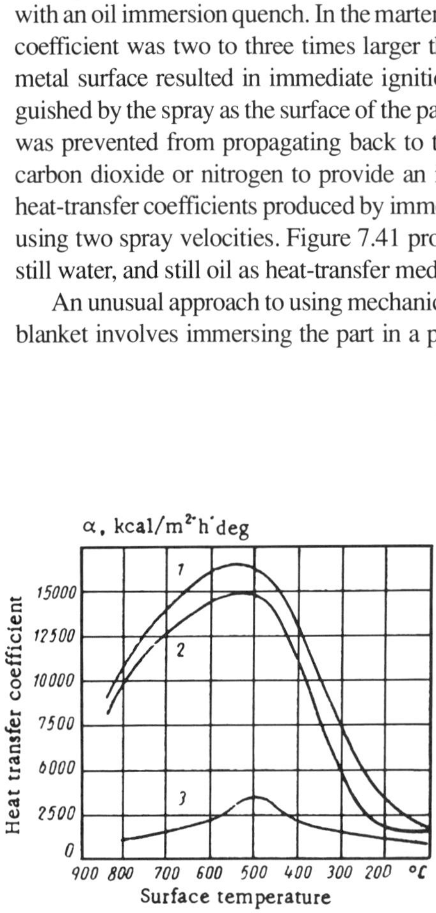

Fig. 7.40 Heat-transfer coefficients for oil spray quenching

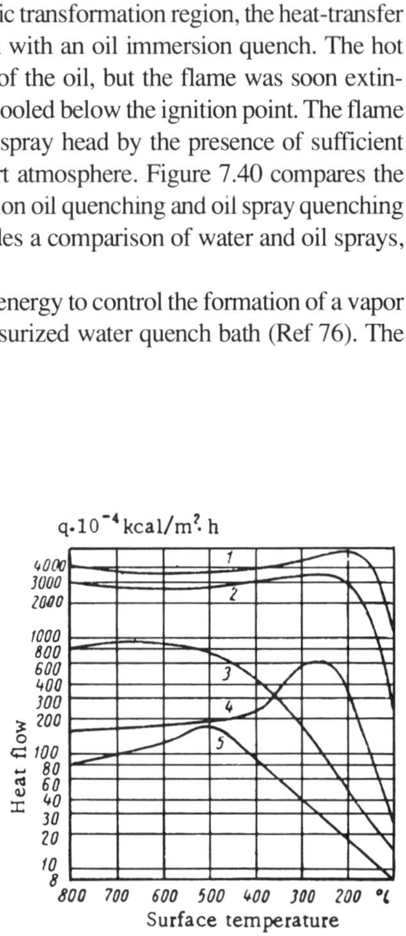

Fig. 7.41 Heat flux versus surface temperature for copper

pressure raises the boiling point of the water, which suppresses the vapor blanket stage of the quench. The nucleate boiling stage of the quench then commences at a higher temperature, resulting in a faster quench. In test parts (notched cylinders of ShKh15 steel), a higher boiling temperature of water was reported to prevent quench cracks as compared with water boiling at its normal boiling point (see Table 7.7) (Ref 76).

The use of pressure to achieve maximum surface compressive stress has been described (Ref 77). This is achieved by raising the pressure of the quench system (bath and part) to raise the boiling point of water and thus significantly increase the quench severity.

Intense quenching has been defined as "quenching wherein the *H* values for water are at least 2.50 and, for oil, 0.70 minimum" (Ref 4). In practice, this involves unusually high agitation using jets of high-pressure quenchant to produce increased heat-transfer rates (Ref 6, 38) and results in very hard as-quenched surfaces. Water, aqueous polymer solutions, or oil is used as the quenchant medium. Intense quenching is reported to have been first applied in the production of the Ford Model T rear axle (Ref 4). As is often the case with induction heating, the objective is to obtain a surface of essentially pure martensite, thus producing a high compressive stress at the surface. Applications include axles, moil points, punches, engine crankshafts, and gears. Hardness distribution results for a gear are shown in Fig. 7.42 (Ref 4).

The use of an induction coil or a plasma torch to heat a surface can produce heat fluxes on the order of 10^2 W/cm^2 (3×10^5 Btu/ft$^2 \cdot$ h). However, fluxes as high as 10^8 or 10^9 W/cm^2 (3×10^{11} or 3×10^{12} Btu/ft$^2 \cdot$ h) can be achieved using high-power lasers or electron beams (Ref 9). The heat transfer to the surface is so fast that conductive heat transfer to the interior is essentially absent. Therefore, with only the surface to cool, the quench rates can also be very high. Heating rates as high as 10^5 °C/s (2×10^5 °F/s) and cooling rates as high as 10^6 °C/s (2×10^6 °F/s) have been reported (Ref 9, 78). The very rapid heating and cooling rates produce unique properties (Ref 79). These high heat rates produce a shock wave that can exceed the yield stress of the steel and result in a plastic flow. This produces a surface with an increase in dislocations (see Table 7.8) (Ref 80).

The results of such high heating and cooling rates on the metallurgy of the surface layer of some steels that were common in the USSR have been discussed (Ref 78-83). These processes produce small grain sizes, surface metallurgy that is very dependent on

Table 7.7 Quench cracking as a function of the boiling point of water

Boiling point °C	Boiling point °F	Parts with cracks, %	Rockwell hardness, HRC
100	212	50	64
107	225	60	65
116	241	10	64.5
130	266	0	64.5
Oil		0	62

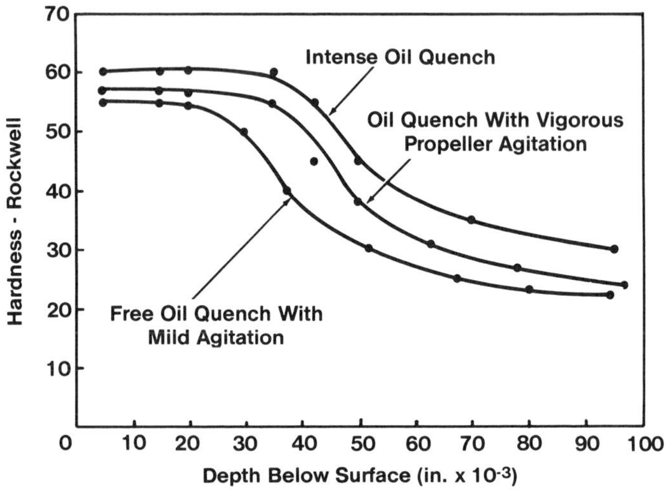

Fig. 7.42 Hardness distribution for intense quenching of a gear

Table 7.8 Laser heating and dislocation density

Steel type	Treatment	Dislocation density
08Yu	Normal	1.3×10^8
	Laser	3.54×10^{10}
08Kp	Normal	2.10×10^9
	Laser	1.4×10^{11}
08Kh118N10T	Normal	4.6×10^8

Table 7.9 Tool lifetimes of laser-hardened parts

Cutting speed			
m/min	ft/min	Treatment	Tool life, h
14	46	Hardened and tempered	1.5-3
		Laser	6-8
20	66	Hardened and tempered	0.5-1.0
		Laser	4-7

alloying elements, hard surfaces, and reduced porosity. Results obtained with an electron beam have also been reported (Ref 84). In this case, a fine structure with unique metastable carbides was produced.

Tools for cutting steel pipe were surface hardened using a continuous laser. The average lifetimes of the tools at two cutting speeds were determined as shown in Table 7.9. The laser surface hardening occurred to a depth of 0.6 to 1.0 mm (0.024 to 0.04 in.) (Ref 80).

Blanking punches that were either conventionally heat treated or surface laser treated were compared (Ref 82). The laser-treated punches had a lifetime two and a half to three times longer than that of the conventionally heat-treated punches. The failure modes also differed. The conventionally heat-treated punches failed by seizing and wear at the cutting edge. The laser-treated punches had a harder surface and generally failed as a result of chipping (Ref 82).

References

1. C.E. Bates, G.E. Totten, and R.L. Brennan, Quenching of Steel, *ASM Handbook,* Vol 4, *Heat Treating*, ASM International, 1991, p 67-120
2. B. Liscic, *Quenching and Carburising*, 3rd International Seminar, Melbourne, Sept 1991, IFHT, p 1-27
3. N.V. Zimin, *Metalloved, Term. Obrab. Met.*, Nov 1967, p 62-68
4. F.K. Kern, *Heat Treat.*, Sept 1986, p 19-23
5. S. Segerberg, *Heat Treatment and Surface Engineering: New Technology and Practical Applications*, Chicago, 28-30 Sept 1988, ASM International, p 177-181
6. F. Moreaux and G. Beck, *Heat and Mass Transfer in Metallurgical Systems*, D.B. Spalding and N.H. Afgan, Ed., Hemisphere Publishing, 1981, p 553-561
7. L.N. Bokanova, V.V. Lebedev, and L.N. Markova, *Met. Sci. Heat Treat.*, Vol 31 (No. 3-4), 1989, p 159-161
8. R.R. Blackwood, *Ind. Heat.*, May 1991, p 46-51
9. G. Beck, *Heat and Mass Transfer in Metallurgical Systems*, D.B. Spalding and N.H. Afgan, Hemisphere Publishing, 1981, p 509-525
10. G. Li, *Proc. 4th Ann. Conf. Heat Treat.*, Nanjing, 25-31 May 1987, Chinese Mechanical Engineering Society, p 171-175
11. V.G. Labeish, *Steel USSR*, Vol 19 (No. 3), Mar 1989, p 134-136
12. N. Hatta and H. Osakabe, *ISIJ Int.*, Vol 29 (No. 11), 1989, p 919-925
13. I. Mudawar and W.S. Valentine, *J. Heat Treat.*, Vol 7 (No. 2), 1989, p 107-121
14. F. Moreaux and P. Archambault, *Quenching and Carburising*, 3rd International Seminar, Melbourne, Sept 1991, IFHT, p 170-176
15. T.A. Deiters and I. Mudawar, *J. Heat Treat.*, Vol 8 (No. 2), 1990, p 81-91
16. M.A. Geller and M.S. Zheludkevich, *Metalloved. Term. Obrab. Met.*, Sept 1989, p 29-30
17. F. Moreaux and P. Archambault, *Prom. Teplotekhnika*, Vol 11 (No. 3), 1989, p 48-55
18. P. Archambault, G. Didier, F. Moreaux, and G. Beck, *Met. Prog.*, Oct 1984, p 67-72
19. T. Fukuda, T. Takayama, N. Hamasaka, K. Ohkawa, K. Tsuda, and H. Ikeda, *Netsu Shori*, Vol 29 (No. 5), 1989, p 296-301
20. B.I. Medovar, A.I. Us, A.I. Krendeleva, N.B. Pivovarskii, N.A. Astaf'ev, and N.M. Shelestiuk, *Probl. Spets. Elektrometall.*, Vol 59 (No. 2), 1987, p 37-40
21. N.I. Kobosko, *Metalloved. Term. Obrab. Met.*, Sept 1989, p 7-14
22. *Surface Hardening*, H.E. Boyer and P.D. Harvey, Ed., American Society for Metals, 1979
23. N.V. Zimin, *Metalloved. Term. Obrab. Met.*, May 1970, p 23-26

24. I.A. Borisov, *Metalloved. Term. Obrab. Met.*, Sept. 1989, p 23-28
25. V.S. Yesaulov, A.I. Sopochkin, V.F. Polyakov, A.V. Nogovitsyn, and V.I. Semen'kov, *Izv. V. U. Z. Chernaya Metall.*, No. 8, 1990, p 82-85
26. M.A. Brich, V.T. Borukhov, M.A. Geller, and M.S. Zheludkevich, *Prom. Teplotekh.*, Vol 12 (No. 6), 1990, p 58-62
27. M. Mitsutsuka, *Tetsu Hagane*, Vol 54 (No. 14), 1968, p 1457-1471
28. V.G. Labeish, *Steel USSR*, Vol 19 (No. 3), Mar 1989, p 134-136
29. R. Jeschar, R. Maass, and C. Kohler, *Proc. AWT-Tagung Inductives Randschichtharten*, 23-25 March 1988, p 69-81
30. M. Mitsutsuka, *Tetsu Hagane*, Vol 69 (No. 2), 1983, p 268-274
31. M. Mitsutsuka and K. Fukuda, *Tetsu Hagane*, Vol 69 (No. 2), 1983, p 262-267
32. R.G. Owen and D.J. Pulling, (Multiphase Transport: Fundamentals, Reactions, Safety, Applications) (Proc. Multi-Phase Flow Heat Transfer Symp. Workshop), 1979
33. K.J. Baumeister, F.F. Simon, and R.E. Henry, paper presented at Winter Meeting of the American Society of Mechanical Engineers, 1970
34. M.S. Jenkins, S.R. Story, and R.H. Davies, *Quenching and Carburising*, 3rd International Seminar, Melbourne, Sept 1991, IFHT, p 161-169
35. C. Kaijiang, Y. Waixia, and C. Kaike, *Acta Metall. Sin. (China)*, Vol 21 (No. 6), 1985, p B297- B302
36. R.A. Mugele and H.D. Evans, *Ind. Eng. Chem.*, Vol 43, 1951, p 1317-1324
37. R.E. Gloss, R.K. Gibbs, and P.D. Hodgson, *Quenching and Carburising*, 3rd International Seminar, Melbourne, Sept 1991, IFHT, p 188-200
38. Yu. P. Gul, V.S. Chmeleva, and V.V. Kirichenko, *Metalloved. Term. Obrab. Met.*, Sept 1989, p 2-6
39. L. Bolle and J.C. Moureau, *Heat and Mass Transfer in Metallurgical Systems*, D.B. Spalding and N.H. Afgan, Hemisphere Publishing, 1981, p 527-534
40. T. Nozaki, J. Matsuno, K. Murata, H. Ooi, and M. Kodama, *Tetsu Hagane*, Vol 62 (No. 12), 1976, p 1503-1512
41. P.D. Hodgson, K.M. Browne, D.C. Collinson, T.T. Pham, and R.K. Gibbs, *Quenching and Carburising*, 3rd International Seminar, Melbourne, Sept 1991, IFHT, p 146-159
42. M. Baumberger and B. Prinz, *Metallwissenschaft and Technik*, Vol 44 (No. 2), 1990, p 154-157
43. M. Baumberger and B. Prinz, *Mater. Sci. Technol.*, Vol 2, April 1986, p 410-415
44. N. Lambert and M. Economopoulos, *J. Iron Steel Inst.*, Vol 208, 1970, p 917-928
45. M. Mitsutsuka and K. Fukuda, *Tetsu Hagane*, Vol 75 (No. 7), 1989, p 1154-1161
46. R.D. Morales, A.G. Lopez, and I.M. Olivares, *ISIJ Int.*, Vol 30 (No. 1), 1990, p 48-57
47. K. Sasaki, Y. Sugitani, and M. Kawasaki, *Tetsu Hagane*, Vol 63, 1977, p S184
48. M. Ishiguro, *Tetsu Hagane*, Vol 60, 1974, p S464
49. T. Nozaki, J. Matsuno, K. Murata, I. Ooi, M. Kodama, *Trans. Iron Steel Inst. Jpn.*, Vol 18, 1978, p 330
50. E. Bolle and J.C. Moreau, *Proc. Int. Conf. Heat and Mass Transfer in Metallurgical Processes*, Dubrovnik, Yugoslavia, 1979, p 304
51. M. Shimada and M. Mitsutsuka, *Tetsu Hagane*, Vol 52, 1966, p 1643
52. E. Bolle and J.C. Moreau, *Proc. Two Phase Flows and Heat Transfer*, Vol III, NATO Advanced Studies Institute, London, 1976, p 1327
53. E.A. Mizikar, *Iron Steel Eng.*, Vol 47, 1970, p 53
54. *Kouzai-no Kyosei Reikyaku,* Special Rep. No. 29, Iron and Steel Institute of Japan, 1977, p 15

55. W.L. McCabe and J.C. Smithe, *Unit Operations of Chemical Engineering*, McGraw-Hill, 1976
56. K.Z. Shepelyakovskii, V.P. Devyatkin, B.K. Ushakov, and V.M. Fedin, *Metalloved. Term. Obrab. Met.*, Sept 1989, p 30-32
57. G.F. Golovin, M.M. Zamyatnin, and N.V. Zimin, *Metalloved. Term. Obrab. Met.*, Nov 1969, p 73-74
58. I.G. Uzlov, *Metalloved. Term. Obrab. Met.*, Sept 1989, p 14-19
59. A.J. DeArdo, *Can. Metall. Q.*, Vol 27 (No. 2), 1988, p 141-154
60. I. Takahashi, Y. Obanya, A. Otomo, N. Akiyama, H. Fukumitsu, and Y. Kobayashi, *Kobe Res. Dev.*, Vol 38 (No. 1), 1988, p 71-74
61. A.H. El-Sawy and C.W. LeMaistre, Paper No. A84-35, The Metallurgical Society of AIME
62. A. Grant, *FWP J.*, Jan 1989, p 13-19
63. G.F. Bobart, *J. Heat Treat.*, Vol 8 (No. 2), 1990, p 148-155
64. O.S. Spark, 4th Annual Meeting of IEEE Industrial and General Applications Group, Detroit, 12-16 Oct 1969
65. G.H. Ledl, *Met. Prog.*, Dec 1967, p 74-78
66. G.F. Bobart, *Ind. Heat.*, Jan 1992, p 37-41
67. O. Longeot, D. Longeot, and J.L. Lebrun, *Heat Treatment and Surface Engineering: New Technology and Practical Applications*, Chicago, 28-30 Sept 1988, ASM International, p 143-150
68. O.S. Spark, *Heat Treat.*, March 1978, p 18-22
69. S.L. Semiatin and D.E. Stutz, *Induction Heat Treatment of Steel*, American Society for Metals, 1986
70. P.D. Harvey, Ed., *Tempering of Steel*, American Society For Metals, 1980
71. K. Ishii, M. Iwamoto, T. Shiraiwa, and Y. Sakamoto, *Heat Treatment and Surface Engineering: New Technology and Practical Applications*, Chicago, 28-30 Sept 1988, ASM International, p 1765-1771
72. N.A. Adamova, Yu.N. Andreev, Yu.V. Yudin, V.G. Sorokin, and E.Z. Chernyakhovskii, *Metalloved. Term. Obrab. Met.*, Sept 1989, p 19-23
73. T. Scoular, *The Heat Treatment of Steel* (conf. proc.), Glasgow, 5 Sept 1989, Scottish Association of Metals, p 33-55
74. C. Hou, *Heat Treat. Met. (China)*, Sept 1989, p 53-56
75. P.D. Jenkins, *Metallurgia*, April 1978, p 196-199
76. N.I. Kobasko, *Metalloved. Term. Obrab. Met.*, Feb 1970, p 64-65
77. N.I. Kobasko, *Metalloved. Term. Obrab. Met.*, Sept 1989, p 7-14
78. S.I. Gubenko, *Metalloved. Term. Obrab. Met.*, Oct 1989, p 2-4
79. I.V. Artamonova, A.A. Nikitin, and I.A. Kyzhkov, *Metalloved. Term. Obrab. Met.*, Oct 1989, p 5-7
80. M.N. Kryanina, A.M. Bershtein, and T.P. Chuprova, *Metalloved. Term. Obrab. Met.*, Oct 1989, p 7-12
81. V.K. Narva, N.S. Loshkareva, M.N. Kryanina, and T.A. Belokonova, *Metalloved. Term. Obrab. Met.*, Oct 1989, p 16-18
82. E.I. Tesker, V.Yu. Mitin, A.P. Karpova, and Yu.V. Bondarenko, *Metalloved. Term. Obrab. Met.*, Oct 1989, p 18-20
83. V.S. Kraposhin, K.V. Shakhlevich, and T.M. Vyaz'mina, *Metalloved. Term. Obrab. Met.*, Oct 1989, p 21-29
84. N.A. Pakhomova, I. Artinger, O.A. Bannykh, and V.A. Ermishkin, *Metalloved. Term. Obrab. Met.*, Oct 1989, p 13-15

CHAPTER 8

Other Quenching Processes

Oil and polymer quenchants have been discussed in detail in Chapters 4 and 5, respectively. Spray and induction quenching processes were discussed in Chapter 7. However, a number of other quenchants and quenching processes are used in the heat treating industry. This chapter presents an overview of the following quenchants and processes:

- *Air*
- *Water*
- *Brine*
- *Molten salt*
- *Fluidized bed*
- *Gas*
- *HIP quenching*
- *Aus-bay quenching*
- *Ultrasonic quenching*
- *Quenching in an electric or magnetic field*

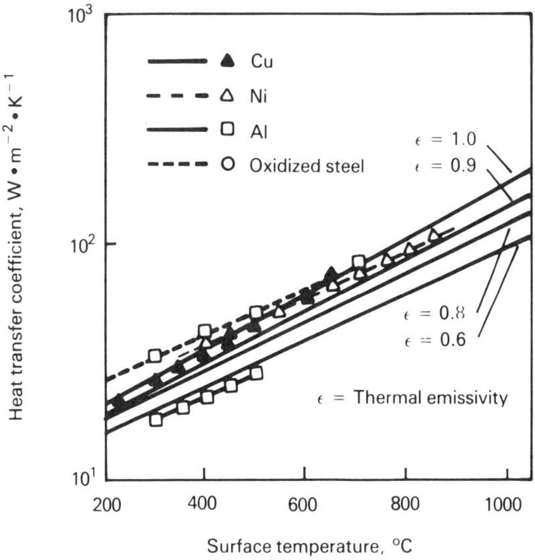

Fig. 8.1 Heat transfer coefficients for air cooling as a function of surface temperature

Air

Perhaps the oldest, most common, and certainly least expensive quenchant medium is air. Air, being a gas, cools by extended A-stage cooling (Ref 1). The heat-transfer coefficients of air with respect to surface temperature for steel, nickel, aluminum, and copper are shown in Fig. 8.1 (Ref 2).

Like other quenching media, the heat-transfer rates of air cooling are dependent on the flow rate of air past the cooling metal. This effect is illustrated in Fig. 8.2, which shows the cooling rate of an instrumented 20 mm (0.8 in.) diam solid silver sphere cooled in still air and compressed air (1 kg/cm^2) (Ref 3). Although increasing the velocity of air flow past the cooling surface does accelerate the cooling rate, it is insufficient to quench-harden many steels; for example, Fig. 8.3 shows that the ability of air to quench-harden plain carbon steel dramatically decreases with increasing carbon content (Ref 4). Attainment of optimal hardness for many steels often requires a more severe quench medium, such as brine or oil.

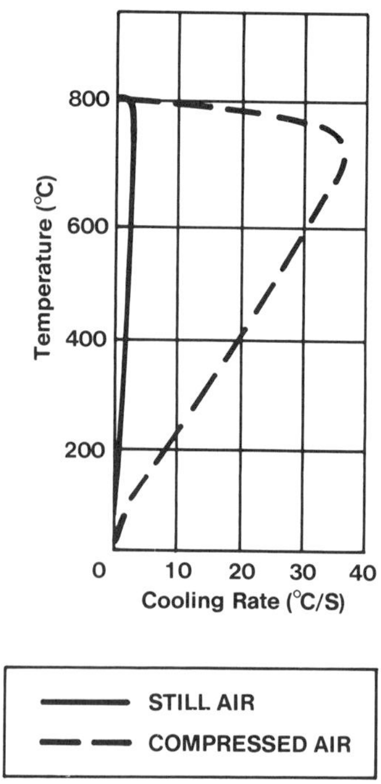

Fig. 8.2 Comparison of the cooling capacity of still and compressed air

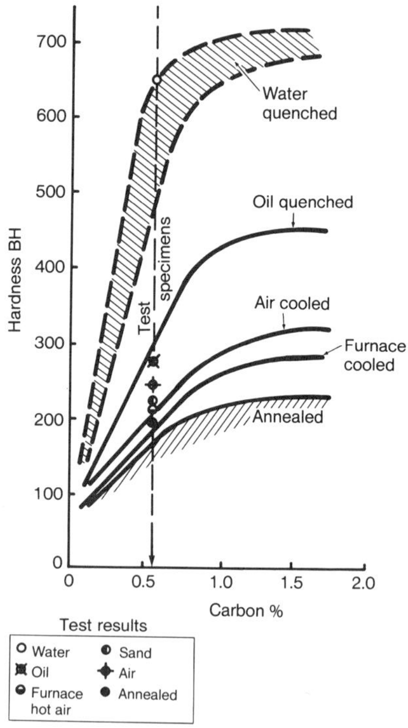

Fig. 8.3 Hardness values for various quenched specimens compared with previous findings

Water

Straight Immersion Quenching. The oldest, most common, and least expensive quenchant medium is water. Water, particularly cold water, is one of the most severe quench media available. In 1935, Speith and Lange (Ref 1) conducted one of the earliest studies utilizing cooling curve characterization techniques to elucidate the mechanism of water quenching. By recording schlieren patterns formed during quenching, it was possible to show that the heat-flow pattern in a still-water quench was turbulent within the boundary layer at the hot metal interface.

This boundary-layer turbulence was proportional to the viscosity of the quench medium. For example, oil, which is substantially more viscous than water, cools via a laminar boundary layer instead of the turbulent layer observed with water.

When these now-classic studies were conducted, it was found that a substantial difference existed between the quench severity produced by distilled water (which was used to model rainwater) and tap water that contained "hard" metal salts. The results of these studies are summarized in Fig. 8.4. In general, tap water exhibited substantially reduced A-stage or vapor blanket cooling profiles at bath temperatures of 20 to 60 °C (70 to 140 °F). (This work was the precursor to more rigorous studies demonstrating the advantages of brine quenching, which will be discussed later.)

In 1940, Rose (Ref 3) demonstrated that the cooling rate during quenching was only minimally affected by the specific heat and thermal conductivity of steel and was primarily affected by the heat-transfer properties of the quenchant itself (Ref 3). One of the ways that this was illustrated was by showing that cooling rates are affected by both

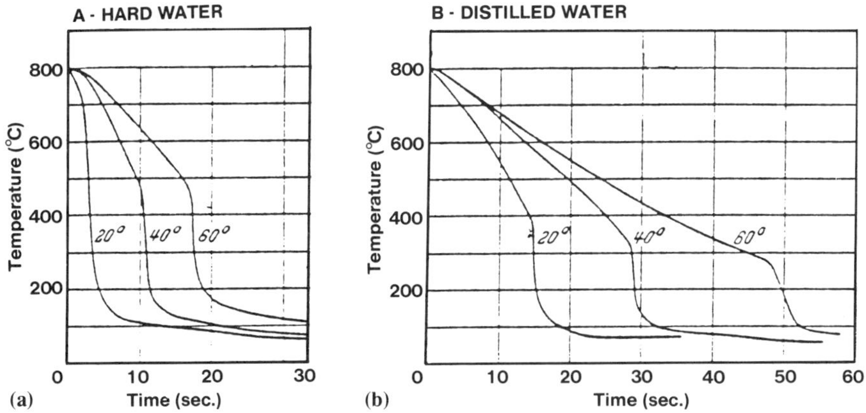

Fig. 8.4 Effect of hard water metals on the cooling rate of water. Cooling curves were obtained with a silver sphere. **(a)** Hard water. **(b)** Distilled water

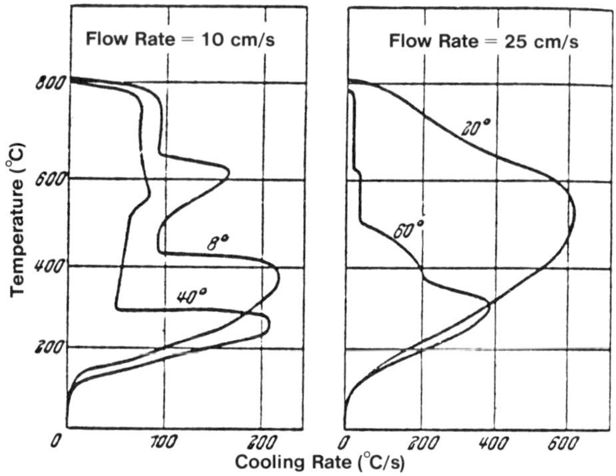

Fig. 8.5 Effect of bath temperature and agitation on the cooling rates of water. Cooling curves were obtained with a silver sphere.

Table 8.1 Effect of water temperature on cooling rate(a)

Water temperature		Time to cool from 732-260°C	Maximum cooling rate (CR_{max})		Maximum cooling rate temperature		Cooling rate, °C/s (°F/s)					
°C	°F	(1350-500°F), s	°C/s	°F/s	°C	°F	704°C	(1300°F)	343°C	(650°F)	232°C	(450°F)
40	105	4.6	153.0	275.4	535	995	60.3	(108.5)	96.6	(173.9)	50.9	(91.6)
50	120	5.6	137.4	247.3	542	1008	31.5	(56.7)	94.3	(169.7)	50.5	(90.9)
60	140	8.0	114.8	206.6	482	900	19.7	(35.5)	87.4	(157.3)	46.3	(83.3)
70	160	11.8	98.8	177.8	448	838	16.8	(30.2)	84.0	(151.2)	46.7	(84.1)
80	175	20.4	79.0	142.2	369	696	14.9	(26.8)	76.6	(137.9)	47.2	(85.0)
90	195	40.6	48.2	86.8	270	518	11.9	(21.4)	26.4	(47.5)	42.9	(77.2)

(a) Data were obtained with a 12.5 × 100 mm (½ × 4 in.) Inconel 600 cylindrical probe with a type K thermocouple inserted into the geometric center. The probe was quenched into distilled water (4 L, or 1 gal) at the specified temperature. The turbulent flow past the probe was linear and parallel to the longitudinal direction of the probe. Cooling curves are shown in Fig. 8.6, and results are illustrated in Fig. 8.7.

water temperature and agitation (Fig. 8.5). In general, the vapor blanket formed during A-stage cooling in water was unstable up to 20 °C (70 °F). Apparently, the turbulence of the boundary layer was sufficient to rupture the vapor blanket.

Rose also reported that test pieces quenched in water at 20 °C (70 °F) produced nonuniform results because of surface irregularities that favored the "adhesion" of the vapor blanket. Because cooling through the vapor layer produced slower cooling rates than those achievable with nucleate boiling, the potentially large thermal gradients at the surface during quenching produced increased cracking and distortion.

At 60 °C (140 °F), the vapor blanket cooling was more stable and required mechanical agitation for film breakage. However, although cracking was avoided, it was more difficult to harden the test pieces.

A more recent study (Ref 5) was conducted to examine the cooling properties of water over a broader range of temperatures, from 40 to 90 °C (10 to 195 °F). The cooling curves obtained (Fig. 8.6) were analyzed; results are illustrated in Fig. 8.7 and are tabulated in Table 8.1. These data show that:

- Cooling times increase with increasing water temperature.
- The maximum cooling rate decreases with increasing water temperature.
- The temperature where the maximum cooling rate occurs decreases with increasing water temperature.

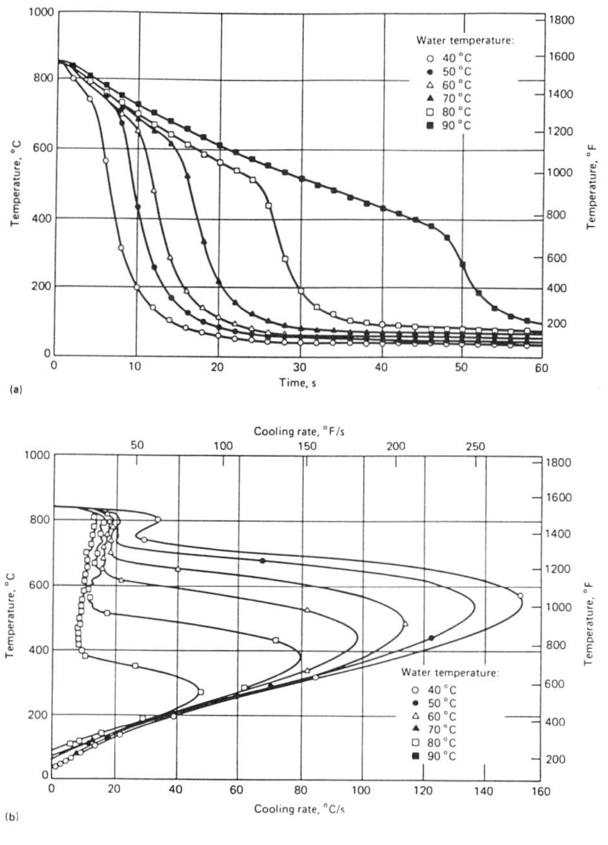

Fig. 8.6 Effect of bath temperature on heat removal in a Wolfson probe. Quenchant is water with a velocity of 0.25 m/s (50 ft/min).

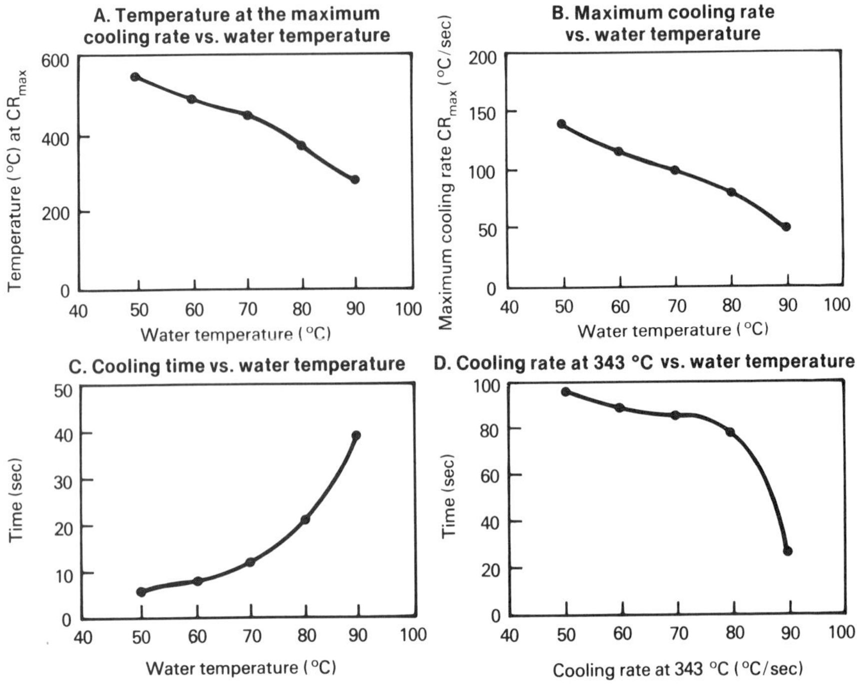

Fig. 8.7 Analysis of cooling curve data obtained with a 12.5 × 100 mm (0.5 × 4 in.) Inconel 600 probe

- The cooling rates at 343 °C (650 °F), with the exception of the 90 °C (195 °F) water (where a dramatic A-phase extension in cooling is observed), decrease only slightly with increasing water temperature.
- The cooling rates at 232 °C (450 °F) are only minimally affected by water temperature.

Chevrier and Beck (Ref 6) studied the heat-transfer characteristics of cold and hot water using a finite-difference calculation technique. This work showed that there is no localized rise in temperature near the surface when quenching in boiling water and that heat transfer occurs exclusively by a vapor blanket (film boiling) mechanism to a surface temperature of approximately 240 °C (465 °F). However, when quenching in cold water, the nucleate boiling and vapor blanket cooling mechanisms coexist on the surface. Therefore, the surface temperature does not vary monotonically with the temperature of the water quenchant.

It was also shown that the surface heat transfer was maximized at approximately 60 °C (140 °F). At lower temperatures, heat transfer is limited by the coexistence of nucleate boiling

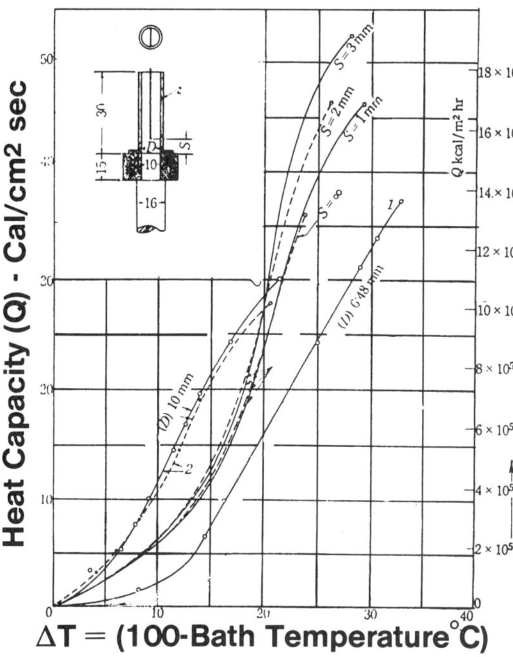

Fig. 8.8 Nukiyama curve of water cooling. Source: Ref 7

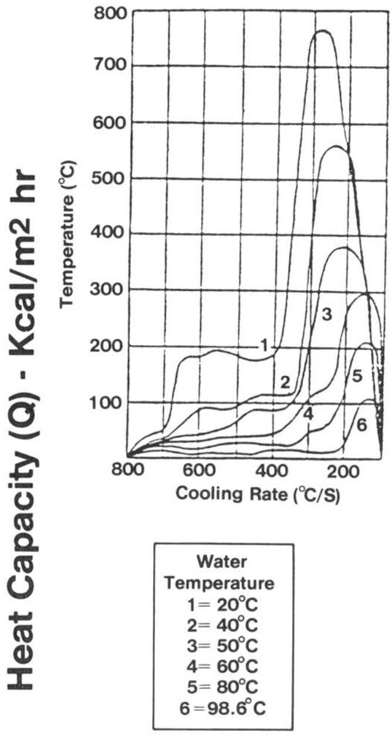

Fig. 8.9 Effect of bath temperature on the quenching performance of water under still-quench conditions

and vapor blanket cooling; above 60 °C (140 °F), heat transfer is limited by vapor blanket cooling, which allows the surface temperature to become more uniform by conduction within the metal. Chevrier and Beck (Ref 6) also showed that the definition of the cooling power of a liquid cannot be more exact than that given by the classic Nukiyama curve (see Fig. 8.8) (Ref 7).

Kobasko (Ref 8) also studied the heat-transfer properties of water during quenching. Nonuniform quenching was most significant when the water temperatures exceeded 50 °C (120 °F). However, the primary difference in this work and that reported earlier by Speith (Ref 1) and by Chevrier and Beck (Ref 6) was that Kobasko focused attention on the effects of salt additions, which provided more uniform cooling by enhancing nucleate boiling (brine quenching). Brine quenching will be discussed later in this chapter.

Petrash (Ref 9) showed that cold water with increasing agitation rates produced lower residual stresses than unagitated hot water. This was attributed to a destabilization of the vapor blanket cooling mechanism, which increased as the water temperature increased (Fig. 8.9). Thermal stresses were attributed to nonuniform cooling caused by nonuniform film rupture during vapor blanket cooling.

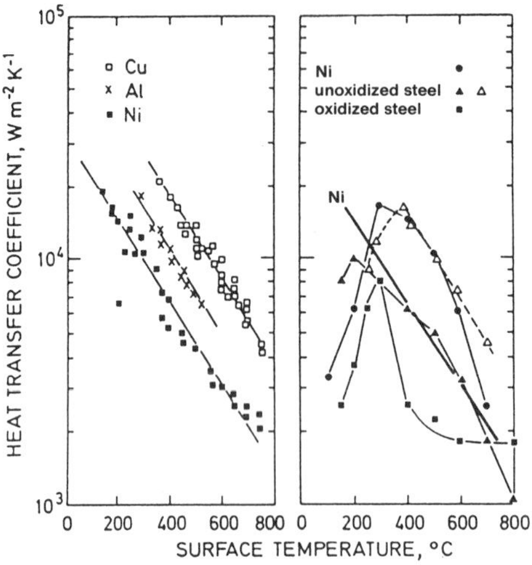

Fig. 8.10 Heat-transfer coefficients for immersion cooling at a water temperature of 20 °C (70 °F)

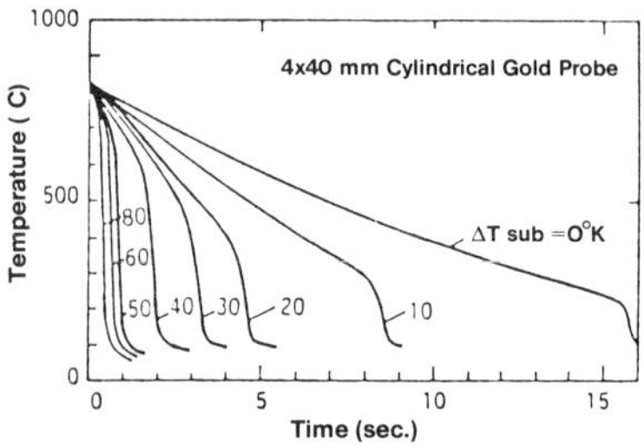

Fig. 8.11 Effect of water temperature on the cooling time and heat-flux curves of water

The film heat-transfer coefficient of a metal increases with the metal conductivity. This is illustrated in Fig. 8.10 (Ref 2) for various metals quenched in water at 20 °C (70 °F). Similar studies, also shown in Fig. 8.10, showed that the film heat-transfer coefficient of oxidized steel was significantly lower than that of unoxidized steel when quenched in water (Ref 2).

Tajima *et al.* (Ref 10-13) evaluated the effect of boundary heat transfer during quenching in water at various temperatures using finite-difference calculation techniques. Figure 8.11 illustrates the cooling curves that were generated using a 4 × 40 mm (0.16 × 1.6 in.) long cylindrical gold probe. The corresponding heat fluxes are shown in Fig. 8.12. At water temperatures greater than 60 °C (140 °F), the vapor blanket cooling was relatively unstable, which is consistent with the data cited above.

These data were then used to calculate the cooling curves attainable with a plain carbon (S45C) steel austenitized in a reducing (nitrogen) atmosphere at 727 °C (1340 °F) (Ref 11). The heat fluxes of steel bars of various diameters quenched in 20 °C (70 °F) water were successfully calculated (Fig. 8.13) (Ref 11), as were the cooling curves (Fig. 8.14) (Ref 11). (Note the correlation between calculated and experimentally obtained cooling curve data in Fig. 8.14.)

The calculated cooling curves of water at various temperatures were then superimposed on the TTT diagram for S45C steel, as shown in Fig. 8.15 (Ref 12). The calculated hardness obtained from this correlation as a function of water temperature is shown in Fig. 8.16 (Ref 12). Interestingly, the cooling rate data obtained at 727 °C (1340 °F), the approximate temperature of the nose of the transformation curve shown in Fig. 8.15,

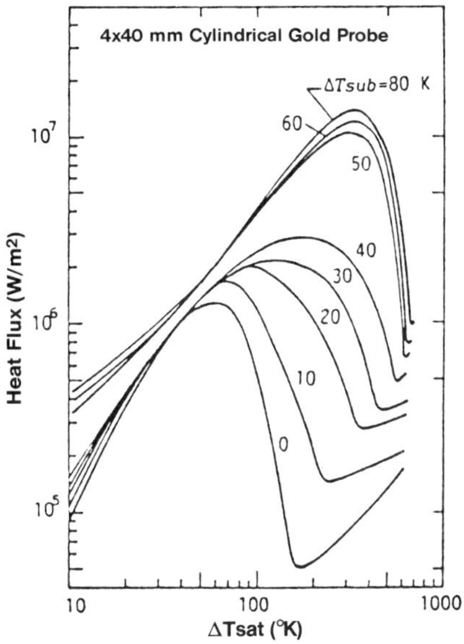

Fig. 8.12 Heat flux as a function of temperature

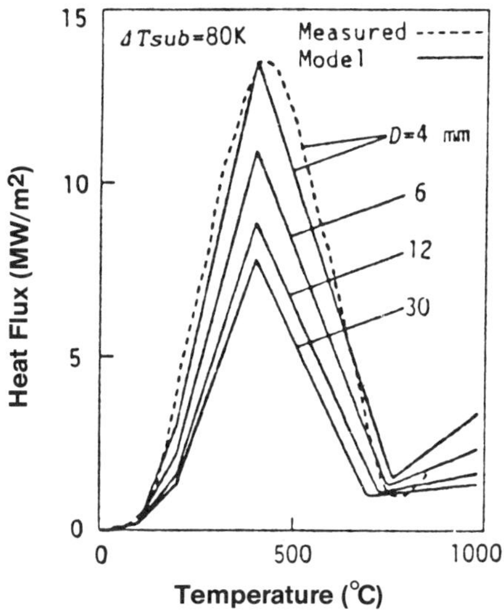

Fig. 8.13 Heat flux of a water-quenched probe as a function of probe diameter

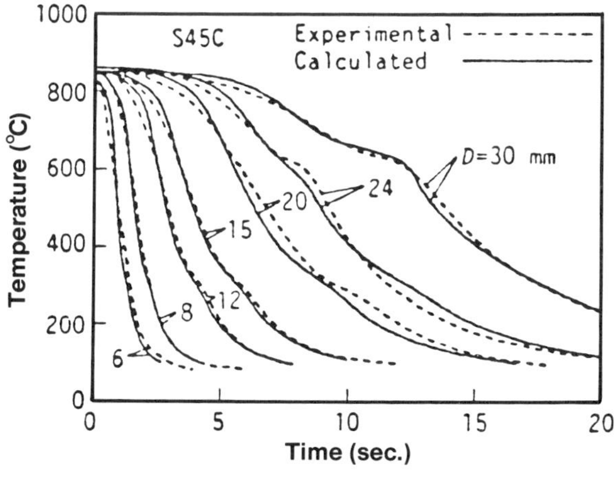

Fig. 8.14 Actual and calculated cooling curves for a gold probe quenched in water as a function of probe diameter

exhibited a large decrease at water temperatures between 40 and 50 °C (105 and 120 °F) (Fig. 8.17).

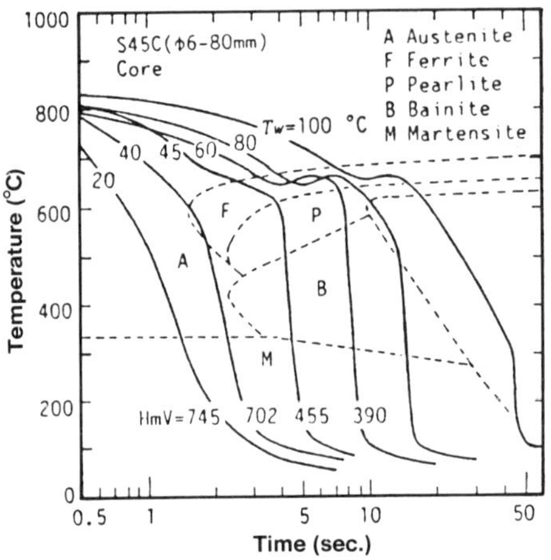

Fig. 8.15 Cooling curves for water-quenched S45C steel probes as a function of temperature (superimposed on a TTT diagram)

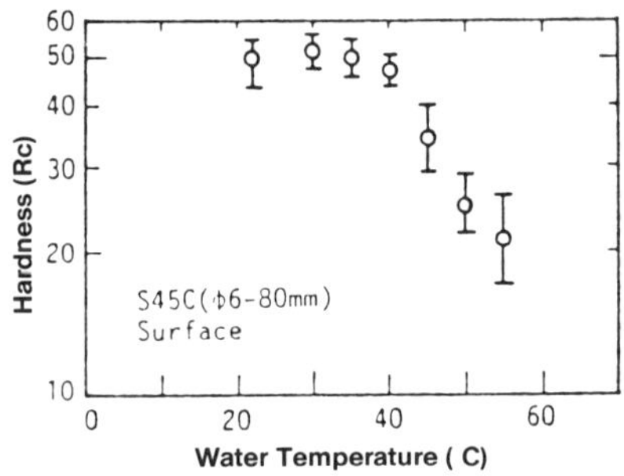

Fig. 8.16 Relationship between surface hardness and cooling water temperature for S45C steel

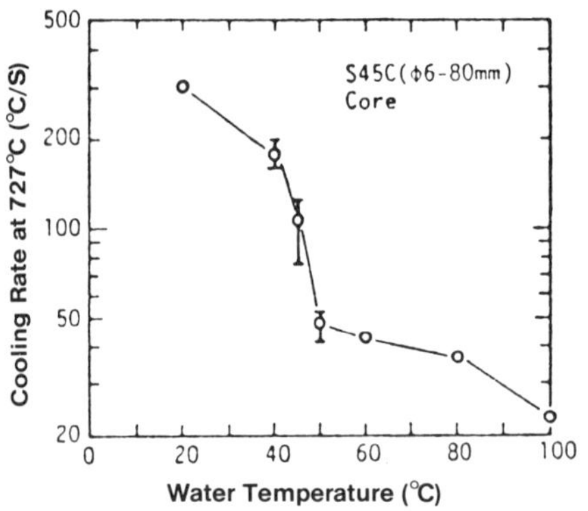

Fig. 8.17 Relationship between cooling rate and cooling water temperature for S45C steel

Maass and Jeschar (Ref 14) demonstrated that the profile of the heat flux for nickel probes quenched in water at different temperatures varied with the shape of the probe, as shown in Fig. 8.18. Interestingly, the transition from vapor blanket cooling to nucleate boiling occurred more quickly for the cylindrical probe than for the spherical probe—evidence that the transition temperature is a function of the shape of a cooling surface.

A similar study was conducted to determine the effect of probe temperature. The heat-flux was found to decrease linearly with increasing water quenchant temperature at all probe surface temperatures for the spherical probe (Fig. 8.19a). The cylindrical probe also exhibited an analogous decrease in heat flux with increasing water temperature, but a change in the slope of heat flux was observed at 70 °C (160 °F) for all surface temperatures (Fig. 8.19b) (Ref 14).

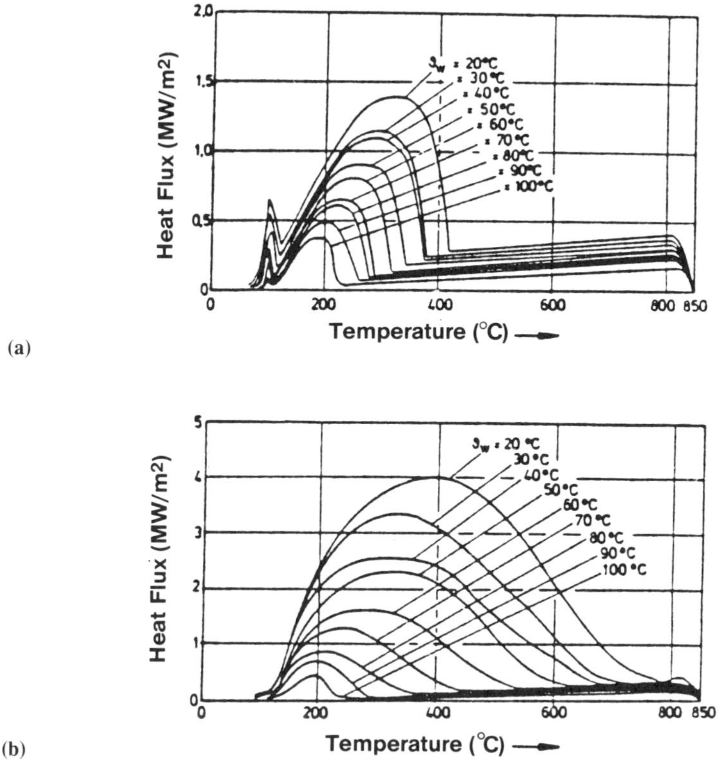

Fig. 8.18 Heat-flux density as a function of surface temperature at different (still-quench) water temperatures. **(a)** Spherical nickel probe. **(b)** Cylindrical nickel probe

Maass and Jeschar (Ref 14) developed equations to predict the temperature at maximum heat flux (T_{BO}) as a function of water quenchant temperature (T_L) for spherical and cylindrical steel probes:

Sphere: $T_{BO} = 1.80 \times T_L^{0.85}$ (Eq 8.1)

Cylinder: $T_{BO} = 6.50 \times T_L^{0.60}$ (Eq 8.2)

Similar studies were conducted with other metals, including aluminum, brass, and copper (Fig. 8.20) (Ref 14).

Double-Step Hot Water Quenching. Hot water quenching has been studied as an alternative to the lead patenting of steel wire. In general, single-step quenching directly into boiling hot water does not produce the desired ductility. However, Yahua (Ref 15)

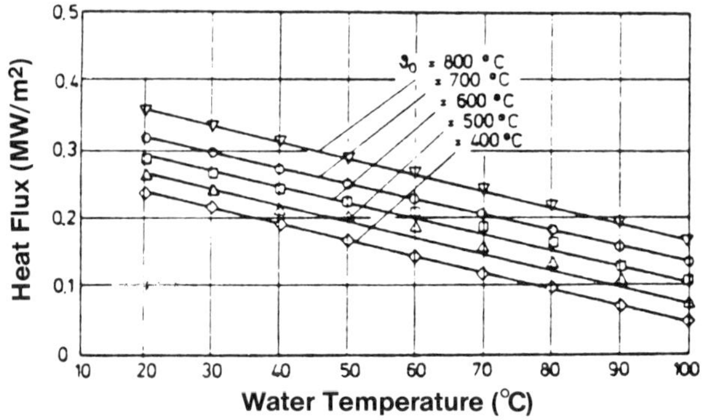

(a)

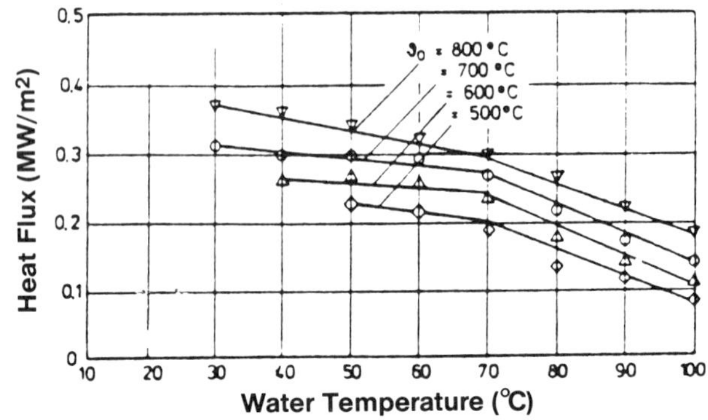

(b)

Fig. 8.19 Heat-flux density in the film phase as a function of water temperature at different surface temperatures and a probe diameter of 40 mm (1.6 in.). **(a)** Spherical nickel probe. **(b)** Cylindrical nickel probe

and Paulus (Ref 16) have found that the following two-step process produces excellent results (Table 8.2):

1. Heat the steel wire to 920 to 950 °C (1690 to 1740 °F) and immerse in boiling hot water for a suitable time (depending on the cross-sectional size of the wire).
2. Remove the wire from the water and air cool.

Use of Polymer Quenchants for Distortion Reduction in a Water Quench. Water, especially at temperatures of <50 to 60 °C (<120 to 140 °F), often produces a nonuniform quench (Ref 1, 4, 8, 9), which may result in spotty hardness, distortion, or cracking. Nonuniformity is caused by relatively unstable vapor blanket formation. Recently, it has been

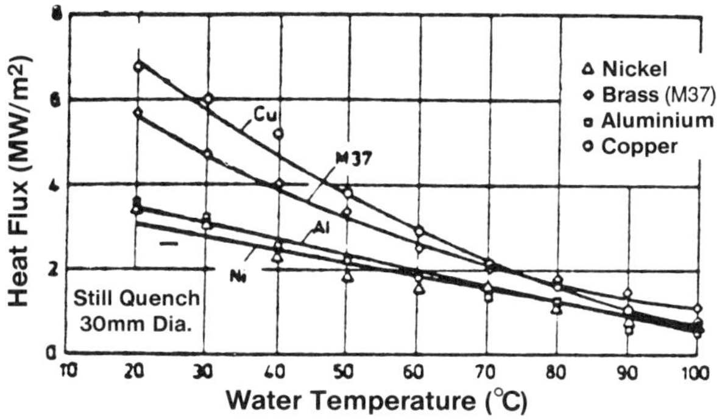

(a)

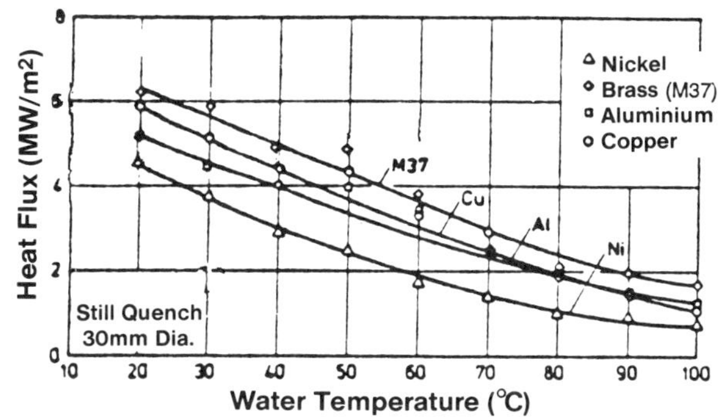

(b)

Fig. 8.20 Maximum heat flux density as a function of water temperature for different materials. (**a**) Spherical probe. (**b**) Cylindrical probe

Table 8.2 Effect of immersion time in boiling water on the mechanical properties and microstructure of 1.6 mm (0.06 in.) diam steel wire

Property	Immersion time, s					
	2	4	5	6	8	~8.5(a)
Tensile strength, MPa (ksi)	~1010 (~145)	1130 (165)	1210 (175)	1170 (170)	815 (120)	940 (135)
Reduction in area, %	55.1	52.5	0	0	0	0
Amount of martensite in microstructure, %	0	0	5	17	36	49

(a) The wire was cooled to 100 °C (212 °F), then air cooled.

shown that the addition of low concentrations (typically 5 to 7%) of an aqueous polymer quenchant dramatically reduces distortion and spotty hardness (Ref 17). (These low concentrations will not hinder the ability of the quenchant to produce the desired hardness.)

It is not unusual for the rejection rate due to cracking to be reduced by at least an order of magnitude. Mukhina *et al.* (Ref 18) have claimed that it is possible to obtain higher cooling rates than water with the addition of low concentrations of an aqueous polymer quenchant.

Aqueous Salt (Brine) Solutions. The term "brine" refers to aqueous solutions containing various percentages of salts, such as sodium or calcium chloride. The cooling rates provided by brine are higher compared with pure water at the same agitation. Although the presence of the salt will decrease the propensity for vapor blanket formation and thus soft spots, higher cooling rates may increase the potential for distortion and cracking.

It has been known for centuries that quench hardening results vary with the quality of the water used for quenching. Speith and Lange (Ref 1), in their now classic work, used cooling curve analysis (see Fig. 8.4) to show that the quench severity of mineral-free rainwater was considerably less than that of tap water, which commonly contains various salts of "hard" metal ions (e.g., Ca^{2+}, Mg^{2+}, Mn^{2+}, Fe^{3+}, Cl^{-}, SO_4^{-2}, and CO_3^{-2}).

Rose (Ref 3) built upon this work by showing that additives such as sodium chloride (NaCl) and sodium hydroxide (NaOH) increase the quench severity of water, especially at higher temperatures, by destabilizing the vapor blanket around the hot metal during A-stage cooling.

Petrash (Ref 9, 19) studied the effects of NaCl and NaOH additions on the vapor blanket stability of water. The addition of as little as 5% of NaCl almost eliminated vapor

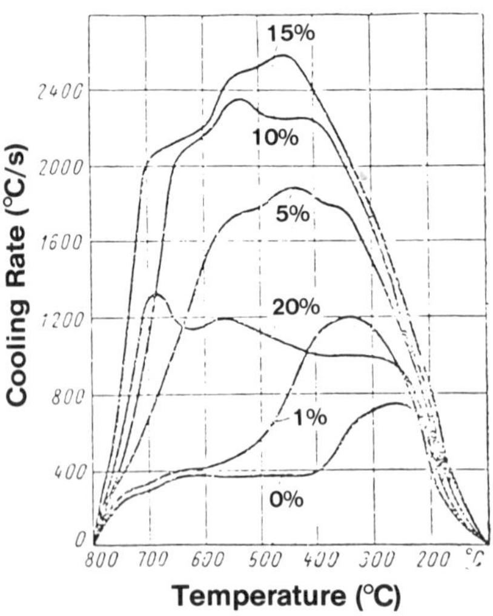

Fig. 8.21 Effect of NaCl concentration and bath temperature on cooling rate

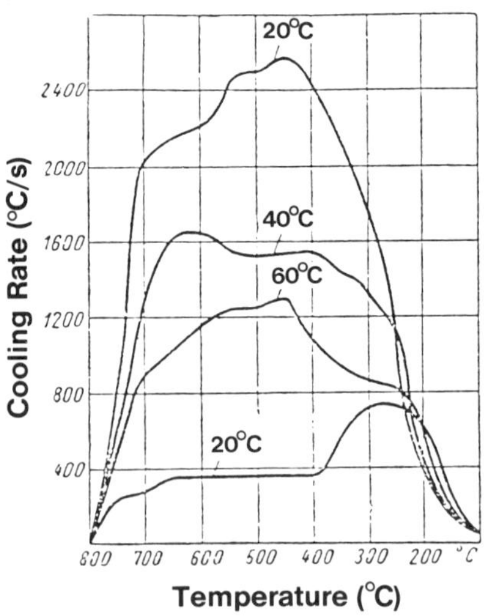

Fig. 8.22 Effect of NaCl bath temperature on cooling rate

blanket cooling. Optimal cooling rate increases were obtained with the addition of 10 to 15% NaCl (Fig. 8.21). Although increasing brine temperatures produce an expected decrease in quench severity, this decrease is not accompanied by the extended vapor blanket cooling observed for water.

The cooling power of brine solutions is not critically affected by small variations in bath temperature. Although a brine can be used near the boiling temperature of water, maximum cooling rates are obtained at about 20 °C (70 °F). The relatively small effect of brine temperature on cooling power is shown in Fig. 8.22 (Ref 9).

The most common alternative to NaCl solutions is aqueous caustic (NaOH) solutions. The addition of NaOH produces cooling rates similar to NaCl at high surface temperatures. The addition of NaOH produces slower cooling rates than NaCl in the martensitic transformation temperatures for many steels (<350 °C, or 660 °F), which would be expected to reduce cracking susceptibility (Ref 9). Figure 8.23 shows the effect of NaOH concentration on cooling rate.

Kobasco (Ref 8) has characterized the relative quenching performance of aqueous NaCl brine and NaOH caustic solutions by determining the relative heat-transfer coefficients using cooling curve analysis. The results of this study, illustrated in Fig. 8.24, showed that the maximum heat-transfer rates of both quenchants are obtained at a salt concentration of approximately 10%.

Jominy end-quench tests have shown that a 10% NaCl solution is an effective hardening medium (Fig. 8.25). It should be noted that although end-quench values were used to construct Fig. 8.25, the specimen was not quenched in the standard end-quench

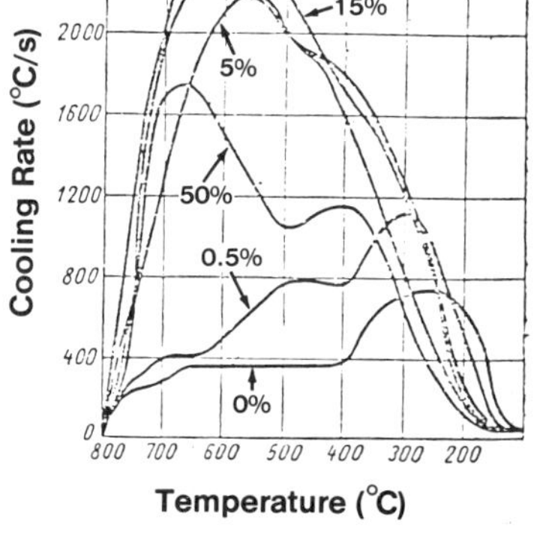

Fig. 8.23 Effect of NaOH concentration on cooling rate. Bath temperature: 20 °C (70 °F)

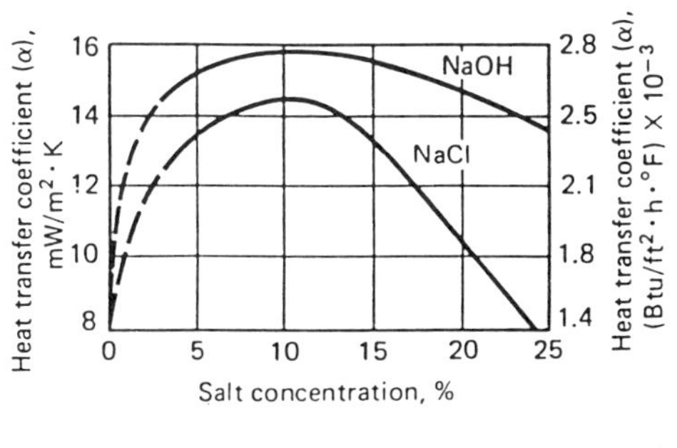

Fig. 8.24 Effect of salt concentration on the heat-transfer coefficients of NaOH and NaCl brine solutions

test but rather in still water, with only the end face contacting the water to simulate surface quenching (Ref 5).

The results of similar Jominy end-quench tests conducted to evaluate the effect of temperature are shown in Fig. 8.26. It is apparent that the cooling power of brine solutions is not critically affected by small variations in bath temperature. Although such solutions can be used at, or near, the boiling point of water, they provide maximum cooling power at about 20 °C (70 °F) (Ref 5). These results are analogous to those reported by Petrash (Ref 9) (see Fig. 8.22).

The concentration of a salt solution is usually quantified by either specific gravity or density. Typical values for various concentrations of NaCl and NaOH are shown in Table 8.3.

Aqueous brine solutions can also be formulated from numerous inorganic and organic salts, including magnesium chloride ($MgCl_2$), lithium chloride (LiCl), and calcium

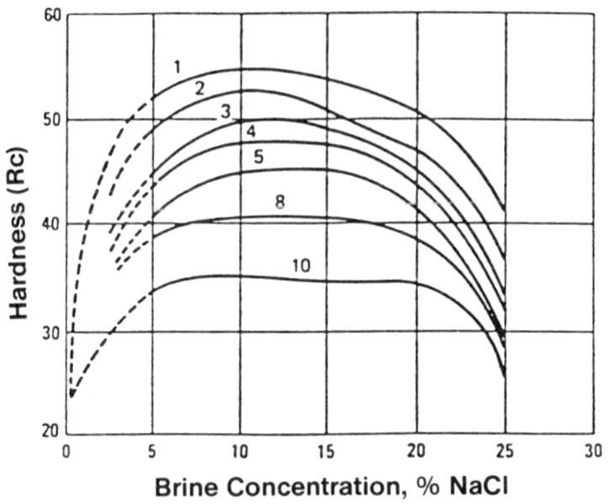

Fig. 8.25 Relationship of hardness to brine concentration during still-quenching of end-quench specimens in a 99 °C (210 °F) brine solution. Numbers above the curves indicate distance from the quenched end of a Jominy bar in units of $^1/_{16}$ in.

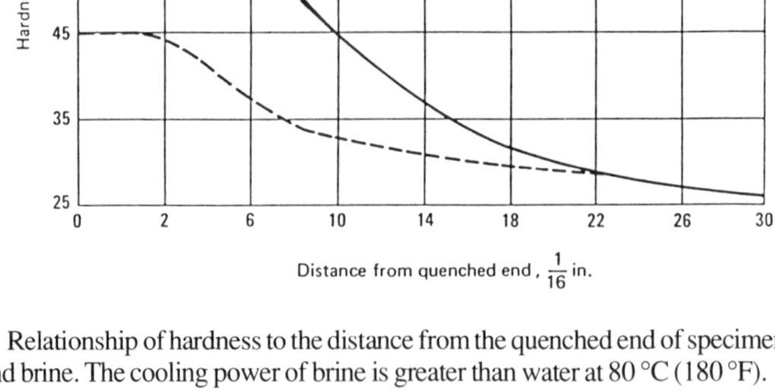

Fig. 8.26 Relationship of hardness to the distance from the quenched end of specimens quenched in water and brine. The cooling power of brine is greater than water at 80 °C (180 °F).

chloride ($CaCl_2$) (Ref 18). The heat fluxes provided by aqueous solutions of these salts at similar densities are compared in Table 8.4.

Experimental studies conducted by Makhina *et al.* (Ref 18) showed that excellent hardness of various plain carbon steels could be obtained with chloride salts, particularly $MgCl_2$ and $CaCl_2$, without cracking (Table 8.5). However, the use of these media suffer from two disadvantages—namely, enhanced corrosion and hydrogen chloride (HCl) contamination of the workpiece via the disproportionation reaction:

$$MgCl_2 + H_2O \rightarrow MgOHCl + HCl \quad \text{(Eq 8.3)}$$

The term D_{max} in Table 8.5 is the maximum diameter of a cylindrical specimen that is completely through-hardened and was calculated from:

Table 8.3 Relationship of brine density to brine concentration

Salt, %	Direct-reading hydrometer	°Be(a)	Salt concentration g/L	Salt concentration lb/gal
NaCl solutions				
4	1.0268	3.8	41.1	0.343
6	1.0413	5.8	62.4	0.521
8	1.0559	7.7	84.5	0.705
9	1.0633	8.7	95.9	0.800
10	1.0707	9.6	107.1	0.894
12	1.0857	11.5	130.3	1.087
NaOH solutions				
1	1.0095	1.4	10.1	0.0842
2	1.0207	2.9	20.4	0.1704
3	1.0318	4.5	31.0	0.2583
4	1.0428	6.0	41.7	0.3481
5	1.0538	7.4	52.7	0.4397

(a) Be = Baume; the specific gravity for liquidus heavier than water is 145/(145 – *n*), where *n* is the reading on the Be scale in °Be.

Table 8.4 Comparison of the heat fluxes of various inorganic salts

Salt	Concentration, %	Density kg/m^3	Density lb/ft^3	Maximum heat flux mW/m^2	Maximum heat flux mW/in.^{2}
NaCl	10	1070	67	13	0.0084
LiCl	23	1138	71	9.5	0.0061
$MgCl_2$	14	1119	70	13	0.0084
$CaCl_2$	10-12	1083.5-1101.5	67.6-68.8	14	0.0090
NaOH	8-10	1087-1109	67.9-69.2	15	0.0097

$$D_{max} = \frac{4\alpha \tau_m}{C\rho \ln T_m} \quad \text{(Eq 8.4)}$$

where α is the heat-transfer coefficient; C is the specific heat capacity of the steel; ρ is the density; τ_m is the time determined from the TTT curve for the alloy of supercooled austenite; and $T_m = (T_o - T_b)/(M_s - T_b)$ where T_o is the austenitizing temperature, T_b is the boiling point of the quenchant at the boundary layer, and M_s is the starting temperature for martensite formation.

Another ionic quenchant medium that has been the subject of considerable research is "sulfite liquor" (Ref 21-24), which is obtained from the noncellulosic component of raw vegetable material and sulfite reaction products of the acids produced by the wood digestion process. Other common names for this material are sulfite yeast concentrate and monosulfite liquor (MSL) (Ref 23).

Table 8.5 Comparison of the hardening power of aqueous chloride and oil

	M_s (a)			D_{max}(a), mm	
Steel	°C	°F	τ(sec)	Oil	Chloride
45	350	660	3	3 (2-4)	16 (14-16)
35 KhM	320	610	30	30	50
40 Kh	350	660	15	15	35
KhVG	245	475	15	8	30
ShKh15	245	475	50	27	56
ShKh15SG	200	390	100	42	71
R18	220	390	400	137	137

(a) The terms M_s, T_m, and D_{max} are defined in text.

Table 8.6 Hardness of powdered steel (0.8% C) quenched with various concentrations of monosulfite liquor

Quench medium	τ_{CO} (N/mm^2)	Hardness, HRC
Water	800	56-58
2% MSL	700-750	55-57
3% MSL	600-650	49-52
4% MSL	450-500	29-36

Table 8.7 Comparison of a plain carbon steel hardened in water, oil, and 6% MSL

Quenching medium	Tetragonal martensite, %	κ-martensite, %	Retained austenite, %
Water	1.046	26.5	21.0
MSL (6%)	1.0425	56.7	17.7
Oil	1.041	67.8	15.3

Solutions of MSL have been used to harden powdered steel (0.8% C) parts produced by dynamic hot pressing (DHP) without cracking. The hardness obtained was inversely proportional to the concentration of MSL, as shown in Table 8.6 (Ref 24). MSL has also been used to harden an unidentified plain carbon steel (Table 8.7).

MSL has a number of disadvantages that preclude its widespread use, including susceptibility to mold formation and rheological instability. Because these solutions are based on emulsions and microemulsions, the particle size of the MSL solution is likely to vary during use, resulting in a concurrent variability in quench severity.

Molten Salts

Two quenching processes that require relatively high quenchant temperatures are martempering and austempering (Ref 25). Martempering requires that the steel part be rapidly cooled to a temperature slightly above the starting temperature for martensitic transformation (M_S), as shown in Fig. 8.27(a), stabilized and then cooled to room temperature. This process is conducted to minimize the potential for cracking.

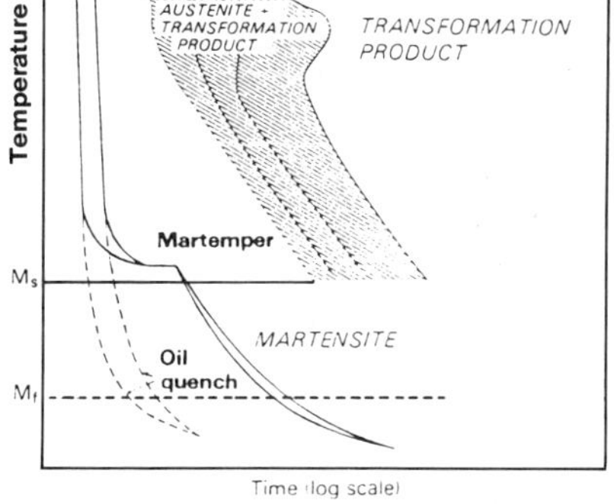

Fig. 8.27(a) Schematic illustration of the relationship between the isothermal transformation diagram and cooling curves during oil quenching and martempering

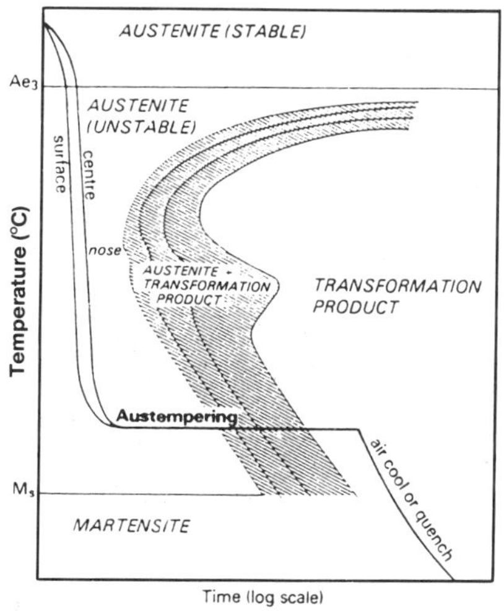

Fig. 8.27(b) Schematic illustration of the relationship between the isothermal transformation diagram and cooling curves during austempering

Austempering is conducted by rapidly cooling to a temperature slightly above the M_S temperature, as shown in Fig. 8.27(b) and holding for a sufficient time to allow austenite to transform to bainite. Austempering may be performed in specially formulated oils (martempering oils) for use at these relatively high temperatures (Ref 20); however, molten salts are usually used. Martempering and austempering processes may eliminate the need for a conventional oil quench and tempering.

Molten salts are usually the medium of choice for high-temperature quenching. Advantages of salt quenching include:

- Highly alloyed steels such as those shown in Table 8.8 (Ref 25) can be isothermally annealed.
- Scaling, distortion, and cracking of high-speed tool steels are minimized (Ref 26).
- The risk of cracking during martensite formation (e.g., spring wire) is reduced (Ref 27).
- The formation of high-temperature transformation products (e.g., bainite, ferrite, etc.) is enhanced.

Most quenching salts are either binary or ternary mixtures of potassium nitrate (KNO_3), sodium nitrite ($NaNO_2$), and sodium nitrate ($NaNO_3$). The minimum quenching temperature depends on the melting point of the salt mixture. The phase diagram shown in Fig. 8.28 illustrates the dependence of the melting point on the composition of the mixture. The ratio of the salt mixture may also affect the viscosity of the medium, which in turn affects the cooling rates (Ref 28). Quenching temperatures of these molten mixtures may range from 140 to 600 °C (285 to 1110 °F). However, melting points as low as 80 °C (175 °F) can be achieved by the addition of up to 10% H_2O (Ref 25). Salt baths are subject to potential explosive degradation at temperatures above 600 °C (1110 °F) (Ref 29, 30).

The selection of a martempering or austempering process also depends on the steel being quenched. Table 8.8 shows the Rockwell C hardness ranges that can be expected for plain carbon steels, alloy steels, and cast iron for each process (Ref 30). Salt bath temperatures in the range of 195 to 350 °C (385 to 660 °F) have only a minimal effect on the Grossmann factor quench severity, as shown in Table 8.9 (Ref 28, 31).

Table 8.8 Martempering versus austempering for various steels

	Cross-sectional size		Hardness, HRC	
Steel class	mm	in.	Martempering	Austempering
Low-alloy steel	<13	<0.5	35-65	35-55
Alloy steel	<100	<4.0	35-65	35-55
Cast iron	<13	<0.5	63	...

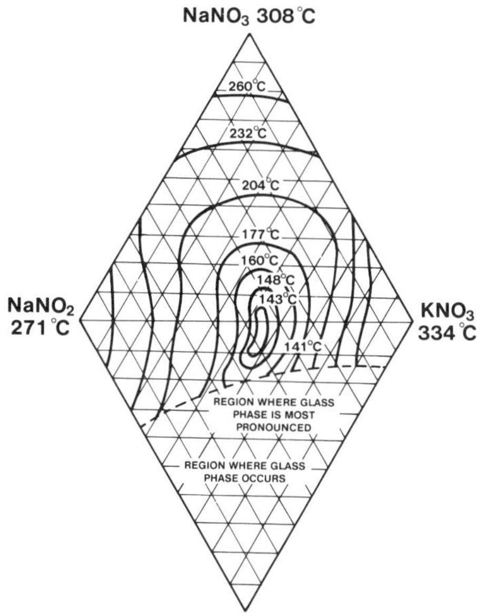

Fig. 8.28 Freezing points of ternary alkali nitrate-nitrite mixtures with composition given in percent

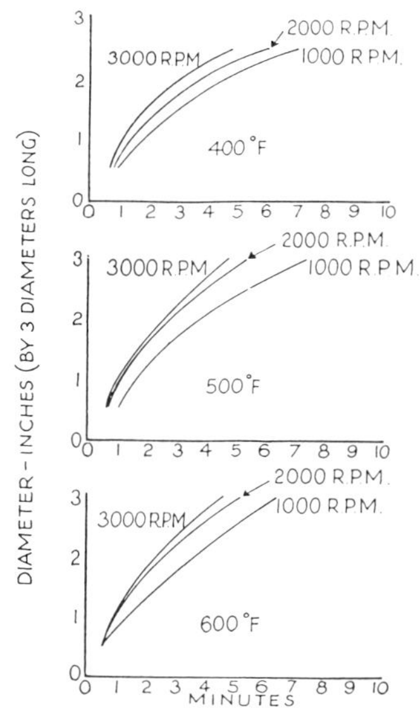

Fig. 8.29 Effect of agitation on the quench severity of molten salt

Table 8.9 Effect of salt temperature on quench severity(a)

Salt temperature		Grossmann *H* factor, $in.^{-1}$	
°C	°F	Center	Surface
195	385	0.46	0.63
200	390	0.45	0.65
230	450	0.40	0.65
270	515	0.45	0.64
295	560	0.41	0.57
350	660	0.43	0.58

(a) A KNO_3-$NaNO_2$ salt with a melting point of 135 °C (275 °F) was used with no agitation. Source: Ref 31

Like all quenchants, the heat-extraction capabilities of molten salt depend on agitation rate. The required agitation rate depends on the ability of a $KNO_2/NaNO_2/NaNO_3$ mixture to through-harden plain carbon steel at various quench temperatures, as shown in Fig. 8.29 (Ref 32).

Schliessmann and Corle (Ref 33) showed that cross-section sizes of less than 38 mm (1½ in.) had faster cooling rates in rapidly agitated molten salt at 205 °C (400 °F) than those attainable with mildly agitated straight mineral oil at 50 °C (120 °F). Conversely,

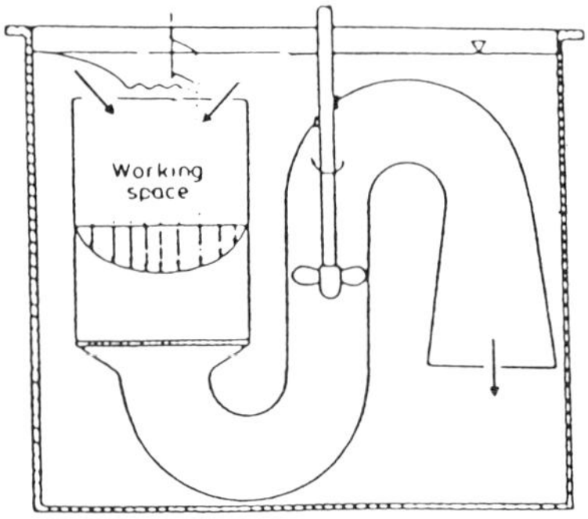

Fig. 8.30 Salt circulation system in a hot salt bath for martempering

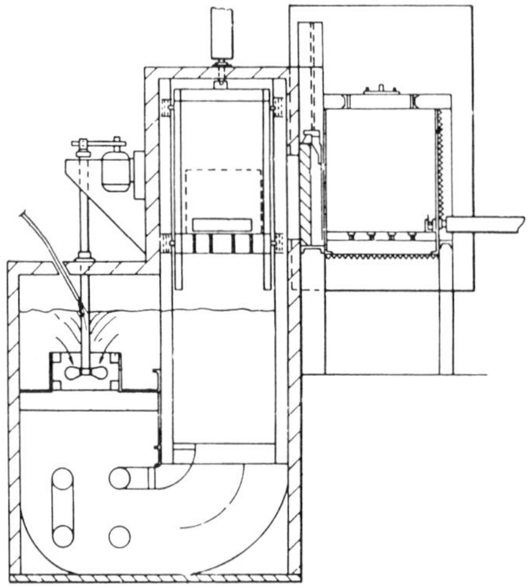

Fig. 8.31 Dual-impeller salt bath agitation system

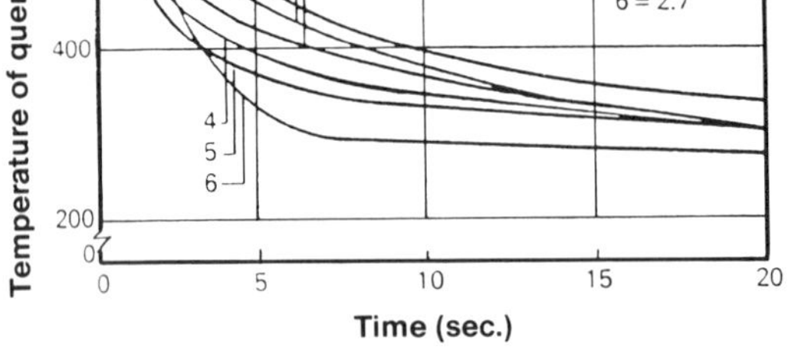

Fig. 8.32 Effect of water additions on cooling curves in a martempering salt bath at 200 °C (390 °F)

the cooling rates for cross-section sizes greater than approximately 50 mm (2 in.) were faster in the mildly agitated oil quench.

Agitation can be applied to a salt bath in various ways. One method that has been described by Liscic (Ref 34) is shown in Fig. 8.30. A dual-impeller system substantially increases quench severity. The impellers are placed in the quench chamber to provide optimal bidirectional flow (Fig. 8.31) (Ref 35).

The quench severity of a molten salt bath may be substantially affected by the presence of water, as shown in Fig. 8.32 (Ref 25) and Fig. 8.33 (Ref 34). Figure 8.34 shows that variations in bath agitation rate and water content can profoundly affect the depth of hardening of AISI 4140 steel (Ref 34). The physical properties of a molten salt bath are also affected by the addition of water (Fig. 8.35) (Ref 36).

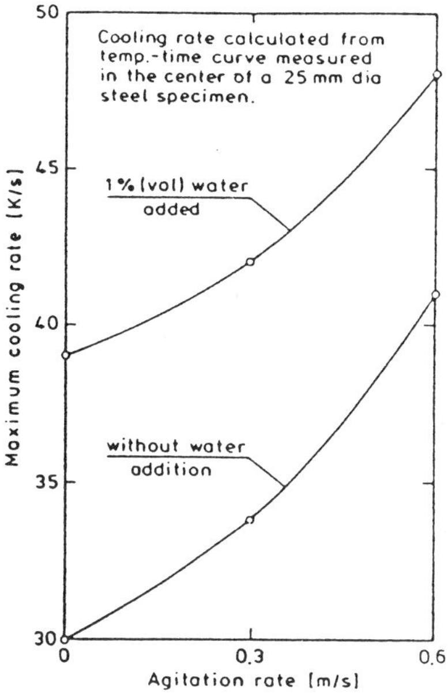

Fig. 8.33 Maximum cooling rate of a hot salt bath (Degussa AS-140) at 200 °C (390 °F) as a function of agitation rate and percentage of water added

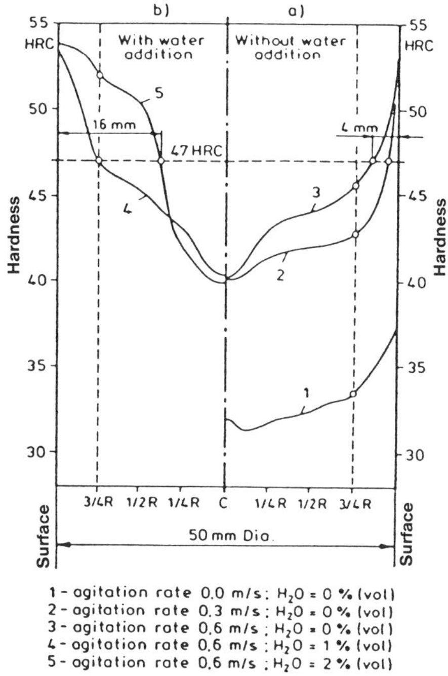

Fig. 8.34 Influence of agitation rate and water addition on hardness distribution in 50 × 200 mm (2 × 8 in.) AISI 4140 steel bars quenched in a hot salt bath (Degussa AS-140) at 200 °C (390 °F)

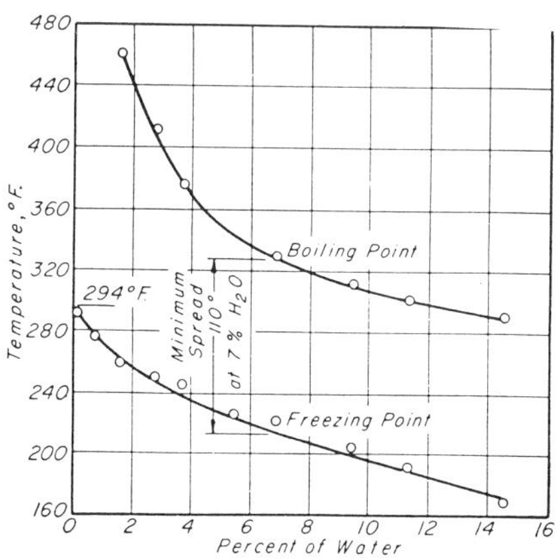

Fig. 8.35 Effect of water on the freezing point and boiling point of a $KNO_2/NaNO_2/NaNO_3$ mixture

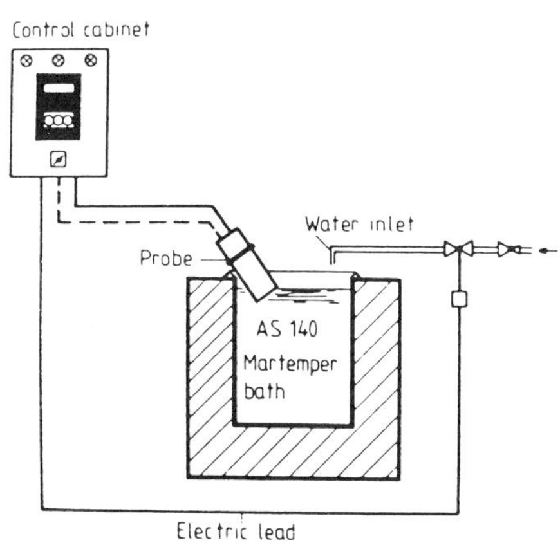

Fig. 8.36 Automatic system developed by Degussa for the addition of water to molten salt

Control of the concentration of added water is important if the maximum benefit of increase in quench severity is to be reproducibly obtained. In practice, however, this has been traditionally difficult to control, because water typically evaporates even during addition. Water addition to molten salt is accompanied by spattering and is therefore can be hazardous. To ensure optimal safety, water should only be added very slowly below 175 °C (350 °F) or using an automatic additive procedure (Ref 36). Degussa has developed an automatic water addition and probe monitoring system (see Fig. 8.36) that permits the maintenance of controlled additions of specific amounts of water to the salt quench in the range of 180 to 250 °C (355 to 480 °F) (Ref 37).

Salts may also absorb water from a humid environment at room temperature when the quench system is not in use. Therefore, before resuming normal quenching operations, it is generally recommended that the salt bath be heated at 95 °C (200 °F) until all of the water is removed (Ref 38).

Howes (Ref 29) has attempted to summarize the effect of bath temperature, agitation, surface area, and quenching time on the ability of a salt bath to harden steel using a

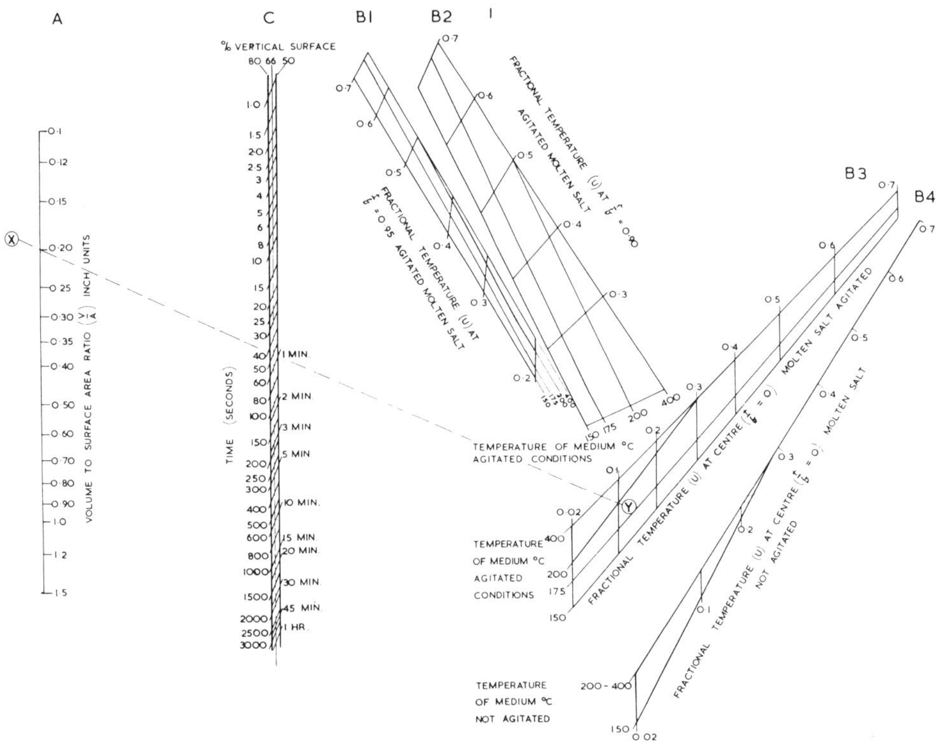

Fig. 8.37 Nomogram for molten salt quenching

"nomogram." Nomograms for a salt bath, with and without water, are illustrated in Fig. 8.37 and 8.38, respectively. These nomograms should be used as follows:

- Select the size of the specimen on scale A.
- Select the appropriate B-scale position and quenching conditions and find the point on the grid corresponding to the desired fractional temperature and quenching medium temperature.
- Connect the points on the A and B scales.
- Read the time on the C scale using the grid line nearest to the actual percent vertical surface of the specimen.

This is illustrated by the quenching of a 25 mm (1 in.) diam cylinder, where the length (L) is equal to two times the diameter (D), into agitated molten salt at 200 °C (390 °F)

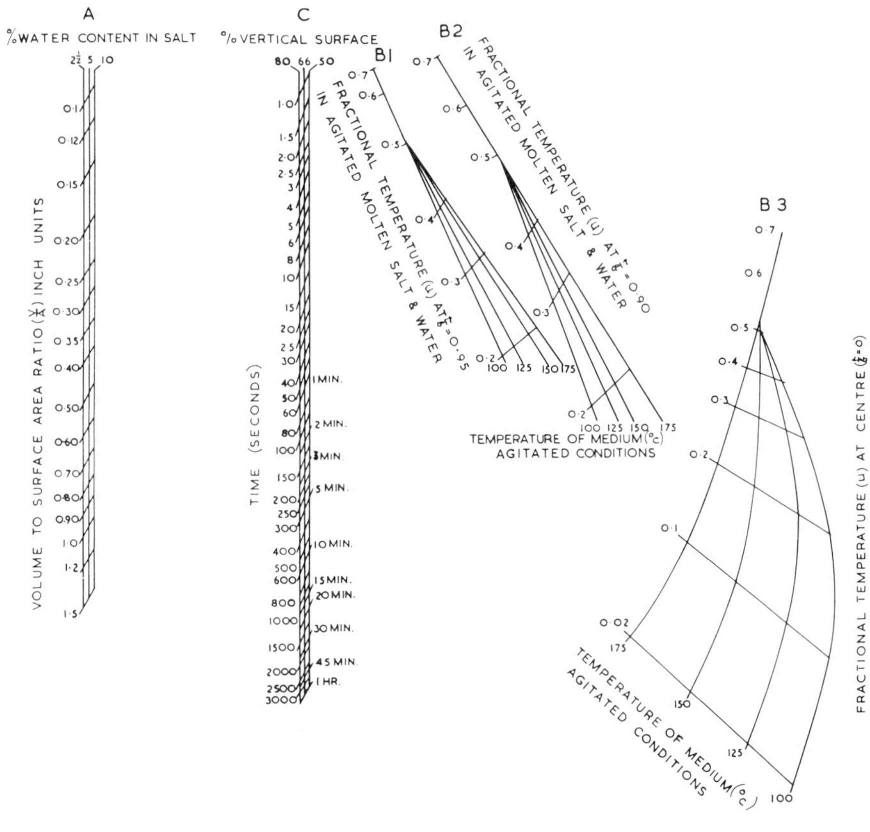

Fig. 8.38 Nomogram for quenching into molten salt and water

(fractional temperature $U = 0.1$) (Ref 39). This dimension ($L = 2D$) corresponds to $V/A = 0.20$. To determine the time required for the center to cool from 900 to 270 °C (1650 to 520 °F), connect a line XY between the A and B3 scales. The time on the 66% vertical surface line scale C shows that approximately 52 s will be required.

In heat treating, it is common for steel to be austenitized in a high-temperature molten salt bath. Such baths are composed of either a binary blend of $KNO_3/NaNO_3$ or a ternary chloride blend, such as KCl/LiCl/NaCl (Ref 40). When austenitization is complete, the steel typically has a salt film on the surface (Ref 41), which is then quenched in a lower-temperature ternary blend of $KNO_2/NaNO_2/NaNO_3$ (see Fig. 8.28).

The adsorbed salt film on the surface of the steel may affect the quench process in two ways: mediation of quench cooling rates and chloride contamination of the molten salt quenchant. The effect of this salt film on the cooling rates of a 40 mm (1.6 in.) diam soft iron sphere was minimal (Fig. 8.39) (Ref 41).

High-temperature chloride salts are only partially soluble in conventional ternary ($KNO_2/NaNO_2/NaNO_3$) salt solutions. Solubility of the chloride salts increases slightly with increasing bath temperatures. However, soluble chloride salts decrease the severity of the molten salt quench (Ref 32). Chloride salt impurities may be separated from the molten quenchant salt mixture by precipitation and subsequent filtration in a separation chamber at approximately 30 to 50 °C (55 to 90 °F) lower than the bath temperature.

Foreman (Ref 28) reported that the presence of chloride salts actually increased cooling rates in a "dry" salt bath. Cooling rate reduction was observed only when chloride was present above its saturation level.

Similar contamination effects are caused by carbonate (from sodium nitrite decomposition) and metallic scale and complexes. However, these insoluble "sludges" may also be removed by filtration and the quench severity maintained by controlled water addition by a system such as that shown in Fig. 8.36 (Ref 30).

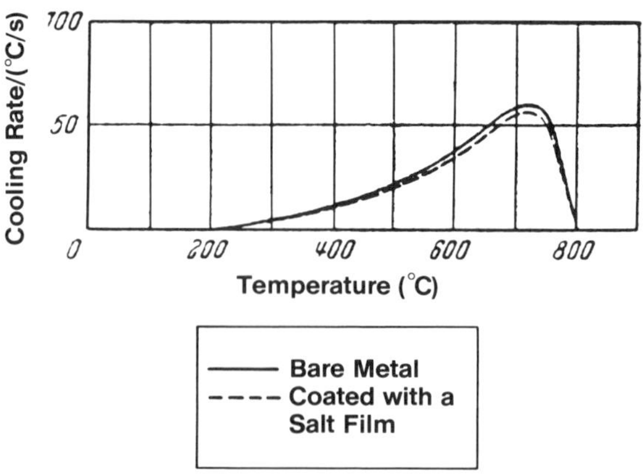

Fig. 8.39 Effect of a salt film on the surface cooling of a 40 mm (1.6 in.) diam soft iron sphere in a molten bath containing KNO_3 salt at 200 °C (390 °F)

Fluidized-Bed Quenching

Fluidized-bed quenching processes may be used as safer and more ecologically sound alternatives to molten metal (Ref 42) and salt bath quenching (Ref 43). Advantages of fluidized-bed quenching include flexibility, improved process control, and cleanliness. One of the disadvantages inhibiting wider use is the cost of the fluidizing gas, which generally is not recycled. Nevertheless, increasing toxicological and environmental regulations are forcing greater consideration of fluidized-bed quenching systems.

A fluidized bed consists of a furnace system (shell, heaters, and insulation) and the quench system (gas diffusion assembly, fluidized-bed solid support, and gas). A typical system is shown in Fig. 8.40 (Ref 44).

A fluidized bed is generated by blowing a gas (e.g., nitrogen, argon, carbon dioxide, helium, or hydrogen) through a solid support (Ref 45). Examples of solid supports include aluminum oxide, iron, copper, and molecular sieves (Ref 45).

Heat transfer, a, within the fluidized bed has three primary components (Ref 46):

- The particle convective component (a_{pc}), which is dependent on heat transfer between the bulk of the fluidized bed and the heat-transfer surface of the particle
- The interphase gas convective component (a_{conv}), where heat is transferred between the particle surface and the interphase gas
- The radial component of heat transfer (a_{rad})

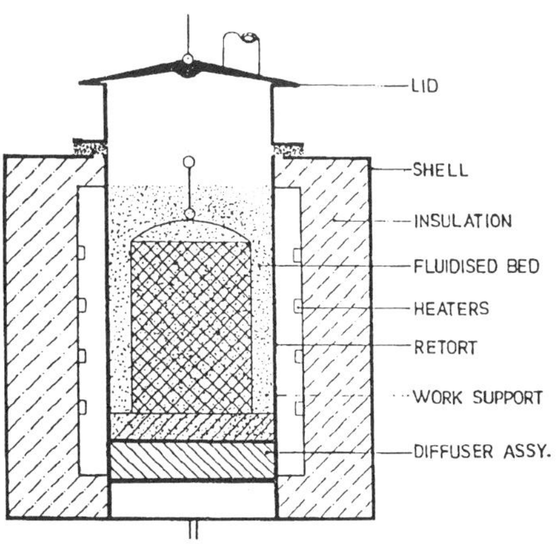

Fig. 8.40 Cross section of an electric fluidized bed furnace

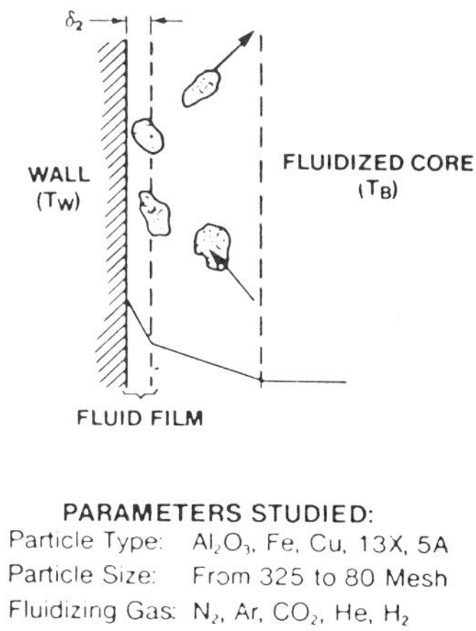

Fig. 8.41 Heat transfer between wall and fluidized bed

$$a = a_{pc} + a_{conv} + a_{rad} \quad \text{(Eq 8.5)}$$

This process can be envisioned by considering the heat-transfer process depicted in Fig. 8.41, where the bubbles of the fluidizing gas rise in the bed, causing movement of the solid particles and thus establishing a mixing process. The particles, or "pockets" of particles disrupt the gas film at the cooling metal surface (Ref 45). The gas flow rate required to fluidize the bed, U, is typically at least three to five times the minimum fluid flow rate 3-5× U_{mff}.

A typical cooling curve shape that is obtained in fluidized-bed quenching is shown in Fig. 8.42. An equation that describes this cooling process is:

$$\frac{T - T_f}{T_i - T_f} = \exp\left[\frac{-Aht}{C_p V \rho}\right] \quad \text{(Eq 8.6)}$$

where T_f is the temperature of the fluidized bed, T_i is the initial temperature of the metal being quenched, A is the surface area of the cooling surface and V is the volume of the metal (probe), ρ is the density of the metal, C_p is the heat capacity of the metal, t is time, and h is the heat-transfer coefficient. Variables that must be considered in the selection of a fluidized bed include medium, particle size, temperature, and gas (Ref 45-47).

Fluidized-bed media can be classified according to three positions on a "Geldart chart" (Fig. 8.43), which characterizes the density difference between the particles and the fluidizing medium (ρp-ρg) and the mean particle size (d_p) (Ref 45). Highest

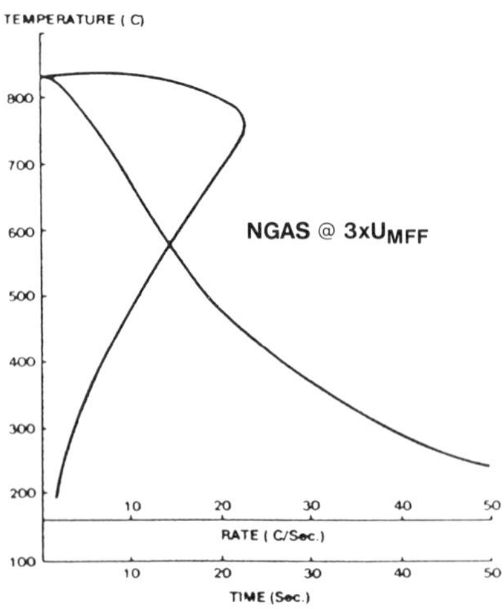

Fig. 8.42 Typical data from fluidized-bed quenching experiment

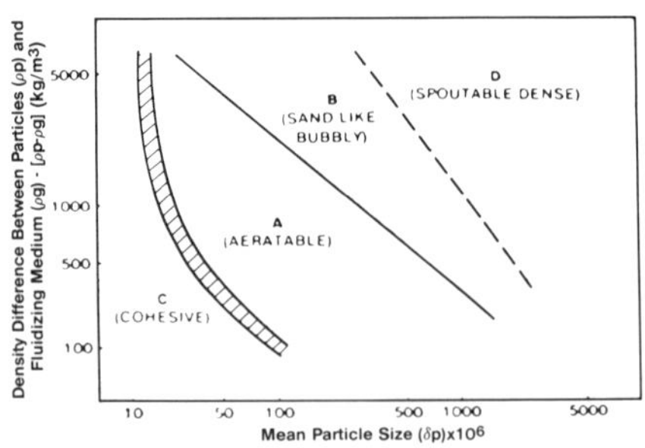

Fig. 8.43 Powder classification diagram (Geldart chart) for fluidization by ambient air

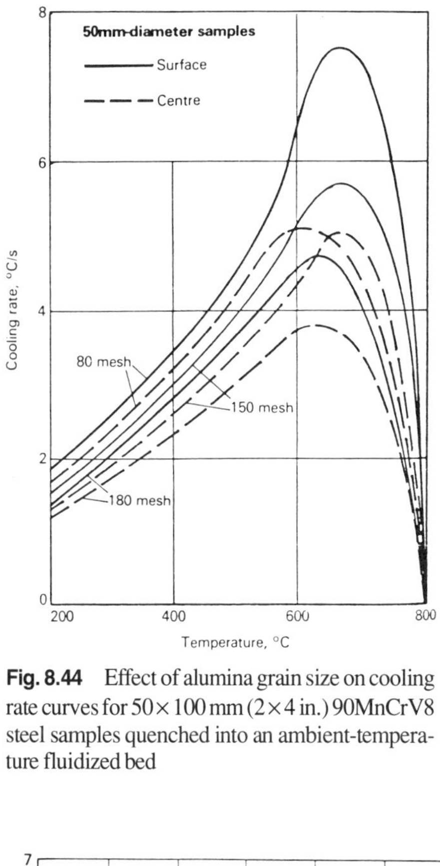

Fig. 8.44 Effect of alumina grain size on cooling rate curves for 50 × 100 mm (2 × 4 in.) 90MnCrV8 steel samples quenched into an ambient-temperature fluidized bed

heat-transfer rates occur with a type B medium, such as aluminum (sand). Aluminum oxide generally provides the best heat-transfer capacity, thermal stability, uniformity, and environmental compatibility compared with other potential candidates (e.g., chromium oxide) (Ref 47).

Particle size of the medium also contributes substantially to cooling properties. In general, cooling rates increase as the alumina grain size increases (Fig. 8.44) (Ref 47). For most applications, 80 mesh aluminum oxide is selected (Ref 47). The lower limit of particle size is approximately 40 mesh. Below this size, the fluidizing region becomes type A or C (see Fig. 8.43) and thus the heat-transfer rate decreases.

The temperature of the fluidized bed can also affect heat-transfer rates during quenching. As expected, cooling rates decrease with increasing bed temperature, as shown in Fig. 8.45 (Ref 47-49).

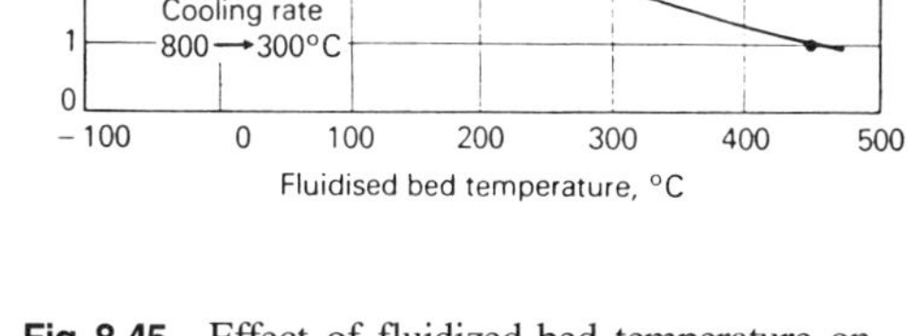

Fig. 8.45 Effect of fluidized-bed temperature on cooling rate between 800 and 500 °C (1470 and 930 °F) and 800 and 300 °C (1470 and 570 °F). 75 × 100 mm (3 × 4 in.) 90MnCrV8 steel samples; temperature measured at center of section

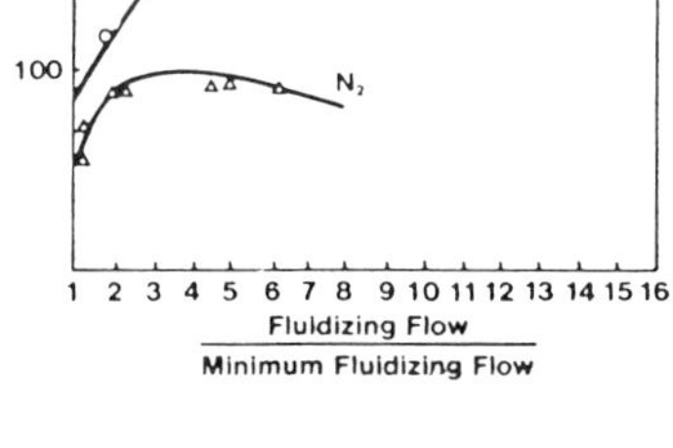

Fig. 8.46 Effect of fluidizing gas on heat transfer

Traditionally, the most common gases for fluidized beds were nitrogen and air. However, other gases, such as hydrogen and helium, provide even greater potential for wider ranges of heat-transfer rates. For example, consider the relative heat-transfer rates of hydrogen, nitrogen, and helium as a function of flow rate (Fig. 8.46). The heat-transfer coefficient at 600 °C (1110 °F) increases with gas flow rate. Higher heat-transfer rates are possible with both hydrogen and helium compared with nitrogen. This work showed that the maximum heat-transfer rates were achieved at approximately $5\times U_{mff}$ for all gases except hydrogen and helium, which exhibited a maximum at approximately $10\text{-}15\times U_{mff}$. Heat-transfer rate decreases after the maximum is achieved because of an increased fraction of gas bubbles in the bed (Ref 45).

Reynoldson (Ref 46) has shown that air-fluidized beds produce cooling rates intermediate between those exhibited by mineral oil and a low-pressure vacuum quench (Fig. 8.47). Other workers have shown that air-fluidized beds are capable of producing the same hardness as oil and salt for various steels (Table 8.10) (Ref 50).

Gas Quenching

The fastest-growing technology in heat treating is gas quenching in a vacuum furnace (Ref 51), a process that produces clean, bright metals. If either ferrous or nonferrous metals are heated under high vacuum (10^{-4} to 10^{-2} mbar) (Ref 52), volatile oxide impurities are readily removed. Under these conditions, vacuum heat treating is a "cleaning" process. However, under high vacuum, the potential for surface reaction with

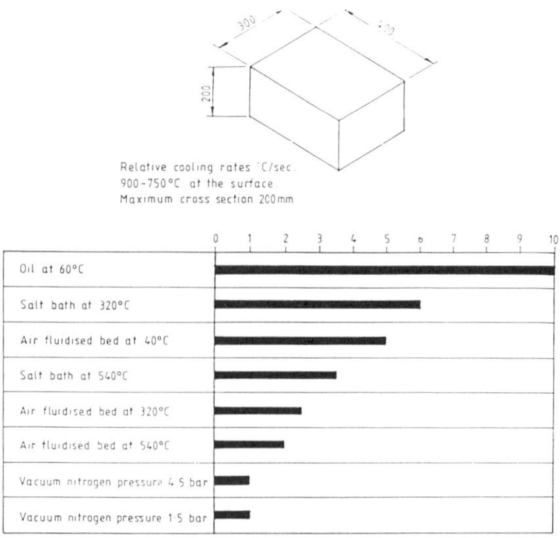

Fig. 8.47 Comparison of the quench severities of various quench media

Table 8.10 Comparison of hardness and distortion properties obtained with oil, salt, and fluidized beds for various steels. Dimensions given in inches.

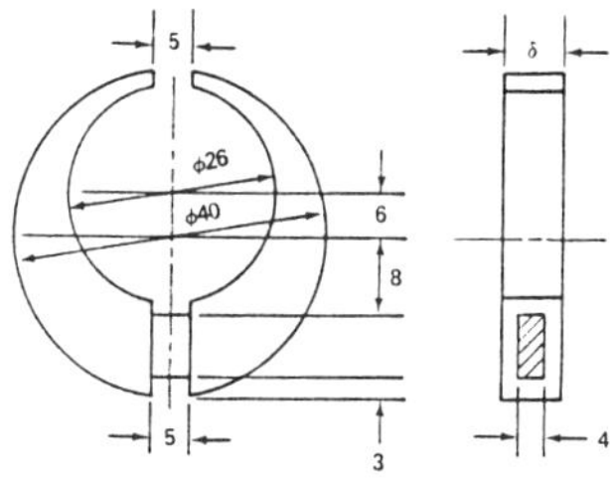

Steel	Quenchant	Outside diameter, in.	Inside diameter, in.	Opening, in.	Hardness, HRC
40Cr	Oil	0.04	0.04	0.12	55.5
	Fluidized bed	0.01	0.01	0.3	55
$60Si_2Mn$	Oil	0.08	0.07	0.06	61.5
	Fluidized bed	0.05	0.03	0.03	61
	Nitrate salt	0.01	0.03	0.06	54
90CiSi	Oil	0.04	0.03	0.04	65
	Fluidized bed	0.02	0.02	0.04	65
	Nitrate salt	0.03	0.05	0.03	65

Table 8.11 Comparison of heat-transfer coefficients of various quench media

Quenchant	Heat-transfer coefficient (λ), $W/m^2 \cdot K$
Gas, recirculated (1000 mbar N_2)	100-150
Gas, overpressure, high velocity	300-400
Salt bath (550 °C, or 1020 °F)	350-450
Fluidized bed	400-500
Stationary oil (20-80 °C, or 70-175 °F)	1000-1500
Recirculated oil (20-80 °C, or 70-175 °F)	1800-2200
Water (15-25 °C, or 60-75 °F)	3000-3500

volatile components (e.g., oxygen) is minimized. Thus, vacuum heat treating is also a "protective" process.

Traditionally, gas quenching was used only where relatively low heat-transfer coefficients, those intermediate between still air and salt bath quenching, were required. Typical heat-transfer coefficients are shown in Table 8.11.

However, high-pressure gas quenching technology (e.g., helium at a pressure of 20 bar), is capable of producing quench severities comparable to conventional recirculated oil, as shown in Table 8.12 (Ref 53). Very high pressure (e.g., hydrogen at a pressure of 50 bar) can produce heat-transfer coefficients even greater than water (3000 to 3500 $W/m^2 \cdot K$, or 530 to 620 $Btu/ft^2 \cdot h \cdot °F$), as shown in Fig. 8.48.

Table 8.12 Comparison of the Lambda (λ) parameters of various quench media and high-pressure helium(a)

Lambda (λ)	Diameter, mm (in.), of the part for cooling in: Oil	20 bar He	6 bar N_2	Air
0.06	10 (0.4)	...	...	...
0.1	18 (0.7)	...	...	...
0.2	30 (1.2)	10 (0.4)	...	...
0.35	45 (1.8)	25 (1.0)	...	...
0.5	60 (2.4)	35 (1.4)	...	...
0.7	70 (2.8)	50 (2.0)	14 (0.6)	...
1.0	90 (3.5)	70 (2.8)	25 (1.0)	10 (0.4)
2.0	130 (5.1)	125 (4.9)	60 (2.4)	18 (0.7)
3.0	170 (6.7)	225 (8.9)	100 (4.0)	25 (1.0)
5.0	220 (8.7)	...	175 (6.9)	40 (1.6)
7.0	280 (11.0)	...	...	...

(**a**) The Lambda (λ) parameter is defined as (Ref 53): $\lambda = [t \times (800\text{-}500\ °C]/100$, where t is time in seconds. For a λ value of 1.0, a 90 mm (3.5 in.) diam bar would be expected to fully harden in oil, or a 70 mm (2.8 in.) diam bar would be expected to fully harden in 20 bar He.

Table 8.13 Definition of degrees of vacuum

Degree	Pressure, torr
Low	Atmosphere-10
Medium	$100\text{-}10^{-2}$
High	$10^{-3}\text{-}10^{-5}$
Very high	$10^{-6}\text{-}10^{-8}$
Ultrahigh	$\geq 10^{-9}$

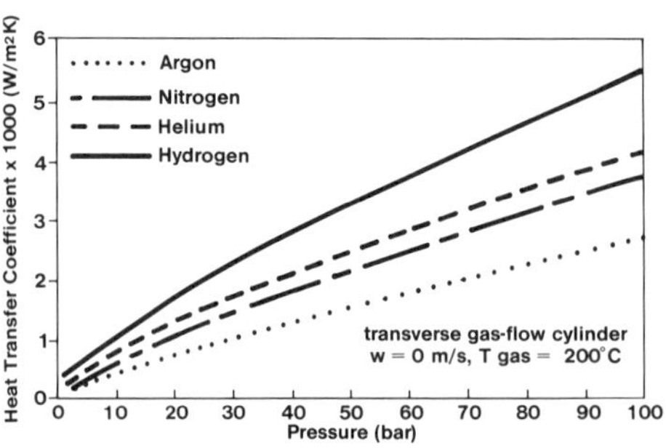

Fig. 8.48 Comparison of various high-pressure gas quenchants. Courtesy of Leybold Durferrit GmbH

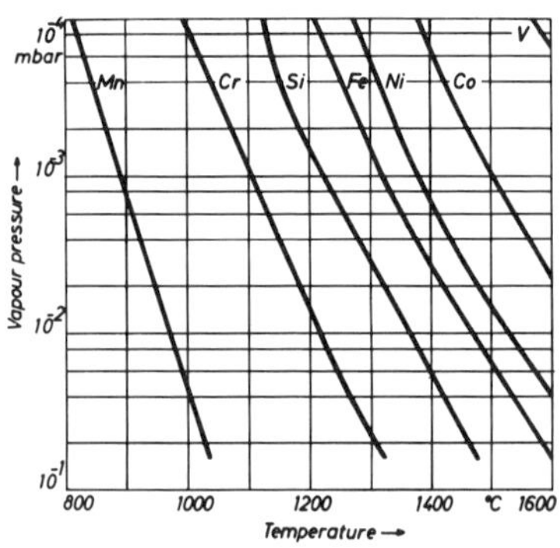

Fig. 8.49 Vapor pressures of different elements. Courtesy of Leybold Durferrit GmbH

There are at least two fundamental aspects of gas quenching: the vacuum heating process and the gas quenching process. Although the primary focus of this section is on gas quenching, a basic discussion of the vacuum heating process is necessary before proceeding.

Vacuum refers to a very low pressure. Various degrees of vacuum have been defined (Ref 54) and are summarized in Table 8.13.

A torr is the standard unit of pressure for vacuum systems. One torr is equal to one millimeter of mercury (mm Hg). For very low pressures, the term micron is often used. One micron is equal to 10^{-3} mm Hg or 10^{-3} torr.

Under high vacuum (e.g., 10^{-4} torr), many metals such as copper are somewhat volatile and thus will exhibit a vapor pressure. Figure 8.49 shows the vapor pressures of various alloying elements as a function of temperature.

In an actual vacuum system, various components exhibit their own vapor pressures, such as residual air, metal oxides, and alloying elements. The fraction of the total system pressure due to each component is called the partial pressure. The sum of the partial pressures will equal the measured system pressure.

Vacuum heat treating is fundamentally different from conventional atmospheric heat treating. In vacuum heat treating, metal oxides that may be present on the surface of a metal will be volatilized by dissociative reactions. On the other hand, conventional heat treating in an atmosphere involves chemical reactions. Because vacuum furnaces are easier to control, they provide a means of supplying a stable process environment. For example, the dew point of a vacuum furnace is readily controlled by the degree of vacuum, as shown in Fig. 8.50 (Ref 54).

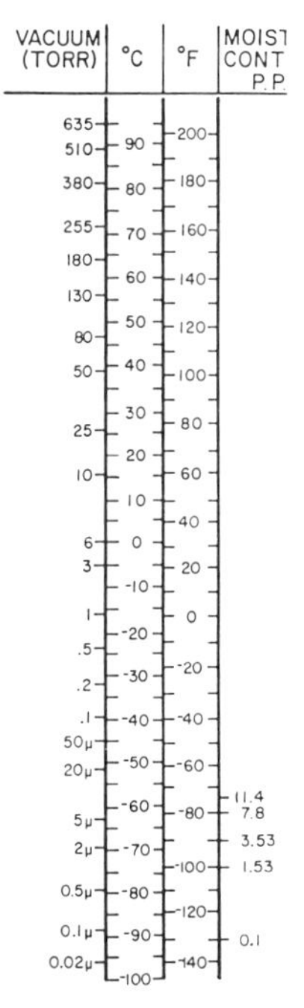

Fig. 8.50 Furnace dew point versus vacuum

Quenching can also be performed in a vacuum furnace. However, relatively slow heat-transfer rates are obtained (5.7 W/m^2 · K, or 3.6 Btu/h · ft^2 · °F) (Ref 55) when compared with other potential quench media (Table 8.12). Therefore, in order to expand the utility of vacuum heat treating, it is essential to provide alternative quenchants for use at the conclusion of the vacuum heat treating cycle.

Various quenchant media used in the vacuum heat treating process include oil (Ref 56), aqueous polymers (Ref 57), and gas (Ref 53, 58). Of these, gas quenching is the most common. Details of liquid quenching processes will be discussed later.

Gas Quenching Variables. Gas quenching is typically performed by pressurizing the furnace (pressure quench) after the heat treating step is complete. This process is called backfilling (Ref 59). Two main factors affect quench severity in gas quenching: gas velocity and gas pressure.

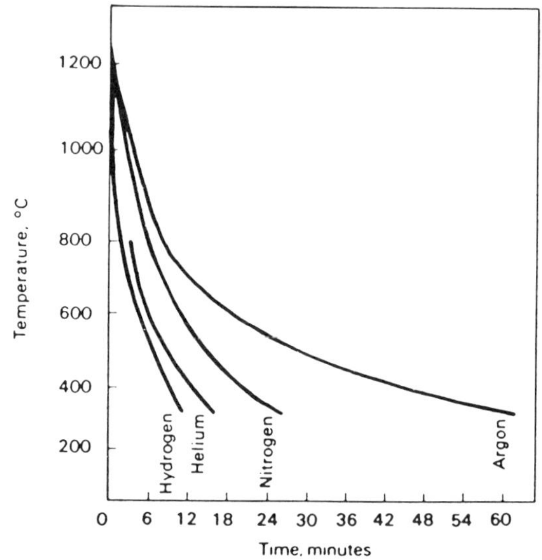

Fig. 8.51 Comparison of the cooling properties of common gases

Table 8.14 Physical properties of quenching gases

Property	Quenching gas			
	Argon	Nitrogen	Helium	Hydrogen
Density, kg/m^3 at 15 °C/1 bar	1.6687	1.170	0.167	0.0841
Density ratio, *w*/*r*, to air	1.3797	0.967	0.138	0.0695
Molar mass, kg/mol	39.948	28.0	4.0026	2.0158
Specific heat(a), kJ/kg · K	0.5204	1.041	5.1931	14.3
Thermal conductivity(a), W/m · K	177×10^{-4}	259×10^{-4}	1500×10^{-4}	1869×10^{-4}
Dynamic viscosity(a), N · s/m^2	22.6×10^{-6}	17.74×10^{-6}	19.68×10^{-6}	892×10^{-6}

(a) Gas conditions: 25 °C, 1 bar. Source: Ref 62

Most gas quenching is done with either nitrogen, argon, helium, or hydrogen (Ref 51). The high cost of argon generally limits its use to situations that demand chemical inertness, such as some aerospace applications employing titanium, tantalum, or niobium (Ref 60). Hydrogen has excellent heat-transfer properties, as shown in Fig. 8.51 (Ref 51), but the explosive potential of this gas requires special safety precautions (Ref 61). The physical properties of quenching gases are summarized in Table 8.14.

Many more recently developed gas quenching applications are based on gas mixtures such as nitrogen/helium. Gas blending allows the user to more cost-effectively calibrate a particular system to higher heat-transfer coefficients than might be achievable with pure helium. Various physical properties of nitrogen/helium blends are summarized in Fig. 8.52 (Ref 62).

Other gas mixtures have also been studied. For example, Lhote and Delcourt (Ref 58) have reported the relative heat-transfer efficiencies when determined at the same pressure (1.5 bar) to be: 70% He + 30% Ar > (N_2 + 10% H_2) > N_2 > Ar.

There are three general approaches to providing gas flow during quenching, as shown in Fig. 8.53 (Ref 52). The first method is to use gas-injection nozzles where the flow is perpendicular to the workpiece (Fig. 8.53a). This process has been used in

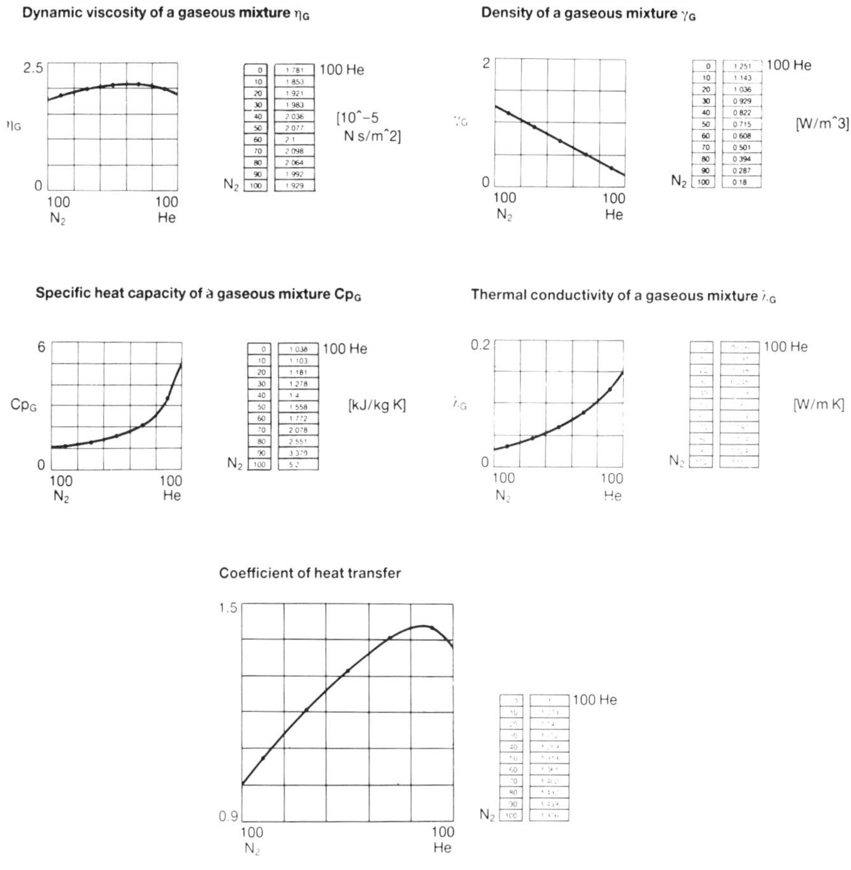

Fig. 8.52 Physical characteristics of nitrogen/helium gas mixtures. Courtesy of Leybold Durferrit GmbH

horizontal vacuum furnaces and is prone to forming cold spots because of nonuniform cooling on the metal surface. The second method is to inject the gas through nozzles parallel to the cooling surface. However, this method may also result in nonuniform cooling (Fig. 8.53b).

The convective heat-transfer coefficient (α) for gas quenching can be represented as (Ref 61, 63):

$$\alpha = \mathrm{A}(V \cdot P)^{\mathrm{m}} \qquad \text{(Eq 8.7)}$$

where A and m are constants that depend on the furnace, component size, and load configuration (Ref 63), V is the flow velocity, and P is the gas pressure.

The rate of heat removal (Q) from the surface during quenching is proportional to the difference in the temperature (ΔT) of the workpiece and the gas (Ref 63):

$$Q = A(\alpha \cdot \Delta T) \qquad \text{(Eq 8.8)}$$

where A is the surface area being cooled. Taken together, Eq 8.7 and 8.8 show the relatively equal importance of both gas flow velocity and pressure (see Fig. 8.54 and 8.55) (Ref 64) and the heat-transfer coefficient of the gas to obtain optimal cooling. These numerical relationships will be analyzed further in a later section.

In addition to gas pressure and flow rate, it is important to consider the uniformity of gas flow around the cooling surface. Gradients of varying heat-transfer rates across the cooling surface will produce corresponding thermal gradients during the quench. These thermal gradients are one of the primary causes of cracking and distortion. To minimize these thermal gradients in gas quenching, a combination of horizontal and vertical flow is often used (Fig. 8.53c). In some cases, the optimal combination of flow patterns is custom designed for the specific part being quenched (Ref 60).

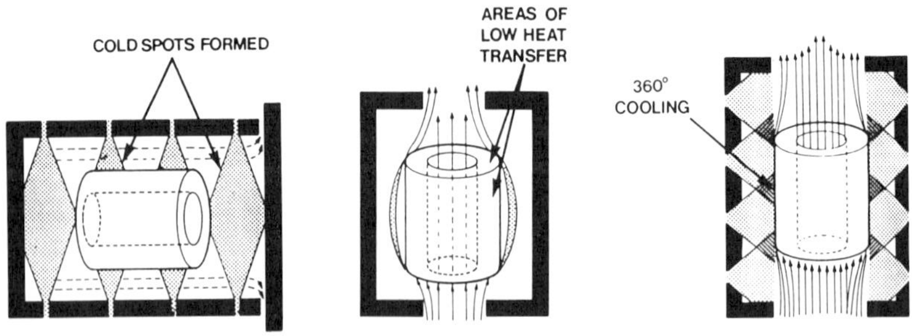

(a) (b) (c)

Fig. 8.53 Effect of gas flow pattern on quench uniformity. **(a)** Gas-injection nozzles, with flow perpendicular to the workpiece. **(b)** Gas-injection nozzles, with flow parallel to the workpiece. **(c)** Combination of horizontal and vertical flow

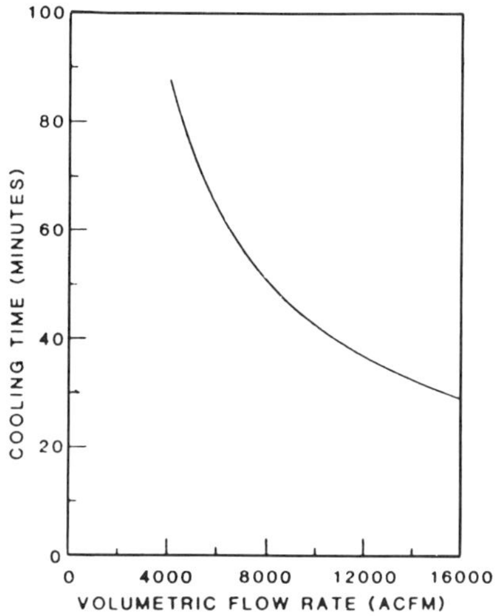

Fig. 8.54 Effect of volumetric flow rate at constant density on cooling time

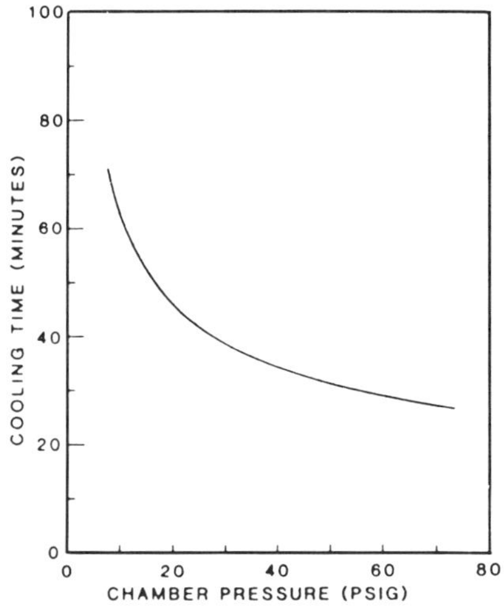

Fig. 8.55 Effect of chamber pressure at constant volumetric rate on cooling time

Gas Quenching Heat-Transfer Fundamentals. Heat transfer in gas quenching consists of two components:

- Convective heat transfer between the part and the gas
- Radiative heat transfer between two parts being quenched, between the part and the quench chamber, and between the part and the gas

Rigorous heat-transfer calculation of a gas quenching system requires analysis of gas flow and temperature, the effect of gas passage through a heat exchanger, the blower, and pressure drops in the recirculating system (Ref 58). This analysis varies for each system and is most effectively performed by the equipment manufacturer; therefore, it is more instructive here to examine the simplified problem of convective heat transfer between the part being quenched and the gas.

The heat-removal rate (Q) from the load to the gas can be represented as (Ref 58):

$$Q = A\alpha\,(T_m - T_g) \qquad \text{(Eq 8.9)}$$

where α is the convective heat-transfer coefficient of the metal part/gas interface, T_m is the temperature of the metal, and t_g is the temperature of the gas.

$$A = \frac{S}{V_o \rho_m C_{pm}} \quad \text{(Eq 8.10)}$$

where S is the surface area of the load, V_o is the volume of the load, ρ_m is the density of the metal, and C_{pm} is the heat capacity of the metal.

The convective heat-transfer coefficient is calculated from three dimensionless characteristics of flow and heat transfer (Ref 58): the Reynolds number (Re), the Nusselt number (Nu), and the Prandtl number (Pr). These dimensionless numbers are defined as:

$$\text{Re} = \frac{\rho_g V_g d}{\mu_g} \quad \text{(Eq 8.11)}$$

$$\text{Nu} = \frac{\alpha_g d}{\lambda_g} \quad \text{(Eq 8.12)}$$

$$\text{Pr} = \frac{\mu_g C_{pg}}{\lambda_g} \quad \text{(Eq 8.13)}$$

where α is the heat-transfer coefficient of the gas, d is the diameter of the part, λ_g is the thermal conductivity of the gas, μ_g is the viscosity of the gas, C_{pg} is the heat capacity of the gas, V_g is the linear velocity of the gas, and ρ_g is the density of the gas. In most cases, the equation for the surface heat transfer to a moving gas will be of the form:

$$\text{Nu} = \text{C}\,\text{Re}^{\alpha}\,\text{Pr}^{\beta} \quad \text{(Eq 8.14)}$$

where C, α, and β are constants that depend on the gas flow pattern and the geometry of the load (Ref 58). Two examples of the application of these equations to different flow patterns are shown in Fig. 8.56 and 8.57 (Ref 58).

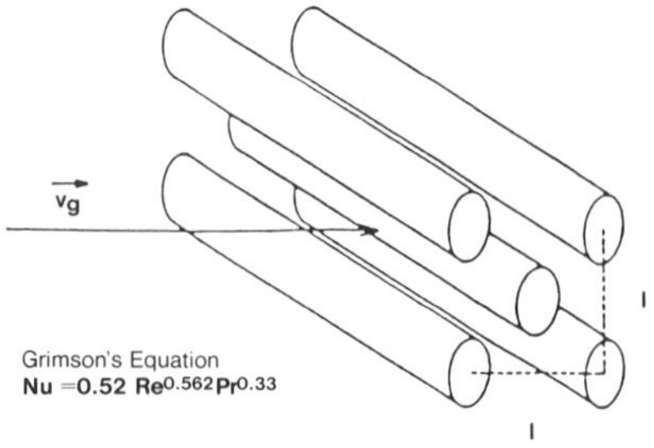

Fig. 8.56 Nusselt number for gas flow perpendicular to the load (Grimson's Eq.)

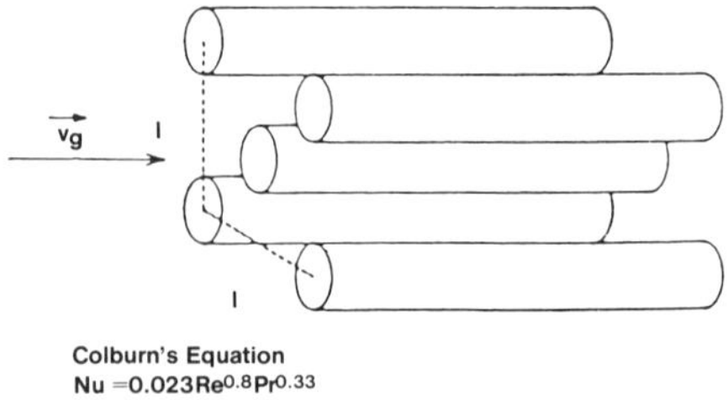

Fig. 8.57 Nusselt number for gas flow parallel to the load (Colburn's Eq.)

The Nusselt number for heat transfer of a recirculating gas (or gaseous mixture) is (Ref 62):

$$\mathrm{Nu} = C\,\mathrm{Re}^{0.8}\,\mathrm{Pr}^{0.43} \quad \text{(Eq 8.15)}$$

Solving Eq 8.15 for the relative convective heat-transfer coefficient (α) yields (Ref 62):

$$\alpha_g = \alpha_{N_2}\left[\frac{\lambda_g}{\lambda_{N_2}}\right]^{0.67}\left[\frac{\gamma_g}{\gamma_{N_2}}\right]^{0.70}\left[\frac{\eta_g}{\eta_{N_2}}\right]^{-0.37}\left[\frac{C_{pg}}{C_{pN_2}}\right]^{0.33} \quad \text{(Eq 8.16)}$$

Using Eq 8.16, it is possible to estimate the heat-transfer coefficient of a gas under various recirculating conditions using the data given in Table 8.14 and Fig. 8.52. For a more fundamental treatment of the heat-transfer properties of a gas, see Ref 59 and 65.

In addition to the fundamental heat-transfer properties of a gas, the effect of the size, shape, and material properties of the load must also be considered (Ref 63). The effects of cross-sectional size and steel alloy type are shown in Fig. 8.58 and 8.59 (Ref 62).

Oil Quenching From Vacuum Furnaces. With the exception of relatively high-pressure (>20 bar) gas quenching, the quench severities achievable with gas quenching are limited to those up to, but usually not including, conventional oil. Therefore, when lower-hardenability alloys are being heat treated in vacuum furnaces, greater quench severity must be provided. Quench oils may be selected for these processes (Ref 51, 56).

A two-chamber vacuum furnace generally is used for oil quenching applications. The two chambers are separated by a vacuumtight door. Before the quench, the quench chamber is activated for equalization with the vacuum in the furnace chamber. The door is then opened and the load transferred into the quench chamber. (The transfer time is usually 10 to 30 s.) The door is then closed and the load quenched with agitation. At the conclusion of the quench, the pressure is raised to atmospheric and the load removed (Ref 51).

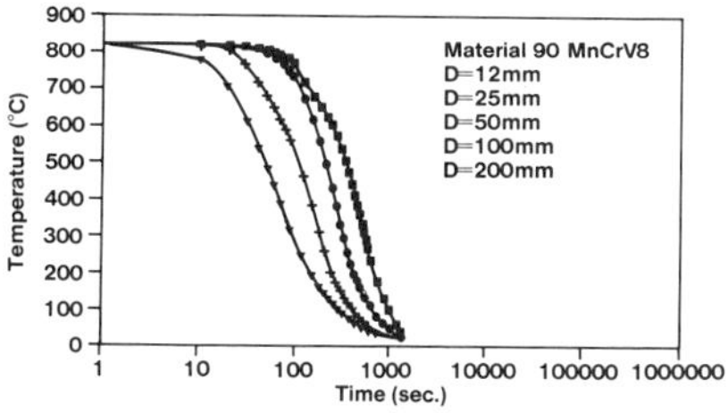

Fig. 8.58 High-pressure gas quenching of 90MnCrV8 alloy. Courtesy of Leybold Durferrit GmbH

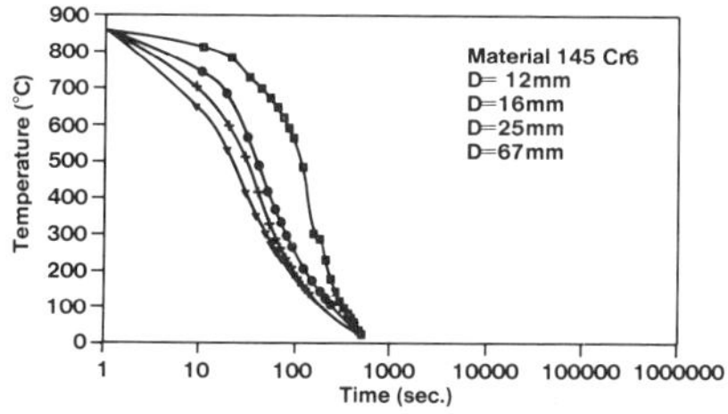

Fig. 8.59 High-pressure gas quenching of 145Cr6 alloy. Courtesy of Leybold Durferrit GmbH

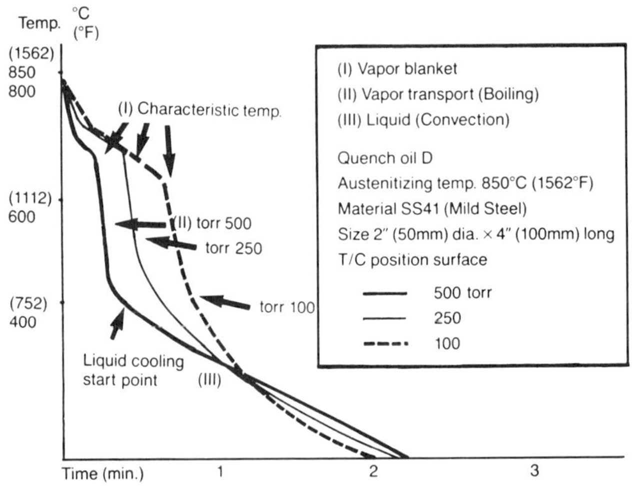

Fig. 8.60 Cooling curves for a 50 × 100 mm (2 × 4 in.) mild steel sample quenched in oil

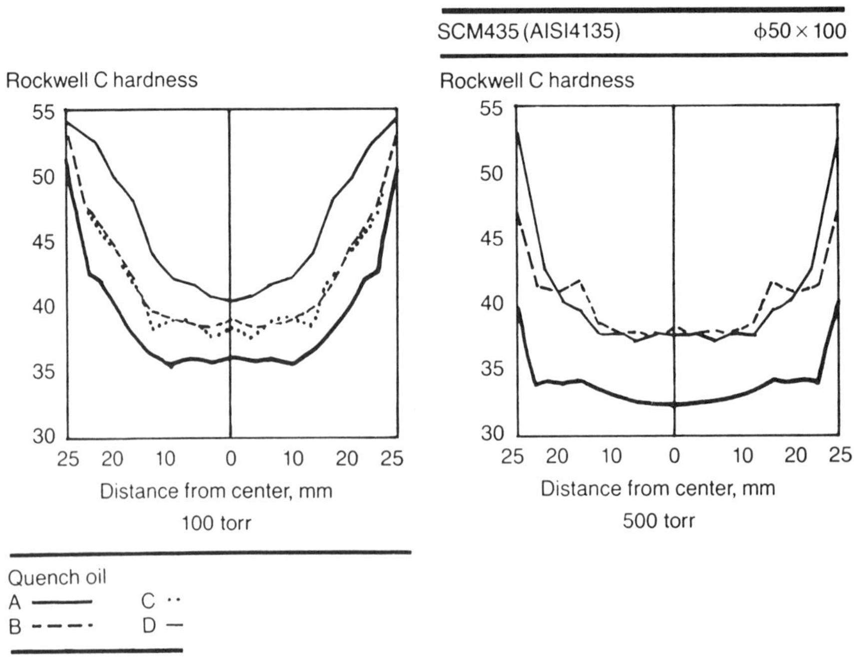

Fig. 8.61 Effect of oil surface pressure on the U-curve of SCM 435 (AISI 4135)

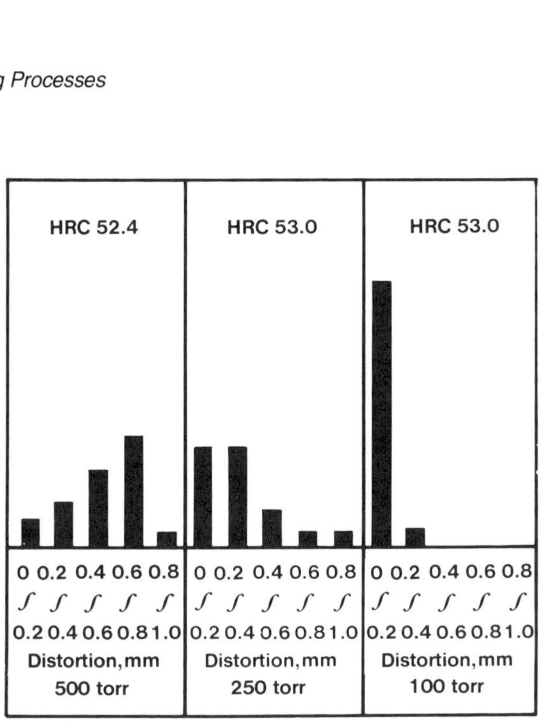

Fig. 8.62 Effect of oil surface pressure on the distortion of an 8 × 130 mm (0.3 × 5 in.) sample of SCM 435 (AISI 4135)

In vacuum quenching applications, the surface pressure above the oil before the quench must be controlled (Ref 51, 56). This is necessary because the surface pressure may exhibit dramatic effects on the A-stage (vapor blanket) cooling properties of the oil, as shown in Fig. 8.60 (Ref 56). In general, A-stage cooling rates *increase* and C-stage cooling rates *decrease* with decreasing surface pressure over the quench oil. The impact of oil surface pressure on the cross-sectional hardness of AISI 4135 is shown in Fig. 8.61.

The surface pressure over the quench oil may also affect distortion, as shown in Fig. 8.62. These data show that distortion decreases as the pressure is decreased, while the hardness level remains constant. These data are different from those obtained when quenching at atmospheric pressure, where hardness reductions often accompany distortion reduction. Thus, the vacuum furnace can be used as a distortion-control device (Ref 56).

Hot Isotactic Process (HIP)

Hot isotactic processing (HIPing) is a high-pressure process that is used to produce more dense parts. HIP quenching combines solution heat treating with extremely high-pressure gas quenching with argon gas at 1000 to 2000 bar and high temperature (typically 2000 °C, or 3630 °F). Under these conditions, cooling rates of approximately 500 °C/min (900 °F/min) are achieved. Advantages of HIP quenching include (Ref 57):

- Combining solution heat treatment with HIPing eliminates the need for port heat treatment.
- Distortion by nonuniform cooling is minimized.
- The use of high-purity gas minimizes surface contamination.

In HIP quenching, argon is used both for pressuring and as the quenchant. The physical properties of argon at extremely high pressures (1000 to 2000 bar) are shown in Table 8.15. The heat-transfer properties of argon vary as a function of pressure and agitation. Figure 8.63 shows that the temperature difference between the center and the surface of a part varies as a function of thickness with increasing argon pressure and flow. As the pressure increases, the temperature lag at the surface decreases. Also, the temperature difference is subsequently lower when a fan is used, as opposed to "still

Table 8.15 Physical properties of argon gas at 900 °C (1650 °F) as a function of pressure

Property	Pressure, bar 6	1000	2000
Density, kg/m^3	2.5	330	570
Specific heat, W/kg · °C	510	540	540
Kinematic viscosity, m^2/s	30×10^{-6}	0.25×10^{-6}	0.13×10^{-6}
Thermal conductivity, W/m · °C	0.05	0.06	0.08
Coefficient of volumetric expansion, °C	0.86×10^{-3}	0.68×10^{-3}	0.58×10^{-3}
Heat-transfer coefficient at free convection(a), W/m^2 · °C	8	180	280

(a) The heat-transfer coefficient is approximately five times higher with forced convection.

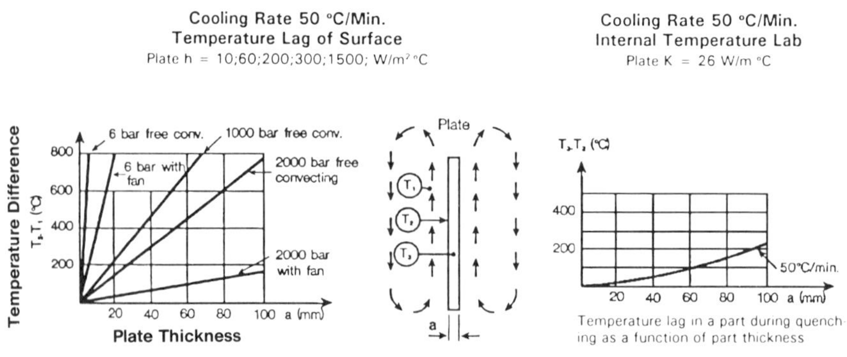

Fig. 8.63 Surface and internal cooling rates of gas quench plate. Courtesy of ABB Autoclave Systems, Inc.

convection" (Ref 57). Figure 8.64 compares the differences in cooling rates as a function of cross-sectional thickness between conventional gas quenching at 6 bar and HIP quenching at 2000 bar (Ref 57).

One of the advantages of HIP quenching is that argon gas is chemically inert; thus, surface reactions and environmental pollution are avoided. Quenchants such as oil, salt, and others produce faster cooling rates than HIP quenching (see Fig. 8.65 for a comparison to oil quenching).

Aus-Bay Quenching

Aus-bay quenching has been reported to offer advantages in distortion and residual stress reduction compared with conventional oil quenching for various steels such as D-6A and D-6AL alloys and has recently been evaluated for 300M alloys (Ref 66). Aus-bay quenching can be illustrated by using the TTT diagram for 300M alloy steel shown in Fig. 8.66. The steel is first quenched from 870 to 540 °C (1600 to 1000 °F) using a vacuum or gas quench. From the TTT diagram, it is evident that the formation of carbide from ferrite is relatively slow. The alloy is held at this temperature for a time sufficient to stress relieve the part and is then quenched in conventional oil.

Ultrasonic Quenching

It has been shown that one of the major causes of quench distortion is nonuniform heat removal from the surface of the part during quenching (Ref 67).

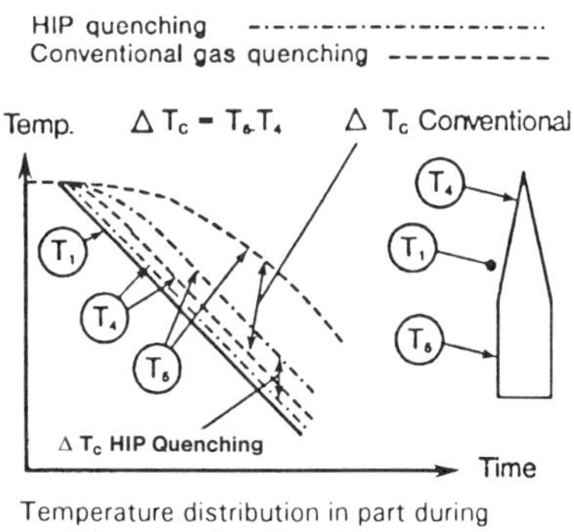

Fig. 8.64 Temperature difference in a workpiece with thin and thick parts. Courtesy of ABB Autoclave Systems, Inc.

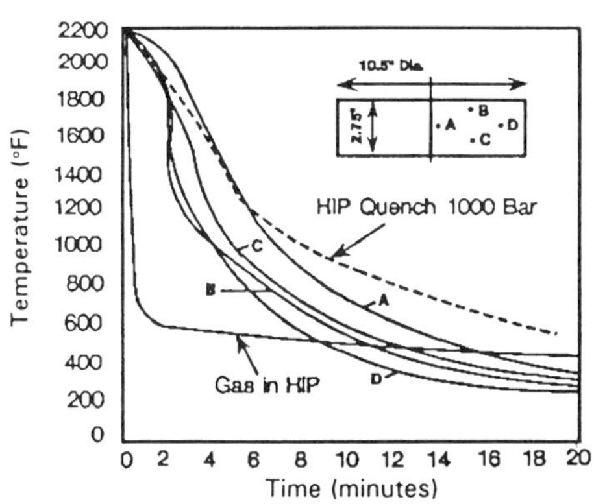

Fig. 8.65 Comparison of cooling curves produced by HIP quenching and by oil quenching. Courtesy of ABB Autoclave Systems, Inc.

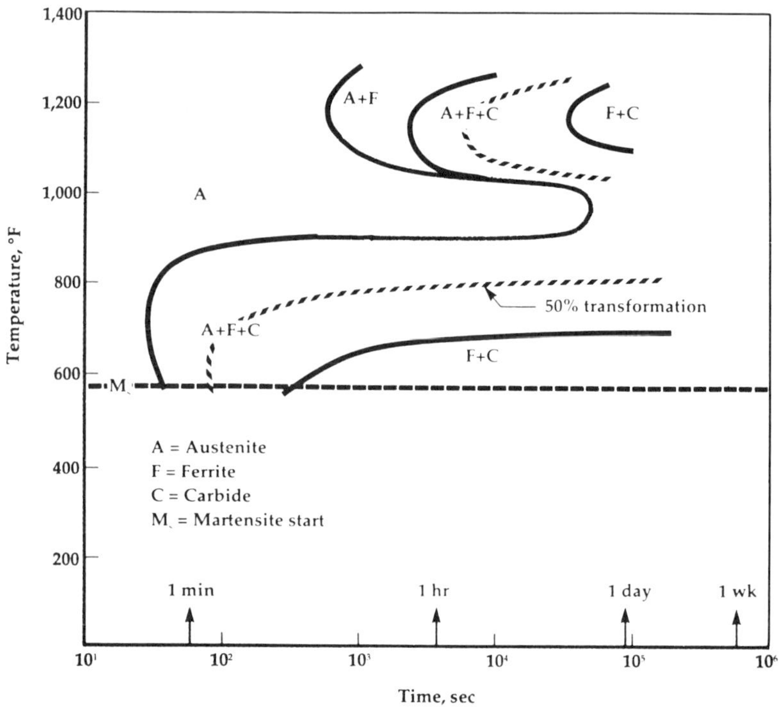

Fig. 8.66 TTT diagram for AISI 300M alloy steel

Table 8.16 Grossmann *H* values of various quenchants with and without ultrasonic energy

Quenchant	Grossmann *H* value
Oil	
Still quench	0.25/0.30
Violent agitation	0.80/1.10
Ultrasonic agitation	1.65
Brine	
Still quench	2.0
Violent agitation	5.0
Ultrasonic agitation	7.5
Hot salt at (400 °F)	
Still quench	0.30
Violent agitation	1.20
Ultrasonic agitation	1.80

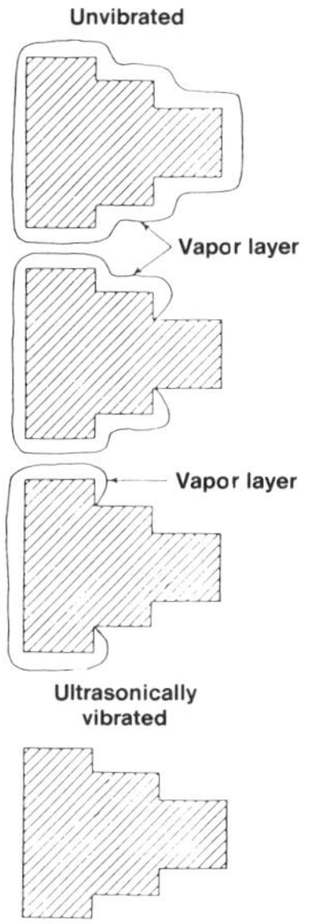

Fig. 8.67 Comparison of potential vapor blanket phases during oil quenching versus ultrasonic quenching

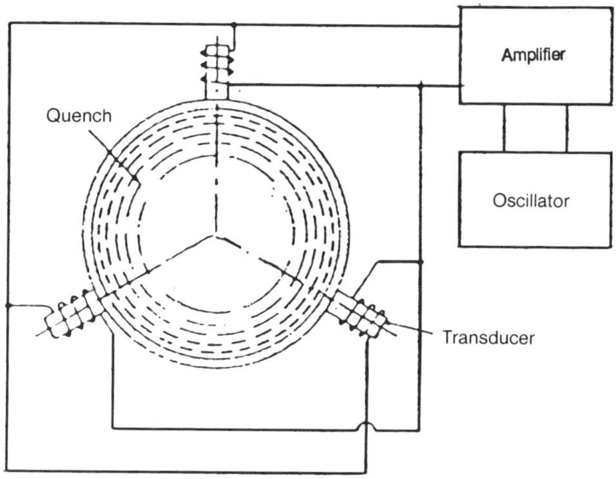

Fig. 8.68 Typical ultrasonic quenching systems, with 20 kc/s transducers in the configuration of a circular quench tank. The transducers are welded at a stress antinode.

Nonuniform heat removal facilitates the formation of thermal gradients, which then result in an increase in the undesirable formation of stresses.

Nonuniform heat removal is caused by uneven wetting of the metal surface, often accompanied by unstable vapor blanket (A-stage) cooling (Fig. 8.67) (Ref 68). The vapor blanket formation can be readily disrupted by the use of vibrational energy, such as ultrasonic energy. A typical ultrasonic system is illustrated in Fig. 8.68; it can be used in virtually any liquid quenchant medium (Ref 68).

Although ultrasonic agitation will produce substantial increases in quench severity (see Table 8.16), the cracking and distortion produced by oil, water, and brine quenchants are often eliminated. As expected from the relatively large H values shown in Table 8.16, the reduction in cracking and distortion is typically accompanied by an increase in hardness (Ref 68).

Quenching in an Electric or Magnetic Field

Another method of enhancing the uniformity of surface heat transfer during quenching is to destabilize vapor blanket cooling by passing an electrical current through the

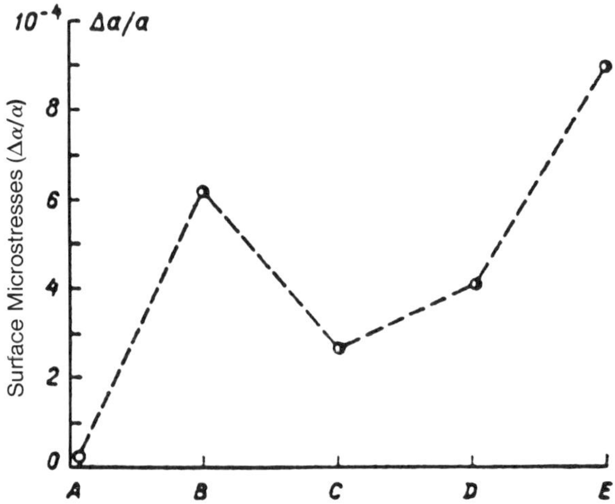

Fig. 8.69 Microstresses on the perimeter of 45C steel specimens quenched in different media

part during quenching (Ref 69-71). It has been shown that hardness can be increased by 10 to 18 HRC for a 0.45% C steel, while virtually eliminating quench-induced microstresses (Fig. 8.69) (Ref 69, 70).

A similar effect in stress reduction with a concurrent increase in as-quenched hardness has been demonstrated by quenching steel into an aqueous suspension of 10 nm magnetite particles (Ref 72). Cooling rates throughout the quench can be controlled by the concentration of the magnetite and the force of the magnetic field.

References

1. K. Speith and H. Lange, *Mitt. Kaiser Wilhelm Inst. Eisenforsch.*, Vol 17, 1935, p 175-184
2. M. Bamberger and B. Prinz, *Mater. Sci. Technol.*, Vol 2, 1986, p 410-415
3. A. Rose, *Arch. Eisenhüttenwes.*, Vol 13, 1940, p 345-354
4. M.O.A. Mokhtar and M.A.E. Radwan, *Heat and Mass Transfer in Metallurgical Systems*, D.B. Spalding and N.H. Afgar, Ed., Hemisphere Publishing, p 547
5. C.E. Bates, G.E. Totten, and R.J. Brennan, *ASM Handbook*, Vol 4, *Heat Treating*, ASM International, 1991, p 91
6. J.C. Chevrier and G. Beck, *Mem. Sci. Rev. Metall.*, Vol 69, 1972, p 623-632
7. S. Nukiyama, *J. Soc. Mech. Eng. Jpn.*, Vol 37, 1934, p 367-374

8. N.I. Kobasko, *Metalloved. Term. Obrab. Met.*, No. 3, 1968, p 2-6
9. L.V. Petrash, *Izv. V.U.Z. Chernaya Metall.*, No. 1, 1958, p 153-159
10. M. Tajima, T. Maki, and K. Katayama, *Nippon Kikai Gakkai Ronbunshu (B-hen)*, Vol 53, 1987, p 3383-3388
11. M. Tajima, T. Maki, and K. Katayama, *Nippon Kikai Gakkai Ronbunshu (B-hen)*, Vol 54, 1988, p 3491-3496
12. M. Tajima, T. Maki, and K. Katayama, *Nippon Kikai Gakkai Ronbunshu (B-hen)*, Vol 54, 1988, p 3497-3500
13. M. Tajima, T. Maki, and K. Katayama, *JSME Int.*, Vol 33, 1990, p 340-348
14. R. Maass and R. Jeschar, *Gas Warme Int.*, Vol 38, 1989, p 142-150
15. T. Yahua, *Wire Ind.*, Vol 57, 1990, p 44-48
16. P.A. Paulus, *Steel Technol. Int.*, 1988, p 310-314
17. G.E. Totten, unpublished results
18. M.P. Mukhina, N.I. Kobasko, and L.V. Gordeeva, *Metalloved. Term. Obrab. Met.*, No. 9, 1989, p 32-36
19. L.V. Petrash, *Metalloved. Term. Obrab. Met.*, No. 3, 1958, p 56-61
20. K. Boiko, *Heat Treat.*, Sept 1985, p 22-24
21. V.N. Oustovoit, N.Yu Churukin, and F.L. Ernandes, *Metalloved. Term. Obrab. Met.*, No. 10, 1986, p 31-35
22. S.A. Grishin and N.Yu Churukin, *Metalloved. Term. Obrab. Met.*, No. 10, 1986, p 36-37
23. A.V. Tolstousov and O.A. Bannykh, *Metalloved. Term. Obrab. Met.*, No. 91, 1989, p 37-39
24. V.A. Blinovskii and K.K. Shugai, *Metalloved. Term. Obrab. Met.*, No. 10, 1986, p 35-36
25. C. Skidmore, *Heat Treat. Met.*, Vol 2, 1986, p 34-38
26. G.W. Dexter, *Heat Treat.*, Aug 1989, p 21-23
27. A.T. Bykadorov, V.E. Vinogradova, T.V. Kamarova, and I.L. Kheifers, *Steel USSR*, Vol 19, 1989, p 134-136
28. R.W. Foreman, paper presented at ASM International National Heat Treating Conference, Chicago, Sept 1988
29. M.A.H. Howes, "The Cooling of Steel Shapes in Molten Salt and Hot Oil," Ph.D. thesis, London University, 1959
30. R.W. Foreman, *Heat Treat.*, Oct 1980, p 26-29
31. M.J. Sinnott and J.C. Shyne, *Trans. ASM*, Vol 44, 1952, p 758-774
32. L. Rosseau, *Metallurgia*, Vol 49, 1954, p 27-33
33. J.A. Schliessmann and R.R. Corle, *Met. Treat.*, Vol 23, Dec-Jan 1971-72, p 3-8
34. B. Liscic, *Quenching and Carburizing*, 3rd International Seminar, International Federation for Heat Treatment and Surface Engineering, Melbourne, 1991, p 1-27
35. J.A. Lincoln and A. Keogh, U.S. Patent 4,431,464, Feb 1984
36. E.N. Case and A.M. White, *Met. Prog.*, Oct 1953, p 122-124
37. G. Wahl, *Carburizing—Processing and Performance*, ASM International, 1989, p 41-56
38. B. Thomas, *Heat Treat. Forg.*, Dec 1931, p 1111-1112
39. H. Scott, *Trans. ASM*, Vol 22, 1934, p 68-96. Note: The term U (temperature ratio) is defined as: $U = (T - T_o)/(T_i - T_o)$, where T_i is the initial quenching temperature, T_o is the final bath temperature, and T is the temperature at a given position and time after the start of the quench.
40. R.W. Foreman, U.S. Patent 4,647,318, 1987

41. W. Peter, *Arch. Eisenhüttenwes.*, Vol 21, 1950, p 395-402
42. R. Branders, *Wire Ind.*, Feb 1990, p 89-91
43. W. Krebs, *Giesserei*, Vol 77, 1990, p 337-344
44. Product brochure, Fluidtherm Technology Ltd., Madras, India
45. M.A. Delano and J. Van den Sype, *Heat Treat.*, Dec 1988, p 34-37
46. R. Reynoldson, *Quenching and Carburizing*, 3rd International Seminar, International Federation for Heat Treatment and Surface Engineering, Melbourne, 1991, p 78-103
47. P. Sommer, *Heat Treat. Met.*, Vol 2, 1986, p 39-44
48. W. Luty, *J. Heat Treat.*, Vol 3, 1983, p 108-113
49. W. Luty, *Härt. Tech. Mitt.*, Vol 36, 1981, p 194-199
50. Z. Kulin and H. Genlian, Proc. 4th Int. Cong. Heat Treatment of Materials, Vol II, 3-7 June 1985, Berlin, p 1293-1320
51. T.W. Ruffle and E.R. Byrnes, *Heat Treat. Met.*, Vol 4, 1979, p 81-87
52. D. Devenny, *The Heat Treatment of Steel* (Proc. Conf.), Glasgow, Sept 1989, p 71-86
53. M. Price, Leybold Durferrit GmbH, personal communication, 1991
54. "Introduction to Vacuum Heat Treating," product brochure, Fiber Materials, Inc.
55. D.L. Cocks, Tool Materials for Molds and Dies: Applications and Performance, CSM Press, Colorado School of Mines, 1987, p 340-350
56. D.H. Herring, M. Sugiyama, and M. Uchigaito, *Heat Treat.*, July 1987, p 34-39
57. F.X. Zimmerman and C. Bergman, "Hip Quenching—A Heat Treating Method," ABB Autoclave Systems, Inc., 1992
58. B. Lhote and O. Delcourt, *Quenching and Carburizing*, 3rd International Seminar, International Federation for Heat Treatment and Surface Engineering, Melbourne, 1991, p 104-127
59. D.H. Herring, *Heat Treat.*, Sept 1985, p 36-37
60. W.R. Jones, *Heat Treat.*, Sept 1985, p 34-35
61. M.H. Jacobs, *Heat Treat.*, Sept 1985, p 30-33
62. P. Heilmann, "*Universal and Economical—The New Vacuum Furnace with Convective Heating and Gas High-Pressure Quenching*," Leybold Durferrit Technical Information Bull. No. D 184-1-3-390 Ki
63. E.J. Radcliff, *Ind. Heat.*, Nov 1987, p 34-39
64. S.R. Heubner, *Heat Treat.*, Sept 1985, p 28-29
65. K.W. Doak, *Heat Treat.*, Sept 1985, p 38-41
66. A.B. Smith, *Adv. Mater. Proc.*, Vol 140 (No. 2), 1992, p 51-52
67. R.K. Zhelokhovtseva, *Steel USSR*, Vol 15, 1985, p 238-239
68. R.F. Harvey, *Met. Treat.*, Aug-Sept 1966, p 3-5
69. A.A. Skimbov, I.A. Kozhukhar, and N.N. Morar, *Sov. Surf. Eng. Appl. Electrochem;*, Vol 2, 1989, p 136-138
70. A.A. Skimbov, I.A. Kozhukhar, and N.N. Morar, *Elektronnaya Obrab. Mater.*, Vol 2, 1989, p 87-88
71. L.I. Bershtein, A.S. Siper, and Yu.L. Bershtein, *Met. Sci. Heat Treat. Met.*, June 1990, p 414-416
72. S.N. Verkhovskii, L.I. Mirkin, and A.Ya. Simonovskii, *Fiz. Khim. Obrab. Mater.*, Vol 2, 1990, p 127-132

CHAPTER 9

Quench System Design

Optimal quench uniformity is essential if the potential for cracking, distortion, residual stress and spotty hardness is to be minimized. This means that heat transfer (i.e., film formation and breaking) during A- and B-stage cooling must be as uniform as possible throughout the quenching process.

One of the most important factors affecting quench uniformity is the design of the quench system. Deficiencies in system design have frequently been inadequately addressed by both heat treating engineers and equipment suppliers, often with disastrous results. With the exception of a few little-known company specifications, there are no industrywide guidelines for quench system design. Therefore, there is no extensive compilation of state-of-the-art design criteria to assist the engineer in the design and construction of a quenching system that will provide optimal heat transfer and quench uniformity.

Agitation is one of the most critical areas of system design. This chapter provides an overview of the impact of agitation on quench uniformity, followed by a general discussion of the selection and use of various types of agitators, including recirculation pumps, jet mixers, forced air (sparging), and impellers. (A detailed discussion of impeller agitation is presented in Chapter 10.) A brief overview of heat-exchanger types and their selection criteria is also provided, along with simplified calculations for approximating heat-exchange requirements. Finally, the use of membrane separators in various aqueous quench systems is discussed.

Effect of Agitation on Quenching Properties

The effect of agitation on the performance of various quench oils has been studied in considerable detail. Liscic (Ref 1) reported that the ability to through-harden AISI 4135 steel increased with increasing agitation (Fig. 9.1). Although decreasing oil temperature provided some improvement in through-hardening, it was considerably less effective than increasing agitation rate. Kweon *et al.* (Ref 2) similarly showed that

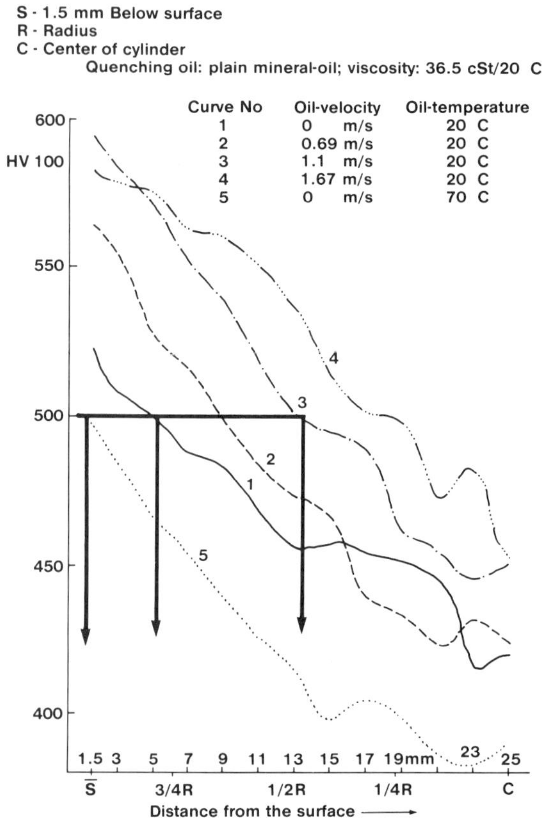

Fig. 9.1 Effect of agitation rate and oil temperature on the through-hardening of AISI 4135 steel

increasing agitation increased cooling rates and the through-hardening ability of both oil and water quenchants, as shown in Fig. 9.2 for SNCM 21 steel.

Vivas and Tardio (Ref 3) studied the influence of both agitation and temperature on the distortion of various carburized steels. They showed that agitation of a quench oil was necessary to destabilize A- and B-stage cooling if uniform heat transfer throughout the quench was to be achieved. Quenchant temperature was found to be an important covariable affecting distortion during the quench (see Fig. 9.3).

The relative efficacy of using immersion or spray quenching with an aqueous polymer and a mineral oil quenchant was studied by Moreaux *et al.* (Ref 4, 5). Their results showed that depth of hardness increased with increasing agitation (Fig. 9.4).

Moreaux *et al.* (Ref 4, 5) also showed that the as-quenched hardness for a complex part quenched in either an aqueous polymer or a mineral oil was highest for the portion of the part subjected to the greatest agitation. These data, presented in Fig. 9.5, clearly

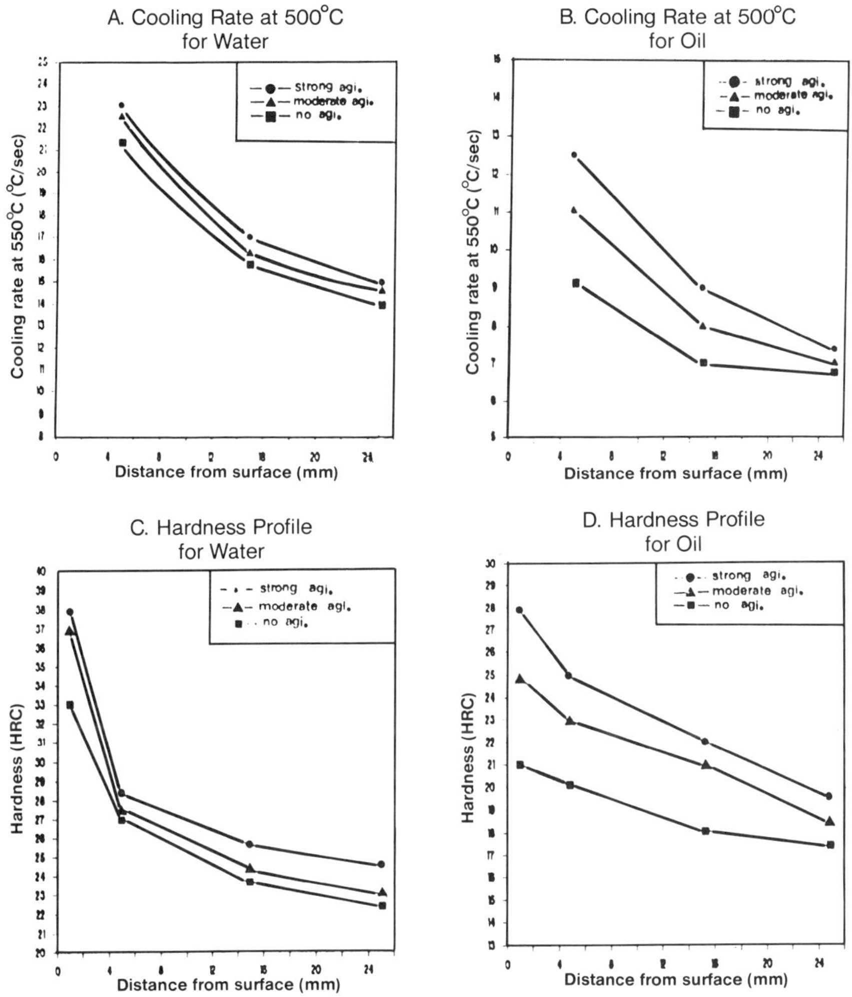

Fig. 9.2 Effect of agitation on the through-hardening of SNCM 21 steel using oil and water quenchants

show that proper racking to ensure optimal quenchant flow around all surfaces is necessary for quench uniformity. The effects of racking on quench uniformity will be discussed in more detail later in this chapter.

Propeller agitation of both oil and water under commercial quenching conditions has been reported to significantly improve quench uniformity, producing less cracking and distortion (Ref 6-8). It has already been shown that the through-hardening capability, or quench severity, of a quenchant is linked to agitation rate. (It will be shown subsequently

Nº	MAT. NUM.	C	Mn	Cr	Ni	Mo	Si	Al	Cu	MAT. SIM.
1	F - 1550	0,17	0.81	1,19	–	0,21	0,33	–	–	18CrMo4
2	F - 1522	0.18	0,75	0,60	0,52	0,16	0,24	0,027	0.02	20NiCrMo2
3	F - 1515	0,20	1.15	0,175	0.23	0.03	0.24	0.022	0.23	20Mn6
4	F - 1516	0.18	1,16	1.10	0.26	0,03	0,15	0,019	0.17	16MnCr5

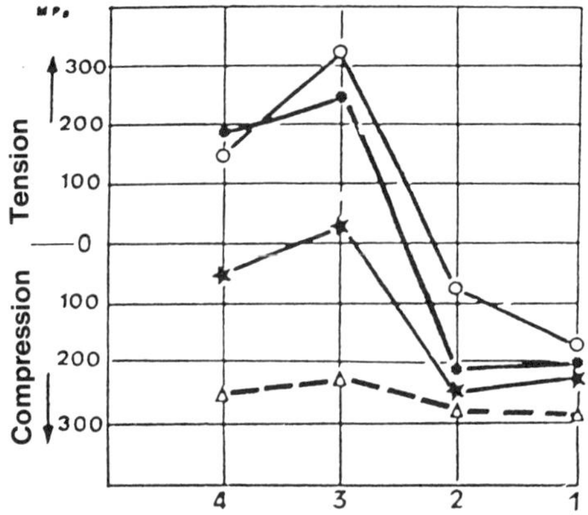

Oil Viscosity	Temp.°C	Symbol
Salt Bath	30	○
Salt Bath	50	●
Salt Bath	160	★
(NO_2 + NO_3)	250	Δ

Fig. 9.3 Comparison of surface stresses formed during quenching of various carburized steels in systems with agitation

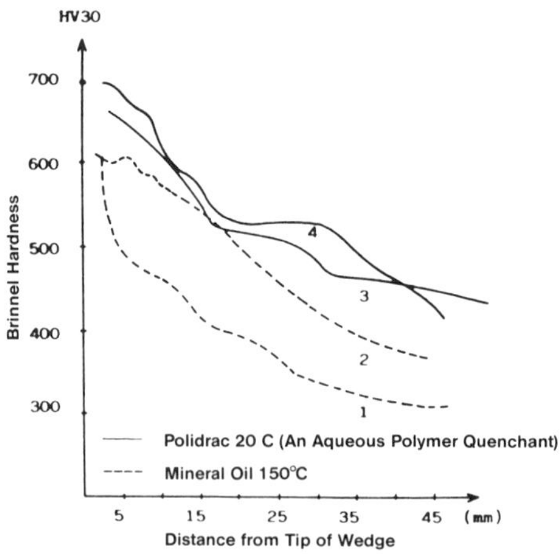

Fig. 9.4 Comparison of immersion and spray quenching with a mineral oil and an aqueous polymer Note: (1, 3) Immersion Quench = 0.25 m/s, (2, 4) Spray Velocity = 1.2 m/s.

that quench severity varies with the size of the tank, viscosity of the quenchant, type and placement of agitators, and so on.) Because the quench severity obtained with a given

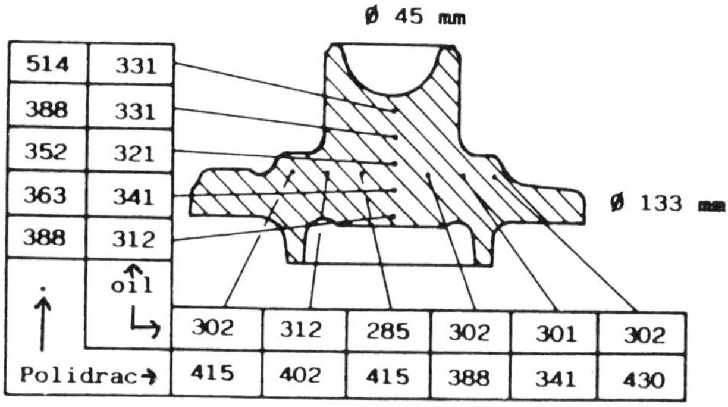

Fig. 9.5 Comparison of the through-hardening capabilities of a quench oil and an aqueous polymer (Polidrac) with agitation. Note: Hardness penetration (HB5/750) in 462C steel. The Polidrac solution was at 20 °C and the mineral oil at 70 °C.

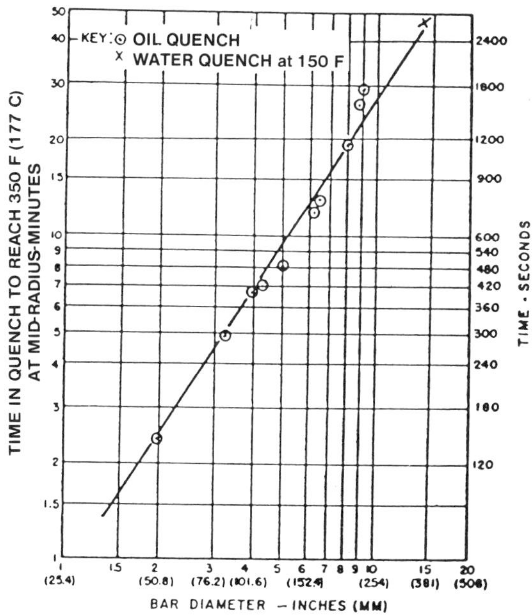

Fig. 9.6 Correlation of time to through-harden AISI 4340 steel as a function of bar diameter with agitation

quenchant is system dependent, it is recommended that a quenching time versus bar diameter correlation plot, such as that shown in Fig. 9.6, be used to guide the heat treater in the selection of the proper immersion time for a particular system.

Depth of hardening and quench uniformity for a quench oil are substantially greater in an impeller-agitated tank compared with either a still quench or the "mild" agitation achievable with a recirculation pump (Ref 9). The depth of hardening for the steel alloy shown in Fig. 9.7 increased for all cross-sectional sizes up to 38 mm ($1^1/_2$ in.) for (vigorous) propeller-agitated oil versus a still-quench oil (Ref 10). Agitation clearly affects the hardness and depth of hardening during the quench because of the mechanical rupture of the relatively unstable A- and B-phase cooling regimes that typically occurs in vaporizable quenchants such as oil, water, and aqueous polymers.

Khavskii and Zelokhovtseva (Ref 11) proposed that cracking and deformation are predominantly affected by the uniformity of the quench and not the quenchant itself. Thus, the design of the agitation system for a particular process is perhaps the largest single contributor to the uniformity of the quench process.

Small-Scale Laboratory Determination of Agitation Effects

This section will review selected methods of assessing the effects of agitation on quench severity. Several laboratory quenchant testing systems, the effects of directional flow by a heated surface, and the role of turbulence on quench severity will be discussed.

Traditionally, quench severity was evaluated by quenching a heated probe, such as those shown in Fig. 3.4 to 3.14, under still-quenching conditions. The most common method in current use (although not universally accepted) is to still-quench a 12.5 × 60

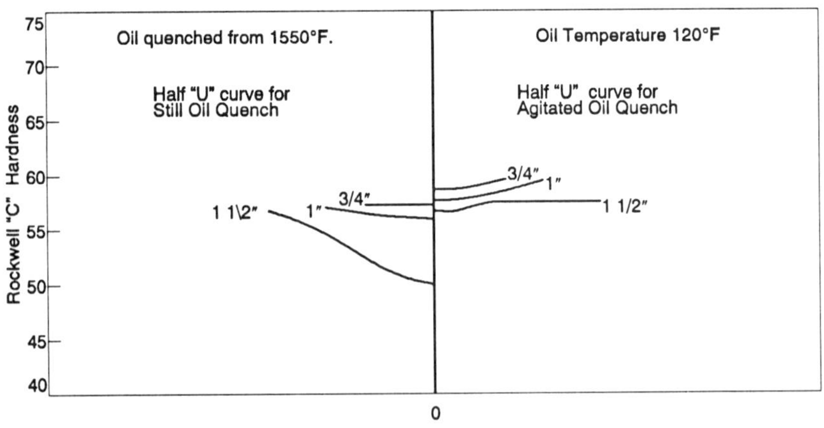

Fig. 9.7 Effect of agitation on the through-hardening capability of a conventional quench oil

mm (0.5 × 2.4 in.) Inconel 600 probe (see Fig. 3.12) into a 2 L (0.5 gal), tall-form, stainless steel beaker containing the quench oil (Ref 12, 13).

However, this system is inadequate for an understanding of agitation effects. Traditionally, numerous systems have been used to study agitation, such as a rectangular tank with a propeller stirrer in the corner or a 208 L (55 gal) drum equipped with either a recirculation pump or a propeller stirrer. In some cases, agitation has been applied either by moving the part in the tank manually (by "swirling" or "dunking") or by using a paddle after the part is immersed. However, such test methods clearly do not allow careful, reproducible study of agitation effects.

Determination of the flow and turbulence of a particular quenchant medium required to achieve the desired degree of through-hardening for a particular alloy may necessitate quenching of probes or parts with cross sections greater than 25 mm (1 in.). Such work requires the use of an "intermediate-scale" system that is larger than the systems used for routine quality control testing of the quenchant and smaller than the quench system itself. Examples of two intermediate-scale systems (40 to 500 L, or 10.5 to 130 gal) are shown in Fig. 3.64 (Ref 1) and Fig. 9.8 (Ref 14). However, these systems are rather large for routine evaluation of quenchants in the laboratory. Therefore, a number of smaller-scale systems have also been developed.

One small-scale system (Ref 12) that has been used to study agitation effects is based on a still-quench system modified by the insertion of an H-baffle (Fig. 9.9) (Ref 15). The agitation rate is quantified by the speed of the propeller (in revolutions per minute). However, as the data in Table 9.1 show, this experimental apparatus exhibits considerable data scatter.

1. Centrifugal circulating pump.
2. Heating coil for elevating quenchant temperature.
3. Flow control valve.
4. Orifice meter and manometer (calibrated) for flow measurement.
5. Induction coil for heating test specimen.
6. Induction power source.
7. Guide retaining fixture for heated test specimen.
8. Flow-through quench fixture where flow from pump is directed against specimen (quenchant flow subsequently overflows into quenchant reservoir).
9. Heat exchanger for control of quenchant temperature.
10. Thermometer for measuring quenchant temperature.

(Total Volume = 10 Gallons)

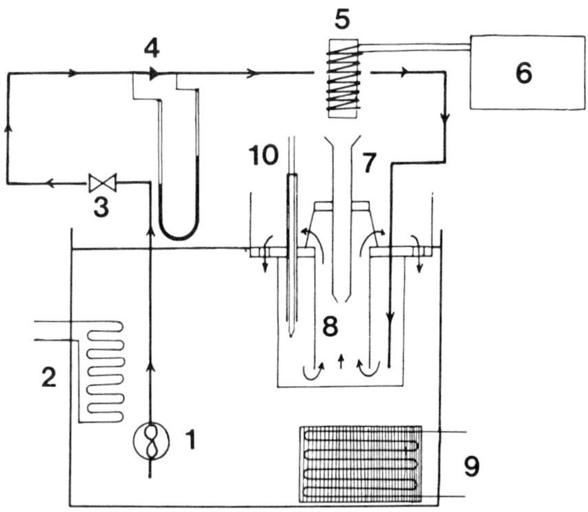

Fig. 9.8 Heins-Mueller quenchant testing apparatus

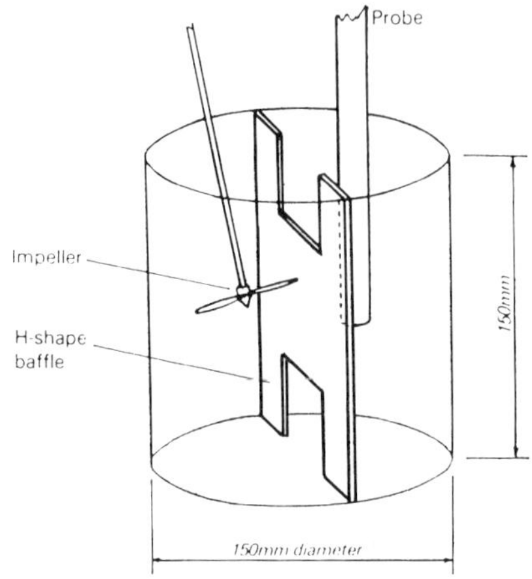

Fig. 9.9 H-baffle quenchant testing apparatus

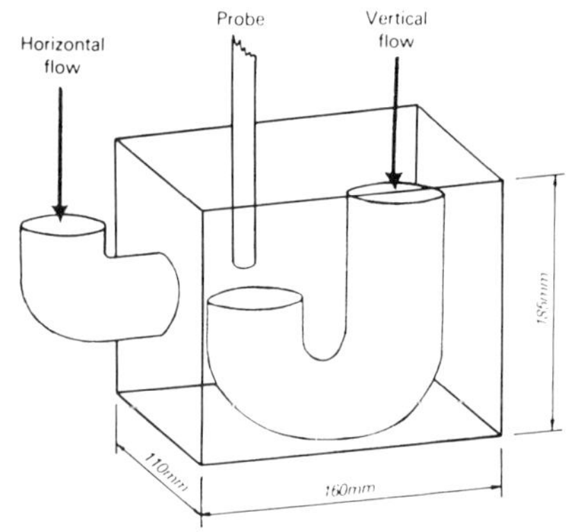

Fig. 9.10 J-tube pumping system for measurement of vertical and horizontal flow effects. Total volume, 2.5 L (0.66 gal)

Table 9.1 Effect of propeller speed on cooling curve performance using an H-baffle agitation system

	Propeller speed, rev/min			
Cooling curve parameter(a)	**0**	**500**	**1000**	**1300**
Time, s, to cool to:				
600 °C (1100 °F)	11 ± 2	9 ± 2	8 ± 2	8 ± 1
400 °C (750 °F)	12 ± 2	11 ± 2	10 ± 2	10
200 °C (390 °F)	18 ± 2	16 ± 2	15 ± 2	15 ± 1
Maximum cooling rate, °C/s	101 ± 7	104 ± 9	102 ± 6	94 ± 7
(°F/s)	(182 ± 13)	(187 ± 16)	(184 ± 11)	(169 ± 13)
Temperature at maximum rate, °C	556 ± 41	599 ± 25	577 ± 23	549 ± 43
(°F)	(1033 ± 74)	(1110 ± 45)	(1071 ± 41)	(1020 ± 77)
Rate at 300 °C (570 °F), °C/s	57 ± 2	56 ± 2	55 ± 2	56 ± 2
(°F/s)	(103 ± 4)	(101 ± 4)	(99 ± 4)	(101 ± 4)
Film boiling transition				
Temperature, °C (°F)	723 ± 3 (1333 ± 5)	746 (1375)	724 ± 14 (1335 ± 25)	702 ± 38 (1295 ± 68)
Rate, °C/s (°F/s)	17 ± 1 (31 ± 2)	21 ± 2 (38 ± 4)	28 ± 5 (50 ± 9)	31 ± 1 (56 ± 2)

(a) Determined using a 20% aqueous polymer quenchant solution

A "J-tube" pump agitation system has also been reported. The pumping action is provided either by a recirculation pump or by a propeller stirrer in a draft-tube arrangement (Fig. 9.10) (Ref 15). This apparatus has been used to evaluate the effect of directional flow by the cooling surface. Cooling rate data obtained using an aqueous polyalkylene glycol (PAG) polymer solution at various concentrations showed that horizontal (radial) flow had a greater impact on the maximum cooling rate of the part than vertical (axial) flow (Fig. 9.11) (Ref 15). These data suggest that more uniform quenching properties in production quenching systems should be obtained with axial flow through the load.

Another important difference between axial and horizontal flow by the cooling surface is the critical flow rate required for the A-stage (vapor blanket cooling) to B-stage (nucleate boiling) transition temperature. In general, lower fluid velocities are required to effect the transition for horizontal flow.

The results presented in Fig. 9.11 and 9.12 show that fluid velocity by the cooling surface is a more informative parameter than the speed of an impeller placed elsewhere in the system. The data in Tables 9.2 and 9.3 are significantly more reproducible than those reported for H-baffle systems (Table 9.1) (Ref 15). These data also show that the flow pattern is another critically important parameter if the impact of agitation on quenching performance is to be clearly defined.

A variant of the J-tube pumping system that is currently marketed commercially is the pump agitation system shown in Fig. 9.13. This system was used to quench various 12.5 mm (½ in.) diam 0.45% C (plain carbon) steel bars to calculate surface residual stress and hardness with maximum cooling rate and time to cool to 360 °C (680 °F). Correlation data obtained for various quenchants are shown in Fig. 9.14 (Ref 16). Decreasing the maximum cooling rate and increasing the time to cool reduce residual stress but also decrease hardness. Thus, the technical challenge of quenching is to identify the specific cooling properties of a particular quenchant for a given quenching system that will provide minimal stress and still produce the required hardness.

Mason and Capewell (Ref 17) used the agitation system shown in Fig. 9.15 to compare the relative impact of flow rate on the cooling rate of a conventional quench oil and a 20% PAG quenchant solution. They found that whereas both quenchants exhibit agitation rate performance, the aqueous polymer quenchant exhibited significantly greater agitation rate dependence. (It should be noted that different PAG polymers, as well as other polymers, would be expected to exhibit different degrees in performance, although all would be expected to exhibit greater agitation rate dependence relative to oil.) These data illustrate why many commercial oil quenching systems must be retrofitted with additional agitation equipment to provide directional flow to maximize quench uniformity.

A common error is to fail to account for flow variation throughout the agitated quench vessel. This includes both linear flow rates and the degree of flow turbulence. Tensi and Stitzelberger-Jakob (Ref 18) have reported the use of a closed-loop pump agitation system (Fig. 9.16) that exhibits very little, if any, flow variation within the quench

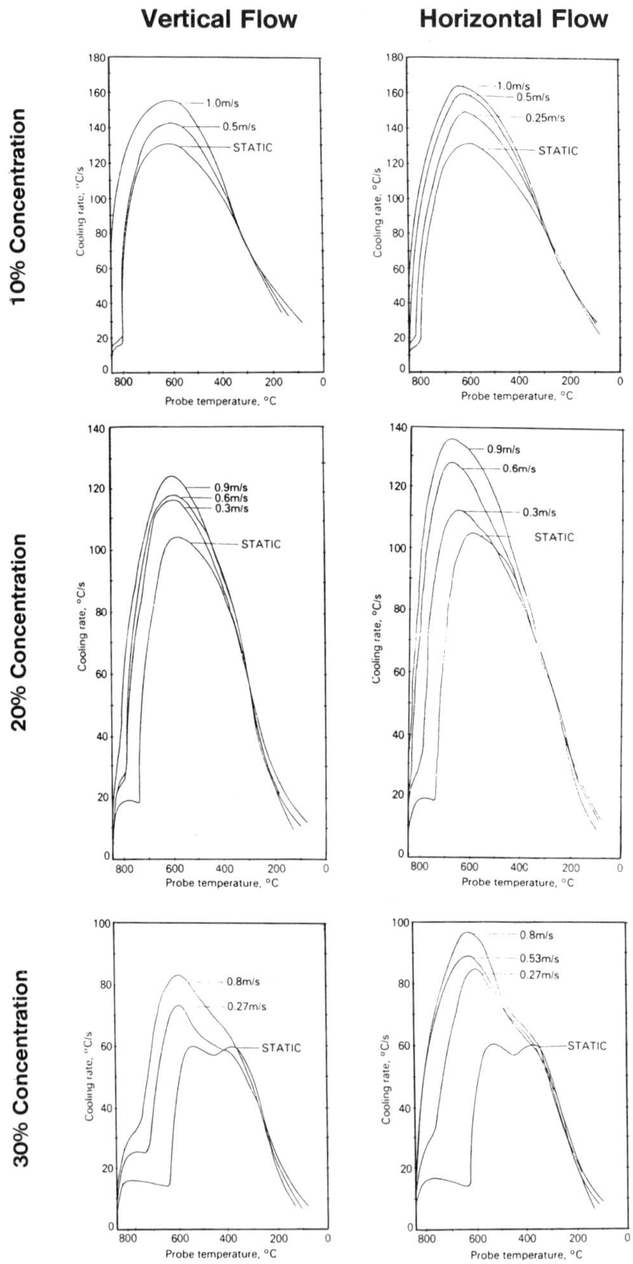

Fig. 9.11 Effect of vertical versus horizontal flow on the maximum cooling rate of an aqueous PAG solution. The agitation system is shown in Fig. 9.10.

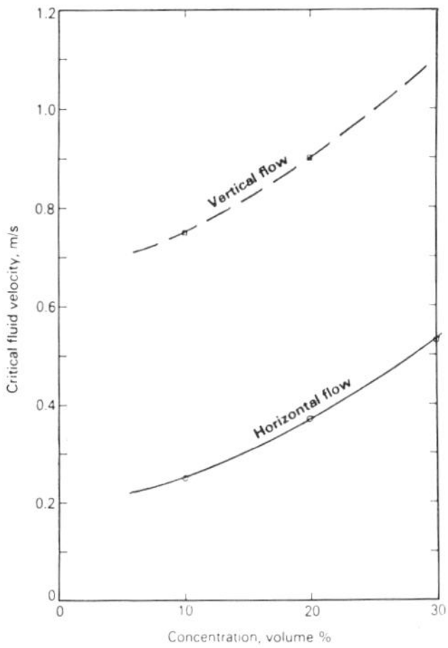

Fig. 9.12 Effect of horizontal versus vertical flow on the velocity required for A- to B-stage cooling transition temperature (PAG quenchant)

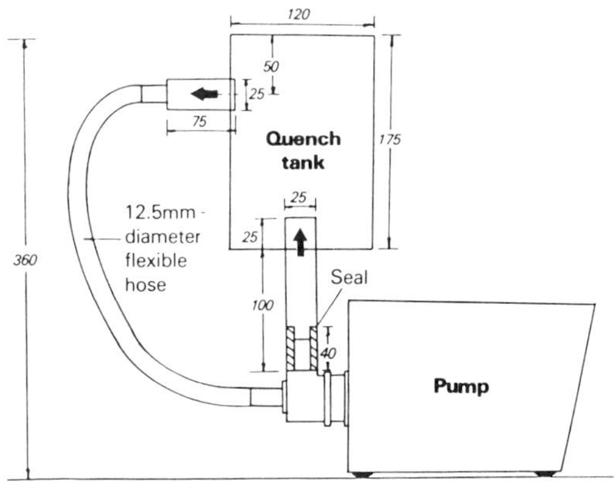

Fig. 9.13 Centrifugal pump apparatus for quenchant evaluation. Courtesy of Drayton Ltd.

Table 9.2 Effect of vertical flow rate on cooling curve performance using J-tube pump agitation

Cooling curve parameter	Fluid velocity, m/s (ft/s) 0 (0)	0.2 (0.7)	0.3 (1.0)	0.6 (2.0)	0.9 (3.0)
Time, s, to cool to:					
600 °C (1100 °F)	94 ± 0.2	6.6 ± 0.1	5.3 ± 0.2	5.1 ± 0.1	4.5
400 °C (750 °F)	11.4 ± 0.4	8.0 ± 0.2	7.1 ± 0.1	6.9 ± 0.3	6.3
200 °C (390 °F)	15.4 ± 0.4	12.1 ± 0.2	11.2 ± 0.2	11.0 ± 0.3	10.3
Maximum cooling rate, °C/s	101.7 ± 2.9	111	118	119.5 ± 2.1	126 ± 6
(°F/s)	(183.1 ± 5.2)	(200)	(212)	(215 ± 3.8)	(227 ± 11)
Temperature at maximum rate, °C	589 ± 1	644 ± 18	623	628 ± 1	610 ± 2
(°F)	(1092 ± 2)	(1191 ± 32)	(1153)	(1162 ± 2)	(1130 ± 4)
Rate at 300 °C (570 °F), °C/s	58.4 ± 0.6	56.8	58.8	56.6 ± 0.3	57.6 ± 1.1
(°F/s)	(105.1 ± 1.1)	(102.2)	(70.6)	(101.9 ± 0.5)	(103.7 ± 2)
Film boiling transition					
Temperature, °C (°F)	741 ± 7 (1336 ± 13)	799 ± 5 (1470 ± 9)	810 (1490)	815 (1500)	850 (1560)
Rate, °C/s (°F/s)	17.4 ± 0.3 (31.3 ± 0.5)	23 ± 2 (41.4 ± 4)	26 ± 1 (47 ± 2)	25 ± 2 (45 ± 4)	Nil

chamber (Fig. 9.17). This apparatus was used to show that optimal uniformity is achieved with turbulent flow.

Table 9.3 Effect of horizontal flow rate on cooling curve performance using J-tube pump agitation

Cooling curve parameter	Fluid velocity, m/s (ft/s)				
	0 (0)	**0.2 (0.7)**	**0.3 (1.0)**	**0.6 (2.0)**	**0.9 (3.0)**
Time, s, to cool to:					
600 °C (1100 °F)	9.4 ± 0.2	5.9 ± 0.1	5.6 ± 0.6	3.8 ± 0.2	3.3
400 °C (750 °F)	11.4 ± 0.4	7.9 ± 0.1	7.6 ± 0.6	5.8 ± 0.4	5
200 °C (390 °F)	15.4 ± 0.4	12.2 ± 0.2	11.8 ± 0.6	9.7 ± 0.3	8.8
Maximum cooling rate, °C/s	101.7 ± 2.9	113.8 ± 1.4	114.5 ± 3.5	125.4 ± 0.8	113.4 ± 2.0
(°F/s)	(183.1 ± 5.2)	(204.8 ± 2.5)	(206.1 ± 6.3)	(225.7 ± 1.4)	(240.1 ± 3.6)
Temperature at maximum rate, °C	589 ± 1	647 ± 9	649 ± 10	665 ± 5	675
(°F)	(1092 ± 2)	(1197 ± 16)	(1200 ± 18)	(1229 ± 9)	(1247)
Rate at 300 °C (570 °F), °C/s	58.4 ± 0.6	57.0 ± 0.3	55.6 ± 1.1	56.8	58.8
(°F/s)	(105.1 ± 1.1)	(102.6 ± 0.5)	(100.1 ± 2)	(102.2)	(105.8)
Film boiling transition					
Temperature, °C (°F)	741 ± 7 (1366 ± 13)	812 ± 12 (1494 ± 22)	838 ± 18 (1540 ± 32)	850 (1560)	850 (1560)
Rate, °C/s (°F/s)	17.4 ± 0.3 (31.3 ± 0.5)	23.4 ± 0.8 (42.1 ± 1.4)	26.8 (48.2)	Nil	Nil

A simpler propeller-driven system has also been reported by Tensi and Stitzelberger-Jakob (Ref 18) and is illustrated in Fig. 3.28. This system possesses a number of advantages over the H-baffle system shown in Fig. 9.9. Besides being inexpensive and easy to construct, there is relatively little variation in flow velocity across the quench chamber, as shown in Fig. 9.18 (Ref 20). Another advantage is that it is possible to construct a linear flow velocity versus rev/min calibration curve, such as that shown in Fig. 9.19 (Ref 18). Thus, it is possible to relate the nature of the flow surrounding the cooling metal that would be applicable in any size quench system with similar flow properties.

Heat transfer depends not only on the direction of the fluid flow relative to the cooling surface (e.g., axial versus radial) and on velocity gradients within the quench chamber, but also on the degree and variation of turbulence throughout the quench tank. Fluid turbulence is defined by the Reynolds number (Re). For fluid flow in a pipe, the Reynolds number is defined as (Ref 20):

$$\mathrm{Re} = \frac{D_p V \rho}{\mu} \qquad \text{(Eq 9.1)}$$

where V is the linear fluid velocity, D_p is the inside pipe diameter, ρ is the fluid density, and μ is the fluid viscosity. Reynolds numbers of less than 2000 represent laminar flow;

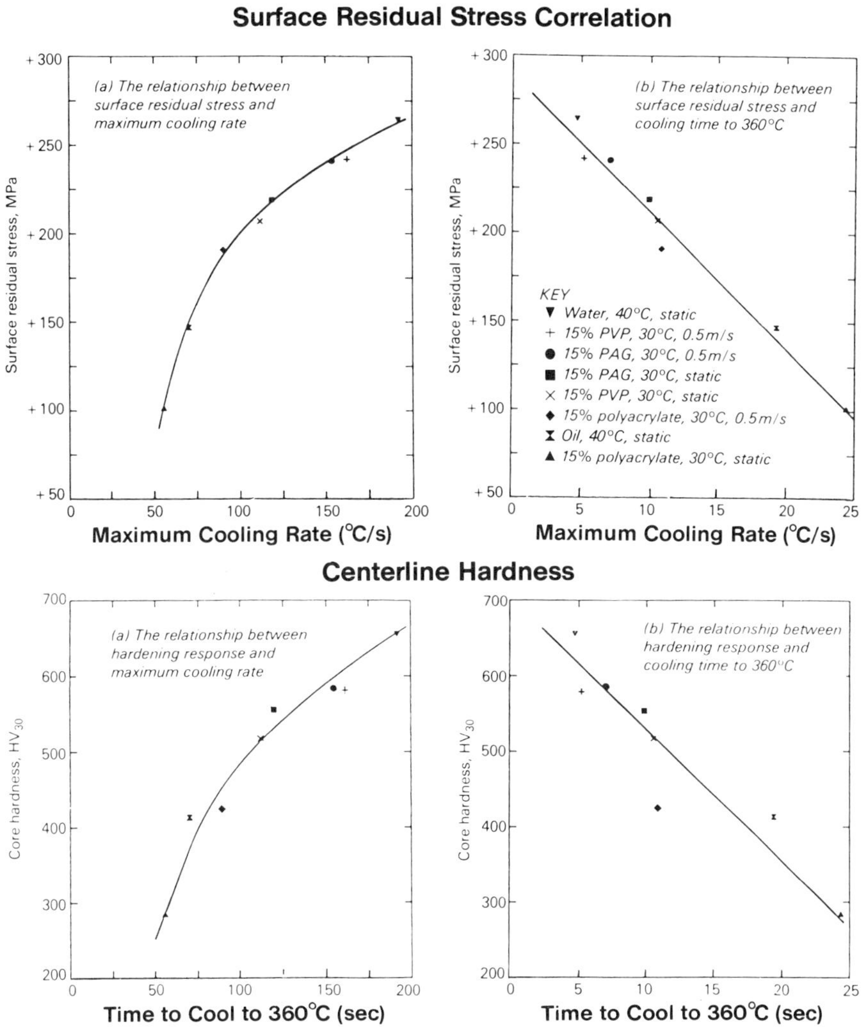

Fig. 9.14 Correlation of surface residual stress and as-quenched hardness. Data obtained with 12.5 mm (½ in.) diam 0.45% C steel bars quenched with various quenchants using the agitation system shown in Fig. 9.13

greater than 10,000, turbulent flow. Reynolds numbers between 2000 and 10,000 are considered to be intermediate-flow regimes.

The Reynolds numbers for a 12.5 mm (½ in.) diam cylindrical probe inserted into pipes of various diameters at two fluid viscosities and a constant volume rate of flow are summarized in Table 9.4. These data also show that the flow of the lower-viscosity fluid

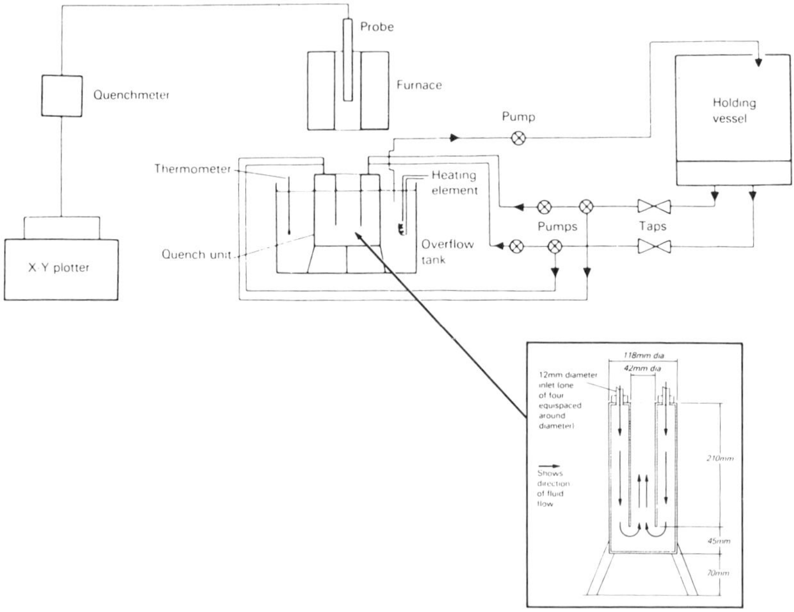

Fig. 9.15 Quenchant testing system used at West Bromwich College of Commerce and Technology

(0.667 cSt) is turbulent for most of the pipe diameters calculated. However, laminar flow would be obtained if the higher-viscosity fluid (6.67 cSt) were used. Thus, when considering turbulence, viscosity, linear flow rate, and fluid density must be taken into account.

The Reynolds number for fluid flow near an impeller mixer (N_{Re}) can be similarly calculated from (Ref 20):

$$N_{Re} = \frac{10.754 N D^2 \rho}{\mu} \quad \text{(Eq 9.2)}$$

where N is the impeller speed (rev/min), D is the impeller diameter (in.), ρ is the specific gravity of the fluid, and μ is the fluid viscosity (cP). In SI units, N_{Re} is defined by:

$$N_{Re} = \frac{1.667 \times 10^{-5} N d^2 \rho}{\mu_o} \quad \text{(Eq 9.3)}$$

where d is the impeller diameter (mm) and μ_o is the dynamic viscosity (Pa · s).

Hilder (Ref 21) studied the effect of fluid turbulence on the maximum cooling rate for a 12.5 × 60 mm (0.5 × 2.4 in.) cylindrical probe (see Fig. 3.12) quenched into a 20% aqueous PAG solution using the system illustrated in Fig. 9.15 and found that the maximum cooling rate increased with increasing turbulence (increasing Reynolds number), as shown in Fig. 9.20. This is explained by considering the stability of the

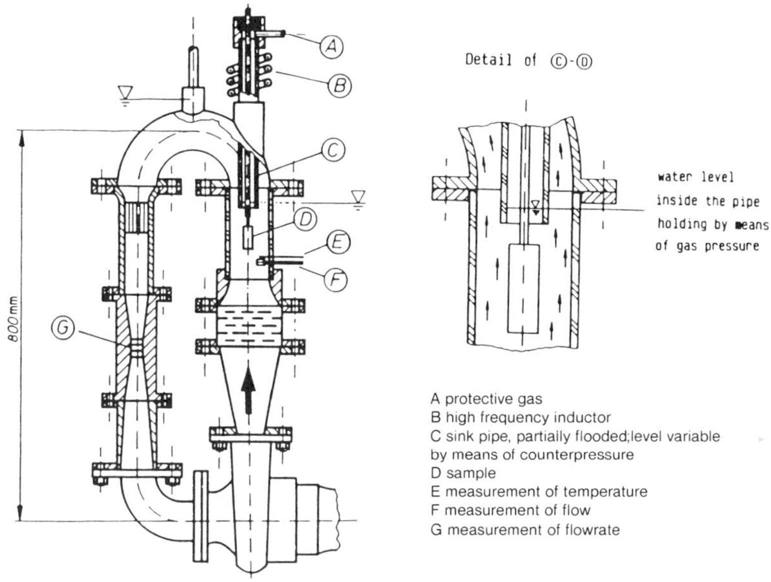

Fig. 9.16 Convection system for measuring quench severity under conditions of uniform flow

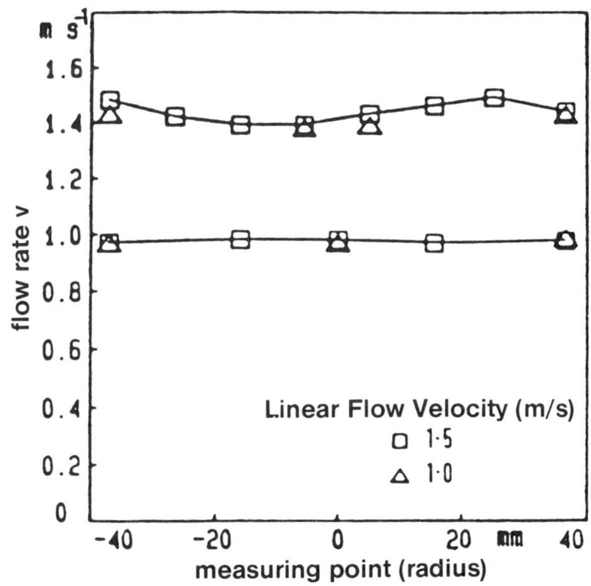

Fig. 9.17 Distribution of flow velocity across the quench zone in the apparatus shown in Fig. 9.16

polymer film that is formed around the metal during quenching (see Fig. 5.12) with respect to the increasing strength of the eddy currents as the flow regime progresses from

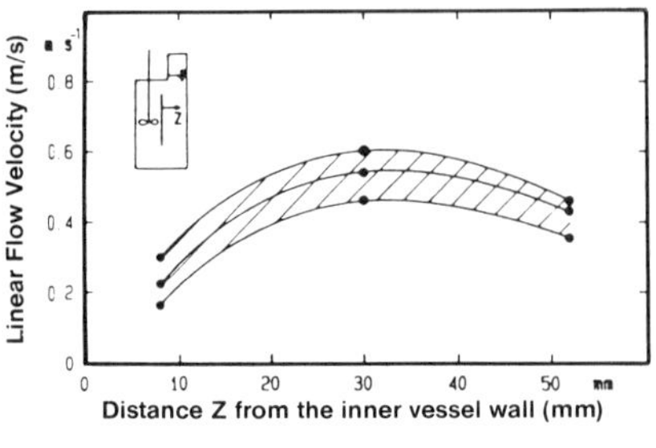

Fig. 9.18 Flow velocity variation across the quench chamber in the apparatus shown in Fig. 3.28

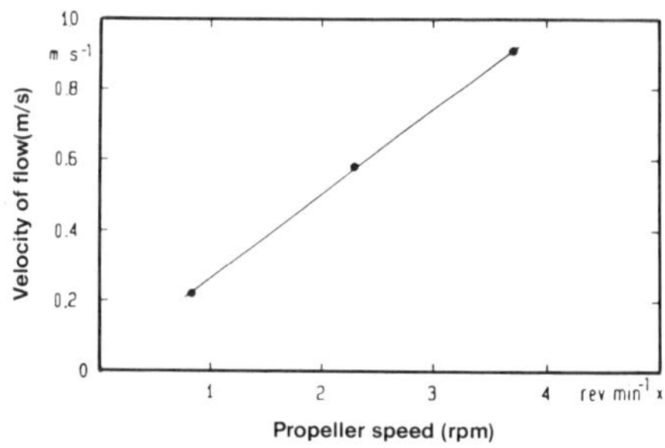

Fig. 9.19 Linear flow velocity calibration for the quench system shown in Fig. 3.28

Table 9.4 Effect of fluid viscosity, velocity, and pipe diameter on Reynolds number

Pipe diameter		Linear velocity by a 12.5 mm (½ in.) probe		Reynolds No. at viscosity of:	
mm	**in.**	**m/s**	**ft/s**	**0.667 cSt**	**6.67 cSt**
79.2	3.12	0.08	0.258	7801	786
50.8	2.0	0.2	0.654	11383	1138
25.4	1.0	1.0	3.27	18971	1897
19.0	0.75	2.4	7.84	22765	2276
15.9	0.625	5.3	17.43	25296	2530

Note: Fluid flow = 23 L/min (6 gal/min) in all cases.

laminar to turbulent (Fig. 9.21). From this representation, it would be expected that the stability of the film formed to turbulent flow would be proportional to the film strength of the hydrated polymer. A similar analogy can be drawn for quenching in oil or water.

Although turbulence effects during quenching are important, increasing turbulence does not necessarily produce a corresponding improvement in the uniformity of heat transfer across the cooling surface and thus reduce the propensity for distortion (Ref 22). It is important to examine the flow regime surrounding the entire hot surface of the part with respect to linear fluid flow rate and turbulence, because the heat must also be carried efficiently away from the hot surface into the quenching fluid during the cooling cycle.

Fluid Flow Modeling

Some of the earliest published work on quenchant flow modeling was conducted by Rose (Ref 23) and by Speith and Lange (Ref 24). Fluid turbulence at the hot metal

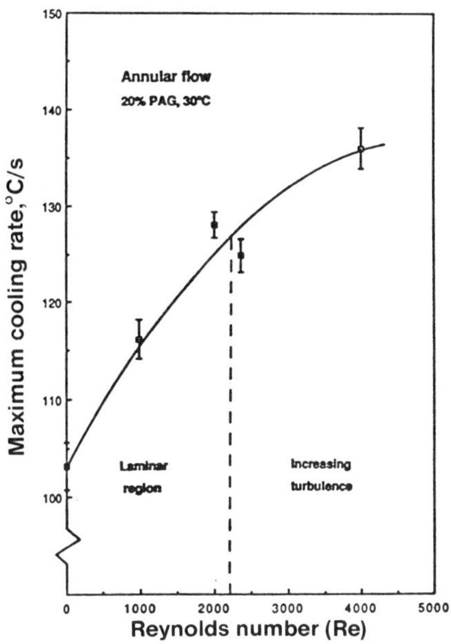

Fig. 9.20 Correlation of maximum cooling rate with Reynolds number. Data obtained with the apparatus shown in Fig. 9.15.

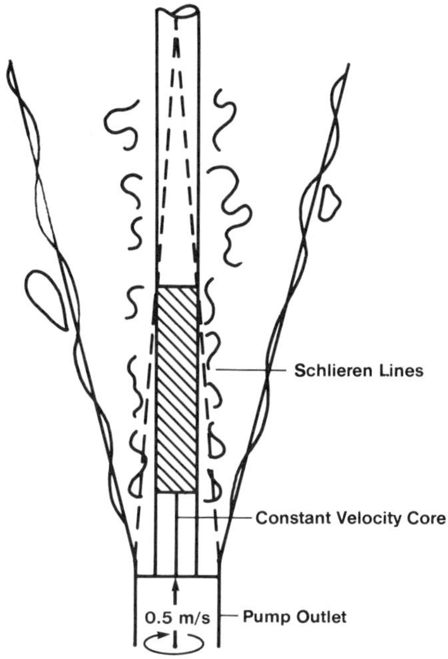

Fig. 9.21 Schematic representation of turbulent flow around a hot metal part during quenching

surface was determined from the schlieren lines of various quenching media, including water, brine, and oil. These authors, in their now-classic work, showed that agitation facilitates heat transfer during quenching by removing the hot quenchant at the heat-transfer interface for replacement by cooler fluid.

Subsequent photographic work performed at U.S. Steel to study schlieren currents during quenching showed that when hot metal is quenched in a fluid with unidirectional flow field, the degree of agitation varies with the position of the metal piece with respect to quenchant flow direction (see Fig. 9.22) (Ref 9, 10, 25, 26). This explains why poor quench uniformity is often achieved in systems where the sole source of fluid agitation is a recirculation pump. It was recommended that either submerged jet mixers or propeller agitation be used to enhance uniform surface wetting and fluid flow.

Fluid flow in a quench tank can also be modeled by the construction of small-scale plexiglass tanks and the use of styrofoam or glass beads or by dye injection. Segerberg (Ref 27) found that the flow pattern in a plexiglass model of an actual quench tank is not uniform throughout, even without a quench load immersed (Fig. 9.23).

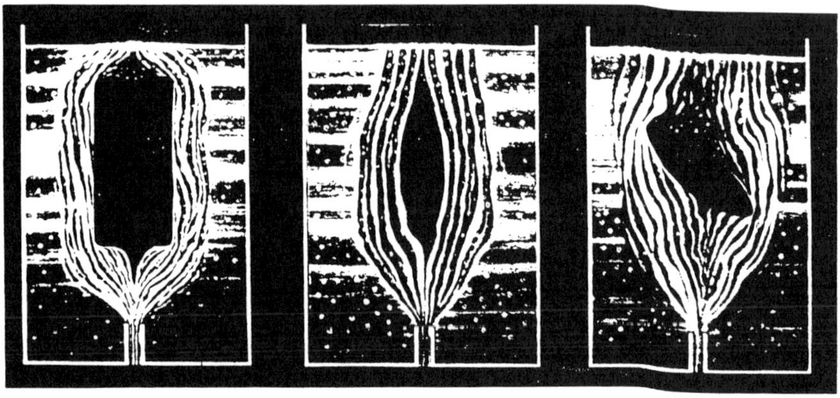

Fig. 9.22 Schlieren cooling lines around a hot metal part during convective cooling in a directional flow field

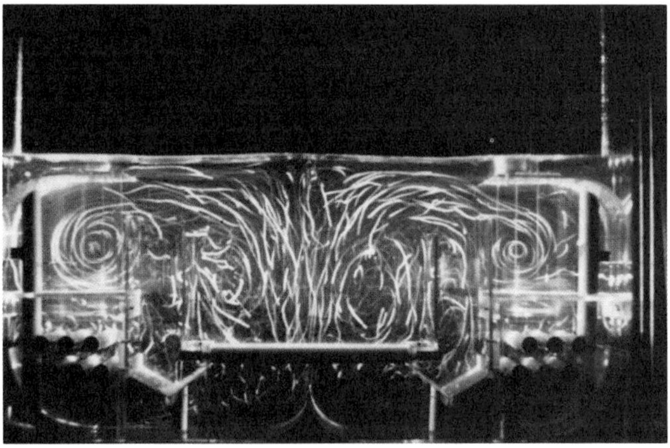

Fig. 9.23 Flow model of an unloaded actual quench tank. Courtesy of the Swedish Institute for Production Research (IVF)

Oldshue (Ref 28) cautions that particular care must be taken in translating flow models obtained with small prototype tanks to larger commercial systems, because the fluid turbulence flow regimes are often considerably different. One precaution to avoid such problems is to account for the Reynolds number increase that often occurs on scaleup.

Computational fluid dynamic (CFD) models such as those shown in Chapter 10 are being used increasingly to determine turbulence, velocity, and viscosity flow patterns in mixing tanks used for heat-transfer operations (Ref 29). However, there is considerable imprecision in these calculations at the present time due to the assumptions that must be made, as much of the required input data is not available.

Tensi and Stitzelberger-Jakob (Ref 18) used "strings" (see Fig. 9.24) to demonstrate the "twistless" flow regime in the quenchant agitation apparatus shown in Fig. 3.28. This flow pattern also is reported to exist in pipes (Ref 18).

Methods of Flow Measurement

Turbine Velocimeter. One of the simplest and most straightforward instruments used to measure flow is the Mead turbine velocimeter (Fig. 9.25). The turbine blade measurement head is attached to a handle and simply immersed at the point of interest. If the fluid viscosity and density are known, the Reynolds number can be easily calculated because the cross-sectional area of the measurement head is known and the linear flow is determined from the experiment.

Streak Photography. If small "particles" of approximately the same density as the fluid are dispersed in a tank and photographed by shining a narrow slit of light onto them,

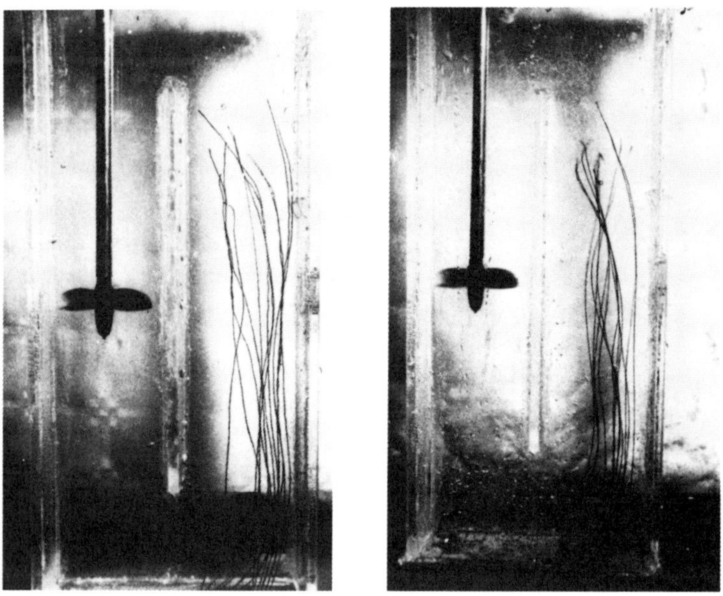

v = 0.25 m/s v = 0.6 m/s

Fig. 9.24 Illustration of "twistless" flow. Courtesy of H.M. Tensi, Technical University of Munich a "streak photograph" is obtained (see Fig. 9.23) (Ref 30). Such photographs show the average motion of the larger eddy currents in the tank.

Pitot-Static Tube. The Pitot tube is useful for measuring unidirectional velocity, such as lamellar or jet stream (Ref 31). It is not suitable for measuring turbulent, multidirectional flow. A Pitot tube can be constructed by drawing a 6.4 mm (¼ in.) length of glass tubing to an inside diameter of approximately 0.4 mm ($^1/_{64}$ in.) and grinding the drawn end flat and square with the tube axis. It can also be made of metal, provided that the tube opening is knife sharp and square with the axis of the tube. A suitable Bourdon gage or manometer is connected to the tube with rubber or transparent plastic tubing. If precise measurements are required, a correction must be made for the quenchant column height in excess of the level of the upper opening of the Pitot tube.

The axis of the Pitot tube must be accurately parallel to, and in the center of, the stream being measured. Alignment is obtained by exploring the stream with the tube and searching for the highest pressure reading. The velocity of the fluid as it strikes the tube is converted to a pressure head that is measured by the pressure gage. The pressure head, h, in feet of water, is converted to velocity, v, in feet per second, by:

$$v = K(2gh)^{1/2} \quad \text{(Eq 9.4)}$$

where K is the Pitot tube constant (most often 1.0, or approximately unity) and g is the acceleration of gravity (980 m/s^2, or 32.2 ft/s^2). The relationships are illustrated in the Pitot tube calibration charts shown in Fig. 9.26.

The electromagnetic current meter (Ref 32) is used to model flow in large systems. It operates by measuring the conductivity of the fluid and the distortion of electromagnetic fields.

The hot-film anemometer detects fluid velocity by measuring the cooling effects of a hot film in the region of interest (Ref 30). These devices are particularly effective for highly turbulent flow measurements.

Laser Doppler Velocimeter (LDV). One of the best instruments for fluid flow calibrations is the laser doppler velocimeter (Ref 30). An overview and example of the use of this instrument are provided in Chapter 10.

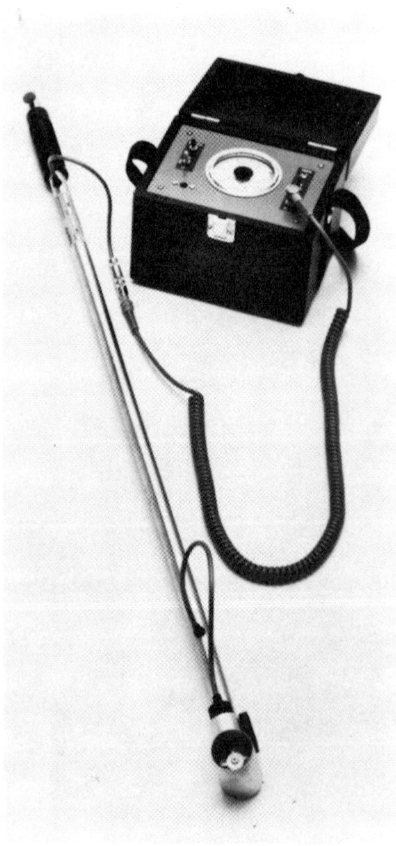

Fig. 9.25 Mead turbine velocimeter. Courtesy of Mead Instruments Corporation

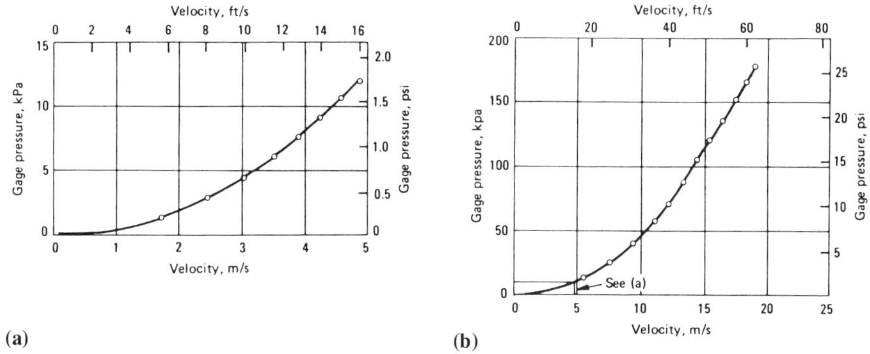

Fig. 9.26 Typical Pitot tube calibration chart. (**a**) Low velocity. (**b**) Full range

Cooling Curve Analysis. Many examples have been shown thus far where cooling curve analysis has been used to measure the effects of agitation on cooling rates in various laboratory tests. Because cooling curve analysis is sensitive to agitation, it is possible to survey a quench tank by measuring cooling rate variation as a function of position of the instrumented part or probe in the tank.

Racking Procedures

Quench severity, as measured by cooling curve analysis, is dependent on linear flow rate, turbulence, quenchant temperature, both interfacial and bulk solution viscosity, uniform surface wetting, and direction of fluid flow impinging on the hot metal surface. Nonuniformity of any of these variables throughout the quench zone and across the boiling surface will result in nonuniform heat removal from the surface of the part during the quench, creating excessive thermal gradients that may cause distortion and nonuniform hardness. Optimization of fluid flow around the part during quenching is necessary for uniform hardness and minimal quench distortion, cracking, and stresses. Therefore, proper part racking for quenching is especially important.

Segerberg (Ref 27) has shown that heat transfer during the quench is a function of the packing density of the parts. Figure 9.27 shows that the degree of A-stage cooling decreases with increasing packing density. Two points must be made regarding this work. First, all of the loads were racked to enhance fluid flow around all of the parts during the quench. Second, the cooling curves were obtained with a 12.5 × 60 mm (0.5 × 1.4 in.) cylindrical Inconel 600 probe that was placed at the center of the load. If the parts themselves were instrumented at various positions across their surfaces and data

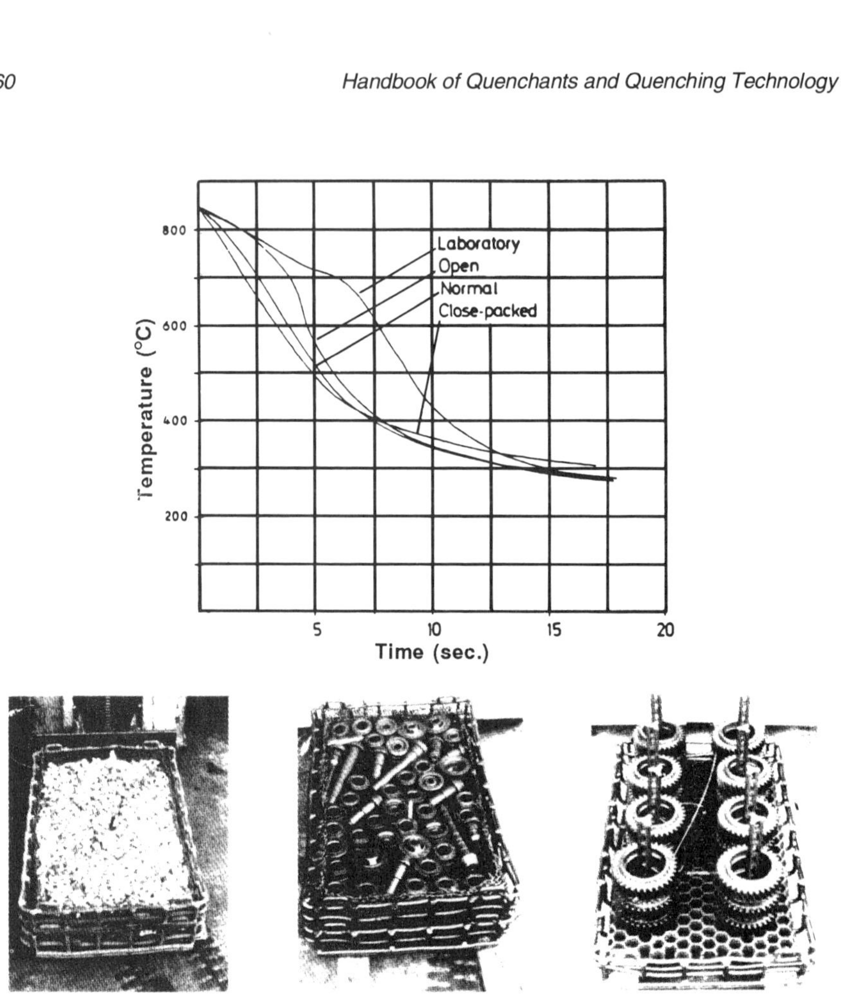

a. "close-packed" b. "normal" c. "open"

Fig. 9.27 Variation of cooling curve performance with packing density. The 12.5 × 60 mm (0.5 × 1.4 in.) Inconel 600 probe was placed at the center of the load.

obtained at different placements within the load, considerable variation in the cooling curve results would be expected.

A number of general recommendations have been made for proper part racking (Ref 33):

- Parts should be racked (placed) to facilitate maximum quenchant flow and contact during the quench.
- Long, slender parts should be suspended.
- Symmetrical parts, such as bearing races and cylinders, can be stacked and supported on a rack or grid.
- Flat parts, such as saw blades, clutch plates, and so on, are best supported on horizontal slotted rods that provide the necessary separation for fluid contact.

- Coils of wire should be supported either vertically on a spider-type grid or horizontally on support rods.
- Small parts can be loaded into a perforated ladle or basket to facilitate quenchant contact upon immersion.
- Fixture design should be simple, free of welds (if possible), and easy to maintain.
- The combined weight of parts and fixtures must be limited to allow for sufficient heat transfer during the quench to minimize temperature rise.

Quench System Design Considerations

To ensure proper quenchant performance, it is essential that the quench system be properly designed (Ref 34). Some of the most important process variables to be considered are:

- Tank sizing with respect to heat evolution during the quench
- Agitation (method and selection)
- Chute design for continuous systems
- Heat exchange for quenchant temperature control
- Filtration

This section and several sections that follow will provide an overview of the importance of each of these variables with respect to proper quench system design.

Tank Sizing and Heat Evolution

The primary function of the quenchant is to mediate heat-transfer rates from the hot metal to the cooler quenchant during the quenching process. Factors to consider here include:

- Nature and thermal properties of the metal
- Bulk properties of the quenching medium, such as heat capacity, density, viscosity, and so on
- Film-forming characteristics of the quenchant at the hot metal interface

The heat-transfer properties of the hot metal/quenchant interface can be described using Grossmann's relationship for quench severity, H:

$$H = \frac{h}{2k} \quad \text{(Eq 9.5)}$$

where h is the film coefficient of the quenchant and k is the thermal conductivity. Thermal conductivity varies with the metal and is a temperature-dependent function.

Figure 9.28 shows the temperature variation of both steel and aluminum as a function of temperature (Ref 34).

The amount of heat (q) transferred from the surface of a hot metal into the liquid quenchant depends on the film (heat-transfer) coefficient (h), surface area (A), and the difference between the initial temperature of the metal before quenching (T_1) and the bath temperature (T_2) (Ref 35):

$$q = hA(T_1 - T_2) \qquad \text{(Eq 9.6)}$$

From Eq 9.6, it is evident that heat flow (q) increases as the ($T_1 - T_2$) value increases. Therefore, faster heat-transfer rates are expected as the temperature of the metal increases or as the quenchant bath temperature decreases. Heat transfer also increases with increasing surface area and as the film coefficient increases.

The effect of varying surface area is illustrated in Fig. 9.29, which shows the temperature of the quench bath for the same size loads of large and small parts throughout the quench (Ref 34). The heat was released more quickly for the load of small parts (higher surface area), as evidenced by the higher temperature of curve F. (The temperature did not rise to the

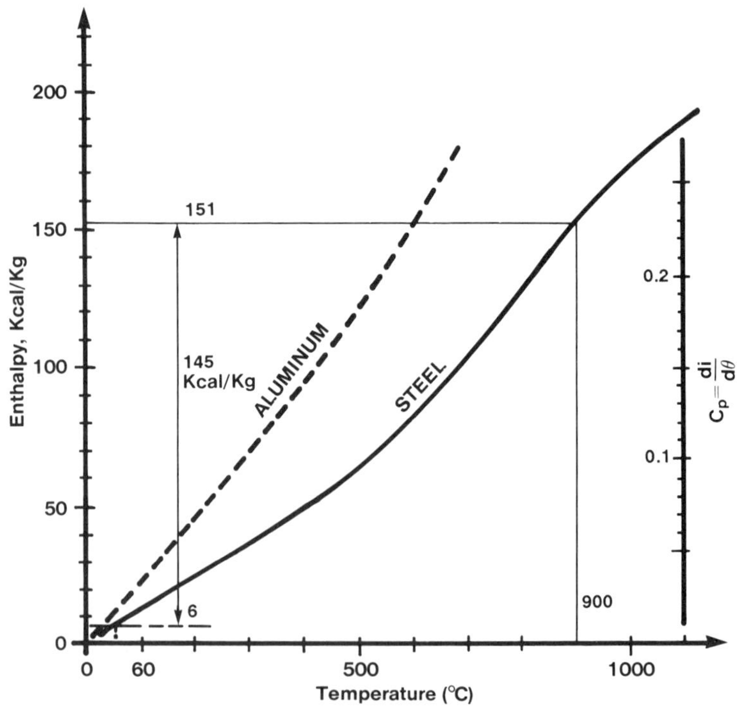

Fig. 9.28 Relative heat exchange during quenching of steel and aluminum

theoretically calculated value of T'' because of environmental heat losses.) The larger parts (lower surface area) caused the bath temperature to rise more slowly over a longer time. Curve B1 shows the importance of adding a heat exchanger to the quenchant circulation loop for greater temperature control.

The amount of heat extracted (given up by the metal to the quenchant), Q, from a part during quenching can be estimated by (Ref 31):

$$Q = C_{pm} W_m (T_1 - T_2) \quad \text{(Eq 9.7)}$$

where C_{pm} is the specific heat of the metal; W_m is the weight (lb) of the metal, including fixtures such as trays; and T_1 and T_2 are the initial and final temperatures of the metal, respectively.

For continuous heat processes, the heat given up by each part being quenched must be removed from the quenchant before the next part is quenched. This is done by the use of an adequately sized heat exchanger. Heat exchangers will be discussed later in this chapter.

Although specific heat is a temperature-dependent function, the simplified relationship shown by Eq 9.7 permits a reasonable estimate of the amount of heat evolved from the metal being quenched. In making this calculation, it is assumed that no heat is lost to the environment during the quenching process.

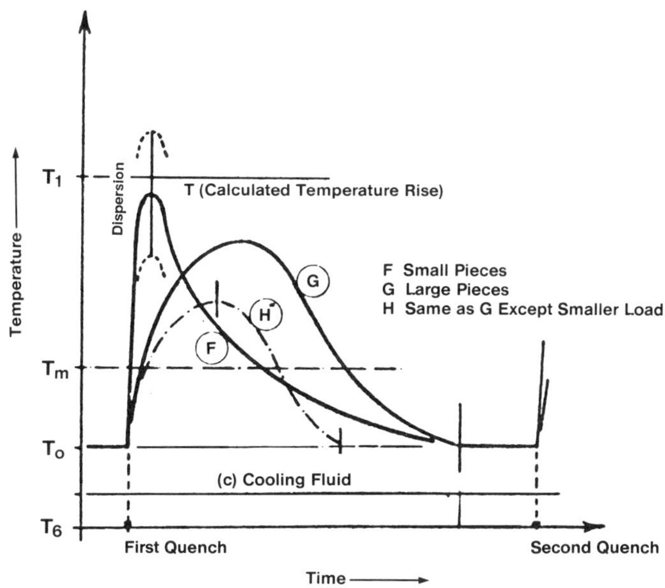

Fig. 9.29 Effect of part size on temperature rise in the bath during quenching

A similar equation permits the estimation of the temperature rise (T_r) of a quenchant bath during the quench cycle (Ref 31):

$$T_r = (T_1 - T_2) = QW_qC_{pq} \qquad \text{(Eq 9.8)}$$

where W_q is the weight of the quenchant, C_{pq} is the heat capacity of the quenchant, and T_1 and T_2 are the initial and final temperatures of the bath, respectively.

From Eq 9.7 and 9.8, it is clear that the temperature rise (T_r) of the bath can be affected by the initial temperature of the metal, the total mass of the load, the volume and temperature of the quenchant in the tank, and the heat capacity of the metal.

Two general rules have been traditionally used for determination of quench tank sizing (Ref 36, 37). The first, the quench load (including fixture weight) should not exceed 0.11 kg/L (1 lb/gal) of quenchant. Second, the temperature rise of the quenchant should not exceed 5 °C (10 °F).

Quench Tank Agitation

As discussed previously, quenchant uniformity is enhanced by the use of agitation. Numerous methods of agitation can be used; however, those most commonly encountered include: impeller, pump, spray (jet), and air sparging.

This section will first discuss the effect of quench tank agitation on heat transfer. A general overview of the various agitation methods will then follow.

Effect of Quench Tank Agitation on Heat Transfer. The previous section showed that heat transfer during quenching is dependent on the surface area of the parts. This is also illustrated in Fig. 9.30, which shows the relative cooling curves obtained with

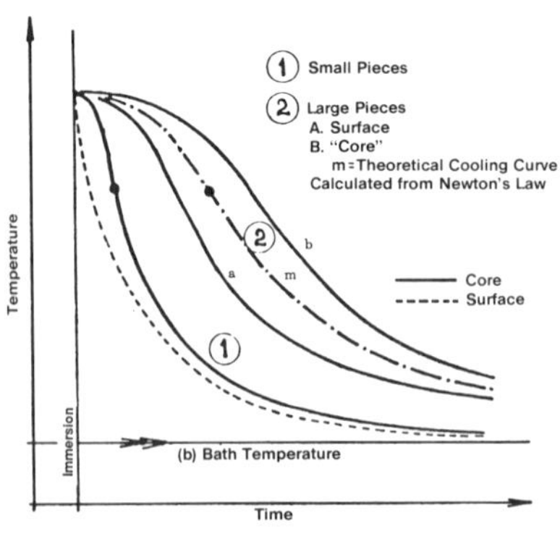

Fig. 9.30 Cooling curve behavior of parts as a function of size

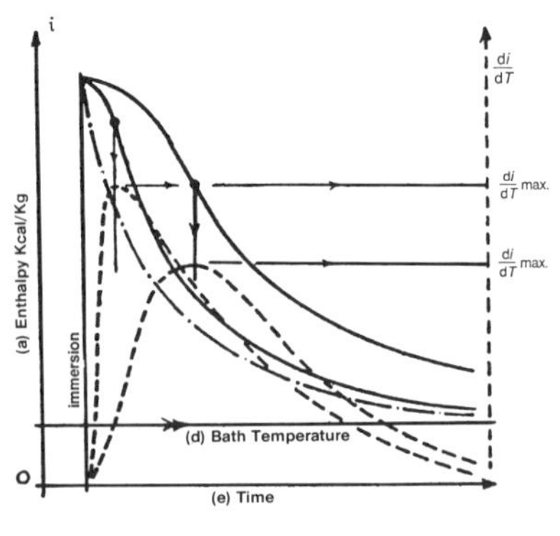

Fig. 9.31 Heat exchange between load and quenchant as a function of part size

agitation for a load of small, rapidly cooling parts with high surface areas (curve 1) and a similarly sized load of larger parts with lower surface areas (curve 2). The center (curve b) and surface (curve a) cooling curves also are shown. The dotted line in Fig. 9.30 represents the theoretical cooling curve (by Newton's law of cooling) that would be obtained were there no increase in the bath temperature during the quench (Ref 34).

Figure 9.31 illustrates the effect of two levels of agitation on heat evolution during quenching (Ref 34): relatively "weak" agitation (curve B) and "rapid" agitation (curve C). The dotted curves are the corresponding cooling rate (dT/dt) curves. Note that the maximum heat transfer occurs at the transition point of A-stage (vapor blanket) to B-stage (nucleate boiling) cooling.

The maximum heat flux (F) of this process is described by:

$$F = P(di/dT)_{\text{max}} \qquad \text{(Eq 9.9)}$$

where F is the heat flux (kcal/h), P is the total mass (kg), and $(di/dT)_{\text{max}}$ is the maximum enthalpy change with respect to temperature where the maximum cooling rate occurs.

The quenchant tank volume (R_{v}) for an acceptable temperature rise is defined as (Ref 34):

$$R_{\text{v}} = \left(\frac{P\Delta i}{1.000V \times \overline{\omega}\, C_{\text{p}}} \right) \qquad \text{(Eq 9.10)}$$

where V is the quenchant volume (m^3), $\overline{\omega}$ is the density of the quenchant (kg/m^3), and C_{p} is the specific heat of the quenchant (kcal/kg · °C).

The maximum instantaneous temperature rise (R_{u}) of an agitated tank can be defined as (Ref 34):

$$R_{\text{u}} = \left[\frac{P(di/dT)_{\text{max}}}{Q_{\text{u}}\, \overline{\omega}\, C_{\text{p}}} \right] \qquad \text{(Eq 9.11)}$$

where Q_{u} is the minimum acceptable agitation rate at the surface of the part being quenched. Assuming that the instantaneous temperature rise is within the same limit of the maximum temperature rise, then:

$$R_{\text{v}} = R_{\text{u}} \qquad \text{(Eq 9.12)}$$

resulting in Eq 9.13 for minimum useful flow (Ref 34):

$$Q_{\text{u}} = \left(\frac{1.000V}{\Delta i} \right) (di/dT)_{\text{max}} \qquad \text{(Eq 9.13)}$$

Using the above relationship and calculating the heat exchange occurring from the parts to the bath, Terrier (Ref 34) was able to calculate the actual flow rate across the load. To calculate the heat-exchange properties of a liquid, it is first necessary to calculate the Prandtl (Pr) and Reynolds (Re) numbers. The Prandtl number is dependent on the fluid and is calculated by:

$$\mathrm{Pr} = \frac{C_{\mathrm{p}}\mu}{\lambda} \qquad \text{(Eq 9.14)}$$

where μ is the viscosity and λ is the thermal conductivity of the quenchant. The Reynolds number defines the flow turbulence of the quenchant:

$$\mathrm{Re} = \frac{dG}{\mu} \qquad \text{(Eq 9.15)}$$

where d is the hydraulic flow diameter and G is the flow mass per unit cross section.

These relationships are used to calculate the surface-exchange coefficient (α):

$$\alpha = \frac{A\lambda \mathrm{Re}^{n} \mathrm{Pr}^{1/3}}{d} \qquad \text{(Eq 9.16)}$$

where A and n vary with the quenchant flow at the surface being quenched. This relationship is valid when $0.5 < n < 0.8$. The surface-exchange coefficient can also be defined as (Ref 34):

$$\alpha = \frac{A\lambda^{2/3} G^{n} C_{\mathrm{p}}^{1/3}}{Q^{1-n} \mu^{n-1/3}} \qquad \text{(Eq 9.17)}$$

For a given fluid at a particular temperature, these equations are reduced to:

$$\lambda = B\left(\frac{G^{n}}{d^{1-n}}\right) \qquad \text{(Eq 9.18)}$$

These equations show that the heat exchange λ occurring at the surface during quenching:

- Increases with flow rate
- Increases with the reduction in hydraulic diameter
- Increases with decreasing viscosity

When considering the amount of agitation required, it is important to determine the pressure loss, or "head loss" (ΔP), that occurs with flow. This loss is defined by a head-flow capacity curve; several are shown in Fig. 9.32. The pressure loss that occurs with agitation can also be defined by (Ref 34):

$$\Delta P = ZQ^{2} \qquad \text{(Eq 9.19)}$$

where Z is the "impedance" of the load (see Eq 9.20) and Q is the effective flow rate across the load.

The impedance of the load is defined by (Ref 34):

$$Z = \left(\frac{H_u}{S^2}\right)\left[\frac{C \cdot r^{1/3}}{p^{1/3}(1 - r^{2/3})^2}\right] \qquad \text{(Eq 9.20)}$$

where H_u is the height of the load, S is the frontal cross section of the load, r is the coefficient of filling (<1), p is the unit weight of the parts, and C is a shape-dependent constant that is a function of the quenchant and the units used for calculation.

Equation 9.20, although difficult to use, permits estimation of the quenchant flow *across* the load ("stirring flow"). However, all quenchant flow does not occur across the surface. Some flow impinges on the surface and then flows around a part (or load); this is called "escape flow."

Terrier (Ref 34) showed a head-flow capacity curve (Fig. 9.32) for agitation provided by a centrifugal pump, an open impeller, and a propeller pump (draft tube). Comparison of the intersection points of the head-flow and total impedance flow curves to determine useful flow (Q_u) reveals several points about each of these methods. (Draft tube and open impeller agitation is described in detail in Chapter 10.) The Q_u point for a draft tube is shown by point 3. The flow (Q_u) is shown at the intersection of the impedance curve at point 3. The centrifugal pump for the same flow requires a much higher head (ΔP). This explains why considerably greater power, and therefore energy, is required for centrifugal pumps to provide the same flow rate attainable with an impeller in a draft tube. The impedance curve crosses the head-flow curve for an open impeller at point 7.

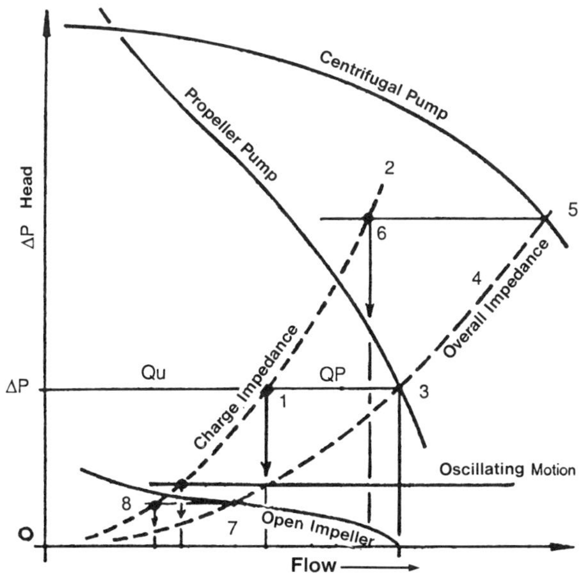

Fig. 9.32 Head-flow curves for various mixers

This shows that substantially lower useful flow rates (Q_u) are attainable with open-impeller stirrers than with a draft tube.

In some cases, it is necessary to calculate the agitation that is independent of the load being quenched. Terrier (Ref 34) reported the following relationship:

$$\text{Agitation rate} = \frac{Q\,\Delta P\,\overline{\omega}}{367V'} \qquad \text{(Eq 9.21)}$$

where ω is the density of the quenchant (kg/m^3), Q is the stirring flow rate (m^3/h), U is the real injection velocity of the flow in the quenching zone (m/s), ΔP is the injection boost ($m = U^2/2g$), and V' is the volume in the quenching zone (m^3).

Impeller Agitation. Impeller stirrers are probably the most common method of employing directional flow in a quench system. The now-classic reference describing the selection, sizing, and use of impeller stirrers was published in 1954 by U.S. Steel (Ref 10). With the exception of the paper by Terrier (Ref 34), no recent publications have dealt with this technology. Chapter 10 provides an extensive updated review of impeller selection and use, including draft-tube design.

Centrifugal Pumps. Although impeller stirrers are commonly employed for directional flow around the load during quenching, the quenchant must be circulated through a heat exchanger or chiller to control temperature during the quench. Centrifugal pumps are most often selected for this application.

There are numerous guides to centrifugal pump selection, use, system design, and so forth. Three particularly useful books on this subject are listed in Ref 38 to 40. Extensive design information on centrifugal pumps is beyond the scope of this chapter.

The first step in determining the correct size of a centrifugal pump is to determine the total system head. The term head is defined as the pressure at any point in a fluid.

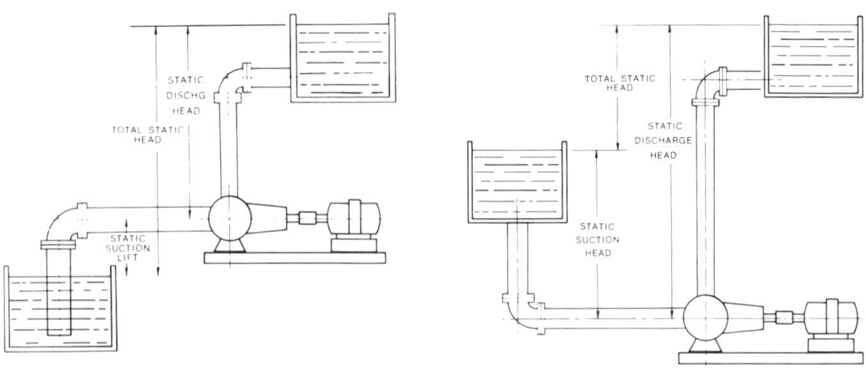

(a) (b)

Fig. 9.33 Static heads for centrifugal pumps. **(a)** Static suction head. **(b)** Suction head

Four types of heads constitute the total system head: static head, friction head, velocity head, and pressure head (Ref 40).

Static head is defined as the vertical distance in feet between the liquid level of the reservoir and the system, as shown in Fig. 9.33. Static head (h_s) is calculated according to:

$$h_s\ (\text{ft}) = \frac{\text{pressure (psi)} \times 2.31}{\text{Specific gravity}} \quad \text{(Eq 9.22)}$$

Friction head (h_f) is another component of total system head. This is the head required to overcome the frictional forces in the piping system for moving the fluid. Friction head depends on the type and size of pipe, flow rate, and fluid, and can be calculated from nomograms relating all of these factors (see Ref 38 to 40).

Velocity head (h_v) also is related to fluid flow and can be calculated as:

$$h_v = \frac{V^2}{2g} \quad \text{(Eq 9.23)}$$

where V is the fluid velocity (ft/s) and g is 32.2 ft/s^2. V is determined according to:

$$V = \frac{rpm \times D}{229} \quad \text{(Eq 9.24)}$$

where *rpm* and D are the revolutions per minute and diameter, respectively of the pump impeller.

Suction lift, if the fluid is below the centerline of the pump, or suction head, if the fluid is above the centerline of the pump, is calculated from Eq 9.22 using the relationship shown in Fig. 9.33. Dynamic head (lift) is the sum of static head plus velocity head (lift). The total dynamic discharge head (h_d) of a pump is equal to the sum of static discharge head, velocity head, and friction head. The total head (H) is equal to:

$$H = h_d + h_s \text{ (with suction lift)} \quad \text{(Eq 9.25)}$$

$$H = h_d - h_s \text{ (with suction head)} \quad \text{(Eq 9.26)}$$

Pumping capacity (Q) is calculated from:

$$Q = A \times V \quad \text{(Eq 9.27)}$$

where A is the cross-sectional area of the discharge line of the pump and V is defined by Eq 9.24. The work performed by the pump (whp), or power delivered by the pump, is defined by:

$$\text{whp} = \frac{Q \times TDH \times \text{sp. gr.}}{3960} \quad \text{(Eq 9.28)}$$

In order to calculate the efficiency of the pump, the brake horsepower (bhp), which is the actual horsepower delivered at the pump shaft, must be calculated:

$$\text{Pump efficiency} = \frac{\text{whp}}{\text{bhp}} = \frac{Q \times TDH \times \text{sp. gr.}}{3960 \times \text{bhp}} \qquad \text{(Eq 9.29)}$$

A pump is characterized by the net positive suction head (NPSH), which is equal to the total suction head-vapor pressure of the fluid in feet absolute. This is a measure of the head required to keep the fluid in liquid form and thus prevent cavitation. NPSH values are dependent on the particular design of a pump. Standard NPSH curves are supplied by various pump manufacturers.

The NPSH available from the liquid on the suction side of the pump is (Ref 39):

For water

$$\text{NPSH} = S + (P_a' - P_{rp}') - h_{SL} \qquad \text{(Eq 9.30)}$$

For liquids other than water

$$\text{NPSH} = S + (P_a - P_{rp})(2.31/\text{sp. gr.}) - h_{SL} \qquad \text{(Eq 9.31)}$$

where P_a' or P_a is the absolute pressure in the vessel on the suction side of the pump, P_{rp}' or P_{rp} is the absolute vapor pressure of the fluid as it is being pumped, h_{SL} is the friction loss from the suction vessel to the pump, and S is the static head or lift in the system.

The NPSH also varies with the pump rotational speed as:

$$rpm \; \alpha \; (\text{NPSH})^{0.75} \qquad \text{(Eq 9.32)}$$

To prevent pump cavitation, the following conditions should be avoided (Ref 39):

- An NPSH that is less than 0.6 m (2 ft) greater than the worst possible recommended condition for operation
- Internal clearance wear inside the pump
- Plugs in the suction piping system
- Entrained gas
- Low liquid level or fluid temperature increases
- Liquid vortexing in the suction vessel, which may cause gas entrainment
- Inadequate sizing of nozzles on the liquid-containing vessel (the suction line usually should be at least one pipe size larger than the pump suction nozzle)

Once the necessary system requirements are calculated, the various pump options can be determined by consulting the manufacturer's literature. The first step is to identify the correct "oak tree curve" (see Fig. 9.34) (Ref 38). The numbers shown on the various "leaves" correspond to pump models. For the purpose of this example, assume that the objective is to pump the quenchant at 19,000 L/min (5000 gal/min) against a total system head of 55 m (180 ft) using a pump operating at 1775 rev/min. Figure 9.34 shows that pump model A1015L would be the best candidate under these operating conditions.

The next step is to consult the performance curve for this particular pump (Fig. 9.35) (Ref 38). Using this curve and the operating conditions cited above, the following conclusions can be drawn:

- The pump impeller diameter should be approximately 380 mm (15 in.).
- An optimal efficiency of greater than 89% will be obtained. (If poor pumping efficiency is indicated, the search for a better pump should continue.)
- Approximately 260 hp will be required to drive the pump.
- Approximately 6.4 m (21 ft) of NPSH must be provided to prevent cavitation. From the curve, it is evident that this is equivalent to approximately 3.7 m (12 ft) of suction.

Submerged-Spray (Jet) Mixing. Although most quenching systems employ either impeller or centrifugal pump agitation, other forms of agitation are available for quench tanks. For example, directional fluid flow can be provided by a submerged-spray, or jet, mixer. This section will provide an overview of the calculations necessary for jet mixing in quench tanks. For a more rigorous discussion of jet mixing, consult Ref 41.

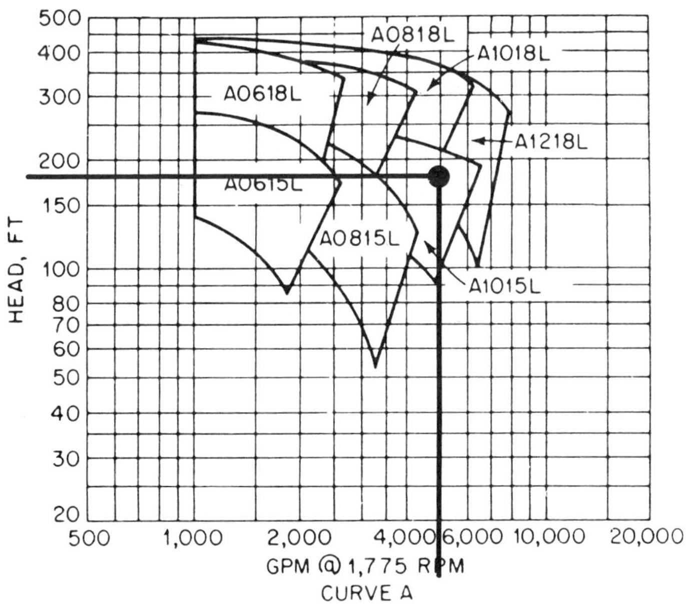

Fig. 9.34 Typical oak leaf chart for centrifugal pumps

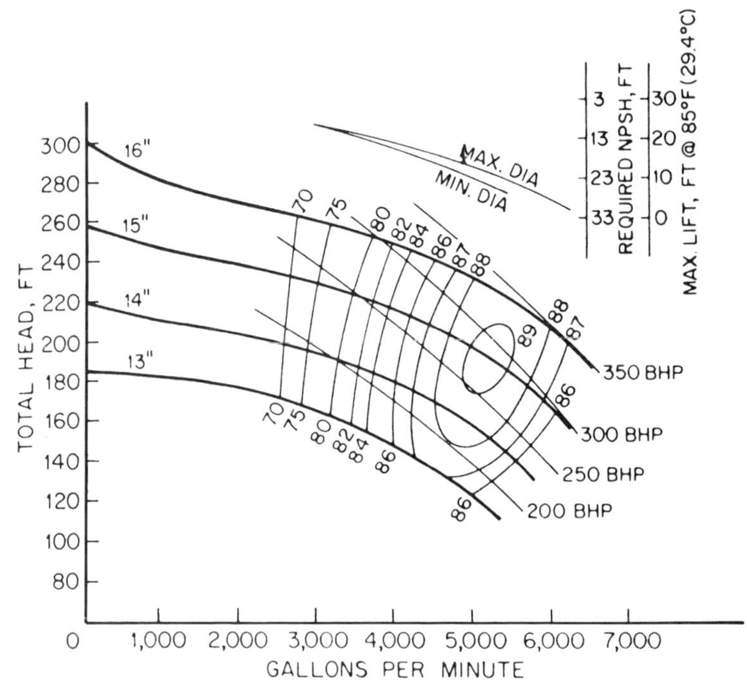

Fig. 9.35 Performance curve for centrifugal pump model A1015L

Jet mixing involves the injection of liquid into the quench tank. The liquid is usually pumped through the jet using a centrifugal pump. Figure 9.36 illustrates a "simple" jet mixing system where the fluid is pumped through a single jet into a rectangular tank. Jet mixers can be used in a variety of arrangements. For example, multiple jets, fed through a center manifold, may be used to obtain more uniform flow over the part surface during the quench. However, for simplicity, the remainder of this discussion will focus on the use of a single jet.

Figure 9.37 illustrates a typical flow pattern of a jet (Ref 42). The radius for the jet (R_j) at any centerline distance (X) from the nozzle can be calculated from (Ref 43):

$$R_j = 0.232X \qquad \text{(Eq 9.33)}$$

The velocity at any centerline distance (U_{cj}) for $X/D_j > 8$ can be calculated from (Ref 43):

$$U_{cj} = 5.13\left(\frac{D_j}{X}\right)U_o \qquad \text{(Eq 9.34)}$$

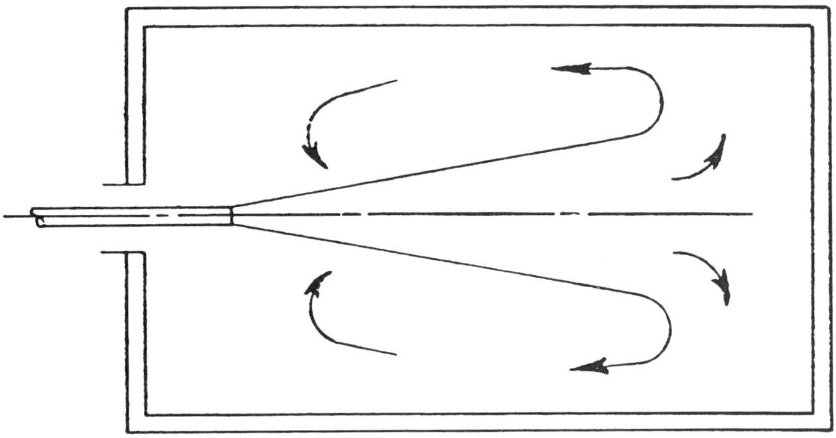

Fig. 9.36 Flow pattern in a rectangular tank with a jet mixer

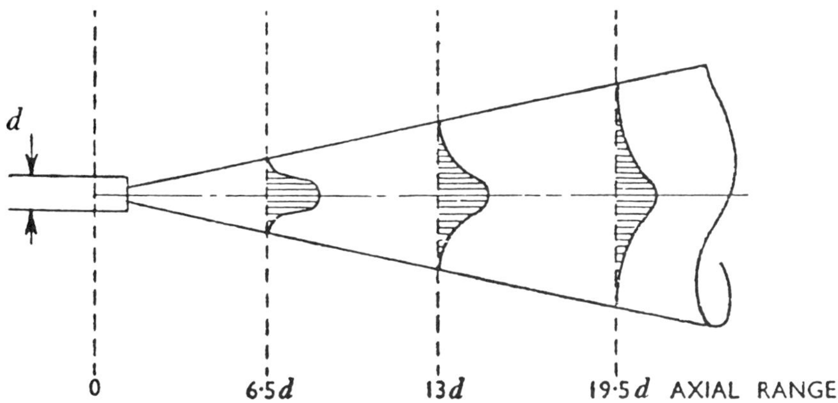

Fig. 9.37 Flow pattern of a jet mixer

where D_j is the diameter of the jet at position j and U_o is the velocity of the jet at the orifice (R_o).

To optimize mixing, the entrainment of the quenchant into the jet should be maximized. The amount of fluid entrained at any position Dj (Q_e) is calculated from (Ref 43):

$$Q_e = \left[0.234\left(\frac{X}{D_j}\right) - 1\right]Q_o \qquad \text{(Eq 9.35)}$$

where Q_o is the volumetric flow rate at the jet nozzle (D_o).

Rushton (Ref 44) showed that the maximum entrainment occurred approximately 17 jet diameters from the nozzle. Therefore, large jets should be used for large tanks, and the jet should be placed within 17 diameters of the surface being quenched.

The power input (P_i) into the jet is (Ref 43):

$$P_i = Q_o \rho g \left(\frac{U_o^2}{g} \right) \quad \text{(Eq 9.36)}$$

where g is the gravitation acceleration constant (32.2 ft/s^2) and (U_o^2/g) is the head of the jet. Tatterson (Ref 41) reported that fluid entrainment decayed to zero when the centerline distance exceeded 100 jet diameters.

The cone angle (θ_j) of the jet varies with kinematic fluid viscosity (η) in units of Stokes:

$$\tan\left(\frac{\theta_j}{2}\right) = 0.238\eta^{0.133} \quad \text{(Eq 9.37)}$$

The jet Reynolds number (R_{ej}) can be calculated from (Ref 41):

$$\text{Re}_j = \frac{\rho U_o D_j}{\eta} \quad \text{(Eq 9.38)}$$

where ρ is the fluid density, U_o is the initial linear flow rate at the jet nozzle, D_j is the position j from the jet nozzle, and η is the fluid viscosity. For jets, turbulent flow begins

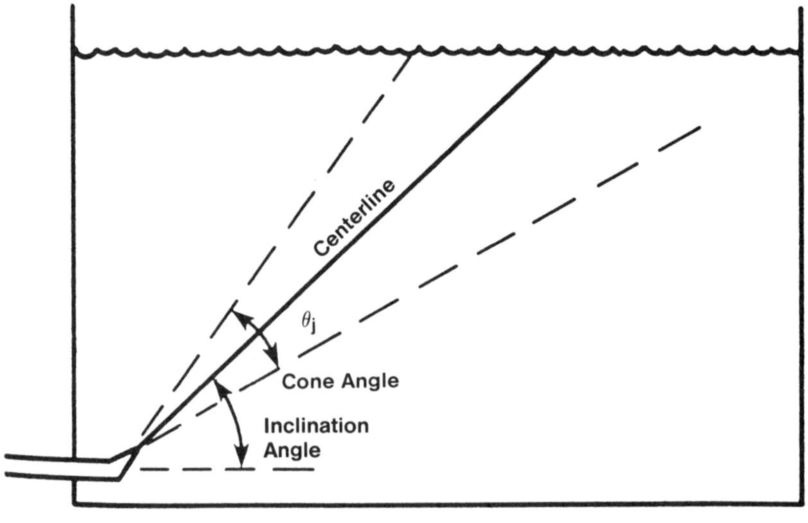

Fig. 9.38 Definition of a jet cone angle

at Reynolds numbers of 2000, and fully turbulent flow occurs at Reynolds numbers greater than 5000 (Ref 41). It is important to recognize that while fluid turbulence facilitates effective and uniform heat transfer, excessive turbulence will prematurely remove the quenchant film, thus negating the heat-transfer mediating effect.

The total volumetric flow (Q_T) of the jet is calculated from:

$$Q_T = Q_o\left[\frac{(0.576\eta^{0.133})\,x}{D_j}\right] \qquad \text{(Eq 9.39)}$$

The mixing time (θ_m) for radial jets is calculated from:

$$\theta_m = \frac{8.7T^2 \sin\alpha}{D_j U_o} \qquad \text{(Eq 9.40)}$$

where T is the tank diameter and α is the angle of inclination (see Fig. 9.38). Using these equations, it is possible to calculate a number of important mixing parameters:

- The size of the jet required to mix a given volume of fluid in a quench tank
- The time required to totally mix the fluid in the tank
- The power of the centrifugal pump required to mix the fluid
- The average circulation rate at any position within the jet
- The fluid turbulence within the jet cone

Characterization of all of these variables is necessary to describe the flow regime being used for quenching.

Fluid mixing by sparging (gas injection) is a relatively little-used method of quench tank agitation. Unfortunately, no literature has been published on the use of gas injection to agitate quenching systems.

Analysis of this form of agitation (Ref 41), when compared with that provided previously for centrifugal pumping, jet mixing, and impeller stirring, is complicated by the fact that quenching in oil and aqueous systems involves boiling and vapor blanket heat transfer. Thus, a similar quantitative treatment will not be provided here. However, some basic engineering data will be reviewed Ref 45).

The pressure required to agitate a system depends on the:

- Head of the fluid
- Friction in the pipe
- Pressure difference required to force the liquid through the pipe orifice (nozzle)

The relationship between linear velocity and flow rate is a function of pipe diameter (Ref 45):

$$Q = 0.327VD^2 \qquad \text{(Eq 9.41)}$$

where Q is the air-flow volume (ft/min), V is the linear velocity (ft/min), and D is the pipe diameter. The pressure loss in the pipe due to friction (P) can be calculated from (Ref 45):

$$P = \frac{Q^2 L}{2690 D^5} \quad \text{(Eq 9.42)}$$

where L is the length of the pipe in feet. When the total length of the pipe is being calculated, it is important to account for the presence of elbows. Table 9.5 can be used to convert elbow size into equivalent straight pipe length for use in the calculation of L (Ref 45).

The volume flow rate of air that will flow through a pipe orifice of known diameter D (in.) is:

$$Q = 21.7\, D^2 \mathrm{C} \sqrt{H} \quad \text{(Eq 9.43)}$$

where H is the head in inches of water and C is a constant that depends on the orifice (determined from Table 9.6).

The agitation provided by gas injection increases (Ref 45):

- With the volume and linear flow rate of the gas from the orifice
- More rapidly than the gas flow rate
- With the depth of the tank (because agitation is caused by the expansion of the gas as it rises in the liquid)

Stognei and Kruk (Ref 46) recently reported on the use of an air-injection technique to supplement the use of centrifugal pump agitation for forgings quenched in oil. They showed that the agitation effect (π) of the injected air was proportional to the growing bubble as it rises:

$$\pi = \frac{3R_{\mathrm{max}}}{g t_{\mathrm{max}}^2} \quad \text{(Eq 9.44)}$$

where R_{max} and t_{max} are the maximum bubble radius and the time required for them to occur, respectively. The π agitation effect is related to the Fourier number (Fo) by:

$$\mathrm{Fo} = \frac{gT^2}{3R_{\mathrm{max}}} \quad \text{(Eq 9.45)}$$

The Fourier number is also related to the heat transfer by (Ref 46):

$$\mathrm{Fo} = \frac{a_{\mathrm{oi}} t_{\mathrm{oi}}}{R^2} \quad \text{(Eq 9.46)}$$

Table 9.5 Pipe equivalent length for elbows

Radius of pipe elbow		Equivalent length of straight pipe	
mm	in.	mm	in.
127.0	5.0	198	7.8
76.2	3.0	213	8.4
50.8	2.0	229	9.0
38.1	1.5	262	10.3
31.8	1.25	323	12.7
25.4	1.0	445	17.5
19.0	0.75	889	35.0
12.7	0.5	3070	121.0

Table 9.6 Orifice constants for gas injection

Orifice type	Constant, C
Conoidal mouthpiece, form of contracted vein	0.97-0.99
Conoidal converging mouthpiece	0.90-0.99
Short, cylindrical mouthpiece rounded at the inner end	0.92-0.93
Short, cylindrical mouthpiece	0.81-0.84
Thin, circular plates	0.56-0.59

where a_{oi} is the thermal diffusivity of the oil (m^2/s), t_{oi} is the mean time between bubble ruptures (s), and R is the characteristic dimension of the tank (m).

The flow rate of the air is determined by (Ref 46):

$$M_a = nV_s\rho_a \qquad \text{(Eq 9.47)}$$

where V_s and ρ_a are the volume and density, respectively, of the air and n is the number of bubbles formed per second. The relationship between the amount of air supplied and quench severity (temperature change on quenching) is shown in Fig. 9.39 and is quantified by (Ref 46):

$$\theta = \frac{T_c - T_{oi}}{T_o - T_{oi}} \qquad \text{(Eq 9.48)}$$

where T_c is the temperature at the center of the quenched part, t_{oi} is the oil quenchant temperature, and T_o is the initial temperature of the quenched part (with all temperatures expressed in degrees Kelvin).

Air injection may be considered for water and brine quenchants, but it is a relatively poor choice for other vaporizable quenchants, such as oil and aqueous polymers. Air injection would be expected to accelerate the oxidative decomposition of any organic material, especially at the temperatures encountered in quenching. The use of an inert gas (e.g., nitrogen, helium, etc.) that would not enhance oxidation of the quenchant would be significantly more expensive.

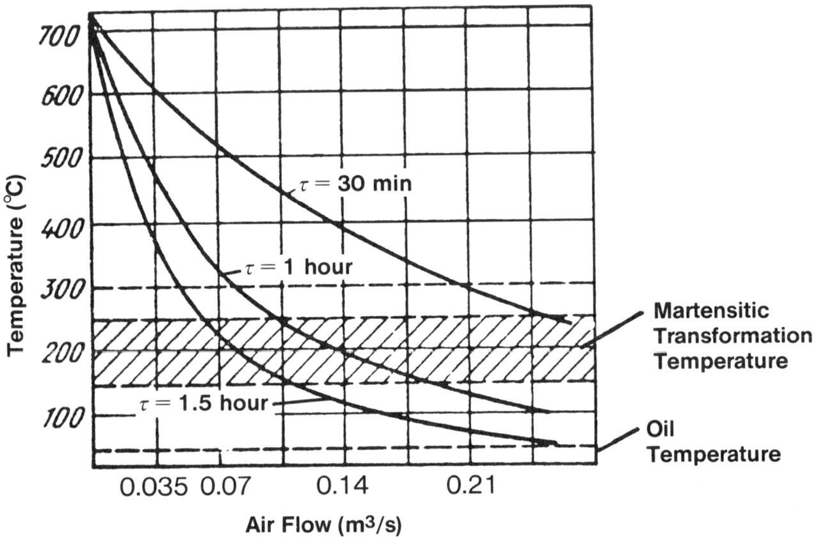

Fig. 9.39 Effect of air-injection rate on cooling curve behavior during quenching

Quench Tank Design

Design Criteria

Quench tank design is based primarily on such factors as the weight of steel to be quenched per hour, part size, part shape, section thickness, steel grade, and final required properties. All of these factors must be considered in any successful design. Practical suggestions include (Ref 31):

- The time required to quench out parts should be measured, and the desired part throughput per hour should be estimated. This information can be used to determine the size of the quench tank. Quenchant volume and production throughput are then used to select the size of the heat exchanger and to determine whether a storage tank is necessary.
- Enough space should be provided around the workpiece to obtain adequate quenchant circulation and maximum heat removal from the part during quenching.
- Hot workpieces should not be allowed to impact on a chute or conveyor until they have fallen through sufficient quenchant to produce a high-strength surface on the part. (Chute design will be discussed later in this chapter.)
- The quench tank should be accessible for maintenance and cleaning.

- Provision for scale removal from the tank must be made, especially if the parts are heated in air. The use of a controlled atmosphere may eliminate scaling; however, in poorly operating furnaces, carbon contamination may occur, which must also be removed. Thus, filtration should be used to achieve optimal quench performance. (Filtration will be discussed later in this chapter.)
- Adequate ventilation should be provided for protection from fumes.
- Adequate fire protection measures should be taken, especially for oil quenchants.
- "Cleanout" plates below the quenchant level should be avoided, because gaskets are prone to leakage and are difficult to maintain.
- Special materials of construction are required for systems using brine or caustic solutions.
- Trays and fixtures of the 25Cr-12N or 35N-15Cr type can be quenched in oil, but not in water, caustic, or brine.
- Wherever possible, the design of the work tank should provide for flexibility and control of such conditions as time cycle, quenchant circulation volume, and uniformity of bath conditions.
- Excessive agitation can blow light parts off conveyors.
- A quenching bath should not be agitated so violently that foaming occurs on the liquid surface. This entrained air reduces the heat-transfer rate, and the foaming of oil could cause fires.

Agitation Placement

Each quench system is uniquely designed for the type of part being quenched. Therefore, the agitator selection and placement will depend on the specific part, type of heat treating process, and so on. However, the common denominator of any well-designed system is uniform, although not necessarily vigorous, agitation around the part surface during the quench (Ref 22, 47). This section will illustrate several successfully designed systems that have been reported previously.

Multiple Open-Impeller Agitation. Multiple agitators should be used in cylindrical tanks when the mixer power requirement (see Chapter 10) exceeds 3 hp (Ref 10). If the power requirement is 3 to 6 hp, then side-entering or top-entering mixers should be used. If the power requirement is in excess of 10 hp, more than two mixers may be required. Depending on the space available, either a side-entering (Fig. 9.40) or top-entering arrangement may be used (Ref 10).

For very deep cylindrical tanks, an agitator arrangement such as that shown in Fig. 9.41 may be used. In this example, a cone at the bottom center of the tank is used to deflect the quenchant upward to provide axial (vertical) flow past the long parts (Ref 10).

For rectangular tanks, one large or two small mixers may be used where the length of the tank is up to twice the width (Fig. 9.42) (Ref 10). However, for tanks where the

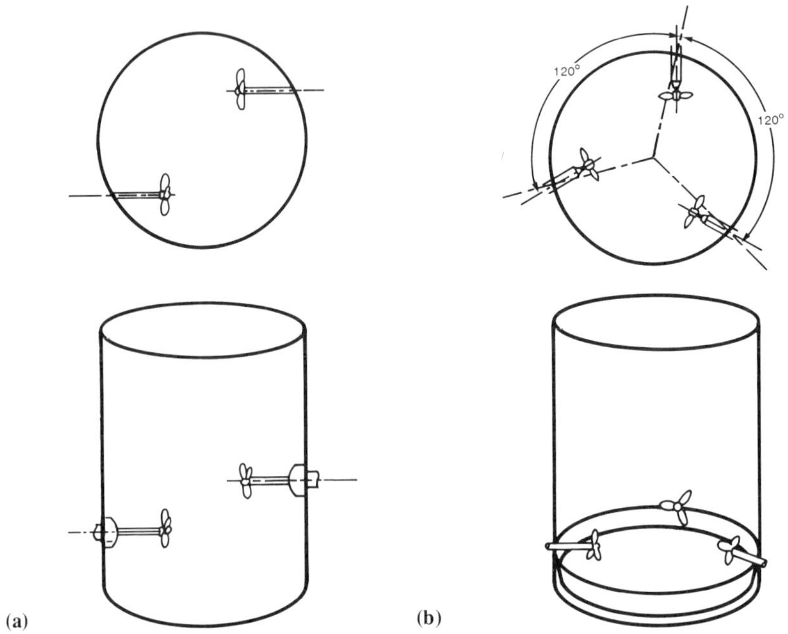

Fig. 9.40 Suggested impeller placement for cylindrical tanks with side-entering mixers

length is greater than twice the width, multiple mixers should be placed in a staggered arrangement (Fig. 9.43) (Ref 10).

Draft-tube mixers may also be used instead of open impellers. Details for draft-tube design and impeller sizing requirements are provided in Chapter 10.

Top-entering open impellers should be placed in the bath at 16° from vertical, with the impeller blade placed so that it is one diameter from the bottom. Side-entering impellers should be one-half of the propeller diameter plus 150 mm (6 in.) from the bottom of the tank (Ref 10).

Figure 9.44 shows a 9500 L/min (2500 gal) quench tank that uses four agitators to develop a total of 100 hp (75 kW) for quenching long bar stock. This system provides an agitation rate of 60 m/min (200 ft/min), with a fluid turnover rate of 114,000 L/min (30,000 gal/min) (Ref 31, 48). The quench tank utilizes a curved bottom to facilitate uniform oil flow around the part during the quench.

Although the agitation rate for this system is probably excessive for most applications, it does illustrate what is achievable with a well-designed impeller agitation system. (For most steel quenching applications, a linear flow rate of 15 to 30 m/min, or 50 to 100 ft/min, is sufficient.)

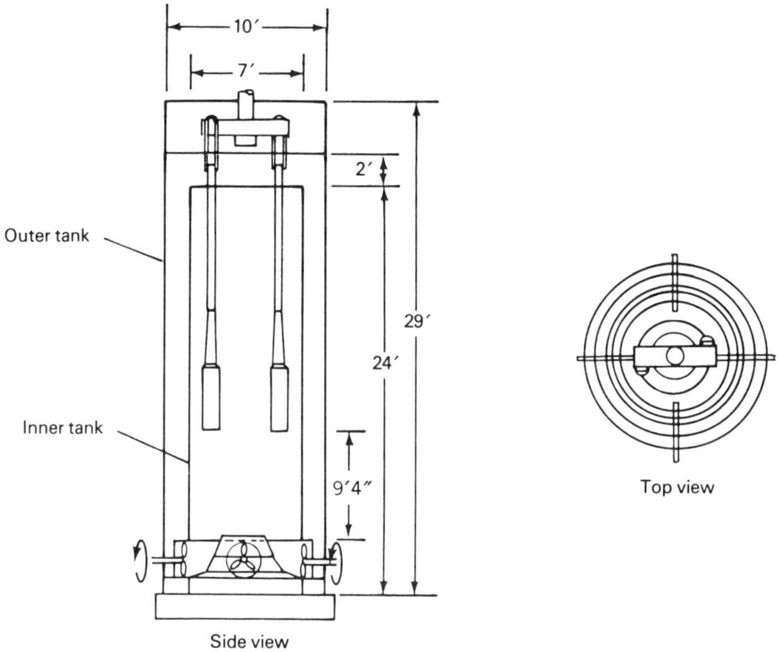

Fig. 9.41 Deep cylindrical tank for quenching long axles and shafts

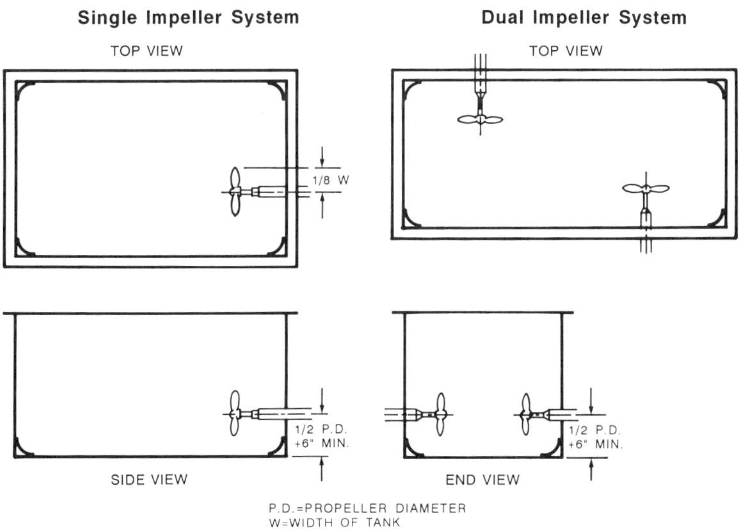

Fig. 9.42 Suggested impeller placement for rectangular tanks

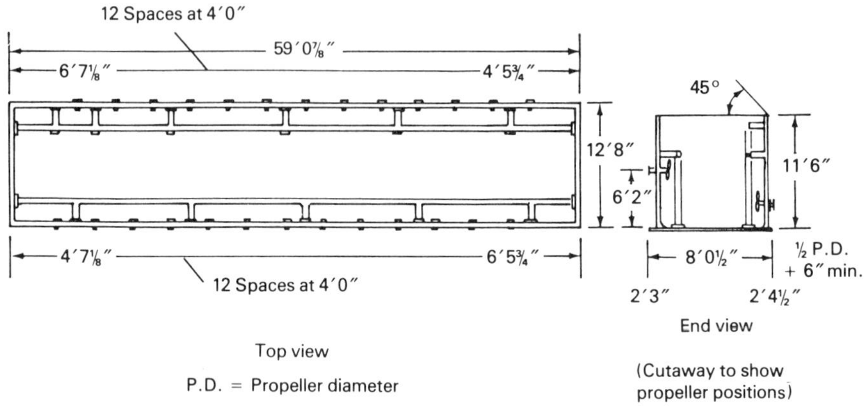

Fig. 9.43 Multiple-impeller rectangular tank

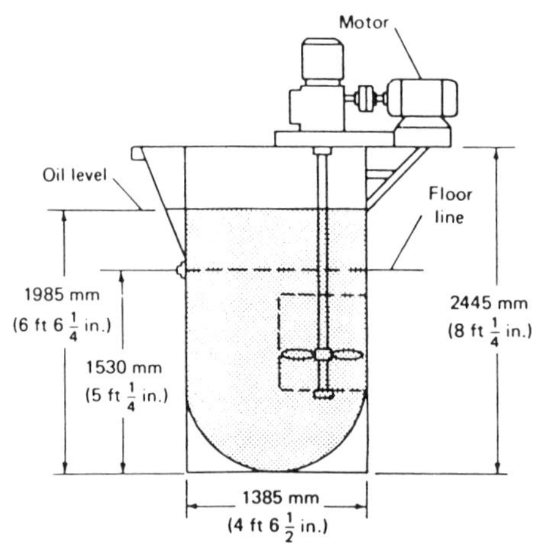

Fig. 9.44 Quench tank for steel bar stock

A system of similar size (10,000 L, or 2660 gal) with side-entering impellers is shown in Fig. 9.45. This system uses three 25 hp agitators that produce a linear flow rate of 20 m/min (66 ft/min) and a fluid turnover rate of 51,100 L/min (13,500 gal/min) (Ref 31, 48). In addition to a rounded bottom to facilitate flow, baffles are used to deflect the quenchant back to the agitators. This arrangement minimizes splashing and vortexing, which may produce cavitation.

A batch-quenching tank with multiple side-entering agitators that can be used to quench trays of smaller pieces is shown in Fig. 9.46 (Ref 31, 48). This system uses baffles at the bottom of the tank to deflect quenchant axially through the quench zone.

Figure 9.47 shows a system design using centrifugal pump agitation. This system is designed to provide axial flow of the quenchant past relatively long axles and crankshafts (Ref 48).

Draft-Tube Agitation. All of the examples of quench tank design shown thus far utilize either open-impeller on centrifugal pump agitation. However, the use of directional flow agitation provides superior results for many applications. One of the most cost-effective ways of providing directional flow is to use draft tubes.

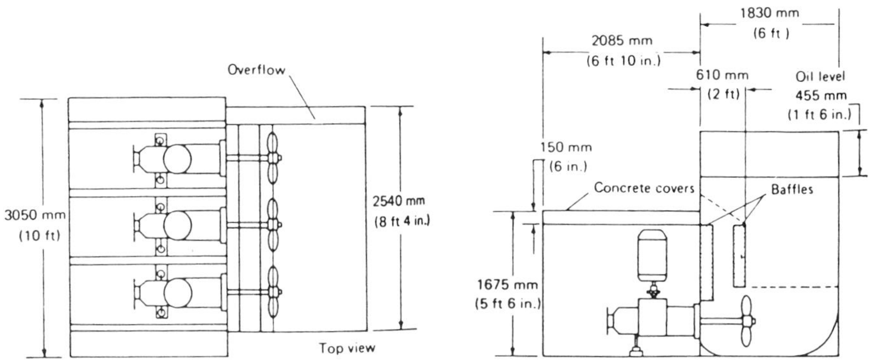

Fig. 9.45 Quench tank equipped with multiple side-entering impellers

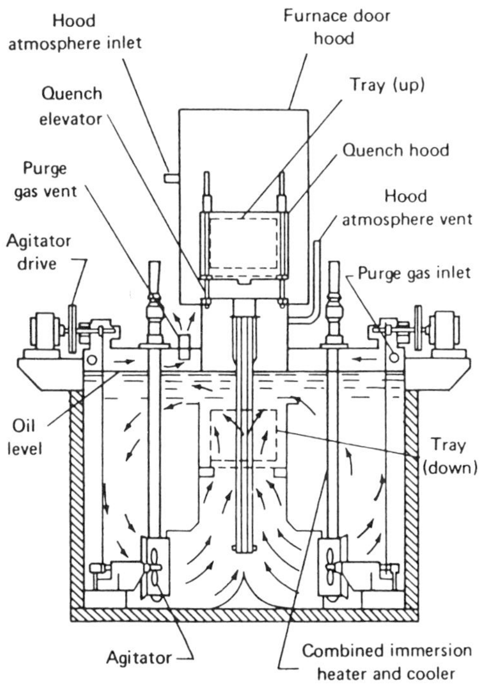

Fig. 9.46 Batch-quenching system with side-entering impellers

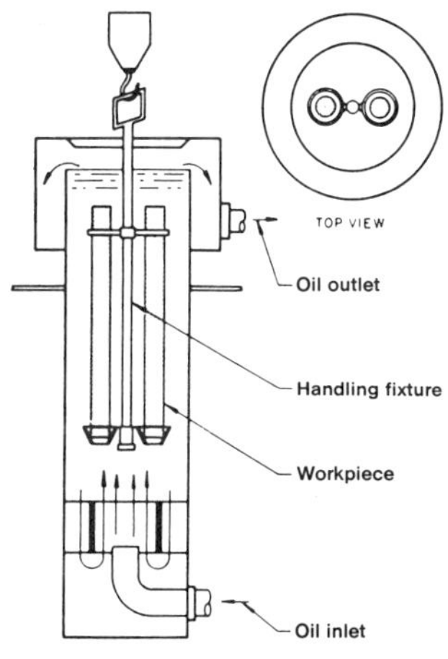

Fig. 9.47 Quench tank with pump agitation for batch quenching axles and shafts

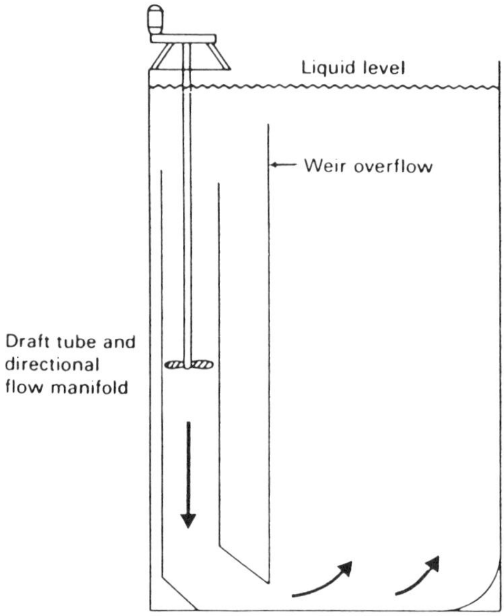

Fig. 9.48 Deep pit quench tank with draft-tube pump

Draft-tube agitation can be used for pit quench systems such as that shown in Fig. 9.48 (Ref 36, 49). In this example, the draft-tube agitator acts as a pump, and fluid flow is directed by the use of rounded corners, baffles, and a weir. The weir overflow is used to ensure that a proper head is maintained above the impeller to prevent cavitation. The draft tube should be at least 300 mm (12 in.) below the minimum liquid level (Ref 36).

Some integral (sealed) quench furnaces also employ quench tanks with draft-tube agitation (Ref 36, 37, 48, 50). The draft-tube agitators are used to provide the desired axial fluid

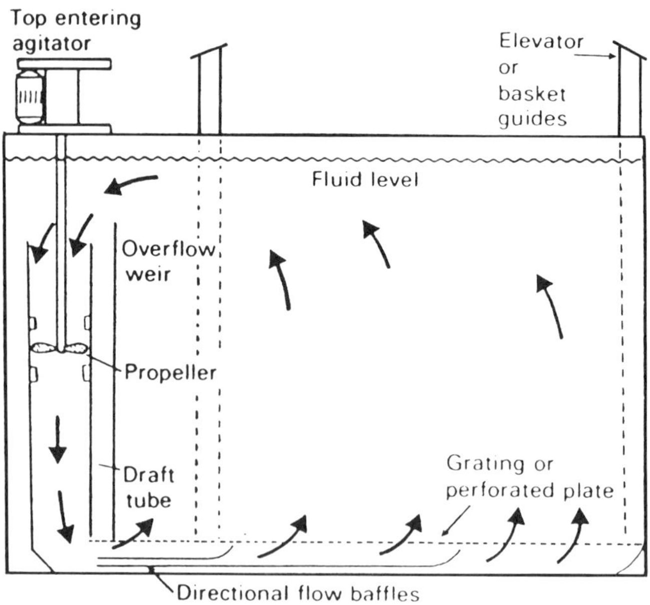

Fig. 9.49 Batch quench tank with single draft-tube agitator

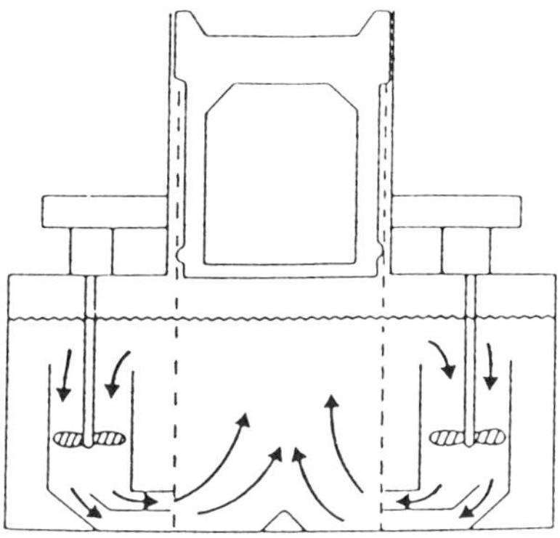

Fig. 9.50 Integral quench furnace with multiple draft-tube agitators

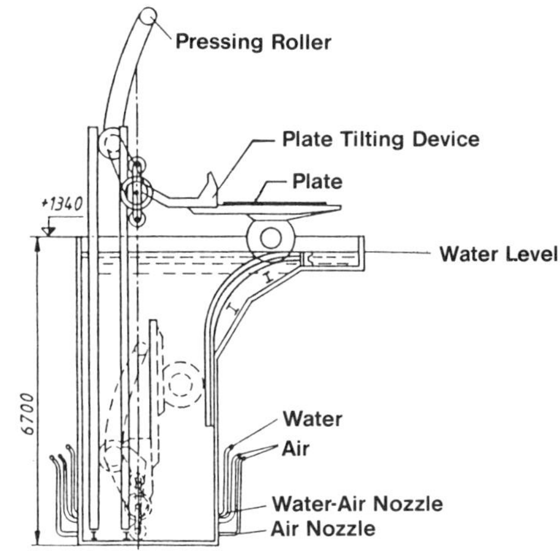

Fig. 9.51 Air-injection agitation system for quenching plate stock

flow through the load during quenching. Single and double draft-tube agitator systems are shown in Fig. 9.49 and 9.50, respectively (Ref 36, 37).

Although integral quench furnaces are most often used with oil quenchants, recent furnace design improvements have permitted the successful use of aqueous polymer quenchants (Ref 51) if the following precautions are taken (Ref 37, 48, 51):

- The workload (including the weight of the fixture) should not exceed 1 lb/gal of quenchant. The temperature rise of the quenchant should be controlled to less than 10 °C (18 °F) with heat exchangers.
- Axial quenchant flow through the load during the quench is recommended.
- There should be a tight inner-door seal to prevent water vapor from entering the furnace vestibule. It is recommended that a slight increase in gas flow (positive pressure) in the furnace vestibule be provided during the transfer of the load into the quench chamber and during the quench.
- The agitators must always be on when the inner door is open to minimize the buildup of water vapor in the vapor spaces in the quench chamber.

Spray Quenching. Some processes have successfully employed a submerged spray (air/water) system for quenching (Ref 22, 52). Typically, such systems (Fig. 9.51) are used for quenching plate stock. Details of various aspects of spray quenching technology are provided in Chapter 7.

Mechanical Agitation. Nearly all of the quenching applications described to this point have employed agitation by fluid-flow control. However, fluid agitation can also be provided by movement of the part (mechanical agitation). Recently, Olivier *et al.* (Ref 47) have reported the use of a mechanical oscillatory device to provide agitation during quenching. Terrier (Ref 34) provided a detailed description of the relative flow head of mechanical agitation versus agitation provided by centrifugal pumps, open impellers, and draft tubes.

Von Hagen (Ref 61) has described a novel quenching system for tube stock (Fig. 9.52). In this system, the tube stock is rotated in a tank containing the quenchant for a time specific to the size and alloy of the stock. The calibration of the required number of revolutions to facilitate optimal martensite formation fit produced the following relationship:

$$N_1 = \left(\frac{6685}{D}\right)\left(1 + \frac{4H}{D}\right) \quad \text{(Eq 9.49)}$$

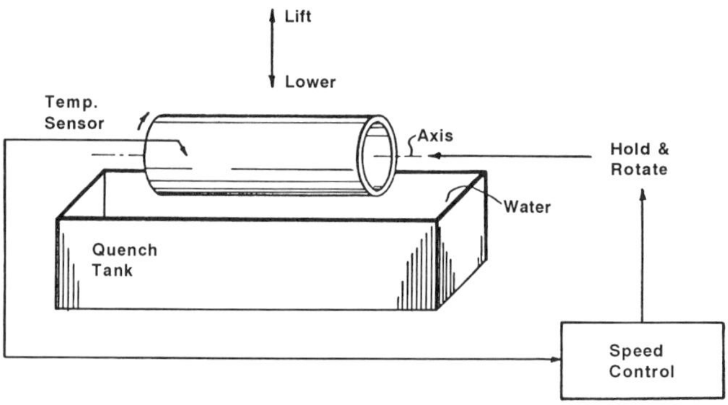

Fig. 9.52 Mechanical agitation system for quenching tube stock

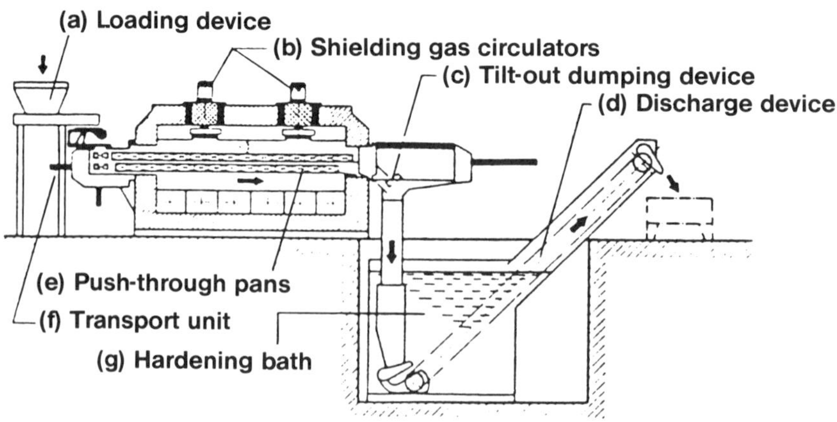

Fig. 9.53 Continuous furnace application with chute quenching

where D is the tube diameter (mm), H is the immersion depth during the quench (mm), and N is the number of revolutions that the tube stock must rotate within the quenchant to produce the desired microstructure. The rotational speed for the quench was reported to be a minimum of 40 rev/min.

Chute Quenching

Continuous heat treating furnace applications typically use a chute to transfer the austenitized steel parts to the quench bath (Fig. 9.53). Because the parts are usually relatively small, it is often assumed that they are "quenched" within the chute zone. Thus, the design of the chute quenching system is critically important to achieve a uniform quench and to minimize cracking and distortion.

Four variables in chute quenching will be discussed:

- Effect of part dropping time
- Agitation
- Heat exchange
- Quenchant contamination

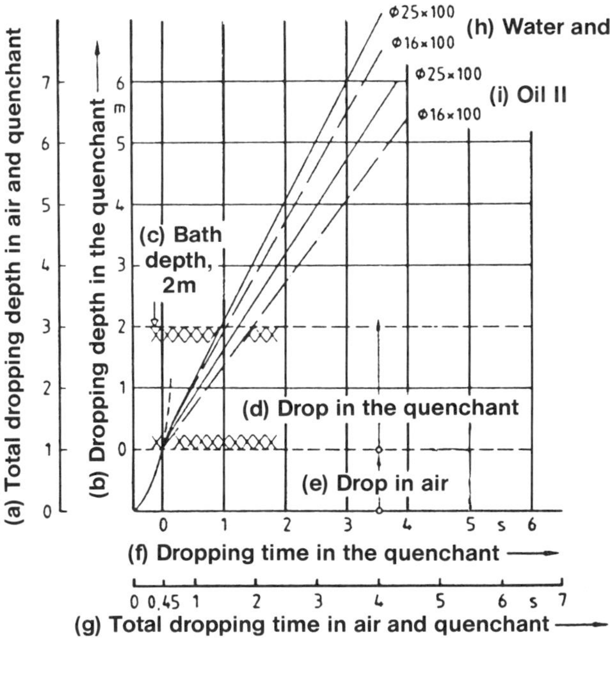

Fig. 9.54 Sinking depth-time relationships for small, cylindrical parts

Effect of Dropping Time. Larger parts generally require longer drop zones distances in the chute, because they require longer times to complete the quenching process (Ref 49). However, until recent studies by Illgner (Ref 53), very little systematic and thorough work has been done to study the effect of dropping time in the chute on quenching properties.

One of the factors studied by Illgner (Ref 53) was the effect of the viscosity of the quenching medium on dropping time. In this study, two sizes of cylindrical steel test pieces—25 × 100 and 16 × 100 mm (1 × 4 and 0.6 × 4 in.)—were allowed to "freefall" through a 2 m (6.6 ft) dropping distance. The quenching media studied were 20 °C (70 °F) water and two oils with viscosity units (U) at 50 °C (120 °F) of 1.6 °U (oil I) and 4.5 °U (oil II). Illgner's study (Ref 53), summarized in Fig. 9.54, showed that:

- The sinking speeds were nearly the same for all quenchants.
- The position of the part when it first contacts the quenchant significantly affects sinking speed. For example, increasing the surface area at contact will produce a proportional reduction in sinking time.
- Although water has the lowest viscosity, it did not produce the fastest sinking speed. The low-viscosity oil often gave faster sinking speeds. See the sinking time data for various quenchants in Table 9.7.
- Even small parts did not reach their martensitic transformation temperature in the chute zone, as shown by the cooling curve data in Fig. 9.55(A).
- Large parts are even less likely to undergo martensitic transformation in the chute zone, as shown in Fig. 9.55(B). Thus, vigorous agitation is necessary to impart the necessary quench severity to complete the quench before part removal from the bath.

Agitation in the Quench Zone. Vigorous agitation in the quench zone, for both oil and polymer quenchants, is necessary to achieve uniform quenching and optimal heat transfer from the part and the quenchant. This factor often dictates the overall success of the quenching process.

Table 9.7 Sinking times for 2 m (6.6 ft) quenchant depth for various parts

Part	Sinking time(a), s		
	Water	Oil I	Oil II
Light cylindrical parts (up to 200 g, or 7 oz)	1.5/1.7	1.0/1.5	1.0/2.5
Heavy cylindrical parts (up to 800 g, or 28 oz)	0.6/1.0	0.6/1.0	0.7/1.1
Nuts, rings, and washers (up to 100 g, or 3.5 oz)	1.8/2.8	1.5/2.2	1.7/2.2

(a) The sinking times shown are the lowest/highest.

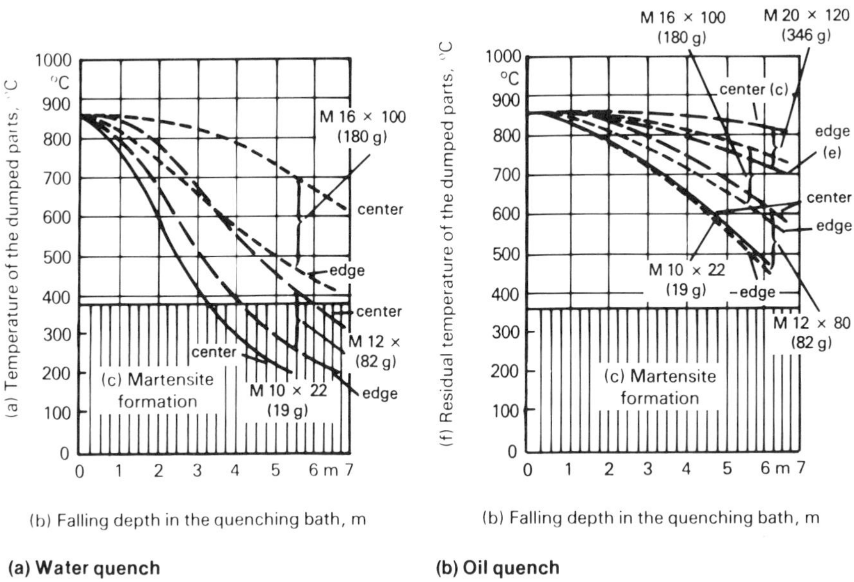

Fig. 9.55 Cooling curve correlation for various cross-section sites in chute quenching

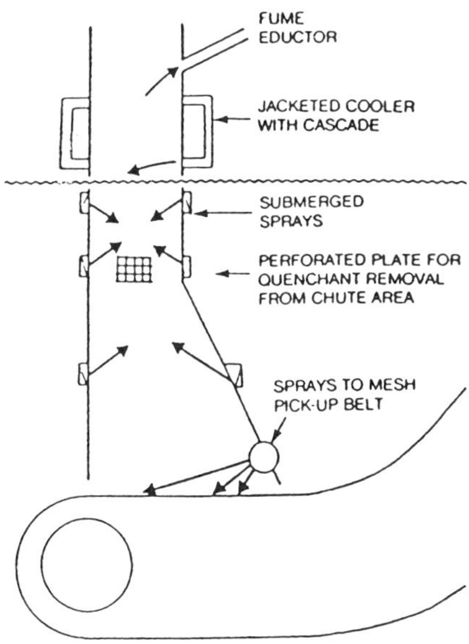

Fig. 9.56 Chute with spray quench system

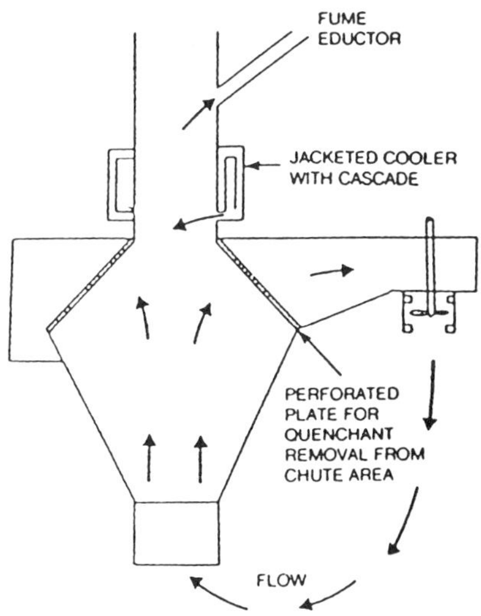

Fig. 9.57 Chute with impeller agitation system

A variety of chute quench agitation designs are possible. Two particularly successful designs are shown in Fig. 9.56 and 9.57. The following design criteria for both oil and aqueous polymer chute quench systems are recommended (Ref 36, 37, 49, 50):

- Adequate quenchant agitation and turnover in the chute zone are critically important to ensure uniform heat transfer from the part during the quench.
- The chute should have a cooling jacket above the quench zone to prevent oil or water vapor from entering the furnace vestibule. Quenchant can be used as the coolant, routed to the cooling jacket upon return from the heat exchanger.
- A fume eductor should be located in the chute area above the cooling jacket to ensure against vapor contamination of the furnace atmosphere.
- The chute area must have openings, preferably screened, to allow the heated quenchant to escape during the quench.
- A mesh belt of sufficient porosity and length should be used to allow for quenchant agitation around the part on the belt in the event that quenching is not completed in the chute zone.

Another recently patented design reportedly provides superior agitation in the chute quench zone (Ref 54). This system is shown in Fig. 9.58. Agitation in the chute zone is provided by rotation of the chute, which has blades attached. For effective circulation, the blades and ribs within the chute reportedly should be at an angle of 5° to 30° to the chute.

Heat Exchange. Because most of the heat removal occurs within the chute zone, and because the quenchant volume is relatively limited within this zone, adequate quenchant turnover for heat exchange during operation is critical. This is necessary to reduce fire potential and atmospheric contamination by oil quenchants and to reduce the potential for water vapor contamination from aqueous polymers.

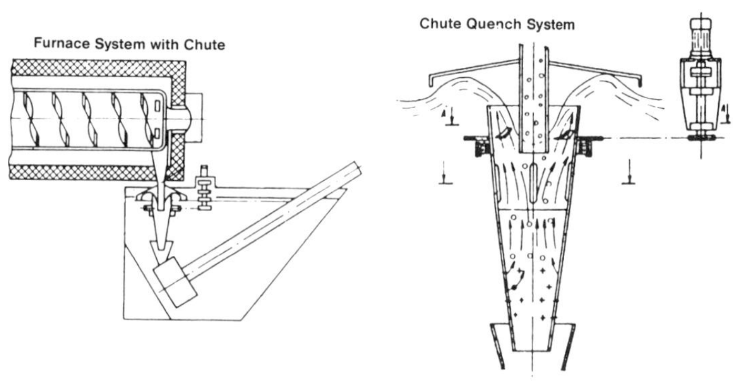

Fig. 9.58 Mechanically rotated chute quench (no pumps or impellers)

Thermal separation (see Fig. 6.24) may occur if the temperature of the quenchant in the chute zone exceeds the separation temperature of the polymer. This will produce a nonuniform quench, and cracking may result.

Continuously Variable Agitation

Agitation during the initial stages of the quenching process generally should be maximized to avoid pearlite formation and to optimize as-quenched hardness (Fig. 9.59) (Ref 55). However, a volumetric expansion accompanies martensite formation, producing increased quenching stresses and thus increased cracking potential. To reduce these stresses, cooling rates should be minimized in the region of the M_S temperature.

Therefore, it is generally desirable to enhance the cooling rates upon initial immersion into the quenchant and to minimize agitation as the M_S temperature is approached. This has recently been achieved using the "Immersion Time Quench System" (ITQS) (Ref 55, 56).

ITQS systems have been developed for both batch and continuous quenching processes. The batch quenching system is similar to that shown in Fig. 9.50. Either two continuously variable impeller stirrers encased in a directional draft-tube arrangement or pump agitation is typically used. The linear flow rate in the quench zone can be quantified using a velocimeter (Fig. 9.25). Typical calibration data are shown in Table 9.8.

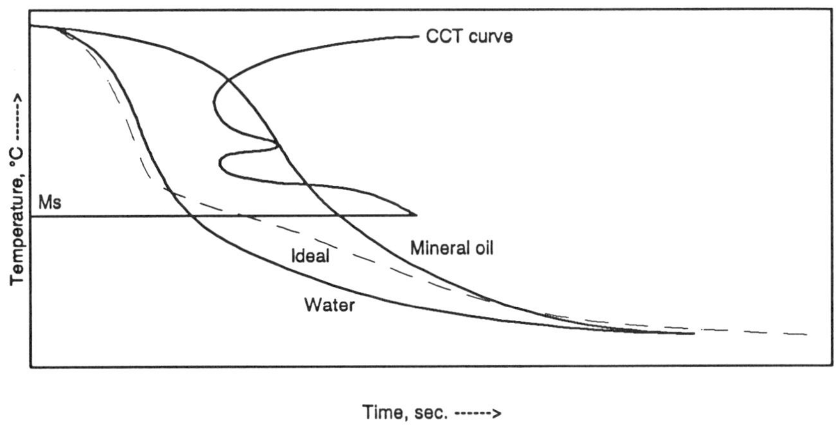

Fig. 9.59 "Ideal" cooling curve

The continuous quench system currently in use is illustrated in Fig. 9.60. The features of this system are:

- The upper conveyor is used during A- and B-stage cooling.
- Cooling rates are controlled on the upper conveyor by both agitation rate and quench time (conveyor speed).
- The lower conveyor is used for C-stage cooling.

Accurate determination of the cooling time for initial cooling (AR I) is critical to ensure the success of the system. Currently this is done by one of two methods. The first involves calculating the H factor to obtain the desired through-hardening and then quantifying the AR I cooling time from a plot that correlates H factor, workpiece

Table 9.8 Flow calibration for a batch ITQS system

Pump meter frequency, Hz	Flow rate m/s	Flow rate ft/s
10	0.03	0.1
20	0.25	0.8
30	0.40	1.3
40	0.50	1.6
50	0.60	2.0
60	0.70	2.3

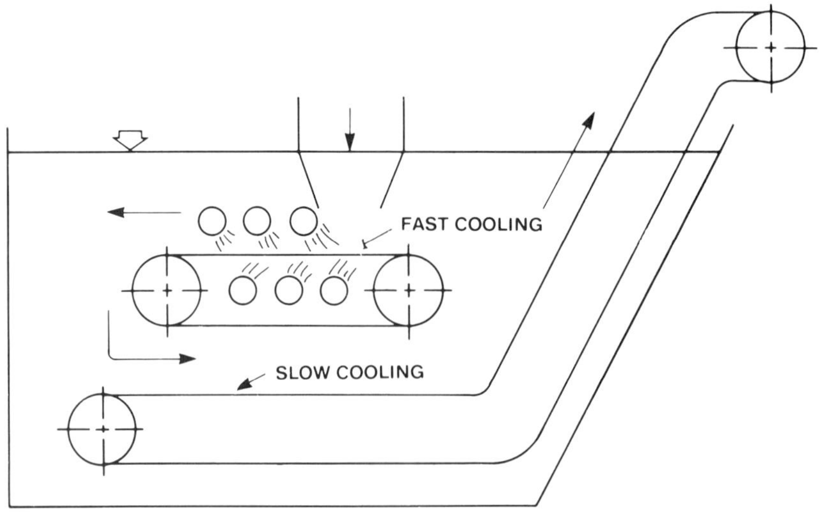

Fig. 9.60 Continuous ITQS system

diameter, and cooling rate (see Fig. 3.27). Once calculated, Table 9.9 is consulted to determine the amount of time for AR I. Experience has shown that the best flow rate in this region is approximately 1.5 m/s (5 ft/s).

Table 9.9 Grossmann *H* factor as a function of A-phase cooling time and cross-sectional size

Diameter		A-phase cooling time, s, for an *H* factor of:			
mm	in.	0.8	1.0	1.5	2.0
15	0.6	5 ± 1	4 ± 1	...	...
20	0.8	9 ± 2	7 ± 1	...	...
25	1.0	14 ± 3	10 ± 2	8 ± 2	...
30	1.2	17 ± 3	13 ± 2	10 ± 2	...
35	1.4	...	16 ± 3	12 ± 3	10 ± 3
40	1.6	...	18 ± 3	15 ± 3	12 ± 3
45	1.8	...	22 ± 4	19 ± 4	17 ± 4
50	2.0	...	26 ± 5	23 ± 5	20 ± 5

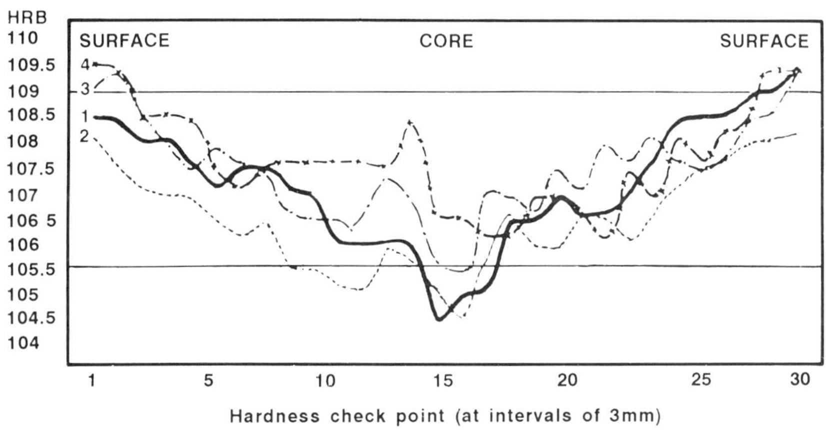

Test Number	Conc. (%)	Temp. (C)	Agitation	Bending (mm)
——— 1	Oil Quench 70	70	0.4 m/sec-Full-0.4m/sec	0.96
·········· 2	Poly(alkylene glycol) 20	45	0.55 m/sec-20 sec-0m/sec	0.4
—·—·— 3	Poly(alkylene glycol) 10	38	0.73 m/sec-60 sec-0.12 m/sec	0.6
-x-x-x-x 4	Poly(alkylene glycol) 7	39	0.73 m/sec-90 sec-0.12 m/sec	1.0

1. Total Quenching Time is 4 Minutes
2. Total Bending Distortion Limit is 1.2 mm

Fig. 9.61 Quenching results for AISI 4140H crankshafts quenched in an ITQS system

The second method to calculate AR I is to determine the initial cooling rate required to avoid undesired transformations and the temperature where this occurs for the steel grade of interest. Thus, it will be necessary to quench the part at a rate equal to, or greater than, the critical value if the maximum amount of martensite is to be obtained:

$$\text{AR I} = \frac{(T_A - T_{CR})}{\text{CR}} \qquad \text{(Eq 9.50)}$$

where T_A is the austenitizing temperature, T_{CR} is the temperature where the critical cooling rate of the quenching medium occurs, and CR is the critical cooling rate, which must be equal to or greater than the critical cooling rate of the process. The critical cooling rate on the temperature where this occurs is determined from the nose of the pearlite transformation curve located on the appropriate CCT diagram.

The ITQS procedure was used to quench AISI 4140H and 1043 steel crankshafts in a polymer quenchant. The hardness and distortion values obtained under various quenching conditions versus an oil quench are shown in Fig. 9.61 and 9.62. These data

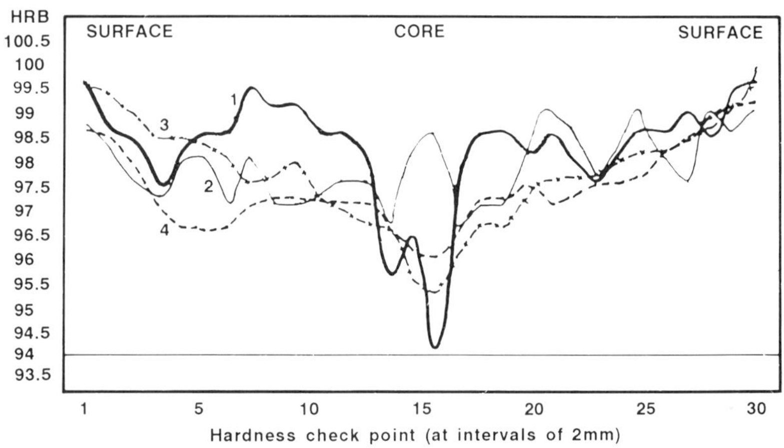

Test Number	Conc. (%)	Temp. (C)	Agitation	Bending (mm)
—— 1	Oil Quench	70	0.4 m/sec-Full-0.4 m/sec	0.6
—— 2	Poly(alkylene glycol) 10	43	0.73 m/sec-110 sec-0.12 m/sec	0.95
—·— 3	Poly(alkylene glycol) 15.75	30	0.73 m/sec-20 sec-0.12 m/sec	0.5
-x—x-4	Poly(alkylene glycol) 15	44	0.55 m/sec-20 sec-0.12 m/sec	0.5

1. Total Quenching Time is 4 Minutes
2. Total Bending Distortion Limit is 1.2 mm

Fig. 9.62 Quenching results for AISI 1043 crankshafts quenched in an ITQS system show that a substantial improvement in distortion reduction is possible with the use of an ITQS process and a polymer quenchant.

Heat Exchangers

When red-hot steel is quenched into a cooler-temperature medium, the heat in the steel will be lost to the medium, creating a temperature rise. The amount of temperature rise will depend on the volume of quenchant and the temperature difference between the hot steel and the quenchant. Temperature typically exhibits a substantial impact on quench severity; therefore, it is important to minimize the temperature increase during the quenching process.

The most important variable in controlling the temperature rise is the ratio of the amount of quenchant to the total weight of hot metal being quenched. Although heat exchangers may be used to cool the fluid, their use does not control the amount of temperature rise during the quench, because the initial amount of heat being lost during quenching will exceed the cooling efficiency of almost any heat exchanger. Therefore, the following points must be considered:

- The ratio of the mass of steel being quenched to the total quenchant volume, to estimate temperature rise.
- Heat-exchanger efficiency, to determine how quickly the temperature of the heated bath can be brought back to the initial, desired temperature.

The variables affecting quenchant temperature rise during the quenching process were discussed earlier. This section will discuss the use of heat exchangers to cool the hot quenchant.

Basic Heat-Exchange Principles*. The operating principle of a heat exchanger involves the transference of heat from the hot quenching fluid to a cooler medium (e.g. air, cold water, or another coolant) via a metal heat-exchanger surface. Figure 9.63 shows that heat exchange involves the transfer of heat from the bulk liquid to the metal surface through a surface boundary layer. The surface coefficient of heat transfer of the boundary layer on the hot side of the metal is written as:

$$H_1 = \frac{q}{A_1\,(T_1 - T_c)} \qquad \text{(Eq 9.51)}$$

where h_1 is the surface coefficient, q is the amount of heat transferred, A_1 is the area of the metal surface on the hot side, T_1 is the average bulk fluid temperature, and T_c is the

*This section was coauthored with Mr. J. Smith, president of SBS Corporation, Rochester, MI.

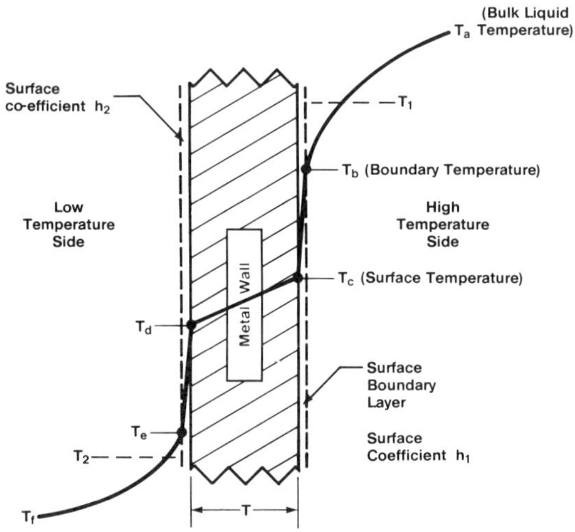

Fig. 9.63 Basic heat-exchange principles

temperature at the hot metal surface. A similar expression for the surface coefficient on the cold side can also be written.

The various surface coefficients for the system can be combined to yield:

$$q = U_1 A_1 \Delta T \quad \text{(Eq 9.52)}$$

Equation 9.52 states that the amount of heat transferred (q) will be determined by the:

- Overall heat transfer coefficient (U), which takes into account the thermal conductivity of the metal wall and the wall thickness
- Temperature drop (ΔT)
- Total area of the cooling surface (A_1)

Equation 9.52 describes the heat-transfer properties of the system. However, the performance of the system depends on such fluid-flow variables as pipe diameter, flow velocity, density, viscosity, thermal conductivity, and specific heat. In practice, the correlation of all of these variables is very complex; therefore, dimensional analysis is often used.

The traditional relationship used to describe heat transfer of a fluid flowing through a pipe is the Dittus-Boelter equation:

$$h = 0.023 \frac{k}{D} \left(\frac{DU\rho}{\mu} \right)^{0.8} \left(\frac{C\mu}{k} \right)^{0.4} \quad \text{(Eq 9.53)}$$

where k is the thermal conductivity of the metal and D is the diameter of the pipe. The term $DU\rho/\mu$ is the Reynolds number and $C\mu/k$ is the Prandtl number, both of which were discussed previously. This equation has been used for water, hydrocarbon oils, and other organic liquids for Reynolds numbers from 10,000 to 500,000 (turbulent flow) and Prandtl numbers from 0.73 to 0.95 (Ref 57).

Thus, heat transfer is also dependent on turbulence and the thermal properties of a fluid. Although turbulent flow is necessary to obtain effective heat transfer and to aid in cleaning the heat-exchange surface, excessive turbulence creates large pressure drops that are costly and inefficient.

The numerical equations describing specific heat-exchange systems in use today may be more complex; however, the fundamental principles of heat transfer involved are similar. An understanding of these relationships is necessary to understand, or troubleshoot, a heat-exchanger system. For example, consider the system shown in Fig. 9.64. In this case, the quenchant flows through a finned tube. It is clear that if the quenchant contains a contaminant, such as sludge, it will plate out on the heat-transfer surface, possibly reducing the total heat-exchange area (A_1) and producing an insulating boundary layer or fouling film (h) that will also reduce heat exchange. If the sludge buildup is sufficient, the fluid flow rate (U) will be decreased. All of these effects will combine to reduce heat-exchange efficiency. Thus, it is important that only clean fluid be circulated through the heat exchanger.

Heat-Exchange Selection. Three primary types of heat exchangers are used in quenching applications: mechanical refrigeration, shell-and-tube (evaporative water-cooled) spray towers, and air-cooled models (Ref 48). Examples of each type are shown in Fig. 9.65.

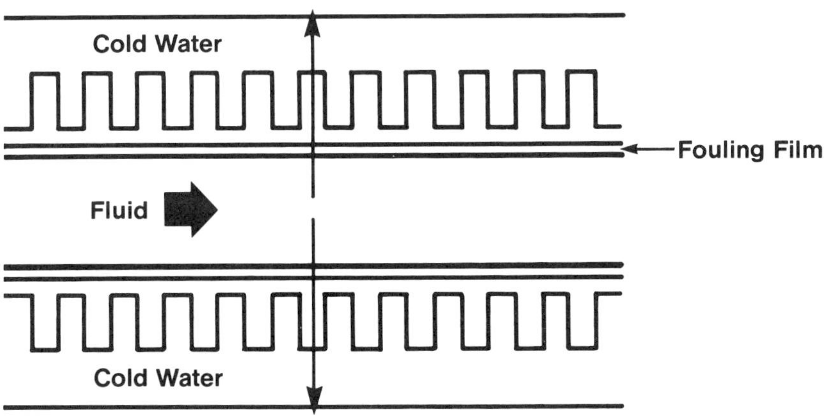

Fig. 9.64 Schematic of a shell-and-tube heat exchanger

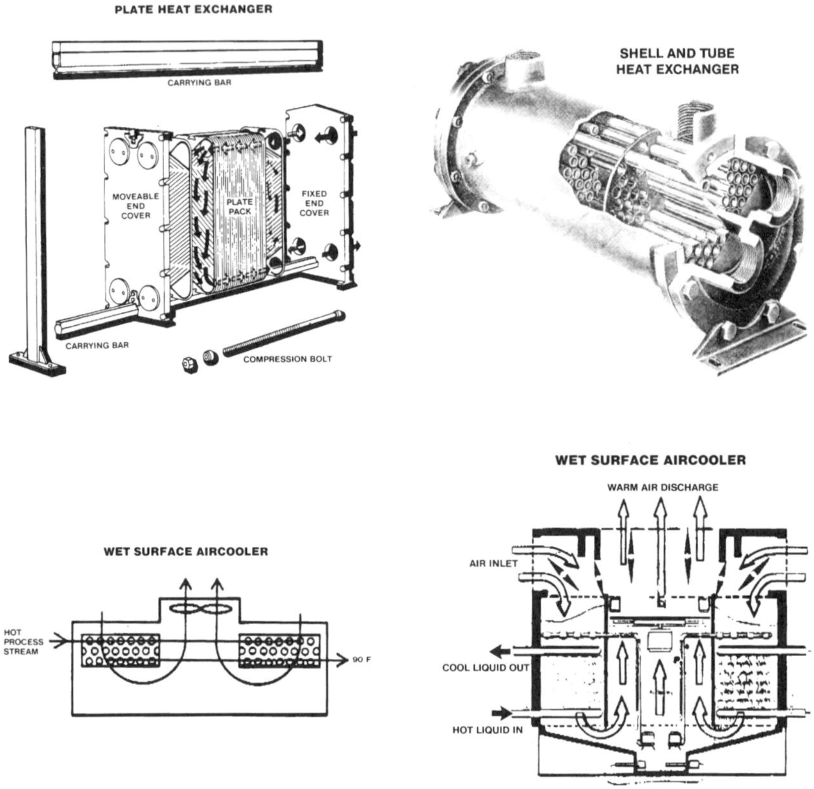

Fig. 9.65 Various types of heat exchangers currently in use

Table 9.10 Guidelines for heat-exchanger selection

Target fluid temperature range		
°C	°F	Heat-exchanger type
< 24	< 75	Mechanical refrigeration
35-45	95-110	Evaporative or water cooled
> 45	> 110	Air cooled

Selection of an appropriate heat exchanger depends on the desired final temperature of the quenchant. Selection guidelines are provided in Table 9.10.

There are two types of evaporative heat exchangers: shell-and-tube and plate-and-frame. The shell-and-tube heat exchanger typically contains a copper tube for the heat-exchange surface. This type of heat exchanger is subject to fouling and plugging. The plate-and-frame heat exchanger, which is primarily used in Europe, contains stainless steel plates as the heat-exchange surface.

Water contamination problems may occur with both types of evaporative exchangers. This is especially critical for oil quenchants and can be avoided by pressurizing the quench oil to prevent water from flowing into the oil. If a leak occurs, the water will go to the sewer. However, precautions must still be taken in the event that the oil pump used for pressurizing wears out. Therefore, it is recommended that a pressure-sensing device (scribe gage) be installed. Such gages have a typical acceptable pressure range of 140 to 280 kPa (20 to 40 psi).

Pumps and Pipe Requirements. Each individual heat-exchanger installation should be reviewed carefully, because it will often have a number of system features unique to the application. However, there are several standard guidelines.

Centrifugal pumps are generally recommended because of their excellent flow capacity at a given dynamic system head. Centrifugal pump selection was discussed earlier in this chapter.

Table 9.11 provides general guidelines for pipe sizing as a function of flow rate. It is essential that the piping not be undersize to produce more total dynamic head when the pump works on a different part of the efficiency curve. Adequate pipe sizing is critical for permitting the turbulent flow necessary to reduce system fouling potential. Also, if the pipe is undersize, turbulent flow will occur within the pipe itself.

Valve and Piping Materials. It is recommended that all pipes, valves, and fittings be constructed of ferrous metals in a heat-exchange system. Cast iron and bronze fittings often undergo corrosion in many quenchant systems because:

- Quenchants may absorb ammonia in some atmosphere heat treating processes.
- Aqueous polymer solutions are typically alkaline.
- Oil sludge formation can be catalyzed with copper in the presence of air.

In some lower-temperature applications, plastic pipe (e.g., PVC) may be used. However, the manufacturer should be consulted to ensure that the plastic is compatible with the quenchant.

Filters

Quench systems may contain solid contaminants, such as carbon, sludge, and scale, from numerous sources. The presence of solid contaminants in a quenchant may affect quenching behavior in various ways, depending on the type and amount. The concentration of these solid contaminants may vary over time; consequently, quenching performance will also vary over time. To achieve statistical process control of a quenching system, the quenchant must remain free of such contaminants (Ref 58).

In addition to its effect on quenching properties, the presence of solid contamination can affect pump and seal wear, heat-exchanger sludging, and so on. Therefore, good

Table 9.11 Recommended requirements for heat-exchange pipe sizing

Flow rate		Pipe size	
L/min	gal/min	mm	in.
190-340	50-90	50	2
340-680	90-180	63	2.5
680-950	180-250	75	3

Table 9.12 Dimensions of standard mesh screen sizes

Meshes per linear inch	U.S. sieve No.	Opening	
		in.	μm
52.36	50	0.0017	297
72.45	70	0.0083	210
101.01	100	0.0059	149
142.86	140	0.0041	105
200.00	200	0.0029	74
270.26	270	0.0021	53
323.00	325	0.0017	44

engineering practice and process control dictate the installation of quenchant filtration systems.

A solid contaminant can be removed from the oil by such methods as centrifugation and filtration. A centrifugal separator suitable for removing carbon and sludge from oil is shown in Fig. 6.6.

Of the various types of solid separation processes for liquid quenchants, the most common is filtration. Filters are divided into two classes. One classification is fixed-pore surface filters, such as mesh or screen filters. Table 9.12 summarizes common mesh filter sizes that are currently available.

The other classification is non-fixed-pore depth filters. These filters typically have higher capacity and enable lower particle size separation than mesh filters. These cartridge filters may be constructed from cotton, wool, cellulose, or synthetic fibers such as polyester, polyethylene, polypropylene, fiberglass, and teflon. (It should be emphasized that paper filters should never be used with aqueous quenchants.)

Filters can be operated in either a full flow or a proportional flow mode. Full flow means that all of the fluid being pumped will pass through the filter. Proportional flow means that only a portion of the stream being pumped will pass through the filter at any one time. However, because this is a continuous process, all of the fluid will be filtered on a time-averaged basis.

The most common rating system used for cartridge filters at present is the β-rating, which is a measure of the ratio of the number of particles greater than a specified size (× in μm) per

Table 9.13 Filter β-ratios and efficiency

β-ratio	Efficiency, %
1	0
2	50
20	95
50	98
100	99
1000	99.9
10,000	99.99

unit volume in the fluid upstream of the filter (N_u) to the number of particles of the same size downstream of the filter (N_d):

$$\beta_x = \frac{N_u}{N_d} \quad \text{(Eq 9.54)}$$

For example, consider that the number of particles 10 μm or greater upstream of the filter is 5000 and the number of the same size particles downstream of the filter is 50. In this case the β-ratio is equal to 5000/50, or 100.

The β-ratio can be converted to percentage filter efficiency (E_x):

$$E_x = \left[\frac{\beta_x - 1}{\beta_x}\right] \times 100\% \quad \text{(Eq 9.55)}$$

Thus, the efficiency is a measure of the ability of a filter to remove the number of particles greater than size x. A summary of β-ratios and efficiencies is provided in Table 9.13.

Filter selection involves two primary criteria: flow rate and fluid composition and temperature. The maximum flow rate and allowable pressure will dictate the filter element size that can be used. The particular filter selected must be compatible with the fluid type at the operating temperature.

Membrane Separation*

The use of membrane separation (molecular filtration) to selectively remove a polymer from aqueous solution has been long known. Croucher (Ref 59) and Foreman (Ref 60) have used membrane separation to vary the concentration of the polymer used to formulate an aqueous quenchant. However, until recently, the technology has not been developed sufficiently to facilitate widespread use in the heat treating industry.

*This section was coauthored with Mr. R. Howard, Marketing Manager of Despatch Industries Inc., Minneapolis, MN.

This section will provide a basic overview of membrane separation and the use of this technology to vary polymer concentration in the heat treating shop. It will also describe the related use of membrane technology for the removal of a polymer quenchant from rinse water.

Basic Overview of Membrane Separation Technology. Semipermeable membranes of known porosity can be used as filters to permit the separation of larger solute molecules from smaller solvent molecules. In the case of aqueous polymers, the larger polymer (P) molecules can be readily separated from the smaller water (W) molecules. A simplistic representation of this process is shown in Fig. 9.66. Either reverse osmosis or ultrafiltration is used for most aqueous polymer quenching solutions. The difference between these two techniques is the porosity of the membrane used for the separation (see Figure 6.22).

As discussed in Chapter 6, many aqueous polymer solutions contain a polymer, additive(s), and water. If the additive is also small enough to pass through the porous membrane, then the aqueous solution will be depleted of additives and concentrated with respect to the polymer. If this occurs, additives must be added again to replenish the solution when the concentrated polymer is diluted for reuse. However, both size separation and ionic rejection are possible. Many semipermeable membranes that will reject salts are available.

A typical membrane separation cartridge is shown in Fig. 9.67. In most large-scale industrial applications, the membrane is spirally wound around a permeate tube. (The permeate is the solution that passes through the membrane.) These membrane assemblies are encased inside a stainless steel cylindrical housing. Each housing may contain multiple membranes, each with permeate tubes connected together and sealed to the

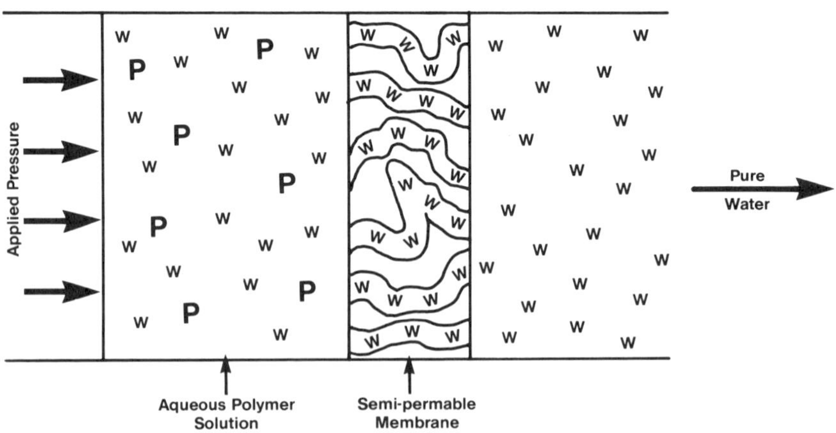

Fig. 9.66 Schematic of the use of semipermeable membranes to separate polymer from water

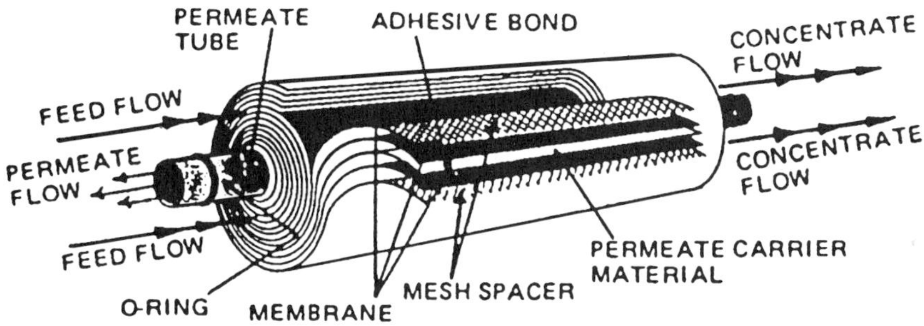

Fig. 9.67 Commercial membrane separation assembly. Courtesy of Osmonics, Inc.

housing with O-rings. The stainless steel housings may be connected ("piped") in series or in parallel, depending on the flow.

The feed solution (the solution to be separated) is pumped at high pressure, typically less than a maximum of 4100 to 4800 kPa (600 to 700 psi), into the membrane assembly by side inlets in order to increase turbulence. Turbulent flow in the membrane inlet is required to prevent plugging of the membrane surface by separated viscous polymer. The retained concentrate (the aqueous polymer that does not pass through the membrane) flows lengthwise through the space between the membranes and exits at the end of the tube opposite from the feed inlet. The solution (permeate) is passed through the membrane into the permeate tube and is discharged from a port located on the centerline of the assembly.

The feed solution is introduced into the high-pressure separation pump by a constant-flow process pump that is sized to satisfy the feed flow rate at an inlet pressure of 140 to 420 kPa (20 to 60 psi). The feed solution passes through a replaceable 5 μm cartridge filter that removes large suspended solids. The filtered solution then flows to an inlet control valve. This solenoid or solenoid-controlled diaphragm valve opens when the system is turned on, allowing the quenchant to flow to the pump inlet. When the system is turned off, the inlet valve closes, preventing nonturbulent flow through the membranes when the pump is not operating.

The high-pressure separation pump feeds PAG solution to the membrane housings at 3100 to 4500 kPa (450 to 650 psi) in either a series or a parallel configuration, depending on the design flow rates. The water is separated from the PAG feed stream and leaves the membrane array as the permeate stream. PAG in the concentrate stream is retained.

Permeate from each membrane housing is collected at a permeate manifold, where a check valve prevents backflow into the membranes. A pressure-relief valve prevents excessive back pressure. The permeate then flows through a flowmeter and a conductivity monitor to the permeate drain outlet on the machine. The concentrate exits the last membrane housing into a concentrate manifold and through a flowmeter before passing through a valve, back to the process container.

Maximum effective polymer reconcentration is in the 55 to 70% range. At 55% concentration, an effective separation pressure of approximately 4500 kPa (650 psi) should be expected. Cellulose acetate membranes, which have been found to be suitable for the separation of PAG quenchants, have a thermal limitation of 40 °C (105 °F). Significant heat is generated, particularly at higher concentrations, because of the viscosity of the feed solution; therefore, a heat exchanger in the feed loop is typically required. At the conclusion of a separation cycle, the membranes should be forward flushed for 3 to 5 min with pure water to remove the gelatinous layer of concentrated PAG from the membranes. The membranes should be flushed periodically with a diluted concentration of phosphoric acid to dissolve and remove calcified deposits that have been trapped in the membranes. When these procedures are followed, extended membrane life can be expected.

Time required for off-line reconcentration of the polymer used to formulate the quenchant removed is a function of the membrane area, which determines permeate flow rate over the range of reconcentration. Solving the equations for volumes of quenchant to be removed can be done manually, but is ideally performed by a PLC (Programmed Logic Controllers) or computer-based controller. Such a controller also performs the transfers, and alarms of nonpermissible conditions. Measurement of actual quenchant concentration can be performed using a hand-held refractometer; however, automated polymer concentration control systems utilize either density- or viscosity-based monitors that provide analog outputs to the system controller and display. Typical density-based concentration monitors provide a measurement accuracy of 0.0001 specific gravity units. Density-based monitors do not distinguish polymer in solution from dissolved salts; therefore, the concentration signal will be distorted by the percentage of salts by weight in solution.

Reconcentration of the removed quenchant to storable concentration is performed off-line in a process tank. Determination as to when sufficient water has been removed

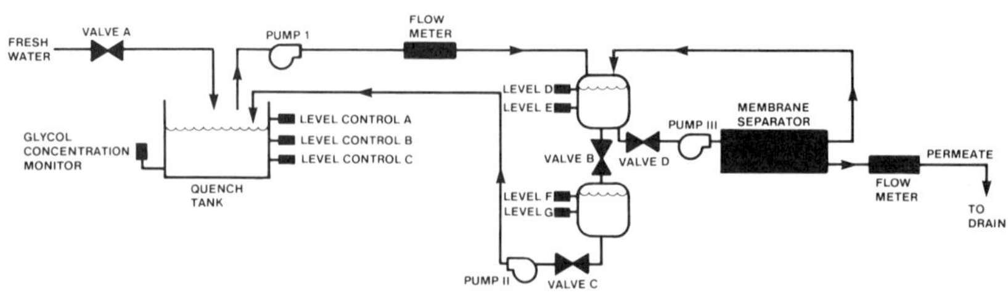

Fig. 9.68 Schematic of polymer concentration control by membrane separation. Courtesy of Despatch Industries

from the quenchant in the process tank is achieved either by measuring total permeate flow or by monitoring process tank concentration with a density- or viscosity-based concentration monitor. Capacities of process and storage tanks are functions of quench tank volume and polymer concentration range. Typical polymer concentration ranges accurately maintained and controlled in the quench tank are 10 to 40%. Figure 9.68 shows a schematic of polymer concentration control.

Definition of Terms. The concentration of the total polymer in aqueous solution is typically expressed as a percentage. The polymer concentrations in the feed, permeate, initial concentrate (as received from the supplier), and average concentration are defined as:

- C_f, feed concentration
- C_p, permeate concentration
- C_c, concentrate concentration
- C_{avg}, average concentration

The concentration ratio of polymer (low molecular mixture fraction) that will pass through the membrane during separation is referred to as "passage" and is defined as:

$$\text{Passage} = \frac{C_p}{C_{avg}} \qquad \text{(Eq 9.56)}$$

The concentration ratio of the polymer that does not pass through the membrane is referred to as "retention" and is defined as:

$$\text{Retention} = \frac{C_s}{C_{avg}} \qquad \text{(Eq 9.57)}$$

The average concentration is calculated from:

$$C_{avg} = \frac{C_f + C_c}{2} \qquad \text{(Eq 9.58)}$$

Salt (ionic) rejection equals the percentage of dissolved salt rejected by the membrane calculated on an average concentration over the membrane. *Salt (ionic) passage* equals 100 minus the percentage of rejection, or the percentage of dissolved salts passed through the membrane.

Salts (calcium, magnesium, sodium, etc.) are introduced into the quench through process salts (salt baths, for example) or from salts in the makeup water. Increasing salt concentration increases the chance for calcium carbonate precipitation and permeate flow loss. As water is separated from the process volume, feed concentration (C_f) increases, concentrate concentration (C_c) increases, permeate concentration (C_p) remains essentially zero, permeate flow rate decreases asymptotically, feed flow rate decreases proportionate to the flow/pressure curve of the pump supplying the membrane array, salt passage as a percentage increases, and the effective pressure required to separate the water from the PAG increases. Figure 9.69 shows representative curves.

Permeate rate (separation water rate) (Q_p) equals the flow rate of purified water that is passed through the membrane. Specified permeate rates are normally at 25 °C (77 °F). *Concentrate rate* (Q_c) equals the flow rate of retained PAG solution. *Feed rate* (Q_f) equals the flow rate of incoming quenchant, or permeate rate plus concentration rate. All of these rates are expressed in liters or gallons per minute.

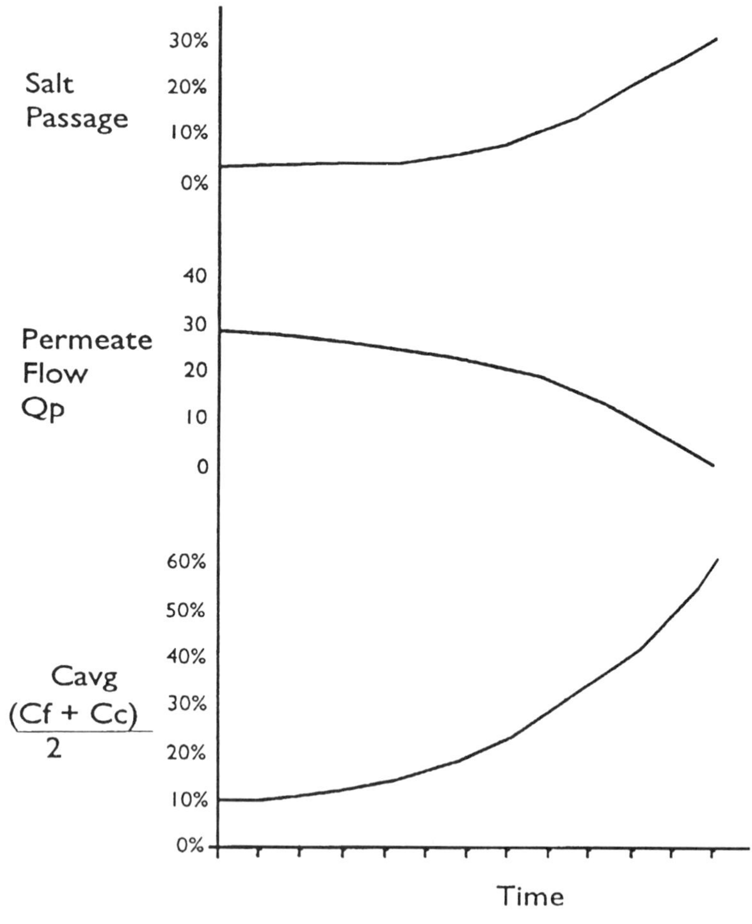

Fig. 9.69 Concentration/flow curves for membrane separation of a PAG copolymer. Courtesy of Despatch Industries

Recovery equals permeate rate divided by feed rate and is expressed as a percentage:

$$\text{Recovery} = \frac{Q_p}{Q_f} \times 100 \quad \text{(Eq 9.59)}$$

For example, 60% recovery means that of a given feed rate, 60% is separated out as pure water.

Membrane Separation for Quenchant Concentration Control. Polymer quenchant concentration control using membrane separation technology is fundamentally a process of removing a calculated volume of a known concentration of polymer and replacing that volume with a known higher concentration for concentration increase, or with the same volume of water for concentration decrease. The original volume of polymer quenchant removed to allow for the concentration change is then reconcentrated to a higher concentration and stored for subsequent quenchant concentration increases. The separation of water from the removed polymer quenchant and subsequent storage of the reconcentrated higher concentration of polymer is typically performed off-line from the quench process in order to allow for rapid quench tank concentration changes. An equation for the volume of quenchant to be removed and processed is as follows:

$$\text{Gallons of quenchant to be removed} = \frac{Q_{wlc} - Q_{whc}}{\%\ \text{polymer added} - \%\ \text{polymer removed}} \quad \text{(Eq 9.60)}$$

where Q_{wlc} is gallons of water at lower concentration and Q_{whc} is gallons of water at higher concentration.

As an example, consider an increase in polymer concentration in a 10,000 gal quench tank from 10% polymer to 20% polymer concentration by replacing a calculated volume of 10% polymer with 55% polymer. The results are shown in Table 9.14. Using Eq 9.61 and data from Table 9.14, the volume of 10% polymer quenchant to be removed and replaced with 55% quenchant concentrate is:

$$\frac{9000\ \text{gal} - 8000\ \text{gal}}{0.55 - 0.10} = 2222\ \text{gal} \quad \text{(Eq 9.61)}$$

Table 9.14 Example of polymer separation stoichiometry for quenchant application

	Volume, gal		
Polymer concentration, %	**Polymer**	**Water**	**Total**
10	1000	9000	10,000
20	2000	8000	10,000

Calculation of the amount of polymer quenchant to be removed to produce a desired lower concentration solution is similar. The volume of quenchant to be removed can be calculated from:

$$\text{Volume of quenchant to be removed} = \frac{Q_{\text{chc}} - Q_{\text{clc}}}{\%\ \text{polymer removed}} \quad \text{(Eq 9.62)}$$

where Q_{chc} is the volume of quenchant at higher concentration and Q_{clc} is the volume of quenchant at lower concentration.

For example, consider the decrease of a polymer quenchant concentration in a 10,000 gal tank from 20 to 10% by replacement with a calculated volume of 20% polymer concentration. Solution of Eq 9.63 provides the volume of 20% quenchant concentration to be replaced by water:

$$\frac{2000\ \text{gal} - 1000\ \text{gal}}{0.20} = 5000\ \text{gal} \quad \text{(Eq 9.63)}$$

Polymer Reclamation From Rinse Water. A polymer quenchant that has been transferred from the quench tank and collected in the rinse water can also be reclaimed by the membrane filtration method. Typically, rinse water should contain no more than 1% polymer (for PAG) in order to be efficient.

Reclamation of polymer from the rinse water is generally achieved by one of two methods. On-line separation involves pumping the rinse water on demand through the membrane separation system. The permeate stream is returned as clean water to the rinse tank, and the concentrate stream is piped to the quench tank as dilute polymer. This method is the simplest, but the elimination of off-line process and storage tanks limits the concentration stream that is being sent to the quench tank to the single-pass recovery capacity of the membrane array. As the separation process progresses and the percentage of polymer decreases in the feed stream from the rinse tank, the concentration of the polymer in the concentrate stream decreases proportionately. Eventually, the concentrate is too diluted to be used in the quench tank. As the permeate is reclaimed in this method, dissolved salts introduced into the quench or rinse water through the process or through ionized makeup water are retained in the quench and rinse water. This problem can be solved by using deionized makeup water or by installing a reverse osmosis water purifier in the makeup water inlet. Figure 9.70 shows a schematic of an on-line rinse water polymer reclamation system.

The second method for polymer reclamation from rinse water utilizes an off-line process tank. Volume of the rinse water is pumped on demand into the process tank and then reconcentrated to any usable concentration. The permeate can be pumped either to the rinse tank for reuse or to drain. The concentrate in the process tank can be pumped either to the quench tank or to storage. As in the on-line system, if the rinse water is reclaimed, dissolved salts will be retained, and some form of deionized makeup water is recommended. Figure 9.71 shows a schematic of an off-line rinse water polymer reclamation system.

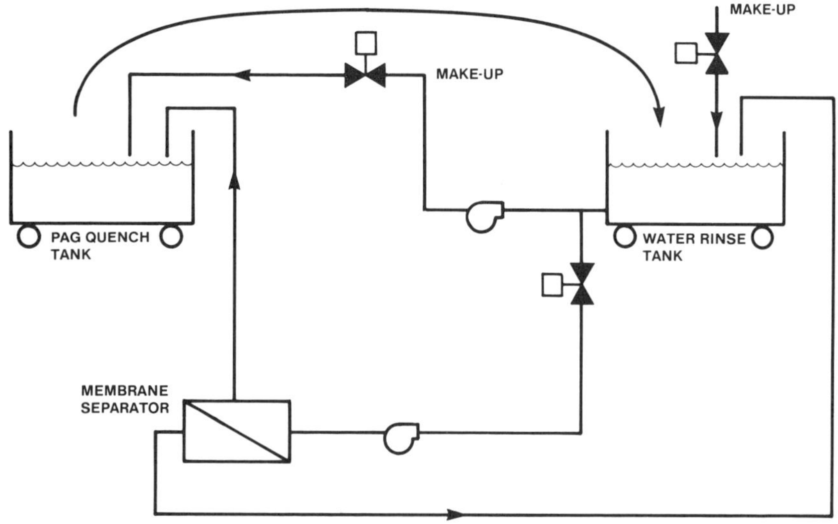

Fig. 9.70 Membrane separation of a polymer quenchant from rinse water (on-line system). Courtesy of Despatch Industries

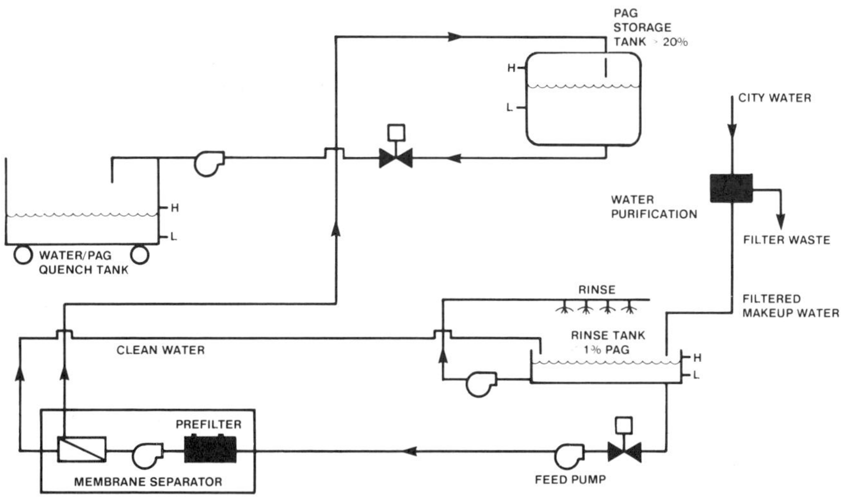

Fig. 9.71 Membrane separation of a polymer quenchant from rinse water (off-line system). Courtesy of Despatch Industries

References

1. B. Liscic, *18th Int. Conf. Heat Treatment of Materials*, ASM Internatiional, 6-8 May 1980, p 51-63
2. D.C. Kweon, C.P. Hong, and M.I. Kim, *J. Korean Inst. Met.*, Vol 26 (No. 3), 1988, p 242-250
3. F. Vivas and J.A. Tardio, *Deform. Met.*, Vol 111, 1985, p 31-47
4. F. Moreaux, G. Beck, B. Clement, J. Olivier, and A. Mulat, *Trait. Therm.*, Vol 225, 1989, p 45-49
5. F. Moreaux, G. Beck, B. Clement, and J. Olivier, *Heat Treat. Surf. Eng.: New Technology and Practical Applications*, 28-30 Sept 1988, p 167-170
6. H.R. Bergmann, *Met. Eng. Q.*, Vol 11 (No. 2), 1971, p 17-19
7. H.R. Bergmann, ASM Tech. Report No. C70-32.1, 19-22 Oct 1970
8. B.D. Wakefield, *Iron Age*, Oct 1970, p 54-55
9. *Ind. Heat.*, Jan 1979, p 14-17
10. U.S. Steel, "Improved Quenching of Steel by Propeller Agitation," 1954
11. N.N. Khavskii and R.R. Zelokhovtseva, *Izv. V.U.Z. Chernaya Metall.*, Vol 3, 1982, p 111-113
12. "Laboratory Test for Assessing the Cooling Curve Characteristics of Industrial Quenching Media," Wolfson Heat Treatment Centre Engineering Group Specifications, 1982. This procedure has been drafted into an ISO specification (ISO/DIS 9950), which is currently in the approval process.
13. D. Close, *Heat Treat. Met.*, Vol 11 (No. 1), 1984, p 1-6
14. R.W. Heins and E.R. Mueller, *Met. Prog.*, Vol 122 (No. 4), 1982, p 33-39
15. N.A. Hilder, *Heat Treat. Met.*, Vol 12 (No. 3), 1985, p 63-68
16. N.A. Hilder, *Heat Treat. Met.*, Vol 14 (No. 2), 1987, p 31-46
17. K.J. Mason and I. Caperwell, *Heat Treat. Met.*, Vol 13 (No. 4), 1986, p 99-103
18. H.M. Tensi and P. Stitzelberger-Jakob, *Mat. Sci. Technol.*, Vol 5, 1989, p 718-724
19. S. Segerberg, Swedish Institute of Production and Engineering Research, personal communication, 1991. Mr. Segerberg reported relatively little, if any, cooling rate variation with respect to probe placement in the quench chamber of this apparatus.
20. J.Y. Oldshue, *Fluid Mixing Technology*, McGraw-Hill, 1983, p 51-53
21. N.A. Hilder, Ph.D. thesis, University of Birmingham, UK, 1988
22. Y.I. Lipunov, K.Y. Eismondt, A.Y. Sitchenko, and A.F. Goncharov, *Frieb. Forschungsh. B, Metall. Werkstofftech.*, Vol B266, 1988, p 37-40
23. A. Rose, *Arch. Eisenhüttenwes.*, Vol 13, 1940, p 345-354
24. K. Speith and H. Lange, *Mitt. Kaiser Wilhelm Inst. Eisenforschung*, Vol 17, 1935, p 175-184
25. R.H. Atkin, A.J. Kemp, and D.G. Paddle, *Metals in the Marketplace*, No. 3, 1978, p 30-37
26. R.H. Atkin, A.J. Kemp, and D.G. Paddle, *Met. Aust.*, Vol 12 (No. 8), 1980, p 12-17
27. S. Segerberg, *Heat Treat.*, May 1988, p 26-28
28. J.Y. Oldshue, 5th European Conference on Mixing, Wurtzburg, West Germany, 10-12 June 1985, p 35-52
29. B.J. Hutchings, R.J. Weetman, and B.R. Patel, ASME Winter Annual Meeting, 1989, p 129-135
30. J.Y. Oldshue, *Fluid Mixing Technology*, McGraw-Hill, 1983, p 162-168
31. C.E. Bates, G.E. Totten, and R.L. Brennan, *ASM Handbook*, Vol 4, *Heat Treating*, ASM International, 1991, p 67-120

32. R.B. Dowdell, Ed., Flow, Its Measurement and Control, in *Science and Industry*, Vol 2, Instrument Society of America, 1974, p 727-733
33. H. Webster and W.J. Laird, *ASM Handbook*, Vol 4, *Heat Treating*, ASM International, 1991, p 67-120
34. A. Terrier, *Trait. Therm.*, Vol 4, 1976, p 3-14
35. C.E. Bates, *J. Heat Treat.*, Vol 6, 1988, p 27-45
36. J.J. Lakin, *Heat Treat. Met.*, Vol 3, 1982, p 73-76
37. G.E. Totten, K.B. Orszak, L.M. Jarvis, and R.R. Blackwood, *Ind. Heat.*, Oct 1991, p 37-41
38. I.J. Karassik, W.C. Krutzch, W.H. Fraser, and J.P. Messina, *Pump Handbook*, 2nd ed., McGraw-Hill, 1985
39. E.E. Ludwig, *Applied Process Design for Chemical and Petrochemical Plants*, 2nd ed., Vol 1, Gulf Publishing, 1984
40. *Goulds Pump Manual—GPMS*. A useful and practical guide on a complete range of pumps. More than 100 pages on calculation procedures, pump maintenance, and engineering tables and figures. Obtain from: Goulds Pumps, Inc., Industrial Products Group, Seneca Falls, New York, 13148
41. G.B. Tatterson, *Fluid Mixing and Gas Dispersion in Agitated Tanks*, McGraw-Hill, 1991
42. H. Fossett and L.E. Prosser, *Proc. Inst. Mech. Eng.*, Vol 160, 1949, p 224-251
43. R.G. Folsom and C.K. Ferguson, *Trans. ASME*, Vol 71, 1949, p 73-77
44. J.H. Rushton, *Petrol. Refin.*, Aug 1954, p 101-107
45. H.L. Kauffman, *Chem. Metall. Eng.*, Vol 37, 1930, p 177-180
46. V.G. Stognei and A.T. Kruk, *Met. Sci. Heat Treat.*, Vol 31 (No. 1-2), 1989, p 69-71
47. J. Olivier, B. Clement, J.J. Debrie, and F. Moreaux, *Trait. Therm.*, Vol 206, 1986, p 29-42
48. H.E. Boyer and P.R. Cary, *Quenching and Control of Distortion*, ASM International, 1988
49. "The Principle of Heat Extraction With the Use of Liquid Single Organic Polymer," technical bulletin, Tenaxol Inc.
50. A.J. Beck, *Heat Treat.*, May 1977, p 55-60
51. D. Diaz, H. Garcia, and B. Bautista, *Proc. 1st Int. Conf. Quenching and the Control of Distortion*, Chicago, Sept 1992, in press
52. R. Jeschar, R. Scholz, U. Reiners, and R. Maass, *Stahl. Eisen*, Vol 107 (No. 6), 1987, p 251-258
53. K. Illgner, *Härt.-Tech. Mitt.*, Vol 42 (No. 2), 1987, p 113-120
54. A.P. Petrukhin, U.S.S.R. Patent 1,247,424, 30 July 1986
55. S.G. Yun, S.W. Han, and G.E. Totten, *Proc. 1st Int. Conf. Quenching and the Control of Distortion*, Chicago, Sept 1992, in press
56. The only supplier for this equipment at the present time is Sam Won Industries Company Ltd., 5-Ho Shinchang Agricultural & Industrial Complex, Shinchang-Myan, Shson-Kun, Chung-Nam, T377-880, Korea (FAX 82-418-44-7509)
57. W.L. Badger and J.T. Banchero, *Introduction to Chemical Engineering*, McGraw-Hill, 1955, p 124-152
58. V. Srimongkolkul, *Heat Treat.*, Dec 1990, p 27-28
59a. T. Croucher, *Heat Treating*, Sept 1984, p 27-30
59b. K. Anderson, T. Croucher and D. Butler, Metals Progress, May 1983, p 37-40
60. R.W. Foreman and A.G. Meszaros, U.S. Patent, 4,251,292, Feb 1981
61. I. von Hagen, W. Nickel, and C. Prasser, U.S. Patent 4,900,376, 13 Feb 1990

CHAPTER 10

Impeller Agitation*

Quenching is an essential element in developing the desired properties of many steel and aluminum alloys. Agitation, or forced circulation of the quenchant, is required to shorten the cooling times. Without agitation, natural convection of the quenchant and quenchant vaporization limit the heat-transfer rate through the fluid film boundary at the surfaces of parts. Under these conditions, changes in the quenchant medium temperature have little effect on the cooling rate. Obtaining a forced convection fluid regime greatly reduces the resistance to heat flow at the fluid film boundary layer. This can be accomplished by mechanically moving the parts through the bath, pumping to recirculate the quenchant, or mechanically inducing agitation/circulation of the fluid. Where control over the cooling rate is important, mechanical agitation provides the best performance at the lowest energy costs. This chapter will describe mechanical agitation with impeller mixers, principles of agitation, and practical recommendations for mixer selection.

Effects of Quenchant Circulation

Three regimes of cooling occur at the surfaces of parts during quenching (see Fig. 5.11). When the hot metal is initially immersed in the quenchant, a vapor blanket forms around the part. This is designated as A-phase cooling; heat-transfer rates are typically slowest in this region. The duration and stability of the A-phase cooling process are dramatically affected by agitation. Generally, A-phase cooling times decrease as agitation rates increase.

The second phase of cooling is called B-stage cooling and is characterized by nucleate boiling; heat-transfer rates are the highest in this region. Further increases in heat transfer during B-stage cooling can be achieved by corresponding increases in agitation rates.

The third stage of cooling, stage C, occurs when the bath temperature is below the boiling point of the quenchant. Heat transfer during C-stage cooling occurs by convective and conductive processes.

Convective heat transfer may be either forced or natural. Forced convective heat transfer occurs when an external force, such as agitation, enhances liquid renewal at the

*This chapter was coauthored with K.S. Lally, manager of the research and development group at Lightnin—A Unit of General Signal, Rochester, New York.

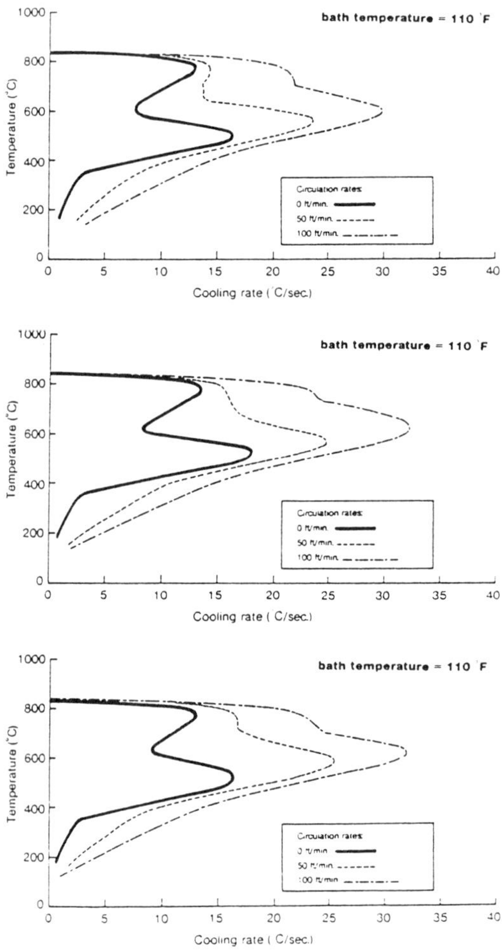

Fig. 10.1 Effect of agitation on the quenching rates of a typical quench oil

cooling surface. Natural convective heat transfer relies on density differences due to thermal gradients to cause liquid surface renewal. Heat-transfer rates are typically higher for the forced convective process. Cooling rates increase with increasing agitation rates.

The effects of agitation on the quenching rates of a typical quench oil are shown in Fig. 10.1 and tabulated in Table 10.1. These results show that:

- Increased agitation rates decrease A-stage cooling rates.
- Increased agitation rates increase B-stage cooling rates and decrease the total time to cool to the martensitic transition temperature for most steels.
- Increased agitation rates increase C-stage cooling rates.

Table 10.1 Effect of agitation on quench oil cooling performance(a)

Bath temperature		Circulation rate		Maximum cooling rate		Temperature at maximum cooling rate		Time to cool from 730-260 °C (1350-500 °F), s	Cooling rate, °C/s (°F/s), at:		
°C	°F	L/min	ft^3/min	°C/s	°F/s	°C	°F		705 °C (1300 °F)	345 °C (650 °F)	230 °C (450 °F)
45	110	0	0	18.4	33.1	519	966	79.8	10.4 (18.7)	3.0 (5.4)	1.6 (2.9)
		1415	50	24.4	44.0	567	1056	39.6	13.9 (25.0)	7.8 (14.0)	4.3 (7.8)
		2830	100	30.4	54.8	600	1110	27.6	22.4 (40.4)	11.6 (20.9)	6.7 (12.1)
65	150	0	0	18.1	32.7	514	958	79.5	11.3 (20.3)	2.8 (5.1)	1.4 (2.6)
		1415	50	25.5	46.0	563	1046	36.2	16.4 (29.5)	8.6 (15.5)	4.4 (8.0)
		2830	100	32.7	58.9	632	1169	26.3	26.8 (48.3)	11.9 (21.4)	6.5 (11.7)
90	190	0	0	18.5	33.4	523	973	84.5	11.2 (20.1)	2.7 (4.8)	1.2 (2.2)
		1415	50	27.0	48.7	578	1072	39.0	17.0 (30.6)	7.2 (13.0)	3.7 (6.7)
		2830	100	32.7	58.9	623	1153	27.9	24.4 (44.0)	11.2 (20.2)	5.6 (10.1)

(a) Quenching perfomance of a "slow" quench oil. All data are based on cooling curves obtained with a 1 × 4 in. cylindrical stainless steel probe instrumented with a Type K thermocouple. Agitation was provided by apparatus shown in Fig. 3-21. The flow by the probe was linear and turbulent.

Agitation increases heat-transfer rates in three ways:

- Agitation minimizes the conditions under which liquid vaporization occurs, because the liquid moves past a hot surface faster and with a greater volumetric flow rate. The result is a smaller temperature rise in the liquid and thus a higher likelihood of the liquid remaining below its vaporization temperature.
- In the C-stage cooling regime, forced convection is achieved throughout the quench bath. Significantly greater rates of cooling, compared with an unagitated system, are achieved.
- Agitation allows better control of the cooling rate of the parts throughout the load during the quench, because temperature uniformity is maintained throughout the bath. Control of the cooling rate is essential for optimizing the quenching process, especially for lower-hardenability steels.

The first two benefits are relatively straightforward; quenching process optimization, however, requires further discussion.

The primary function of a quenchant is to reduce the propensity for distortion and cracking. Cracking potential generally decreases with decreasing cooling rates (especially at or near the martensitic transformation temperature) and with increased uniformity of quenchant wetting on the metal surface. Although decreased cooling rates are desirable, it is essential that the cooling rates of stages A and B be sufficiently fast to avoid the pearlite transformation region of the cooling process (see Fig. 4.15) and thus achieve optimal hardness.

Optimal quenching processes usually require maximum A- and B-stage cooling rates and minimal C-stage cooling rates. The impact of agitation on the cooling rates of various oil and polymer quenchants has been detailed in Chapters 4 and 5. The objective here is to describe design criteria that can be used to construct impeller agitation systems.

Introduction to Mixing Impellers

A mixer follows many of the same operational principles as a pump. The power delivered by the mixer can be analyzed as the product of flow and head:

$$P = Q \times H \quad \text{(Eq 10.1)}$$

where P is the power input of the impeller, Q is the flow rate or pumping capacity, and H is velocity head.

A mixer is used to provide fluid motion and shear. Flow-controlled mixing is strongly dependent on the bulk turnover produced by the impeller. Mixing applications, such as blending, solids suspension, and heat transfer, are strongly related to impeller flow.

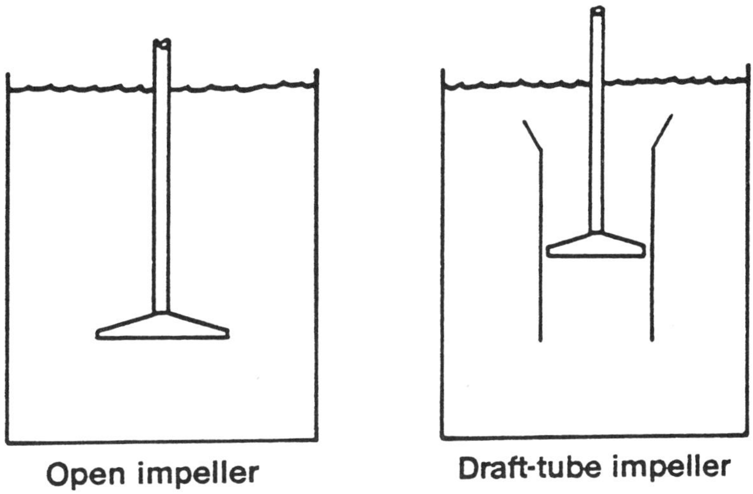

Fig. 10.2 Open-impeller and draft-tube impeller mixers

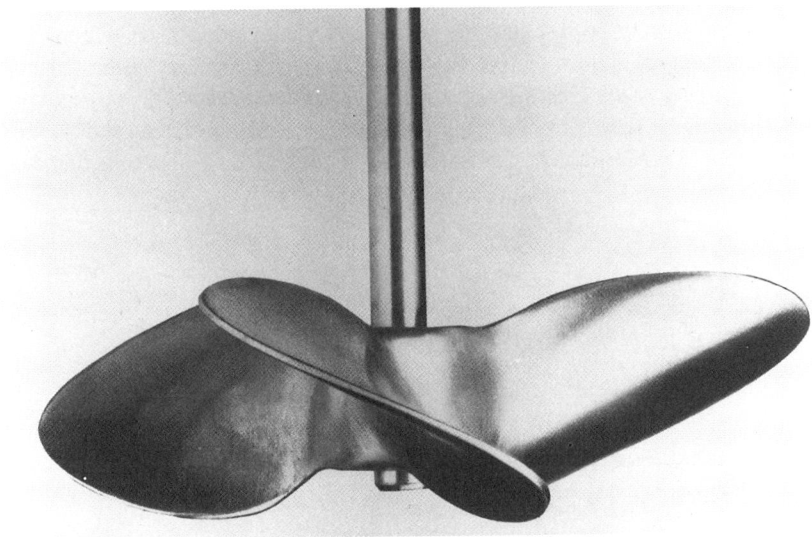

Fig. 10.3 Marine propeller (1.0 pitch ratio). Courtesy of Lightnin—A Unit of General Signal

Mixing impellers used for quenching operations are either "open-impeller" systems or "draft-tube" systems. In open-impeller systems, the mixing impeller does not have a flow-directing surface encasing the impeller (Fig. 10.2). This system relies on the characteristics of the impeller itself to direct flow into the desired region of the tank. In addition, the direct impeller flow induces a significant entrained flow pattern that generally ensures circulation throughout the tank volume.

A commonly used type of open impeller involves axial flow; an example is the marine propeller shown in Fig. 10.3. (Note: The term "impeller" is generic. "Propeller" refers only to a marine-type impeller, although the term is used outside the mixing industry to refer to all types of mixing impellers.) In an axial-flow impeller, the principal direction of flow occurs along the axis of the impeller parallel to the impeller shaft. Axial-flow impellers can be used on top-entering, side-entering, or angled top-entering mixers (Fig. 10.4).

Axial-flow impellers are also used in draft tubes to more closely control the direction of the flow pattern. In this system, the resistance head against which the impeller must pump is higher because of the fluid friction losses in the draft-tube pipe. Velocity-head losses also occur as a result of entrance, exit, and turning losses in the draft tube. The higher head conditions require a slightly different impeller for optimum performance.

Open-Impeller Systems

The first mixers to be installed in quench tanks used three-blade marine-type propellers, with a typical pitch ratio of 1.0 (Fig. 10.3). These impellers were used with either side-entering or top-entering angular-mounted mixers, depending on tank configuration and mixer size. Mixer size and selection guidelines have been documented previously in a U.S. Steel article published in 1954 (Ref 1). Since that time, impeller

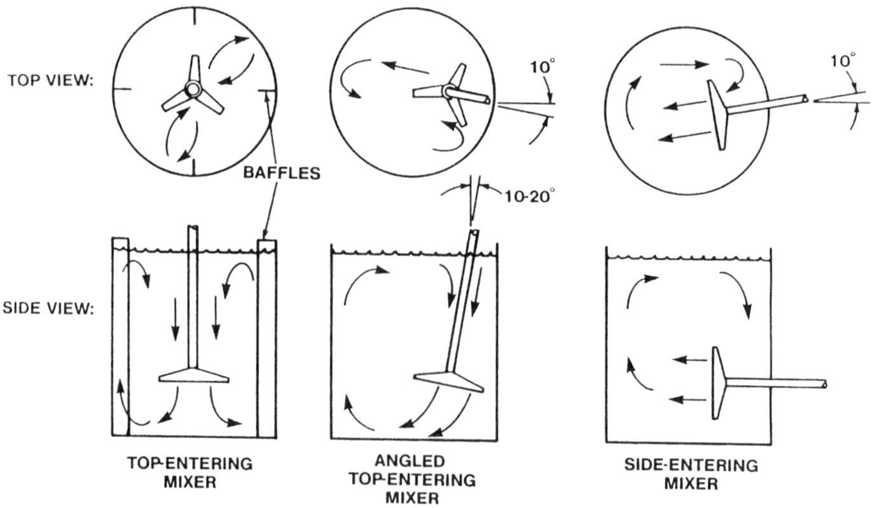

Fig. 10.4 General impeller arrangements and flow patterns

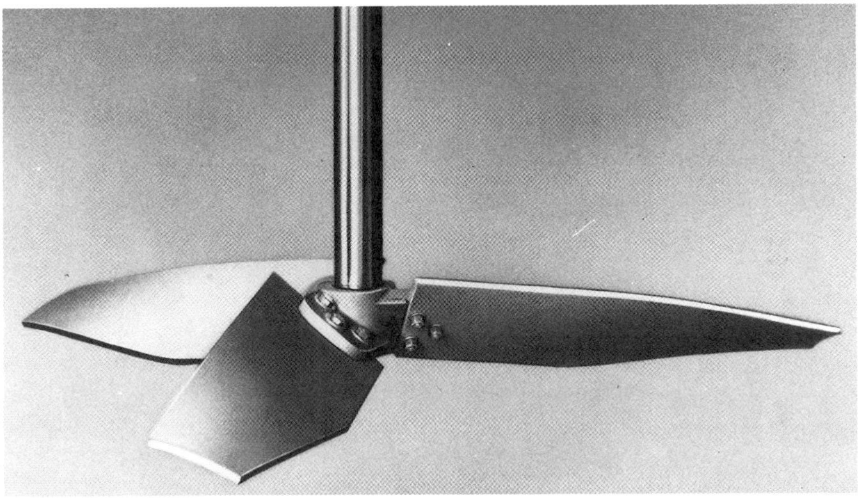

Fig. 10.5 High-efficiency airfoil-type impeller. Courtesy of Lightnin—A Unit of General Signal

development has continued at a rapid rate. With the advent of fluid-flow measurement techniques, such as the laser doppler velocimeter (LDV) (Ref 2), mixer designs have improved dramatically, permitting greater pumping efficiency.

Angled top-entering and vertical top-entering mixers now use the high-efficiency airfoil-type impellers (Fig. 10.5). The flow efficiency (pumping volume per unit power) of such impellers is 40% greater than that of marine propellers and consume the same amount of energy. The new impellers also cost less to construct, largely because of weight reduction and the smaller shaft size needed to carry the impeller loads.

Side-entering mixers use impellers similar to that shown in Fig. 10.6. The blade has a different shape to account for the mechanical forces that act on the impeller because of the higher operating speeds typically required in a side-entering mode of operation. An advantage of this type of impeller over conventional marine propellers is that the bolted blades allow for easier installation.

The velocity profiles of a flow-efficient impeller are shown in Fig. 10.7. This computer-generated plot depicts the velocity vectors produced by an impeller stirrer in a 1220 mm (48 in.) cylindrical tank with a flat bottom. The vertical line at zero is the tank centerline. The vertical line at 20 represents a sidewall baffle, one of four equally spaced. This velocity plot was determined experimentally by scanning the tank with a laser. The dashed horizontal and vertical lines show the traverse that the laser followed. Computer analysis of the scattered light patterns shows the fluid velocities at each point. Emanating from the dashed lines are velocity vectors, showing both magnitude and direction of fluid flow. The line length is proportional to the fluid velocity. Computer analysis of these velocity vectors yields pumping (flow) and shear rates.

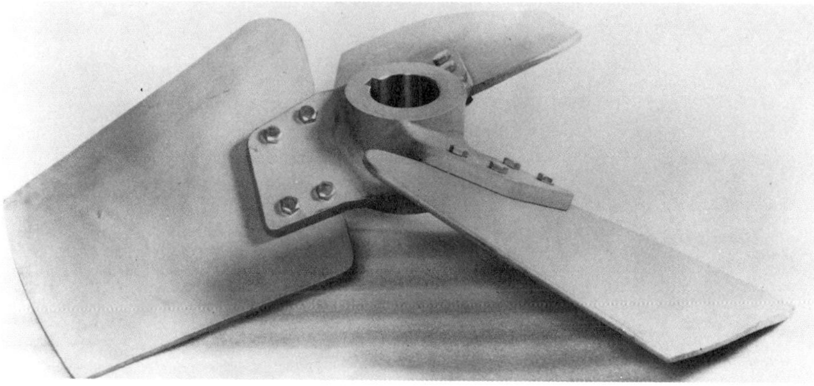

Fig. 10.6 Side-entering mixer impeller. Courtesy of Lightnin—A Unit of General Signal

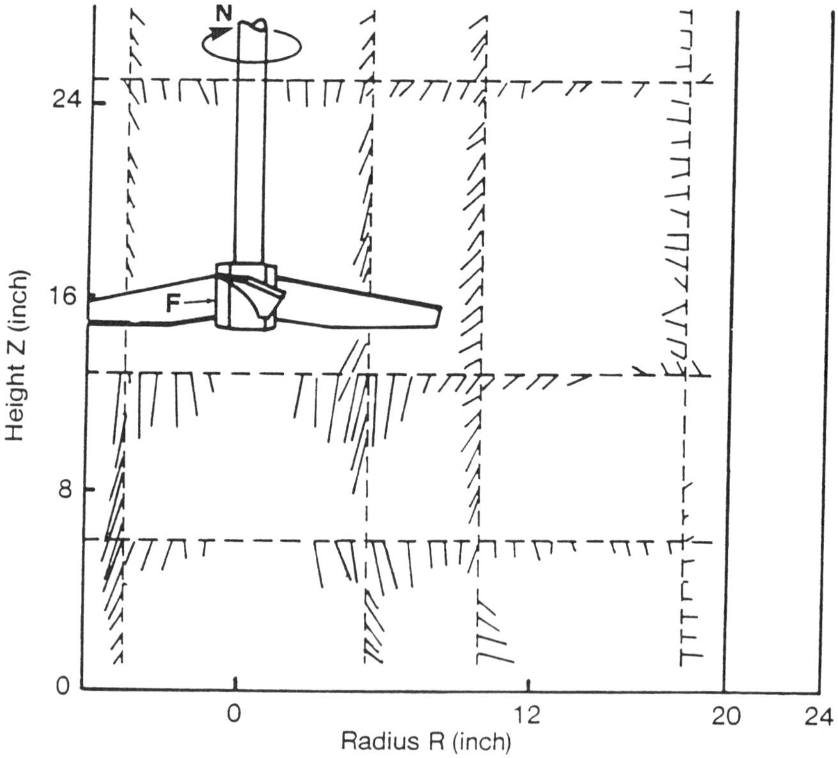

Fig. 10.7 Fluid velocity vectors in the R-Z plane for a flow-efficient impeller

Oldshue (Ref 3) has demonstrated that a given impeller type will provide more flow per unit of power as the mixer speed decreases and the impeller diameter increases at constant power consumption. This is one of the reasons that the typical operating speeds for side-entering mixers have decreased from the 420 rev/min recommended in the earlier U.S. Steel paper (Ref 1) to 280 rev/min today. Only 60% of the power is required to achieve the same flow rate and the same heat-transfer rate for the same impeller type; rotational speed is lower, but blade diameter is larger.

By taking advantage of more efficient impellers, mixer performance can be improved even further. Table 10.2 compares the power levels originally recommended by U.S. Steel (Ref 1) and current requirements based on the advanced technology now available for equivalent agitation. For "violent" agitation, less than half of the original recommended power consumption is currently required. This power level is equivalent to a blend time of approximately 1 min and can be used as a starting point for selecting open-impeller mixer systems.

Draft-Tube Impeller Systems

Draft-tube circulation systems are often used in quench system design. A draft tube is circular duct used to direct flow to and from the impeller. The primary advantage of draft tubes is improved control of the flow pattern within the quench tank.

In an open-impeller system, such as that depicted in Fig. 10.7, the distribution of flow may be such that relatively quiescent or stagnant zones are created in the mixing vessel near the tank bottom, directly beneath the impeller. For example, when the ratio of liquid depth to vessel diameter or side length is greater than 1.0, uniform solids suspension is

Table 10.2 Equivalent quench tank mixer size(a)

	Power level, hp/gal		
Year	Standard quench oil	Water or brine	Comments
1954	0.007	0.005	U.S. Steel publication (Ref 1) recommends a marine propeller speed of 420 rev/min
1970	0.004	0.003	Power levels adjusted for lower mixer operating speeds (280 rev/min)
1990	0.003	0.002	Power levels adjusted for improved impeller pumping efficiency at same speed (280 rev/min)

(a) Tank volume greater than 11,300 L (3000 gal), open-impeller mixtures, "violent" circulation not readily obtained. The use of down-pumping draft-tube circulators produces a strong flow pattern that sweeps the solids away from the draft tube and reduces the tendency for solids deposition.

With a draft tube, it is theoretically possible to approach the condition where the path lengths for most of the fluid particles are the same. When the ratio of induced flow to primary flow is quite small, a "plug-flow" through the tank results. Fort *et al.* (Ref 4) calculated the streamline patterns in a volume located above an open, high-speed impeller in a baffled vessel, as shown in Fig. 10.8. Figure 10.9 represents the flow pattern for the same unit operating within a draft tube. The use of the draft tube significantly improves the spatial uniformity of flow in the system, especially close to the liquid surfacc.

An LDV fluid velocity profile for an operating draft-tube circulator model is shown in Fig. 10.10. The fluid velocity vectors in the upper region of the tank are reasonably

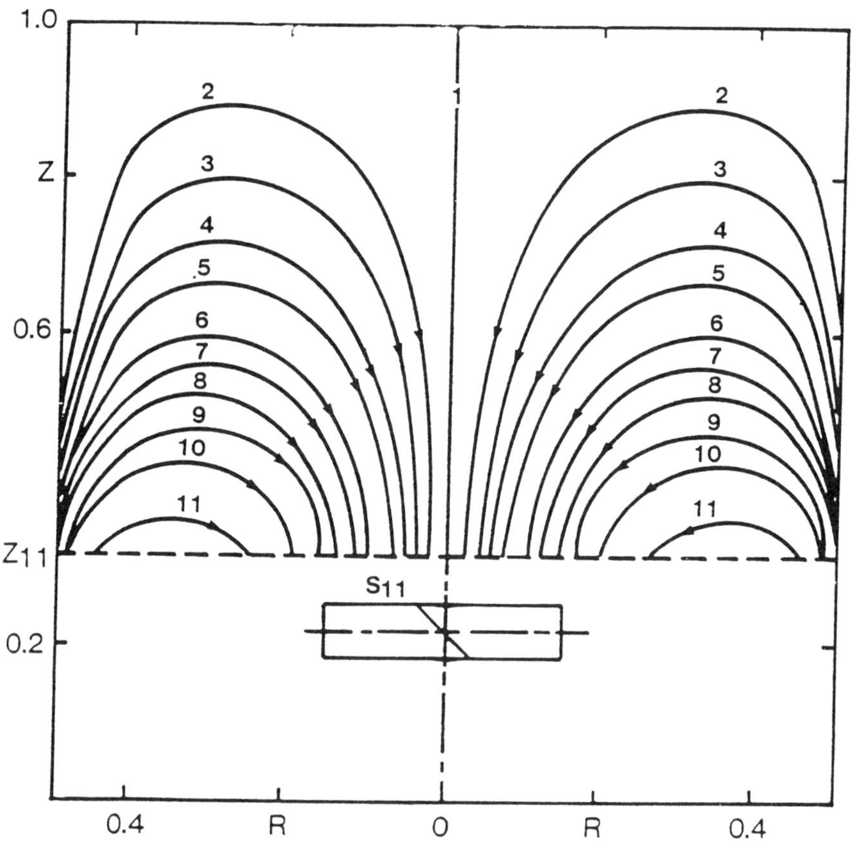

Fig. 10.8 Streamline patterns in a cylindrical vessel with an open-impeller mixer

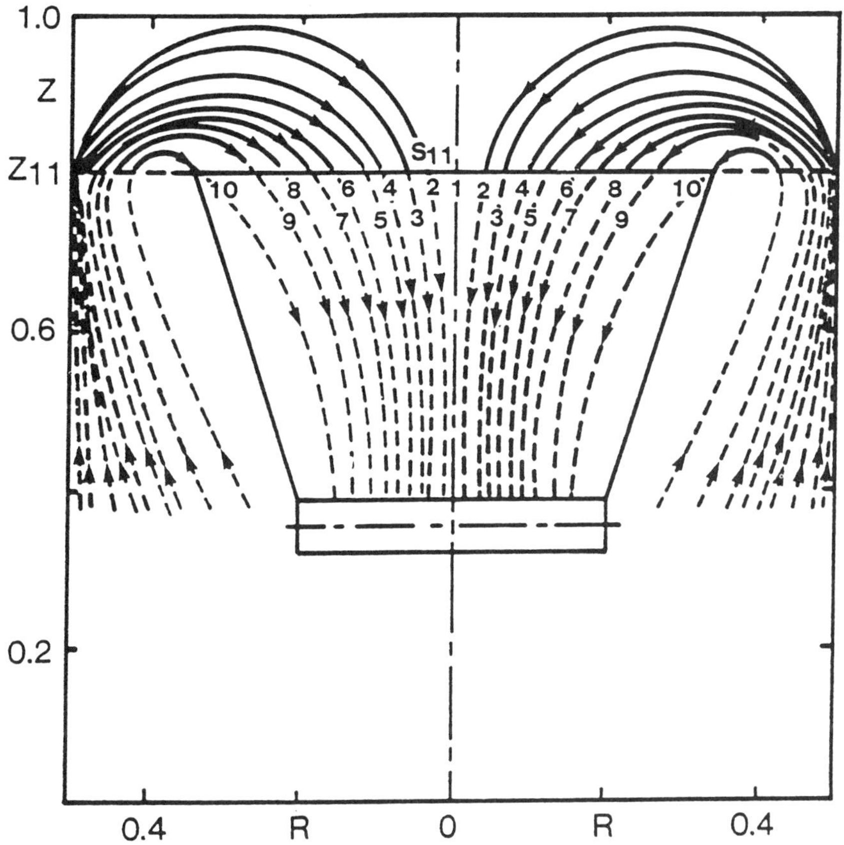

Fig. 10.9 Streamline patterns in a cylindrical vessel with a draft-tube impeller mixer

close to the streamlines depicted in Fig. 10.9. Note that the use of a draft tube eliminated the stagnant zone beneath the impeller. However, "upflow" in the lower portion of the annulus is still not completely uniform. A small recirculation zone exists, as evidenced by "downflow" present at the outer wall of the draft tube.

Impeller performance in a draft tube can be compared with pump operation. In both systems, the volume of flow delivered depends on the resisting head. Draft-tube circulator performance is characterized by using a "head-flow" curve similar to that used for pumps. If a system head curve can be matched to an impeller head and capacity curve, the resulting intersection yields the required unit design.

A typical head-capacity curve for an airfoil-type circulator impeller is shown in Fig. 10.11. In this example, head is plotted as a function of flow for impellers of varying blade-tip chord angles. The region of instability indicated by the upper left portion of the curve denotes the onset of "stall condition." System design should always result in selection of an operating point to the right of this region.

A series of system resistance curves are also plotted in Fig. 10.11. System resistance, K_V, which is a function of system geometry, can be defined as:

$$K_V = 2GH/V_D^2 \qquad \text{(Eq 10.2)}$$

where H is system head (ft), G is gravitational acceleration 32 ft/s^2, and V_D is velocity through the draft tube (ft/s).

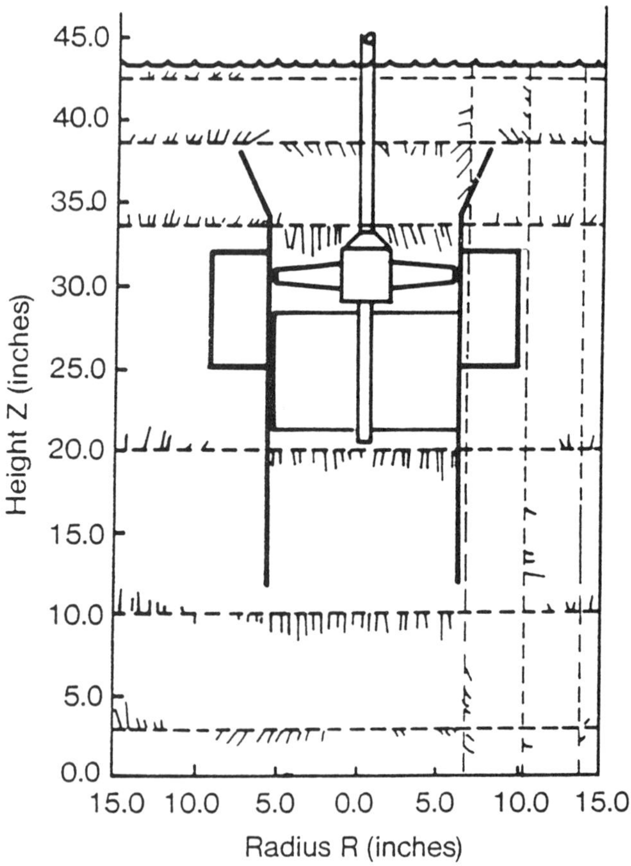

Fig. 10.10 Fluid velocity vectors in the *R-Z* plane for a draft-tube circulator

An appropriate value of K_V can be determined empirically for a given geometry. This, together with the process flow and head requirements, will establish the mixer design. For comparison purposes, a pitched-blade axial-flow turbine performance curve is also plotted in Fig. 10.11. The slope of the curve represents the capacity of the impeller for resisting changes in system head. The data plotted show that, for the airfoil impeller, the pumping capacity is reduced by 30% when the system K_V increases from 1 to 5. The axial turbine, under similar conditions, exhibits a 35% decrease in pumping capacity. At the high system resistance, the airfoil impeller produces nearly 16% more head. Therefore, the advantages of an airfoil type of impeller are:

- Higher head capacity
- Steeper head-flow curve
- More resistance to stall
- Greater operating efficiency

Draft-tube systems normally use the marine propellers for very small mixers and airfoil-type impellers for larger units. In a draft-tube system, an airfoil-type stirrer will

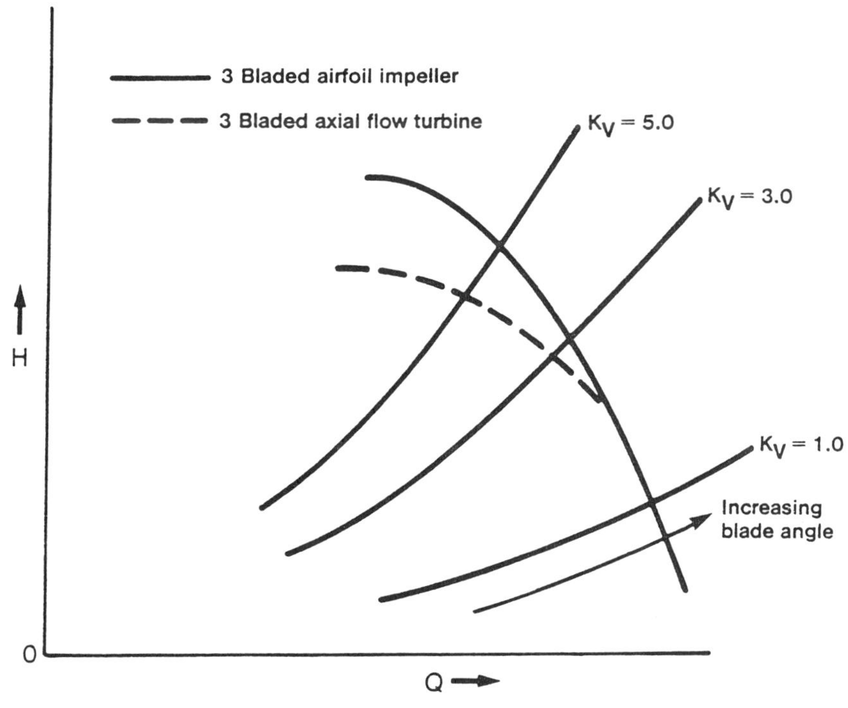

Fig. 10.11 Total head-flow capacity curves for a draft-tube impeller

be designed at a lower tip chord angle than that used on a side-entering mixer in order to avoid stalling under conditions of high head resistance. The lower weight of the airfoil-type stirrer lowers its cost at diameters greater than 610 mm (24 in.).

Based on Fig. 10.11, it is evident that the total system head resistance should be minimized in order to maximize the flow developed by the impeller. Although it is not practical to change most of the characteristics of the quench tank system, the draft tube can be tailored to minimize the head losses. For example, lack of an entrance flare with inadequate liquid coverage over the draft tube can decrease the flow rate by 20%.

As shown in Fig. 10.12, the following characteristics maximize the performance of a draft tube:

- Down-pumping operation takes advantage of the tank bottom as a flow-directing device.
- A 30° entrance flare on the draft tube minimizes entrance head losses and ensures a uniform velocity profile at the inlet.
- Liquid coverage over the draft tube of at least one-half the diameter avoids excessive disruption of the impeller inlet velocity profile.
- Internal flow-straightening vanes prevent fluid rotation or swirl.
- The impeller should be inserted into the draft tube a distance equal to at least one-half the diameter. Inlet velocity profile considerations dictate this dimension.
- A "steady bearing" or limit ring protects the impeller from occasional high deflection. Use of a steady bearing results in a mixer that is lower in cost, but that requires periodic maintenance. For this reason, a limit ring generally is specified as the only protective item. In such case, the mixer manufacturer should be requested to ensure that the total mechanical design of the mixer is safe for operation without a steady bearing. The manufacturer will supply dimensions for the limit ring.
- The impeller normally requires a radial clearance of 25 to 50 mm (1 to 2 in.) between the blade tips and the draft tube. If the size of the draft tube must be minimized, an external notch can be used to reduce the draft-tube diameter by 50 to 74 mm (2 to 3 in.).

All of the geometric parameters of the draft tube combine to determine K_V, the system resistance constant, and thus the flow developed by the impeller.

Impeller Comparison

Because quenching is a flow-controlled mixing operation, and different impeller types produce different values of flow volume per unit of power, it is clear that the simple "power per unit volume" mixer sizing approach currently in use is applicable only to one impeller type at one operating speed. Using the scaleup correlations for flow and power presented below, the equation for flow per power is obtained (Eq 10.5):

$$Q = N_Q N D^3 \quad \text{(Eq 10.3)}$$

$$P = N_P \rho N^3 D^5 \quad \text{(Eq 10.4)}$$

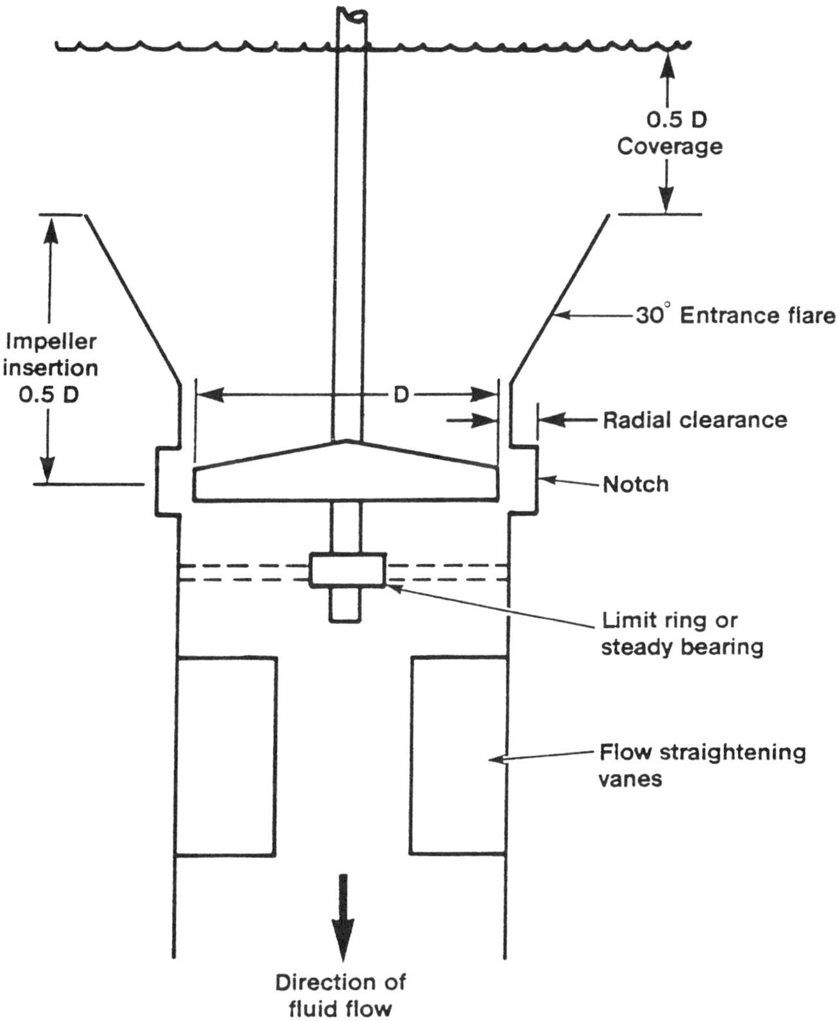

Fig. 10.12 Typical draft-tube impeller system characteristics

$$\frac{Q}{P} = \left(\frac{N_Q}{N_P}\right)\left(\frac{1}{\rho N^2 D^2}\right) \quad \text{(Eq 10.5)}$$

where D is the impeller diameter (in.), N is the impeller speed (rev/min), ρ is fluid density (lb/ft^3), Q is volumetric flow rate (gal/min), and P is power (hp). N_Q is the dimensionless flow number that characterizes the flow-producing capability of the impeller, and N_P is the dimensionless power number that characterizes the power consumption characteristics of the impeller. Equation 10.5 shows that a relative comparison of flow per unit power depends on impeller type, installation geometry, speed, and diameter.

This relationship does not completely specify the mixer, because other significant design elements, such as torque, also are imposed. Equation 10.6 relates mixer torque, T (lbf · in.), to the other parameters of interest:

$$T = N_P \rho \frac{N^2 D^5}{2\pi} \quad \text{(Eq 10.6)}$$

Capital cost of the complete mixer is strongly dependent on torque; therefore, this parameter must be examined as rigorously as flow and power.

Comparing speed and diameter options for one type of impeller is a straightforward task. Comparing two different impellers is much more complex. For flow-controlled processes, probably the best evaluation of different options is to examine mixer parameters on the basis of maintaining constant flow and mixer speed. The equations above can be rewritten as functions of flow (Q) and speed (N) to obtain the following:

$$\frac{Q}{P} = \left(\frac{N_Q^{5/3}}{N_P}\right)\left(\frac{1}{N^{4/3} Q^{2/3} \rho}\right) \quad \text{(Eq 10.7)}$$

Therefore, at constant flow and impeller speed, the flow volume per unit power can be expressed as:

$$\frac{Q}{P} \propto \frac{N_Q^{5/3}}{N_P} \quad \text{(Eq 10.8)}$$

To make use of Eq 10.8, the flow number, N_Q, and power number, N_P, for the impellers of interest must be determined or obtained from the stirrer manufacturer. The power number is readily obtained on a mixing dynamometer by measuring both torque and speed. Because torque is proportional to power divided by speed (Eq 10.6), the power can be calculated. The power number is determined using Eq 10.4.

Measurement of the impeller flow number is more difficult. Flow measurement methods such as streak photography, Pitot-static tubes, and hot-wire anemometers have been described by Oldshue (Ref 3). In addition to being time consuming and difficult to conduct, each of these methods affects flow patterns by its presence.

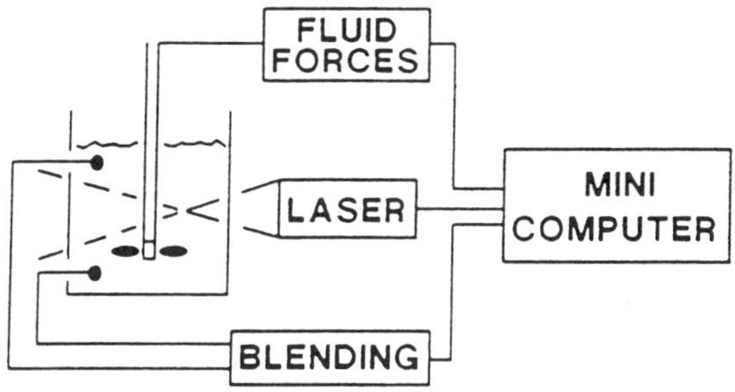

Fig. 10.13 Laser doppler velocimeter system

The laser doppler velocimeter is much more suitable for measuring flows in a laboratory-scale mixing vessel. The LDV uses the doppler frequency shift of light scattered by moving particles to determine particle velocity. Because particles or contaminants travel with the fluid, it is thus possible to measure fluid velocity. (Contaminant particles referred to here are small, up to 10 μm in diameter.)

The LDV consists of a laser, sending and receiving optics with a photomultiplier, and an electronic processor to convert the doppler shift frequency to an analog or digital signal, as shown in Fig. 10.13. Two convergent laser beams are emitted from the sending optics intersecting at a precise point in the body of the fluid (Fig. 10.14). Because of the nature of laser light, an optical fringe or interference pattern is produced at the point of intersection. As contaminant particles move across the interference pattern, they absorb and reflect light at a frequency equal to the doppler frequency difference. Reflected light is collected by the receiving optics. A photomultiplier tube senses the frequency, which is proportional to particle fluid velocity.

An advantage of the LDV is that it does not interfere with fluid flow and thus does not produce any restriction on measuring direction versus sensor orientation. Because great unsteadiness in turbulence (perhaps exceeding ±100%), may occur near a mixing impeller, the LDV is the only instrument identified to date that can effectively measure fluid flow.

Sophisticated devices such as the LDV allow mixing impellers to be characterized in terms of the dimensionless values of flow number (N_Q) and power number (N_P). As illustrated earlier, these parameters can be used to compare impellers directly and to predict impeller performance. They are, however, somewhat dependent on the mixing environment, including the tank geometry and the presence of a draft tube.

Table 10.3 summarizes the flow numbers and power numbers for marine propellers and airfoil impellers. It is important to note that a change in flow number and therefore flow rates (Eq 10.3) will affect flow volume per unit of power when the impellers are used in a draft tube. Impellers operating in a draft-tube develop less flow for the same power consumption, but have the advantage of directional flow control.

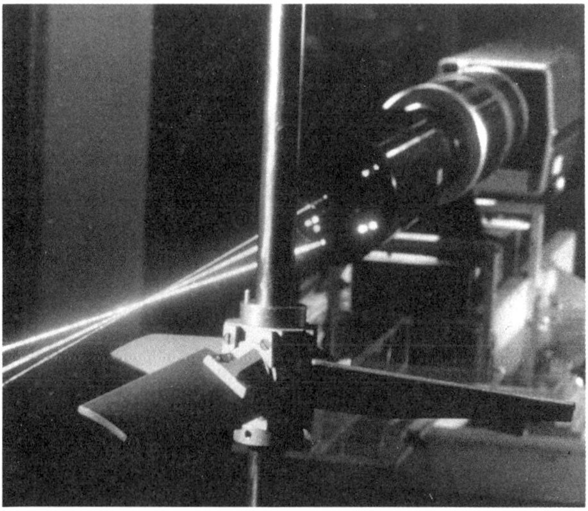

Fig. 10.14 Laser doppler velocimeter measuring flow in a transparent vessel. Courtesy of Lightnin—A Unit of General Signal

A common mistake in the selection of draft-tube mixers is to use flow numbers reported for open impellers. The draft-tube flow number should be used for calculation and will result in the selection of a significantly larger mixer. Draft-tube mixers will be undersize if the mixer manufacturer cannot accurately define the draft-tube flow number. The values shown in Table 10.3 are for a "typical" draft tube. The draft-tube flow number varies considerably, depending on such factors as draft-tube length and geometry. Therefore, it is essential that the draft-tube flow number (and thus K_V) be properly determined. Because of the complexity of this determination, mixer manufacturers should be able to demonstrate that the draft-tube flow number has been properly determined upon request.

Selection of Mixer Size

Fluid flow is a function of both impeller speed and diameter; therefore, specifying mixer power alone will not adequately define the mixer requirements. The procedure used to select the proper mixer size depends on whether the mixer employs an open impeller or a draft tube.

Open Impeller Mixers. The power levels documented by U.S. Steel (Ref 1) have become an industry standard. These power levels were based on a three-blade marine propeller operating at 420 rev/min. Table 10.2 shows how these power levels have been

Table 10.3 Typical mixing impeller flow numbers and power numbers

Type of impeller	Open-impeller system(a)			Draft-tube system ($K_V = 4$)		
	N_Q	N_P	Relative Q/P	N_Q	N_P	Relative Q/P
Marine propeller (three blades, 1.0 pitch ratio)	0.46	0.35	1.0	0.40	0.40	0.69
High-efficiency airfoil-type impeller	0.56	0.33	1.47	0.45	0.36	0.94

(a) Includes both top-entering and side-entering mixers

adjusted to account for changes in mixing technology. Mixer operating speeds have decreased significantly, and impellers that are more pumping efficient have been developed. Through improved mixer design, the power levels required to provide equivalent mixing have decreased by more than 50%.

Continued reliance on the U.S. Steel recommended power levels precludes the benefits of added flow and incrementally better quench performance that have arisen as the result of lower mixer operating speeds. Therefore, some quench operations may have come to expect a higher performance level than the 0.003-0.002 hp/gal power level would develop. For this reason, the following power levels are recommended for open-impeller mixers using high-efficiency impellers at 280 rev/min:

- Standard quench oil: 0.004 hp/gal
- Water or brine: 0.003 hp/gal

Power levels are summarized in Table 10.4. The power levels should be adjusted for other speeds using Eq 10.7, which can be simplified to:

$$P \propto N^{4/3} \qquad \text{(Eq 10.9)}$$

Because some top-entering mixers operate at lower speeds, the required power level can be reduced, as shown in Eq 10.9, for equivalent mixing. These power levels apply equally to top-entering mixers, whether vertical or angular mounted, and to all side-entering mixers. The blend time will be less than 1 min.

Draft-Tube Mixer Systems. The current trend in the heat treating industry is to use draft-tube mixers rather than open-impeller mixers in order to achieve more uniform velocity profiles throughout the quench tank. In many cases, the mixer has been selected using the 1954 U.S. Steel criteria (Ref 1), even though draft-tube mixers require higher power levels for equivalent flow. This has possibly been offset by a combination of lower operating speeds (and thus more flow volume per unit of power) and the high-efficiency impellers now in use.

No true equivalence exists between draft-tube mixers and open-impeller mixers for heat transfer and blending. However, assuming that equivalent impeller pumping

Table 10.4 Recommended quench tank mixer sizes(a)

Mixer type(b)	Power level, hp/gal	
	Standard quench oil	Water or brine
Open impeller, top or side entering, 280 rev/min(c)	0.004	0.003
Draft tube, 280 rev/min(c)	0.006	0.0045

(**a**) Recommended mixer sizes provide higher agitation levels than originally recommended by U.S. Steel (Ref 1). (**b**) Based on use of high-efficiency impellers; increase power levels by 50% for other impeller types. (**c**) Adjust power level for other mixer output speeds using Eq 10.9.

capacity will translate to similar quench performance, the recommended power levels for draft-tube mixers are:

- Standard quench oil: 0.006 hp/gal
- Water or brine: 0.0045 hp/gal

These power levels are appropriate for draft-tube systems where K_V is equal to 4 and for 280 rev/min mixers with high-efficiency impellers. The mixer manufacturer should adjust these power levels to account for the precise system geometry (i.e., if K_V is other than 4). The power levels for other types of impeller mixers should also be adjusted for different operating speeds using Eq 10.9.

A more sophisticated approach to draft-tube mixer selection is to use past experience to maintain constant linear flow velocity past the part. An existing system with a similar tank geometry, but not necessarily the same size, can be used to establish a successful mixer/circulator size. The system geometry of both the new system being designed and the system being used for analogy must be similar to successfully extrapolate quench characteristics provided by a given draft-tube fluid velocity. The as-quenched metallurgical characteristics of a part in the active region of the tank can be correlated with a known draft-tube velocity. The fluid flow velocity will relate to an average flow velocity through the draft-tube and will also relate to an average flow velocity based on the cross-sectional tank area.

The draft-tube velocity of an existing installation can be determined by one of two methods. In the first method, the quenchant supplier measures the velocity of fluid flow through the draft tube. In many cases, however, the precision of the measurement is poor due to both the intrusive nature of the flow measurement device, such as a velocimeter, and the nonuniform velocity profile across the draft tube, especially near the liquid surface.

In the second method, the mixer supplier determines the system resistance, K_V, and then estimates the impeller draft-tube flow number. The draft-tube flow number, in conjunction with the mixer speed and diameter, allows calculation of the flow rate and thus the draft-tube velocity. In this case, it is recommended that mixing impeller literature references that report flow numbers only for open impellers *not* be used. Use of such data may cause the draft-tube velocity to be overestimated by as much as 100%.

Once the draft-tube velocity has been determined, the fluid flow rate through the draft tube is known. This can be used to calculate an average velocity past the part. This should be based on either a horizontal or a vertical cross-sectional area, depending on which makes sense and whether a convenient plane of symmetry exists (Fig. 10.15). The value for average velocity is the one which most closely predicts the cooling characteristics that the part will experience; this should be maintained constant in the new installation. This average velocity value will then translate to a flow rate, Q. Slight tank geometry changes can be accepted as long as the average fluid velocity by the part is maintained.

Knowledge of the average velocity will permit calculation of the draft-tube flow rate for a new installation. Mixer suppliers can then select an optimum combination of mixer speed and diameter to achieve this flow rate after determining the system resistance, K_V. Mixer speed and diameter can be estimated using Eq 10.3 and 10.4 along with the draft-tube flow number and draft-tube power number from Table 10.3. Typical operating speeds are 150 to 280 rev/min. Mechanical mixer design constraints may limit the speed to much less than 280 rev/min, so care must be taken not to finalize the impeller diameter and draft-tube diameter without input from the mixer manufacturer.

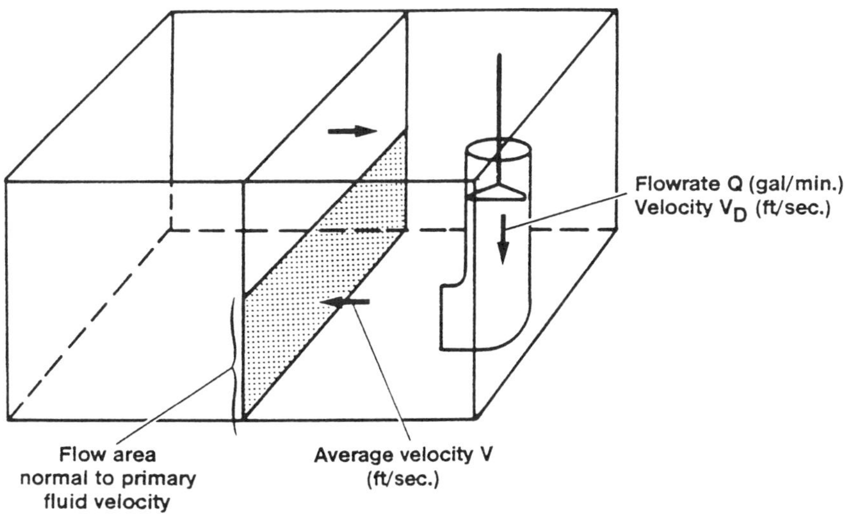

Fig. 10.15 Average linear flow velocity in a quench tank with a draft-tube mixer

Mixer Details

Multiple Mixers. The determination of the number of mixers needed for a quench tank depends primarily on the vessel geometry. The quantity of mixers is selected to:

- Ensure that all quench zones in the tank experience forced convective heat transfer
- Ensure that the fluid velocity will have minimal variation throughout the quench zones to avoid nonuniform or differential cooling rates

Unfortunately, it is not yet possible to fully quantify these objectives. For the most part, understanding of mixer flow patterns will permit the selection of the optimal number of mixers. Observations of successful quench installations and advice from mixer manufacturers are probably the best sources of information.

In general, a cylindrical or rectangular tank where the length-to-width ratio is less than 2:1 can generally utilize a single mixer. However, if the length-to-width ratio is greater than 2:1, multiple mixers are generally recommended. Figure 10.16 shows several possible multiple mixer arrangements for rectangular tanks where the length-to-width ratio exceeds 2:1. These configurations are applicable primarily for draft-tube mixers and side-entering mixers. Top-entering mixers may be arranged differently.

Tank baffles eliminate vortexing and convert swirling motion to productive top-to-bottom fluid motion. In rectangular tanks with properly placed multiple mixers, the combined effects of tank corners and interference between mixer flow patterns generally eliminate the need for baffles.

Vertical cylindrical tanks with top-entering mixers use a standard baffle configuration (Fig. 10.17). Side-entering mixers generally do not need to be baffled because of the asymmetric flow pattern. Draft tubes require internal baffles or flow-straightening vanes (Fig. 10.12).

Cavitation occurs when the pumping action of the impeller creates localized zones of relatively low pressure. If this pressure is below the vapor pressure of the liquid at the fluid temperature, then fluid vaporization may occur. This may lead to erosion of the impeller and unstable operating performance. Properly designed impellers, whether open impellers or impellers operating in a draft tube, do not experience cavitation in quench tank operations. The high vapor pressure of quench oils and the smooth flow profiles around the impeller help to prevent this problem.

Example of Mixer Selection

Solution of a Design Problem. The problem is to select the appropriate agitation system for a 5000 gal quench tank that is 6 ft wide, 12 ft long, and 10 ft deep. Assume

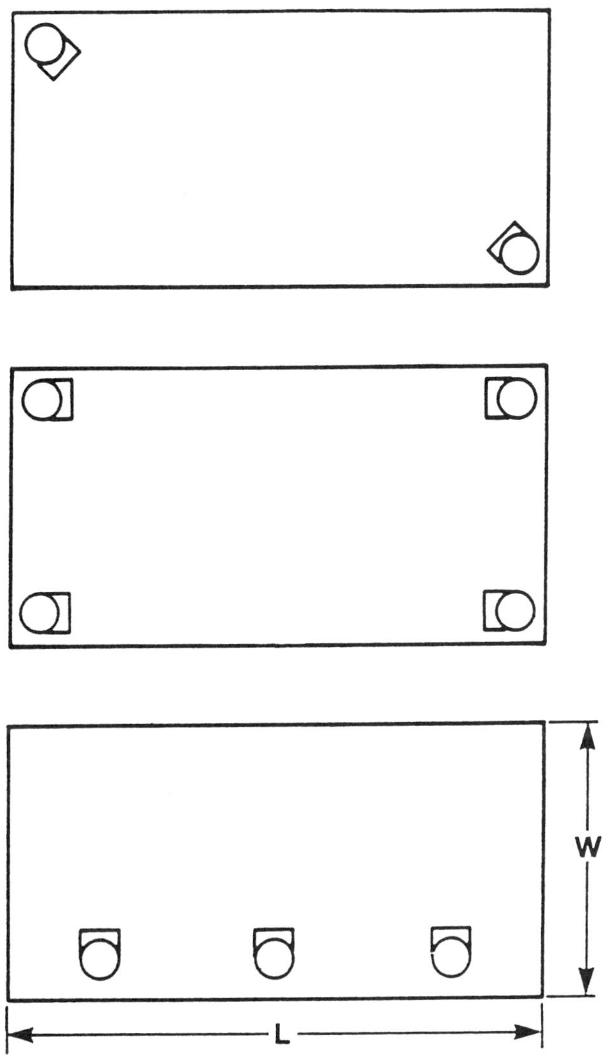

Fig. 10.16 Typical draft-tube mixer arrangements for tanks where the length-to-width ratio exceeds 2:1

that a typical quench oil is being used, with a specific gravity of 0.93, and that draft-tube mixer circulators are desired.

Table 10.4 lists the desired total mixer power level at 0.006 hp/gal. This corresponds to a total power requirement of 30 hp. Figure 10.16 indicates that two mixers will probably provide a more uniform velocity flow field because of the length-to-width ratio of this tank. Therefore, two 15 hp (30 ÷ 2) mixers is the preliminary selection.

In order to refine this mixer selection, note that Table 10.4 assumes the use of high-efficiency impellers. The draft-tube flow number and power number for a typical impeller stirrer are given in Table 10.3, based on a system resistance of $K_V = 4$. This is a reasonable assumption when there is not an excessive number of parts to restrict the flow upward in the main quench zone.

Equation 10.4 relates the power consumption to the impeller diameter and speed. Using this equation and normal mixer design practice, which allows the mixing impeller to be loaded to 80% of the motor size, yields a 30 in. airfoil impeller stirrer operating at 280 rev/min.

Depending on the location of the mixer mounting surface above the quench tank, the mixer mechanical design may be limited by the natural frequency of the mixer. This is highly dependent on the length of the mixer shaft and must be determined by the mixer manufacturer. Assume that the mixer speed must be decreased to 190 rev/min to meet normal design standards. The selection of a different required power level for equivalent flow rates is necessary. Employing Eq 10.9, a maximum of 18 hp, or dual 10 hp mixers, is required. Equation 10.3 can be used to show that each mixer will require a 35 in. airfoil impeller mixer operating at 190 rev/min. The following calculation of the flow generated

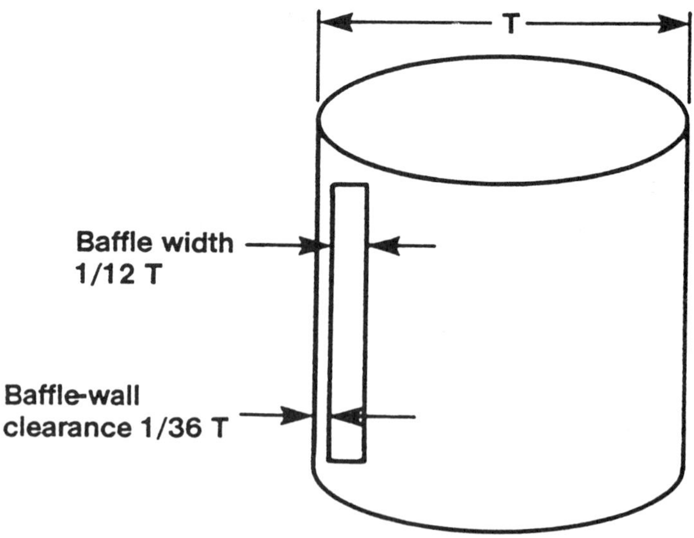

Fig. 10.17 Typical baffle dimensions (one of four equally spaced baffles)

by both the higher-speed and the lower-speed mixers shows that they are approximately equal:

$$Q = N_Q N D^3 \qquad \text{(Eq 10.3)}$$

For mixer 1, Q = 0.45 (280 rev/min) (30 in.^{3})/(231 in.^{3}/gal) = 14,727 gal/min. For mixer 2, Q = 0.45 (190 rev/min) (35 in.^{3})/(231 in.^{3}/gal) = 15,869 gal/min. As a final check, the average cross-sectional fluid velocity through the tank should be calculated for comparison with other successful installations.

Recirculation Pumps. A continuation of the above example problem will illustrate the relative inefficiency of a recirculation pump to provide fluid mixing and circulation. In the example, it was shown that each 10 hp mixer provided a flow rate of 15,869 gal/min. However, it is also possible to select pumps to achieve the same flow rate.

Conventional centrifugal pump design requires that two pumps, each over 300 hp, be used. Centrifugal pumps have much higher head capabilities, which are largely wasted in this application. A more economical alternative is to use a low-head axial propeller pump that requires two 40 hp pumps. If one extends the concept of the axial propeller pump to even lower head requirements, such as those present in a draft-tube circulator, the result is a derivation of the two 10 hp mixers previously selected. If a pump is optimized to the physical system, the optimal pump based on power consumption is the draft-tube mixer previously selected.

A conventional pump usually requires 4 to 30 times as much power to achieve the same flow rate and same degree of agitation or circulation in the tank. Another disadvantage of a recirculation pump is that the large-diameter mechanical seal on the pump requires periodic maintenance.

Future Improvements

Optimal Quenching Performance Through Engineering Design. An obvious but difficult challenge in heat treating research is the development of quenching technology that can be varied with the heat treating process to produce optimal metallurgical properties. In the past, most of the focus has been on the development of quenchants that produce "ideal" behavior. "Ideal" quenchants typically exhibit maximum cooling rates and minimum cooling times during A- and B-phase cooling and minimum cooling rates during C-phase cooling, as shown in Fig. 5.11. Although the continuing development of improved quenchant media is essential, it is also important that even greater research efforts be focused on quench system design as it relates to the impact of flow control during all three stages of cooling.

Part placement, with respect to the mixer, affects the cooling rate. Most of the progress made in determining optimum part placement is empirical in nature and resulted from trial and error. The fluid dynamics and quench tank flow fields are complicated; hence, the design and construction of quench tanks has been handled empirically. This is especially true where quench rates are dependent on micromixing and local fluid velocities. Measurement of these quantities is difficult and is generally limited to scaled models of the mixing tank.

Computational fluid dynamics (CFD) may provide greater understanding of the details of fluid flow in the quench tank and possibly provide insights for determination of optimal location and orientation of a part. Predictions of the quench tank flow field have been made using a commercially available CFD computer program, FLUENT™. This computer program solves the discretized Navier-Stokes equations using a control-volume-based finite-difference technique. Conservation of mass and momentum are solved for steady-time average turbulent flow in three-dimensional coordinates.

An example of the use of CFD to model fluid flow is shown in Fig. 10.18(a) and (b). Using the quench tank geometry from the example problem discussed earlier, the case of a single draft-tube mixer is examined. The draft tube is located in one corner of the tank as shown. CFD uses an input of the flow field in the draft tube near the impeller. This was previously measured using an LDV. The CFD computational code then predicts the three-dimensional flow field throughout the tank.

Both Fig. 10.18(a) and (b) show the three-dimensional velocity vectors predicted through various planes. A limited number of velocity vectors are shown for clarity. The general flow pattern sweeps across the bottom of the tank from left to right. There is a rotational element to the flow around the axis of the general flow pattern. The results also show that relatively quiet zones exist near the corners of the tank, except where the draft tube is located.

CFD flow fields can also be calculated when parts of various sizes and orientations are placed in the quench tank. Fluid velocity near the part surface can then be predicted. Because CFD shows good agreement with actual measurements (Ref 5), this technique

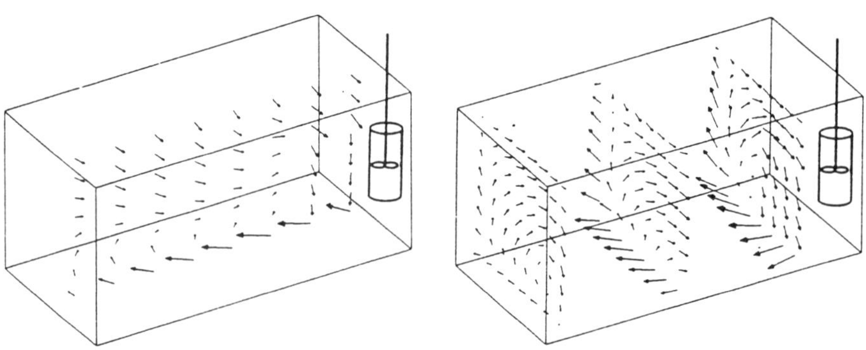

(a) (b)

Fig. 10.18 CFD prediction of a three-dimensional flow field in a quench tank. Fluid velocity increases with arrow length. **(a)** Vectors in x-z axis. **(b)** Vectors in y-z axis may allow determination of optimal location and orientation of the part in the quench tank.

References

1. "Improved Quenching of Steel by Propeller Agitation," U.S. Steel, 1954
2. R.J. Weetman and R.N. Salzman, Impact of Side Flow on Mixing Impeller, *Chem. Eng. Prog.*, Nov 1988
3. J.Y. Oldshue, *Fluid Mixing Technology*, McGraw-Hill, 1983
4. I. Fort, O. Jaroch, and M. Hortalek, Homogenization of Streamlines Field in a Cylindrical Vessel With Axial High Speed Impeller, *Collect. Czech. Chem. Commun.*, Vol 42, 1977, p 3555-3567
5. B.J. Hutchings, R.J. Weetman, and B.R. Patel, "Computation of Flow Fields in Mixing Tanks With Experimental Verification," presented at ASME Winter Annual Meeting, 1989

CHAPTER 11

Residual Stress, Distortion, and Cracking

Quenching is usually performed to optimize properties such as hardness and strength while at the same time minimizing residual stresses, distortion, and crack formation. However, these are often contradictory objectives. For example, although increasing carbon content increases hardness, it also increases the potential for cracking and distortion. Similarly, increasing cooling rates aid in the achievement of high hardness, but increase the potential for cracking and distortion as well. This chapter will discuss the metallurgical basis of these effects.

Definition of Stress

When a force, or load, is applied to a material, a dimensional change results. Fundamentally, this is illustrated by a stress-strain curve (see Fig. 11.1) (Ref 1).

Engineering stress (σ) is defined by the total load (P) divided by the original cross-sectional area (A_o):

$$\sigma = \frac{P}{A_o} \quad \text{(Eq 11.1)}$$

Strain (ε) is determined by the elongation of the sample when a load P is applied by the original length (L_o) according to:

$$\varepsilon = \frac{\Delta L}{L_o} \quad \text{(Eq 11.2)}$$

Stress is related to strain by the bulk modulus (E_{bulk}), if the elastic limit of the material is not exceeded:

$$E_{bulk} = \left(\frac{P}{A_o}\right)\left(\frac{\Delta L}{L}\right) \quad \text{(Eq 11.3)}$$

When a tensile force is applied in the x-direction, the resulting stress (σ_x) not only produces a strain (ε_x) in the same direction, but also in the transverse

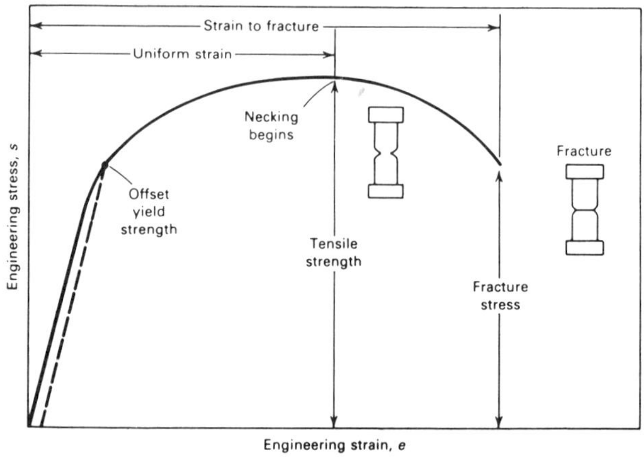

Fig. 11.1 Example of a stress-strain curve. Intersection of the dashed line with the curve determines the offset yield strength.

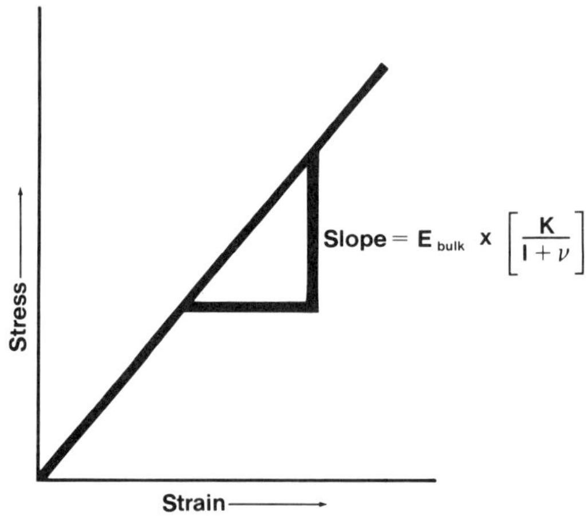

Fig. 11.2 Graphical calculation of bulk modulus from stress-strain behavior

direction. The ratio of transverse strain to longitudinal strain is Poisson's ratio (ν):

$$\varepsilon_y = \frac{-\nu \rho_x}{E} \quad \text{(Eq 11.4)}$$

Because of material variations, the bulk elastic modulus should be determined experimentally instead of using a reference book value, which could lead to substantial errors in stress analysis. The modulus is determined by attaching a strain gage to the specimen and measuring the deflection after applying a series of loads. The E_{bulk} value is determined from the slope (Eq 11.5) by either regression analysis or graphically (see Fig. 11.2).

$$\text{Slope} = E_{\text{bulk}} \times \left(\frac{K}{1 + \nu} \right) \qquad \text{(Eq 11.5)}$$

where K is the stress factor and is dependent on the metal being tested. For steel, K is approximately 1.7×10^{-4}.

Metallurgically, stresses may be developed in a material during quenching by a volumetric change associated with such transformations as martensite transformation from austenite. This process is fundamental to the understanding of stress formation during quenching and will be discussed in more detail.

Effect of Heat Treating Process Factors

Distortion during heat treatment can be classified as either size (movement) or shape (warpage) (Ref 2). Size distortion is affected by steel composition, austenitizing temperature and time, and cooling rates throughout the quenching and tempering process.

Shape distortion (and cracking) may also occur during austenitization or quenching. It is caused by various heat treating process factors. This section will review the most common heat treating parameters that aggravate cracking and shape distortion problems, including (Ref 2-4):

- Part design
- Material selection
- Engineering requirements of the part
- Nonuniform heating
- Nonuniform quenching
- Quenchant selection

Part Design. The propensity for quench cracking is particularly affected by such part shapes as:

- Long parts with small cross sections or thin parts with a large area
- Nonsymmetrical shapes
- The presence of holes, deep keyways, and grooves

Many quench cracking problems can be solved by a design or material change (see Fig. 11.3). However, in many cases design changes are not possible, and modifications of the quenching process are required (e.g., the selection of a less severe quench).

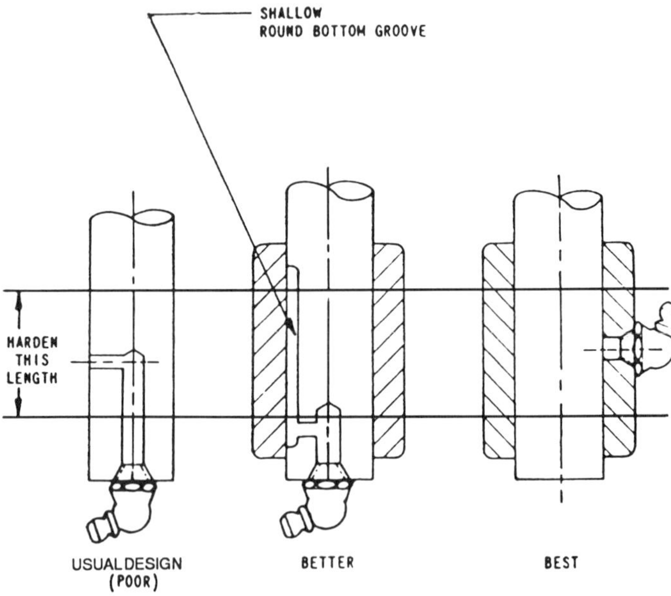

Fig. 11.3 Alternate designs to alleviate cracking when hardening a shaft over a cross hole used for lubrication

Material Selection. Most quench cracking problems are related to nonuniform heating and quenching (Ref 4). However, it is possible to select a grade of steel for a particular part where the composition in the low specification range will not produce the desired hardness and the composition in the high specification range will cause cracking. Kern (Ref 4) has provided the following guidelines and precautions relative to steel selection:

- Check the composition of the steel in relation to the analysis tolerances and ladle ranges.
- Some steel mills may supply product where alloy compositions are at either the high or the low end of the specification range.
- Some steel grades are prone to microsegregation of manganese or gross segregation of chromium (e.g., AISI 1340H, 1345H, 1536, 1541, 4140H, and 4150H).
- Decarburization of up to 0.06 mm (0.0025 in.) per 1.6 mm ($^1/_{16}$ in.) cross-sectional diameter is possible without appropriate precautions.
- Due to scale formation and decarburization, it is not possible to specify a minimum surface hardness for as-rolled or forged steel.

Difficult Engineering Requirements. In many cases, the design engineer will specify a grade of steel without considering the property variations that are possible within the allowed chemistry specification range.

Incorrect Heating. Steel is prone to many surface reactions that lead to scale formation, decarburization, and so on. Therefore, it is essential that well-designed furnaces be used, permitting adequate atmosphere control and uniform heating. Some suggestions for improved furnace design include (Ref 4):

- Atmosphere-hardening furnaces should have purged charge and discharge vestibules. Belt and shaker hearth furnaces should have a purged quench enclosure.
- Excessive furnace loadings should be avoided. Generally, if the steel exceeds 20% of the distance from the discharge to the charge door, it is overloaded.
- If the furnace is too large, excessive heat may be radiated into the work area.

Nonuniform Quenching. None of the above examples is related to quenching. However, one of the most common causes of cracking is the quenching process, although not necessarily the quenchant. One of the primary causes of quench cracking is a nonuniform heat transfer during quenching. Nonuniform quenching creates large thermal gradients between the core and surface of the part (Ref 3-5).

Another cause of nonuniform quenching is part geometry. Parts with both thick and thin cross sections will cool at different rates, leading to different volumes of phase transformation products at a particular point in the quench cycle. Both effects contribute to stress formation, which may result in increased distortion and cracking (Ref 6).

Quenching uniformity may be enhanced by agitation (Ref 4, 5). The purpose of the agitation system is to obtain uniform heat removal over the surfaces of all parts throughout the load being quenched. The batch quench system shown in Fig. 11.4 illustrates a system where axial (vertical) quenchant flow occurs through a load of round bars lying horizontally in a basket. In this case, the bottom surfaces of the bars experience greater agitation than the top surfaces of the bars. Cracks form on the upper surfaces because of nonuniform heat loss. Agitation produces higher heat loss at the bottom, thus creating a large thermal gradient between the top and bottom surfaces.

If a submerged spray manifold is used to facilitate more uniform heat removal, the following design guidelines are recommended:

- The total surface of the part should experience uniform quenchant impingement.
- The largest holes possible (2.3 mm, or 0.09 in., minimum) should be used.

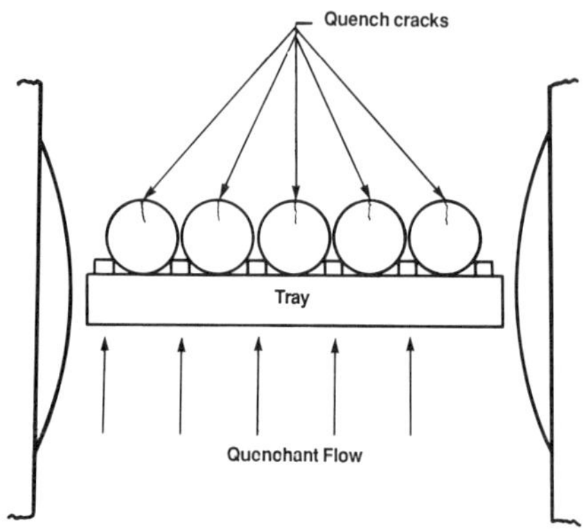

Fig. 11.4 Harmful effects of vertical quench flow

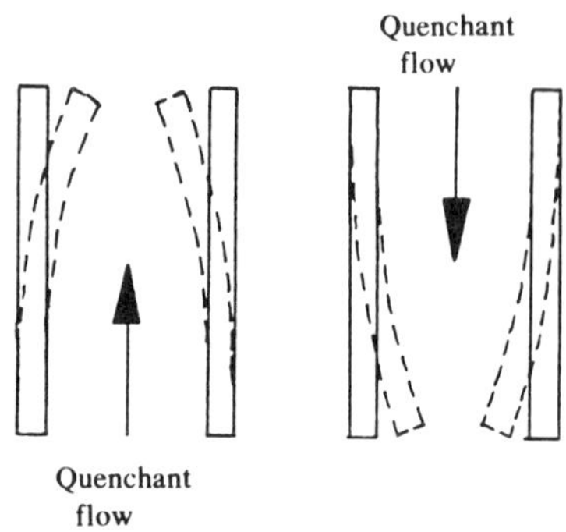

Fig. 11.5 Effect of quenchant flow direction on distortion

- The manifold face should be at least 13 mm (0.5 in.) from the surface of the parts being quenched.
- Rapid removal of hot quenchant and vapor should be possible.

In another case (see Fig. 11.5) recently reported by Von Bergen (Ref 2), excessive distortion resulted when the quenchant flow was either in the same direction relative to the direction of part immersion or in the opposite direction. This problem was corrected by minimizing the quenchant flow to that required for adequate heat exchange during the quench and by moving the part in the quenchant mechanically using an up-and-down action.

Quenchant Selection. Quenchants may reduce distortion and cracking in two ways. The use of a quenchant may increase the uniformity of heat transfer and thus reduce thermal gradients. A quenchant may also be used to reduce the heat-transfer rates at, or near, the martensitic transformation temperature. Achieving reduced cooling rates in the martensitic transformation permits some stress relief associated with the reduction of volumetric expansion of martensite, as shown in Fig. 11.6 (Ref 6).

The selection of a particular quenchant depends on the quench severity desired. Generally, a more severe quench is used for plain carbon steels—for example, water, brine, or caustic solutions and a 5 to 10% solution of a polymer quenchant in water. Accelerated oils are used for crack-sensitive, low-hardenability steels. Conventional (slow) oils or higher concentrations of polymers in water (typically 15 to 30%) are used for higher-alloy steels. Austempering or martempering processes may also be used for particularly crack-sensitive alloys.

Alternative quenchants were described in Chapter 8 and will not be discussed further here.

Metallurgical Sources of Stress and Distortion

The heat treating of steel involves heating the alloy to a temperature sufficient to dissolve and diffuse the carbon and alloying elements in the iron matrix (the austenitizing temperature). The steel is then cooled at a rate sufficient to obtain the desired properties, such as hardness and strength.

When steel cools at relatively low rates, it undergoes a structural change from a more dense face-centered cubic (fcc) structure to a less dense body-centered cubic (bcc) structure, as illustrated in Fig. 11.7. At faster cooling rates, transformation to the bcc structure (ferrite) can be suppressed, allowing a transformation from the fcc structure to a lower-density body-centered tetragonal (bct) structure,

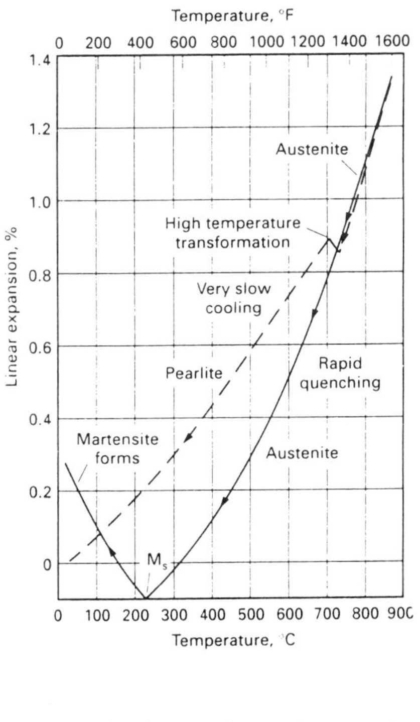

Fig. 11.6 Steel expansion and contraction upon heating and cooling

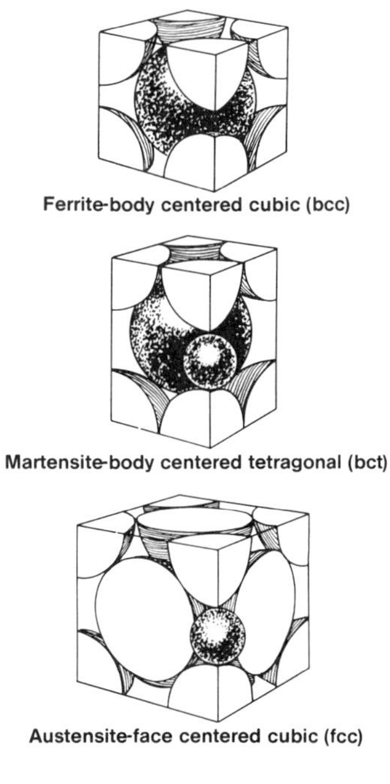

Fig. 11.7 Possible crystal structures for steel

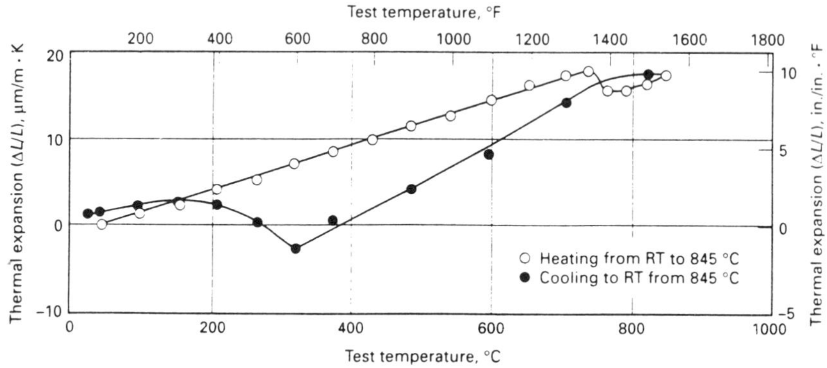

Fig. 11.8 Thermal expansion and contraction curves for AISI 4340 steel. RT, room temperature

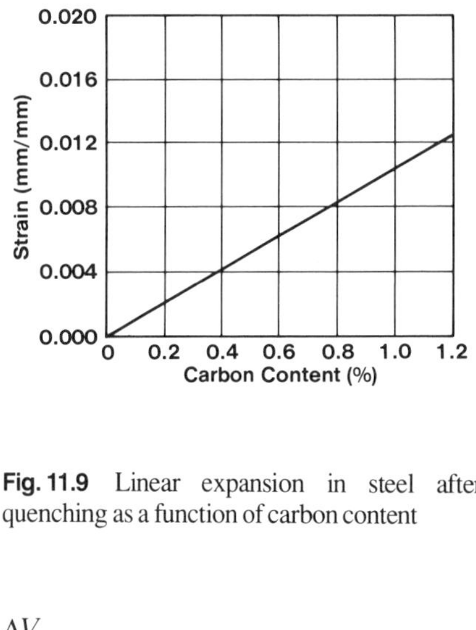

Fig. 11.9 Linear expansion in steel after quenching as a function of carbon content

commonly called martensite. Such a steel undergoes a volumetric contraction during cooling until martensite is formed, at which point expansion occurs (see Fig. 11.8).

The volumetric change can be controlled only by controlling the carbon content of the steel. The data illustrated in Fig. 11.9, calculated from x-ray diffraction data for fully martensitic structures, show that volumetric expansion increases linearly with carbon concentration (Ref 7).

Thelning (Ref 8) reported that volumetric expansion occurring in quenching is described by:

$$\frac{\Delta V}{V} \times 100 = (100 - V_c - V_a) \times 1.68C + V_a(-4.64 + 2.21C) \qquad \text{(Eq 11.6)}$$

where: $(\Delta V/V) \times 100$ equals the percentage change in volume, V_c equals the percentage by volume of undissolved cementite, V_a equals the percentage by volume of austenite, $(100 - V_c - V_a)$ equals the percentage by volume of martensite, and C equals the percentage by weight of carbon dissolved in austenite and martensite, respectively. Kunitake and Susigawa (Ref 9) reported that the tendency for cracking decreases as the start of the martensite transformation temperature (M_s) increases. The M_s temperature was approximated from:

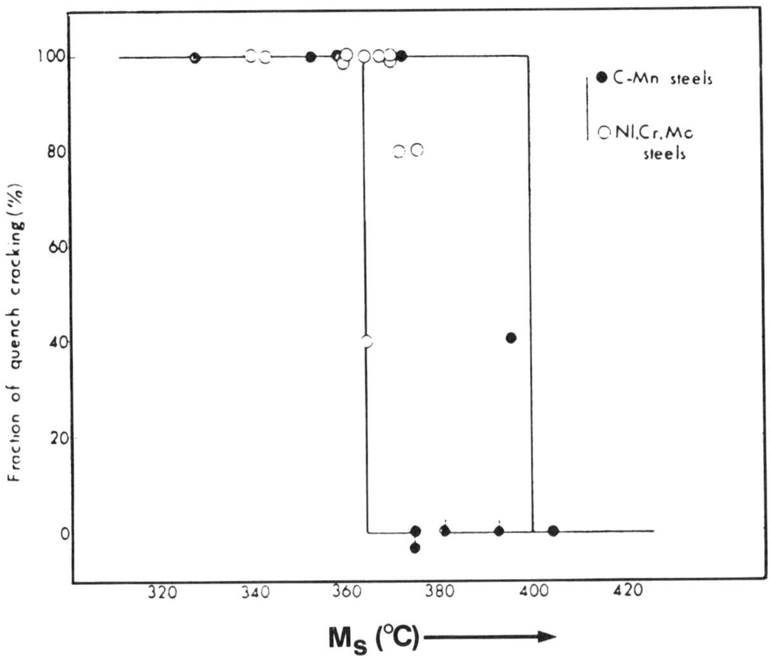

Fig. 11.10 Relationship between quench cracking frequency and M_S temperature

$$M_S(°C) = 521 - 353\,C - 225\,Si - 24.3\,Mn - 27.4\,Ni - 17.7\,Cr - 25.8\,Mo \quad \text{(Eq 11.7)}$$

The correlation between the occurrence of quench cracks and M_S temperature is shown in Fig. 11.10. A similar study produced a poor correlation between austenite grain size and quench cracking, as shown in Fig. 11.11 (Ref 9).

It is known that increasing carbon concentrations increase the propensity for cracking (see Fig. 11.12). However, the correlation of these data is rather poor, indicating that other factors must also be involved in determining the cracking sensitivity of a particular steel.

The relationship between D_I (ideal diameter) and quench cracking is shown in Fig. 11.13. (A detailed description of ideal diameter is provided in Chapter 2.) Although the cracking tendency did increase with D_I, there was considerable data scatter. This indicates that there is a strong interrelationship between cracking tendency and elemental composition, even at a constant hardenability (Ref 6).

Kunitake and Susigawa (Ref 9) developed a relationship to interrelate the combined effect of both carbon content and elemental composition on cracking

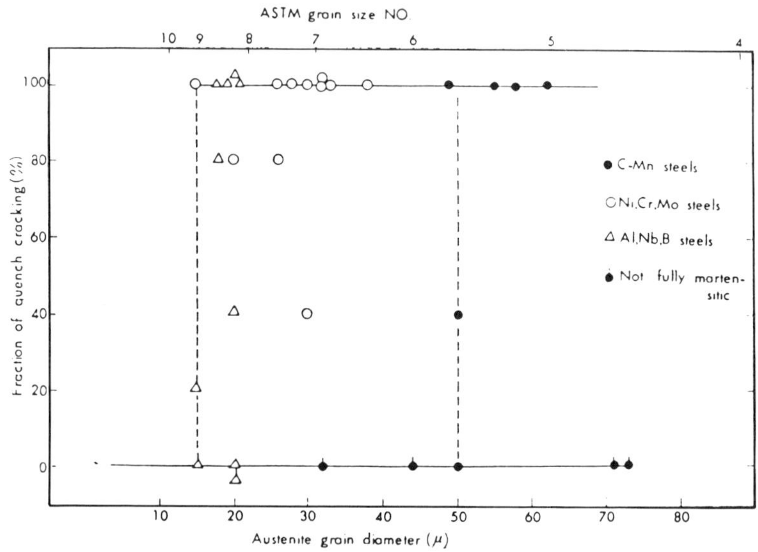

Fig. 11.11 Relationship between quench cracking frequency and austenite grain size

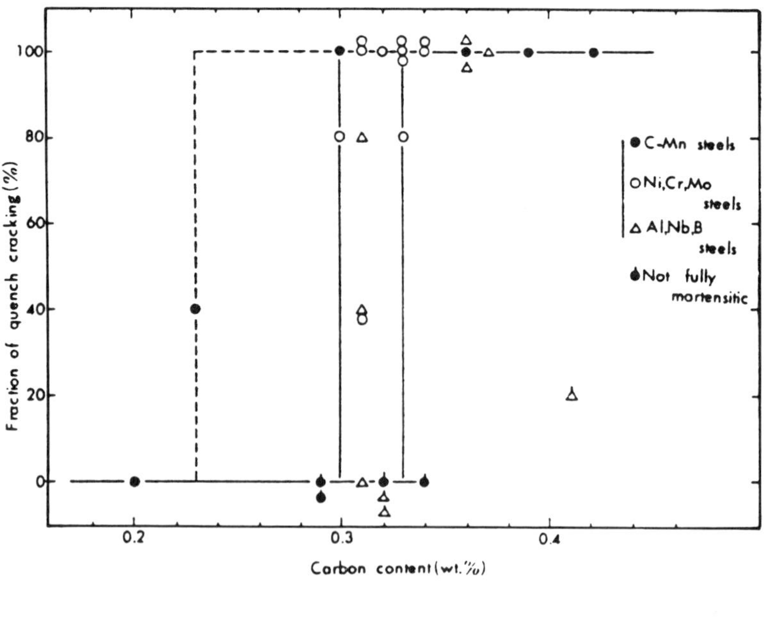

Fig. 11.12 Relationship between carbon content and quench cracking frequency

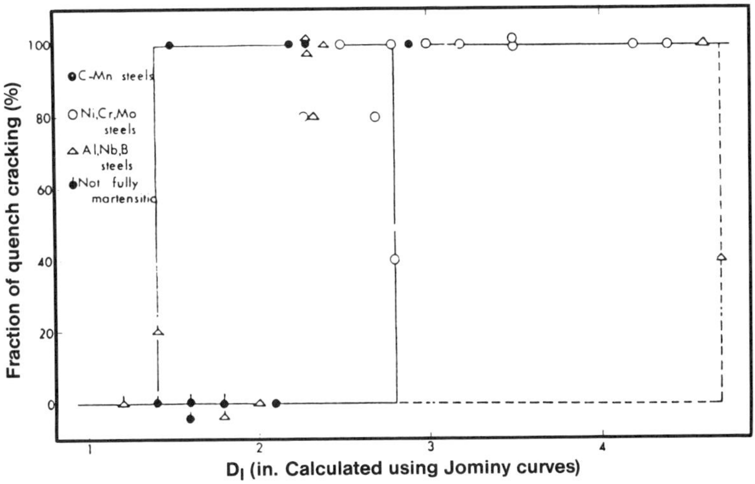

Fig. 11.13 Relationship between quench cracking frequency and D_I

propensity. This was designated as the "carbon equivalent" (C_{eq}) and is calculated by:

$$C_{eq} = C + \frac{Mn}{5} + \frac{Mo}{5} + \frac{Cr}{10} + \frac{Ni}{10} \quad \text{(Eq 11.8)}$$

Figure 11.14 shows a good correlation between carbon equivalent and steel cracking. In general, steels are classified as "crack sensitive" if the C_{eq} value is greater than 0.52 to 0.55% (Ref 9).

Another measure of cracking tendency is the difference in the start and finish temperatures of martensite formation ($M_s - M_f$) (Ref 7). A summary of the M_s and M_f values for some common steels is provided in Table 11.1. The correlation between cracking sensitivity and the transformation temperature range is due in part to the low M_f caused by high-carbon steels (which expand more) and to the fact that wide transformation ranges may result in cracking of the brittle untempered martensite formed at higher temperatures in the transformation range.

Fujio *et al.* (Ref 10) showed that the volumetric expansion caused by martensite formation can be estimated from the maximum cooling rate in a particular type of steel, as shown in Fig. 11.15. Similar correlations were evaluated for both cooling time and cooling rate at the M_s temperature. However, these correlations were dependent on cross-sectional size and thus could not be used for gears or other parts with complex shapes. Volumetric expansion can be estimated for various cross-sectional sizes by a correlation between the volume fraction of martensite versus the cooling rate at the M_s temperature for the steel (see Fig. 11.16).

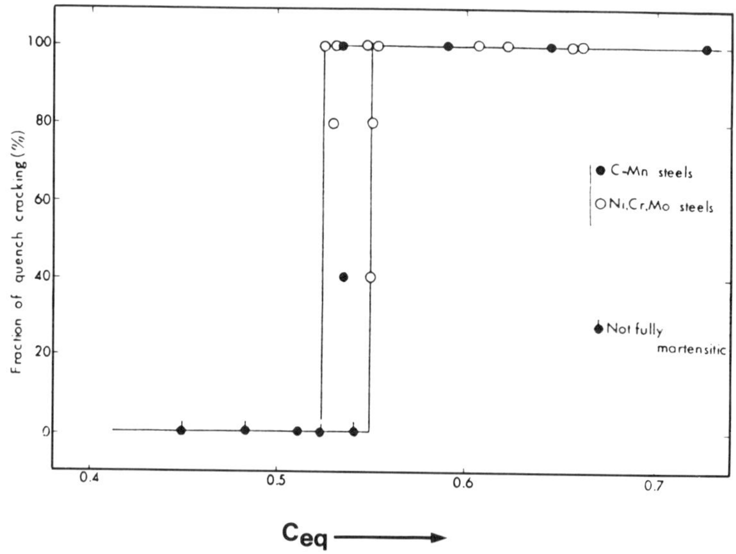

$$C_{eq} = C + \frac{Mn}{5} + \frac{Mo}{5} + \frac{Cr}{10} + \frac{Ni}{10}$$

Fig. 11.14 Relationship between quench cracking frequency and C_{eq}

Table 11.1 M_s and M_f values for selected steels

AISI No.	Austenitizing temperature		M_s		M_f	
	°C	°F	°C	°F	°C	°F
1065	815	1500	275	525	150	300
1090	885	1625	215	420	80	175
1335	845	1550	340	640	230	450
3140	845	1550	330	630	225	440
4130	870	1600	375	710	290	550
4140	845	1550	340	640	220	425
4340	845	1550	290	550	165	330
4640	845	1550	340	640	255	490
5140	845	1550	330	630	240	460
8630	870	1600	365	690	280	540
8695	845	1550	135	275	...	...
9442	860	1575	325	620	−15	4

Nakhimov and Bron (Ref 11) performed studies to separate the volumetric effects due to the steel composition (the first component of deformation) and those effects due to thermal and structural changes (the second component of deformation) and warping (the third component of deformation). To separate

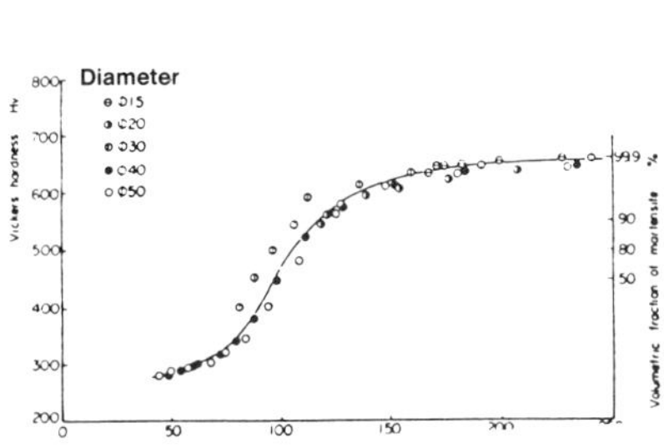

Fig. 11.15 Relationship between maximum cooling rate and volumetric fraction of martensite

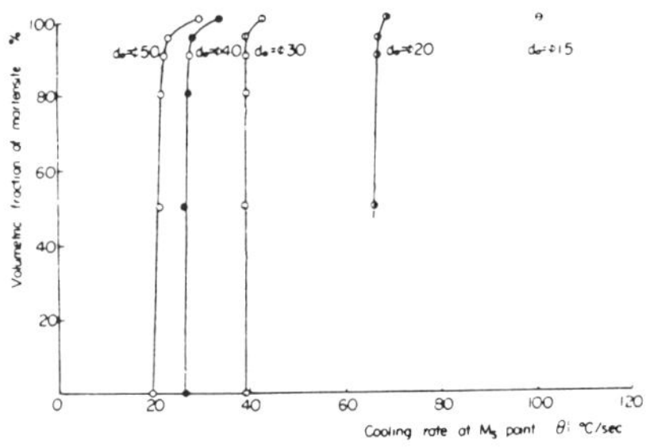

Fig. 11.16 Relationship between cooling rate at the M_S temperature and volumetric fraction of martensite

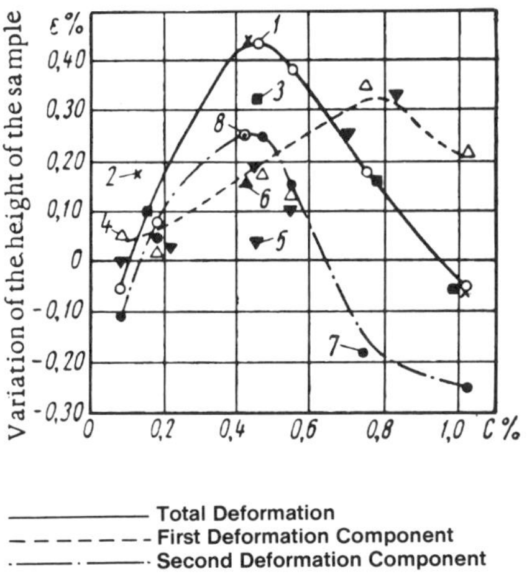

Fig. 11.17 Influence of carbon concentration on the deformation of steel quenched in water

these effects, a cubic steel specimen was quenched three times. The dimensional change after the first quenching (ΔV_1) was considered to be the "total" deformation. The volume changes after the second (ΔV_2) and third quenching (ΔV_3) were considered to be due to the second component of deformation. The deformation due to the first component of deformation was calculated from:

$$\Delta_{vol} = \Delta_1 - \left(\frac{\Delta_2 + \Delta_3}{2}\right) \quad \text{(Eq 11.9)}$$

The influence of carbon content on the first and second deformation products and total deformation is shown in Fig. 11.17. The studies also showed that:

- The first deformation component increases with total carbon content, increasing hardenability, decreasing cross-sectional size, and increasing austenitizing temperature. This component leads to an increase in the size of the workpiece.
- The second deformation component is composed of thermal and structural deformation components of opposite sign. This means that shape deformation can be controlled by controlling heat transfer during quenching.
- The third component, warping, is a function of the uniformity of the quenching process.

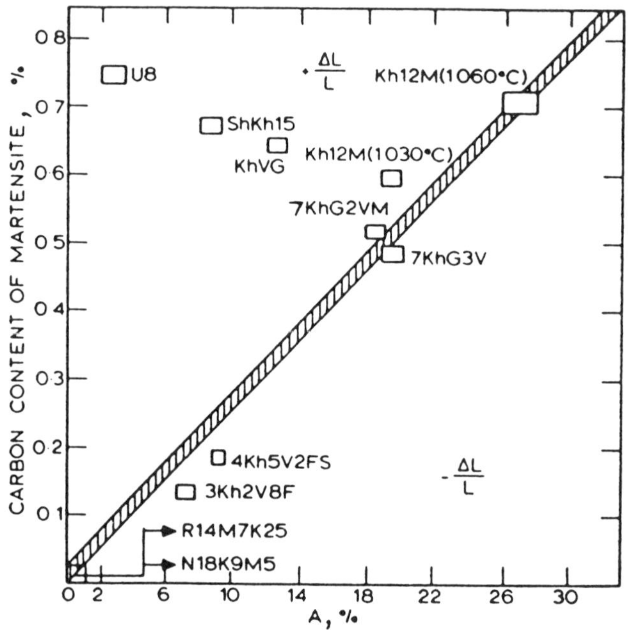

Fig. 11.18 Changes in linear dimensions during quenching relative to carbon concentrations in martensite and retained austenite

It is well known that retained austenite can substantially affect distortion. Geller and Brimene (Ref 12) published a nomogram that can be used to predict dimensional changes caused by total carbon concentration in the martensitic transformation product and the amount of retained ausenite. The following comments will assist in interpreting the nomogram shown in Fig. 11.18:

- The shaded line represents zero distortion. Steels with martensitic carbon contents and retained austenite levels falling on the line will exhibit essentially no distortion.
- Steels with martensitic carbon contents and retained austenite levels that fall below the shaded line will exhibit shrinkage upon quenching.
- Steels that have total martensitic carbon contents and retained austenite levels that fall above the line will exhibit expansion.

This nomogram was developed for various construction and tool steels. Therefore, it should be used with caution for other steel grades (e.g., high-speed tool steels).

Residual Stress, Deformation, and Cracking

Residual Stress

Mayr (Ref 13) determined the influence of carbon content on the specific volume of different phases in plain carbon steels. These results, which are shown in Fig. 11.19, illustrate that the degree of shrinkage upon cooling depends on both the carbon content of the transformation product and the relative amount of various transformation products. Thus, two types of stresses are developed during quenching: thermal and transformation (Ref 13).

The interaction of these stress formation processes is depicted in Fig. 11.20 (Ref 13), which shows three different cooling curves for the surface and core of a solid cylinder. The correspond-

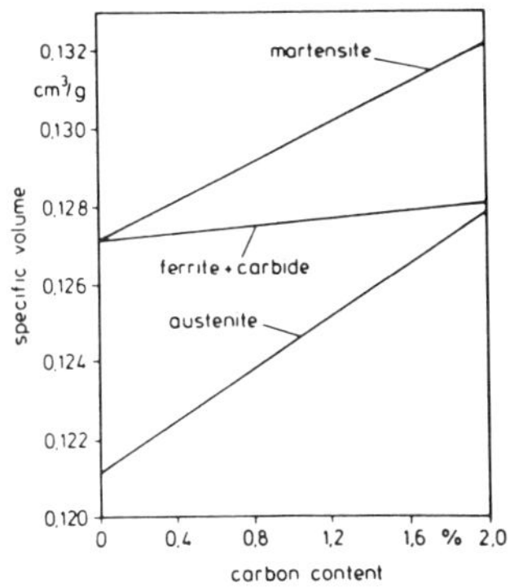

Fig. 11.19 Influence of carbon content on specific volume of different phases in carbon steels

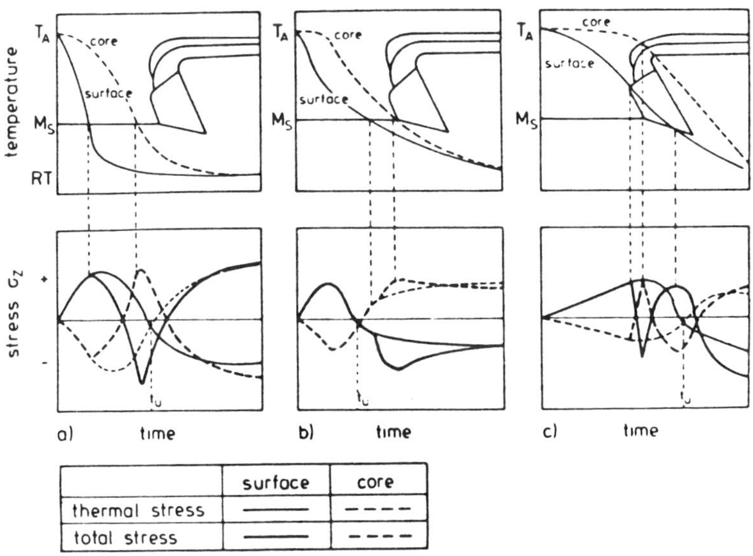

Fig. 11.20 Schematic CCT diagrams with cooling curves for surface and core and calculated behavior of the corresponding thermal and total stresses

ing stress formation curves are also shown. These curves represent three different cooling processes—a, b, and c:

- In case a, the phase transformation occurs at both the surface and in the core before the thermal stresses change sign (t_u). Therefore, the volume increase at the surface will cause an earlier conversion of the total stresses, relative to the thermal stresses. This will cause internal stresses at the core to exceed the hot yield strength, producing a plastic elongation. When the core completes the martensitic transformation, the surface stresses will be tensile and compressive at the core.
- In case b, the transformation begins after the conversion of the thermal stresses. This means that there will be greater compressive stresses at the surface due to the balance of stresses at the center. Thus, there will be little, if any, plastic deformation.
- Case c shows that although transformation begins last at the center, it is completed first. Figure 11.16 shows that the stresses at the surface convert three times, with the final result being a tensile stress at the surface and a compressive stress at the core.

These data show that varying cooling behaviors may exhibit complex effects relative to dimensional stability.

The data in Fig. 11.9 and 11.12 show that increasing carbon content increases both cracking tendency and volumetric expansion upon quenching. Nelson *et al.* (Ref 14) showed that surface residual stress of plain carbon steels is also a function of carbon content, as illustrated in Fig. 11.21. These data show that both hardness and compressive residual stress increase with carbon content.

Wildau and Hougardy (Ref 15) studied the effect of the martensite transformation for a 50CrV4 cylindrical steel specimen (M_S = 270 °C, or 520 °F) on longitudinal, tangential, and radial residual stresses. These results (see Fig. 11.22) show that the stresses became increasing tensile with increasingly case depth.

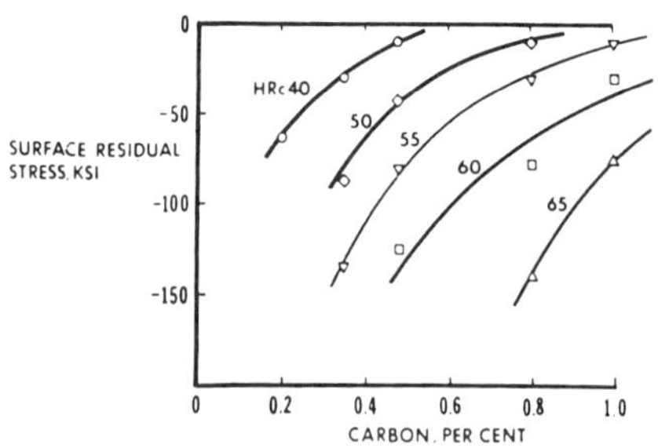

Fig. 11.21 Surface residual stress as a function of carbon content for various steel hardnesses

Figure 11.23 shows the stresses generated at the surface and core of a 30 mm (1.2 in.) diam solid cylinder of 50CrV4 steel quenched in water as a function of time. Under these quenching conditions, there is a 35 s delay in martensite formation at the core rela-

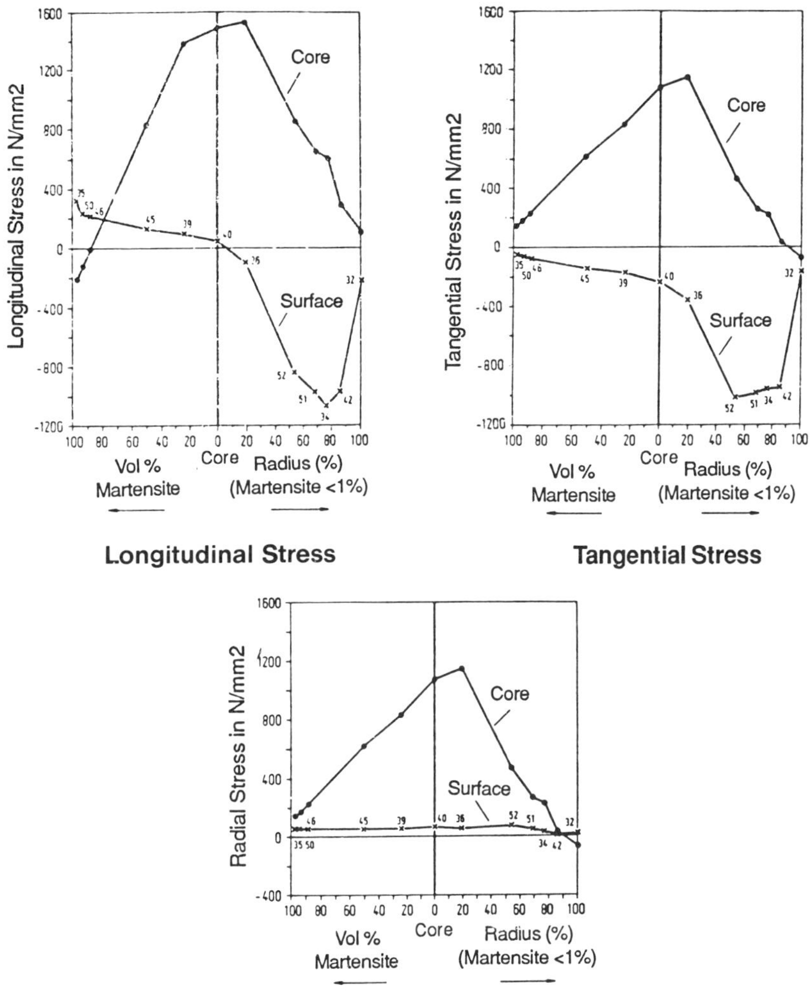

Fig. 11.22 Quench stress generation as a function of cross-sectional size and martensite content for 50CrV4 steel

tive to the surface, leading to the formation of tensile stresses at the core and compressive stresses at the surface.

Okamura and Kawashima (Ref 16) studied stress formation during quenching computationally using a finite-element method. The material properties used for this work are tabulated in Table 11.2 and are illustrated as a function of temperature in Fig. 11.24(a) and (b). The calculated and measured distortions obtained

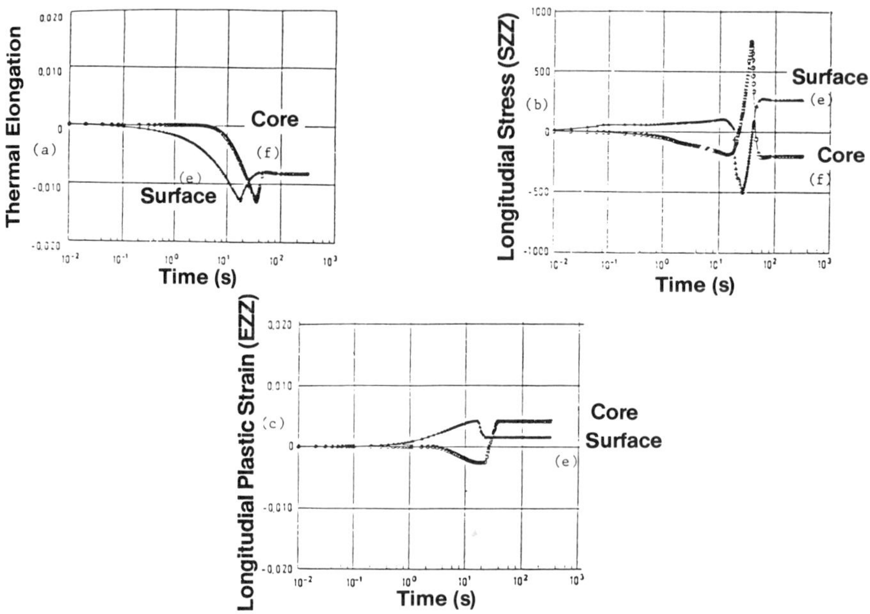

Fig. 11.23 Distortion as a function of residual stress

Table 11.2 Steel material used for stress calculations

Transformation product	Density (ρ), kg/m^3	Heat capacity (C_p), kcal/kg · °C	Thermal conductivity (ε), kcal/m · h · °C	Enthalpy (ΔH), kcal/kg	β, %
Austenite	7.9×10^3	0.1405	$8.67 \times 10^{-3}\,T + 17.5$	...	−0.27
Bainite	7.9×10^3	$1.056 \times 10^{-4}\,T + 0.105$	$-9.72 \times 10^{-3}\,T + 32.1$	18.0	0.30
Martensite	7.9×10^3	$1.056 \times 10^{-4}\,T + 0.105$	$-9.72 \times 10^{-3}\,T + 32.1$	20.0	0.60

after quenching are shown in Fig. 11.25. This work showed that both deformation and residual stress are related to transformation behavior during quenching.

A similar study was performed by Fernandes *et al.* (Ref 17) on a 132 × 65 mm (5.2 × 2.6 in.) diam cylindrical specimen of an AFNOR XC80 steel. The temperature-dependent material properties for austenite and plastic transformation products used for this work are shown in Fig. 11.26. The enthalpy relationship used for the transformation from austenite to pearlite is (Ref 17):

$$\Delta H\ (\text{J/m}^3) = 1.59 \times 10^9 - 1.50 \times 10^6 \times {}^\circ\text{C} \qquad \text{(Eq 11.10)}$$

The total stresses (σ_e) were calculated from (Ref 17):

$$\sigma_e = \left(\frac{1}{\sqrt{2}}\right) \times \sqrt{(\sigma_r - \sigma_\theta)^2 + (\sigma_\theta - \sigma_z)^2 + (\sigma_z - \sigma_r)^2} \quad \text{(Eq 11.11)}$$

where σ_r is the radial stress, σ_θ is the tangential stress, and σ_z is the axial stress.

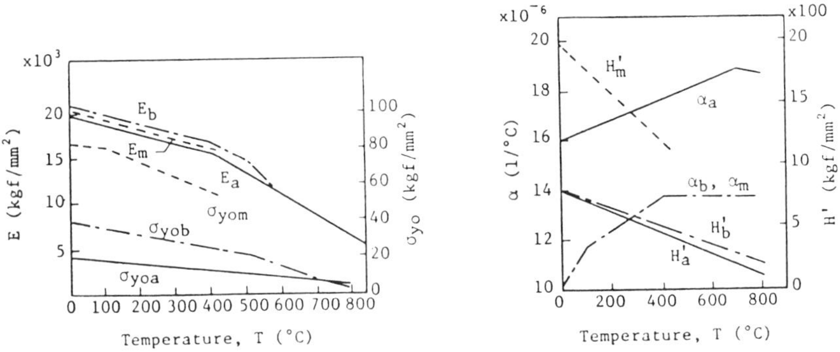

(a) (b)

Fig. 11.24 Material properties of austensite, bainite, and martensite as a function of temperature. **(a)** Modulus and stress versus temperature. **(b)** Thermal conductivity and density versus temperature

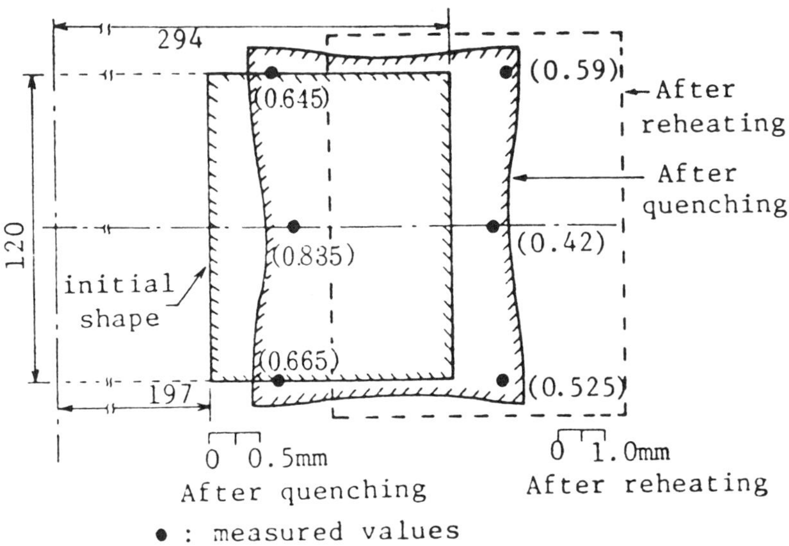

Fig. 11.25 Dimensional changes after quenching

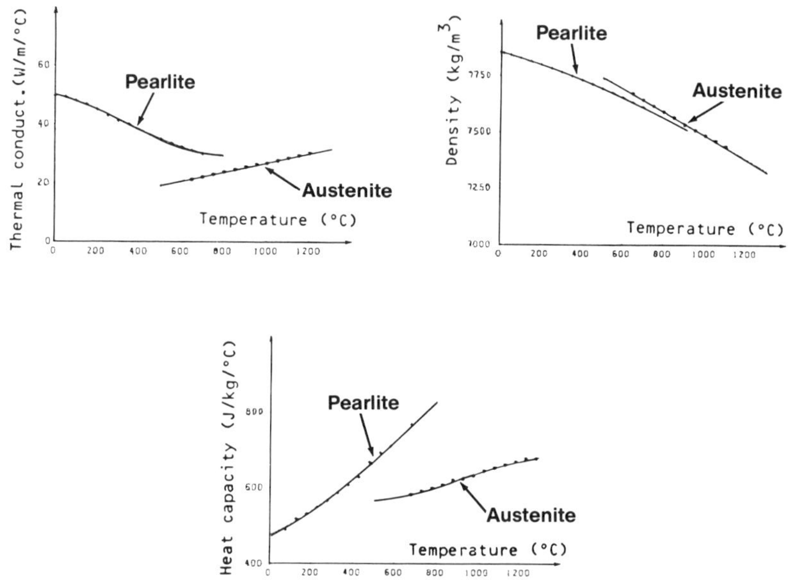

Fig. 11.26 Thermophysical properties of austenite and pearlite as a function of temperature

Table 11.3 Total stress as a function of mean cooling rate

Specimen location	Total stress, MPa (ksi), at a mean cooling rate, °C/s (°F/s), of: 4.6 (8.3)	7.4 (13.3)	14.3 (25.7)	17.0 (30.6)
Center	10 (1.5)	19 (2.8)	35 (5.1)	40 (5.8)
Surface	20 (2.9)	36 (5.2)	62 (9.0)	70 (10.2)

Note: Using 13 × 65 mm (0.5 × 2.5 in.) cylindrical test specimens

The stress data in Table 11.3 were obtained for the point just prior to the austenite to pearlite transformation. These data show that the total stresses and the stress difference between the surface and core increase with cooling rate. The data in Table 11.3 also indicate that the internal stresses during quenching may affect the transformation behavior of steel and are a function of cooling rate. Therefore, appropriate corrections should be made for residual stress calculations (Ref 17).

It has been shown that both thermal and transformation stresses are encountered during quenching. It is often assumed that each of these stresses has opposite signs. Kuzmin (Ref 18) studied the effects of these stresses on the residual stresses produced as martensite forms during quenching. The results of

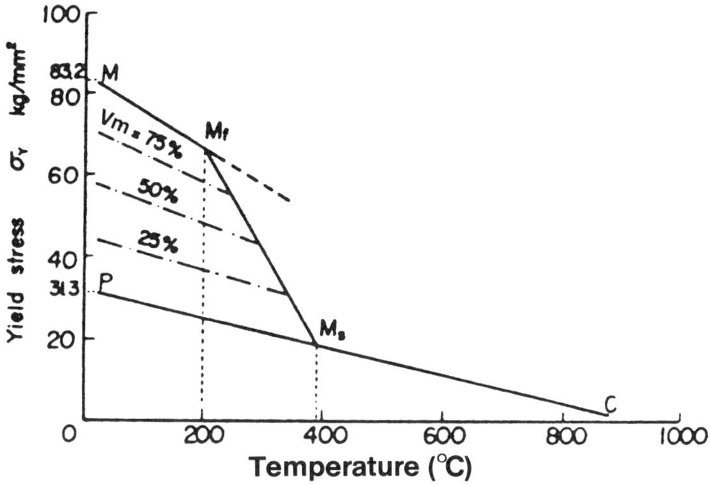

Fig. 11.27 Temperature dependence of yield stress

this study showed that during martensitic transformation, elastic deformations become residual deformations. In general, tensile stresses stimulate martensite formations, and compressive stresses retard them. It was also shown that during quenching, compressive stresses formed in the core increase with cooling rate because of a relatively large volumetric expansion in the core. To reduce residual stresses, rapid cooling during the initial martensitic formation at the surface, followed by slow cooling, is desirable. (Higher surface cooling rates produce larger martensite aggregates, which have lower surface tensile stresses.) This may be done by a two-step quench in water and oil (Ref 18). The variation in yield stress with respect to temperature for a plain carbon S45C steel is dependent on the degree of martensitic transformation, as shown in Fig. 11.27 (Ref 19).

Although martensitic transformation occurs with a volumetric expansion, there is a net shrinkage of about 1% during cooling from the austenitizing temperature to room temperature. During rapid cooling, the outer surface is in tension and the center is in compression before the transformation to martensite begins. A practical example of this is illustrated by the quenched bar in Fig. 11.28 (Ref 20).

In another example, a bar 75 mm (3 in.) long and 22 mm (⅞ in.) in diameter was quenched in water from 775 °C (1425 °F). Deformation occurred in the center of the bar resulting in a 0.04 mm (0.0015 in.) increase in diameter (see Fig. 11.29). If the hardened surface of the bar were wet-ground to relieve the internal stress and produce a final diameter of 7.6 mm (0.3 in.), the final length of the bar would be 0.2 mm (0.008 in.) less than the as-quenched length. A

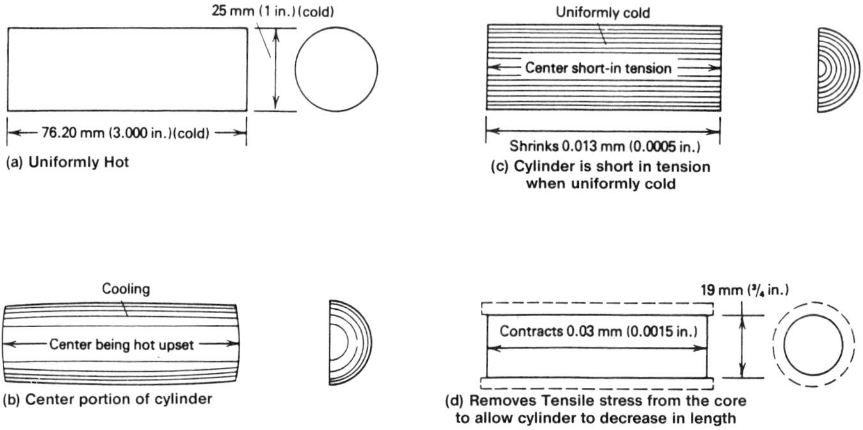

Fig. 11.28 Effect of quenching on a steel cylinder quenched in water from 775 °C (1425 °F)

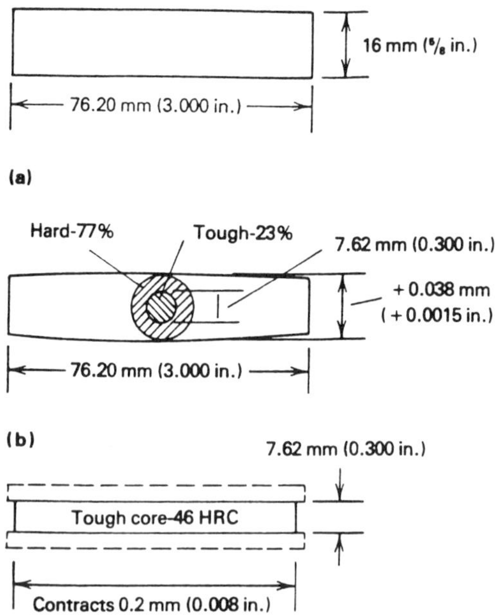

Fig. 11.29 Strains produced by hardening a 1.1% C tool steel cylinder. **(a)** Dimensions of an annealed bar before quenching. **(b)** Hardness. **(c)** Core hardness after water quenching; cylinder diameter reduced from 16 to 7.62 mm (0.6 to 0.3 in.) by wet grinding

variety of similar experiments have been conducted to demonstrate that tensile and compressive stresses developed during cooling and martensite formation can result in significant deformation of steel parts (Ref 20).

The stress states in a through-hardened 0.6% C steel, an unhardened 0.25% C steel, and a case-hardened 0.3% C steel after quenching are shown in Fig. 11.30. The corresponding length and diameter changes after quenching 50 mm (2 in.) bars into water from 850 °C (1560 °F) are also shown (Ref 21).

Distortion

Two components produce distortion during quenching: the presence of thermal gradients and volumetric expansion during martensite formation. Deformation resulting from thermal gradients produced during cooling is illustrated in Fig. 11.31 (Ref 22). If a part is initially uniformly hot and is rapidly quenched, the outer surface shrinks while the center remains relatively hot. This process

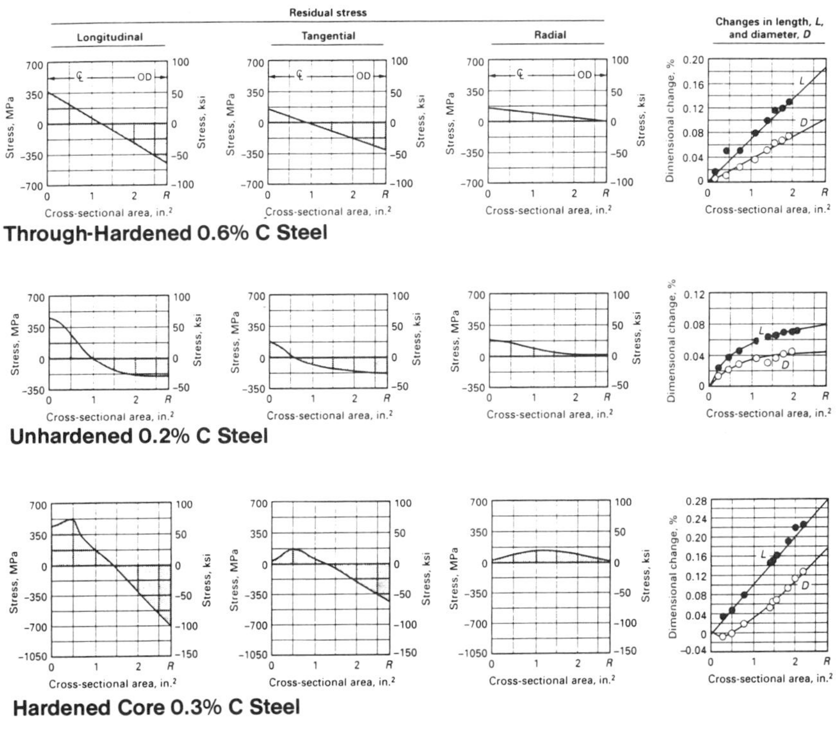

Fig. 11.30 Residual stress distribution and distortion in 50 mm (2 in.) diam carbon steel bars quenched in water from 850 °C (1560 °F)

puts the outer surface into tension and the inside into compression, causing internal plastic flow. As the center of the part cools and the temperature reaches a uniformly low value, contraction in the center of the bar occurs, which reverses the stress state and places the center in tension (Fig. 11.31c) (Ref 22). For this reason, it is generally desirable to minimize the thermal gradient between the surface and the center of a part during quenching to minimize the potential for cracking and distortion. For the same reason, it is also important to minimize thermal gradients across the cooling surface. This is done by providing optimal agitation throughout the quench.

Another example of distortion caused by thermal gradients is illustrated in Fig. 11.32 (Ref 22), which shows a bar uniformly quenched on all surfaces and a bar quenched on only one side (Ref 23). When the bar is quenched on only one side, the faster-cooling side will contract earlier and at a more rapid rate than the opposite side. The rapidly cooled side becomes shorter and stronger (harder) as it cools, causing plastic deformation on the hot side. The deformation is followed by cooling and contraction of the more slowly cooled side.

Quenchant temperature can also significantly affect distortion. Murzin *et al.* (Ref 24) found a substantial reduction in distortion of a gear rim and splined

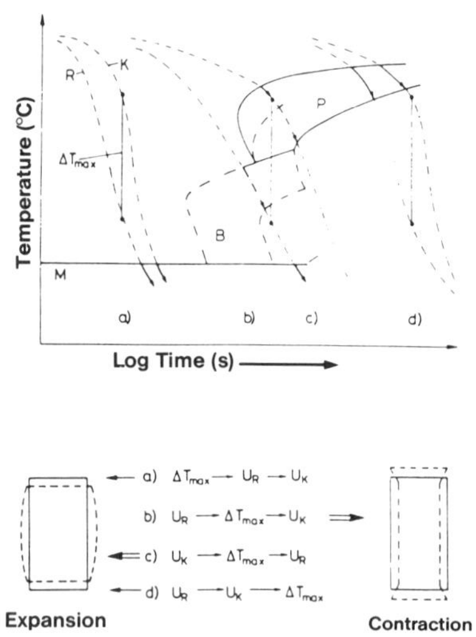

Fig. 11.31 Size change by phase transformation and temperature difference during quenching. U, transformation; R, surface; K, core

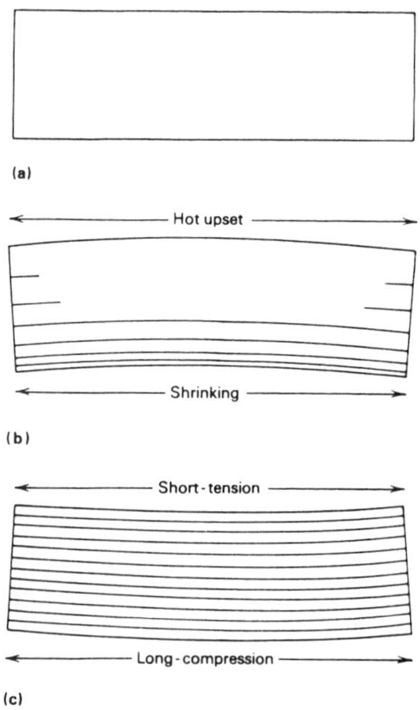

Fig. 11.32 Warpage caused by nonuniform quenching of a steel block. **(a)** Uniformly hot. **(b)** Uniform cooling. **(c)** Uniformly cold

holes by quenching in oil at 170 °C (340 °F) compared with quenching in oil at 70 °C (160 °F). On the other hand, the distortion of spline grooves was greater in the hot oil quench. However, the relative amount of distortion in either quenchant was critically dependent on racking procedures. On the basis of this work, the following recommendations were made for the quenching of gear wells used in heavy-duty transmissions:

- Accelerated air cooling of the forged blocks
- High-temperature tempering (4 h at 620 to 640 °C, or 1150 to 1185 °F) after rough machining
- Carburizing in the horizontal position
- Heating in a protective atmosphere and quenching in hot oil in the horizontal position

Pavee (Ref 25) found that a substantial reduction in distortion in carbonitrided stepdown gears could be achieved by 30 s delayed agitation after immersion in the cold (35 to 40 °C, or 95 to 105 °F) quench oil.

Kakiuchi (Ref 26) examined the effect of water temperature on the distortion of hollow cylindrical test specimens for the four steel grades shown in Table 11.4. These results, illustrated in Fig. 11.33, showed that:

- Low-carbon steels (SUYB1 and SNCM420) produced decreasing distortion with increasing water temperature.
- Quench-hardenable steels (SUP9 and SKS3) exhibited greater distortion with increased martensite formation.

Toshioka (Ref 27) examined the relative distortion in plates with holes that were quenched into both hot and cold oil. The results of this study, shown in Fig.

Table 11.4 Steel compositions used by Kakiuchi (Ref 26)

Element, %	1 (SUYB1)	2 (SNCM420)	3 (SUP9)	4 (SKS3)
Carbon	0.01	0.20	0.58	0.86
Silicon	0.21	0.27	0.26	0.27
Manganese	0.20	0.67	0.82	1.20
Phosphorus	0.007	0.010	0.023	0.012
Sulfur	0.009	0.024	0.018	0.007
Copper	0.01	0.09	0.01	0.04
Nickel	0.02	1.55	0.02	0.02
Chromium	0.01	0.62	0.84	0.47
Molybdenum	...	0.20	...	0.12
Tungsten	...	...	...	0.40
Lead	0.16	...	...	...
M_s temperature, °C (°F)	538 (1000)	417 (783)	297 (567)	188 (370)

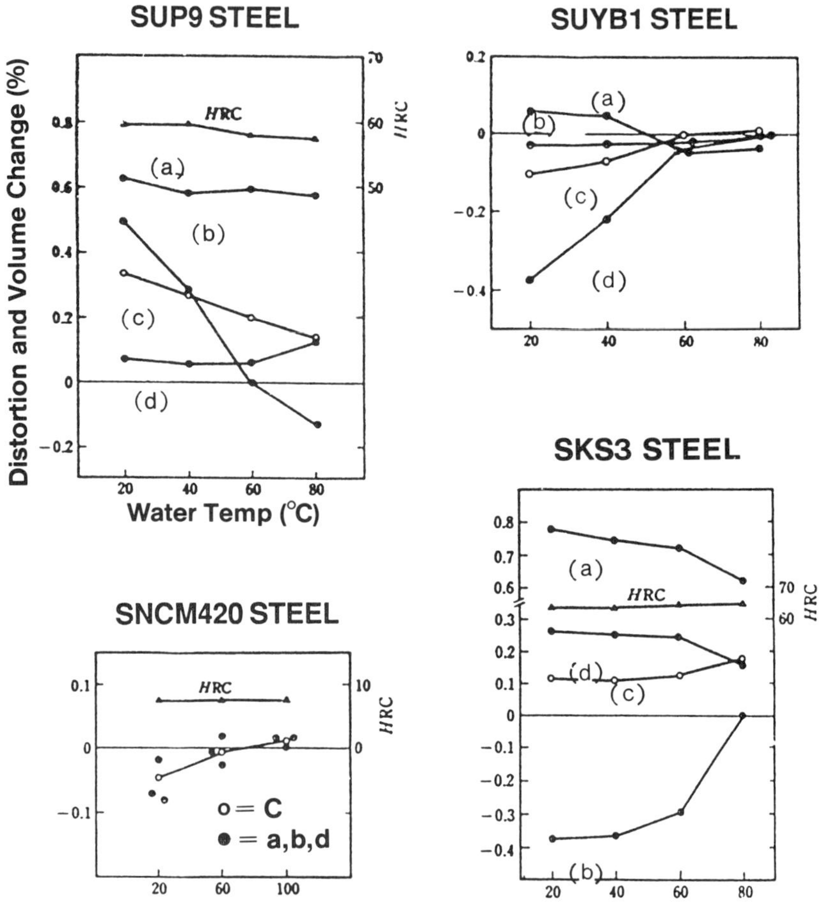

Fig. 11.33 Distortion of four steels as a function of water quenchant temperature. a, thickness; b, volume; c, outer diameter; d, inner diameter

11.34, demonstrate that an increase in longitudinal dimension was greater for the hot oil, although hole distortion was observed due to lower thermal shock during quenching, which produces less contraction.

Kakiuchi (Ref 28) also reported that for quench-hardenable steels, tempering up to 600 °C (1110 °F) returns the specimen volume to approximately the initial condition. However, tempering did not restore the outer diameter, inner diameter, or thickness to the original dimensions. Approximately 50% of the residual stress was removed by tempering at 200 to 300 °C (390 to 570 °F); 100% was removed by tempering at 500 to 600 °C (930 to 1110 °F).

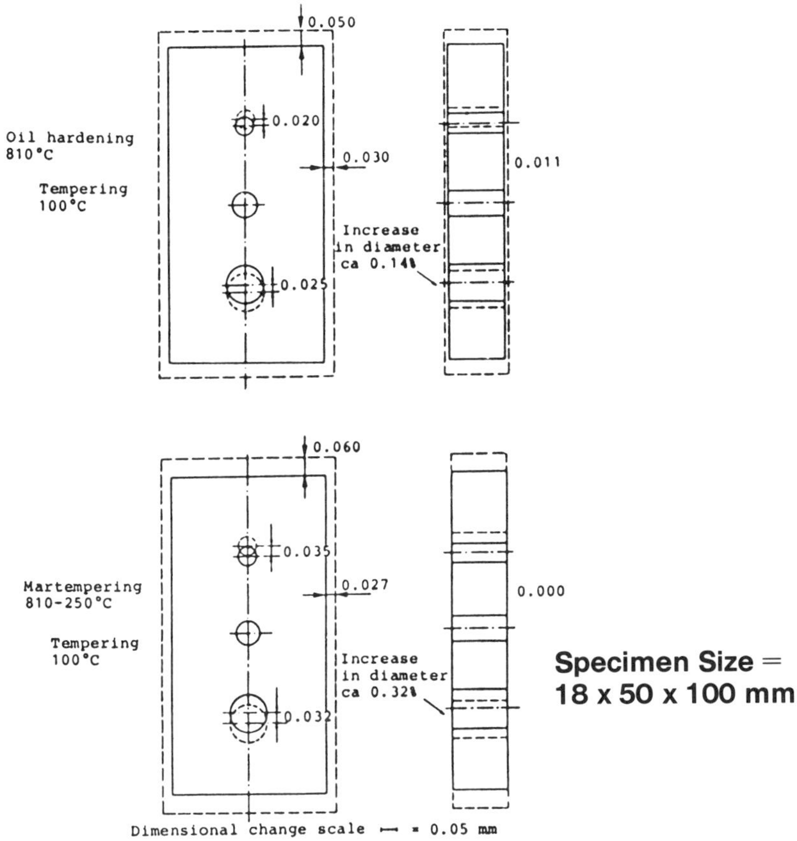

Fig. 11.34 Dimensional changes during hardening of plates made from O1 steel after quenching in oil and martempering

Uzlov and Danchenko (Ref 29) proposed that the optimal heat treatment conditions for a steel could be determined from a multilevel correlation of cooling rates from the austenitizing temperature with various properties, including residual stress and hardness. Representative correlations for three steels whose compositions are given in Table 11.5 are shown in Fig. 11.35 (Ref 29).

In general, the distortion that occurs during quenching depends on the size and shape of the bar, bar composition, and the quenchant. Parts with section size ratios greater than 1 to 4, large parts with relatively thin cross sections, and parts containing slots, keyholes, drilled holes, or grooves cause problems because of the difficulty in achieving uniform heating and cooling rates. Some of these difficulties are depicted in Fig. 11.28, 11.32, and 11.34 (Ref 30).

Table 11.5 Steel compositions used by Uzlov and Danchenko (Ref 29)

	Steel		
Element, %	**30**	**50**	**67**
Carbon	0.34	0.51	0.74
Silicon	0.33	0.30	0.25
Manganese	0.72	0.60	0.27
Sulfur	0.012	0.009	0.011
Phosphorus	0.018	0.010	0.015
Chromium	0.09	0.05	0.06
Nickel	0.06	0.07	0.07
Copper	0.06	0.06	0.05

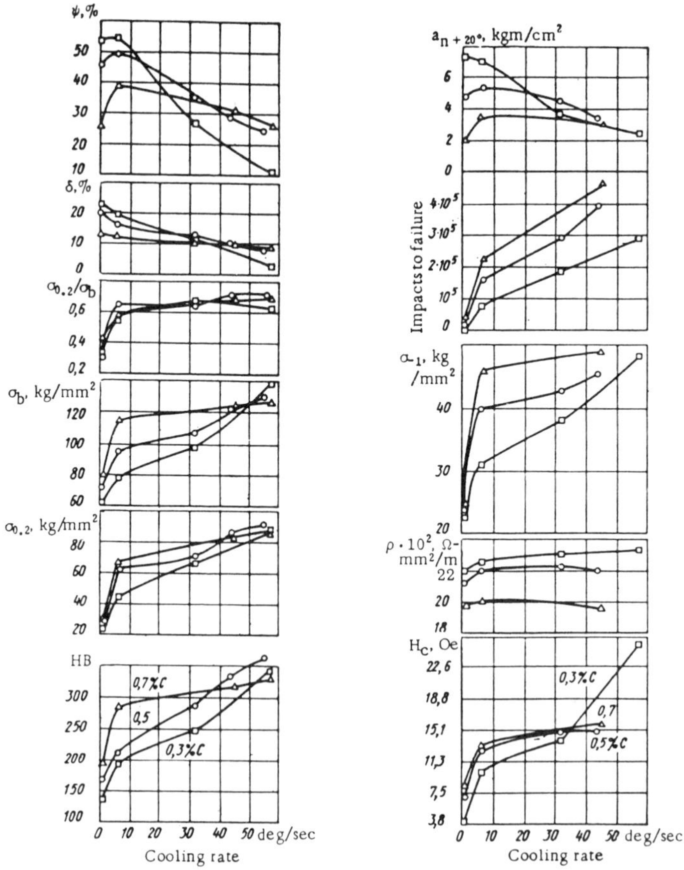

Fig. 11.35 Effect of cooling on mechanical and physical properties of steel

Figure 11.36 illustrates the distortion that can occur in a track shoe as a result of the presence of a lightening groove. Flanges and bosses also create conditions that make it difficult to achieve uniform cooling rates (Fig. 11.37). A large-diameter flange is very difficult to cool at a rate consistent with the bar, and this almost invariably produces cracks.

Keyholes and grooves often cause a shaft to warp during heat treating, and keyways with sharp corners often initiate cracks. The keyway cracking problem can be minimized to some extent by designing the part with generous radii (Fig. 11.38).

Cracking

It is difficult, if not impossible, to obtain a "uniform" quench when parts with a variety of cross-sectional sizes are subjected to an immersion quench. One

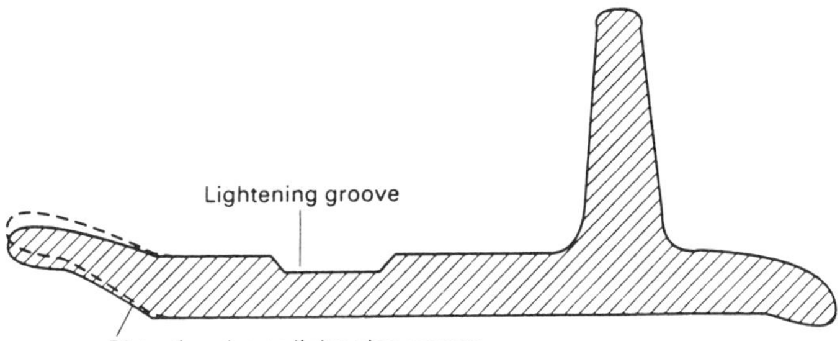

Fig. 11.36 Cross section of a water-quenched SAE/AISI 1037 steel track shoe with 0.25 mm (0.01 in.) distortion caused by a lightening groove

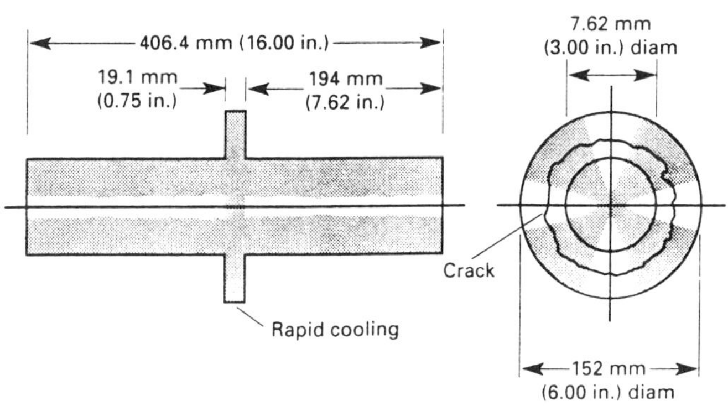

Fig. 11.37 Steel flange design prone to quench cracking

method of addressing this problem is to use a submerged spray, or jet cooling (Ref 31). Uniform quench severity can be obtained by varying the spray intensity of the jets at critical impingement points on the part.

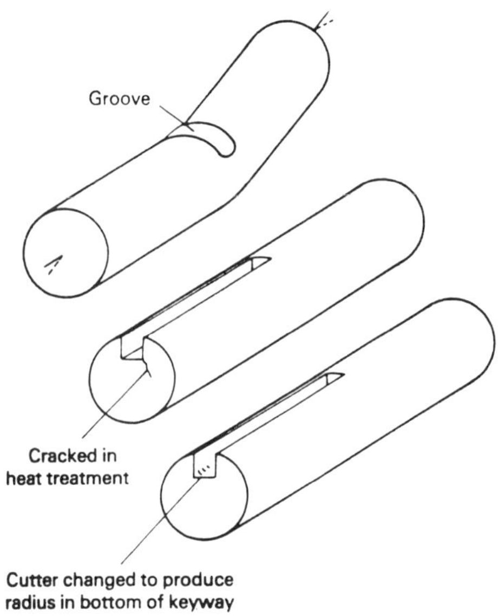

Fig. 11.38 Grooves as a cause of distortion or cracking

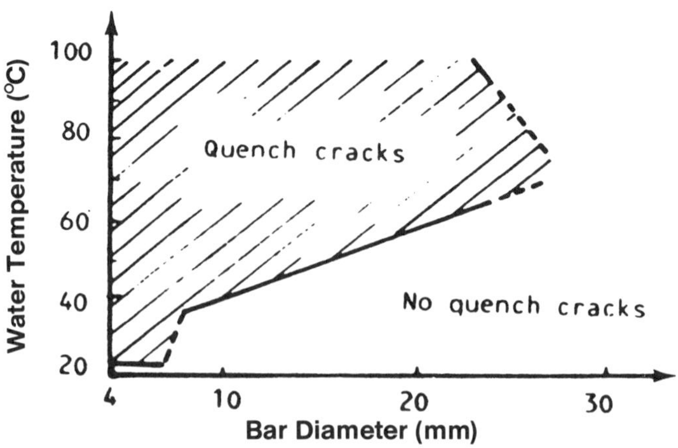

Fig. 11.39 Quench cracking as a function of water quenchant temperature and bar diameter (0.6% C, 1.6% Si, 0.5% Cr)

Traditionally, it has been proposed that the cooling rate in the region of the M_S-M_f transformation should be minimized to decrease the potential for cracking. Recently, Zhelokhovtseva (Ref 32) has shown that increasing the cooling rate in the M_S-M_f range either reduced or eliminated cracking for the steels listed in Table 11.6. The test specimens used to generate the data in Table 11.6 were prismatic in shape (10 × 10 × 55 mm, or 0.4 × 0.4 × 2.2 in.). Heat exchange was provided by a hydrodynamic radiator, and agitation was provided by an ultrasonic vibrator. (The use of ultrasonic agitation provides far more uniform destabilization of the vapor blanket during cooling, as well as more uniform bubble nucleation during the nucleate boiling stage.)

The data in Table 11.6 show that when sufficient ultrasonic agitation was used to produce a cooling rate for the quench oil that was approximately equivalent to water, no cracks were observed. (It should be noted that the water quenchant was used without agitation.) These data suggest that the cooling rate in the M_S-M_f range is not critical, provided that it is uniform (Ref 32).

Cracking is actually a stress-relieving process (Ref 33). If stress reduction by plastic deformation is impeded, the potential for cracking increases. The relative

Table 11.6 Dependence of cracking on cooling rate during martensitic cooling

Steel grade	Quenchant	USV(a) pressure, atm	Cooling rate of center at 300-200 °C (570-390 °F) °C/s	°F/s	Number of cracked specimens out of 100
45	Water	...	150	270	25
	Oil	...	10.4	18.7	0
	Oil	5	21	38	0
	Oil	10	61	110	0
	Oil	15	76	137	0
	Oil	20	90	162	1
	Oil	25	144	259	0
U8	Water	...	65	117	75
	Oil	...	6	11	2
	Oil	5	25	45	2
	Oil	10	29	52	3
	Oil	15	35	63	2
	Oil	20	44	79	3
	Oil	25	52	94	2
U10	Water	...	66	119	65
	Oil	...	6.3	11.3	2
	Oil	5	33	59	3
	Oil	10	45	81	2
	Oil	15	52	94	1
	Oil	20	64	115	2
	Oil	25	60	108	2

(a) USV, ultrasonic vibrator

Table 11.7 Size changes and residual stress during heat treatment

Process	T(c)	$\Delta V/V$(d)	σ_E(e)	Example
Microstructure change	Small	No	0	Tempering precipitation hardening
Thermal stress	Large	0	<0	Quench austenitic steel
Thermal transformation and stress(a, b)				
Transformation R + K	Large	>0	>>0	Through-hardening in air
after thermal stress max	Small	>>0	>>0	Through-hardening in water
Transformation R before K *after* thermal stress max	Large	>0	>0	Medium hard penetration
Transformation K before			<0	Shallow hard penetration
R *after* thermal stress max	Large	0 to >0		
Transformation R + K	Small	No	No	
before thermal stress max	Large	No	<0	Normalizing, no quench

(**a**) The metallurgical transformations corresponding to these processes are shown in Fig. 11.31. (**b**) R = surface, K = core. (**c**) T, temperature difference between the surface and the core ($T_K - T_R$). (**d**) $\Delta V/V$, relative volume change. (**e**) σ_E, remaining residual stress at the surface

effects of thermal and volumetric changes on distortion and cracking are summarized in Table 11.7. The metallurgical transformation processes that correspond to the cooling processes indicated in Table 11.7 are illustrated in Fig. 11.30. The data in Table 11.7 indicate the potential for cracking (Ref 33).

It should be noted that carbon concentrations higher than 0.4% will result in the formation of plate martensite. The danger of microcracking rises as the concentration of carbon increases above 0.4%. Other potential causes of cracking include: high localized stress due to part geometry, nonuniform quenching, and surface decarburization. Decarburization produces a higher M_S temperature, thus causing martensite formation sooner than the regions below the decarburized surface (Ref 33).

Moreaux *et al.* (Ref 34) studied the effect of cross-section geometry and quenchant (water) temperature on the potential for crack formation for cylindrical test specimens with a length three times the diameter. A steel containing 0.60% C, 1.6% Si, and 0.5% Cr was used. The results given in Fig. 11.39 (Ref 34) show that the transition temperature from film to nucleate boiling will affect the thermal stresses of austenite formation. The nucleate boiling to convective cooling transition will primarily affect transformation stresses.

Beck (Ref 35) studied the effect of the interrelationship between quench severity and average cooling rates on the severity of crack formation. This work was conducted using a 34 mm (1.3 in.) diam cylindrical specimen of 60SC7. Generally, the propensity for cracking increased with the severity of the quench (Fig. 11.40) (Ref 35).

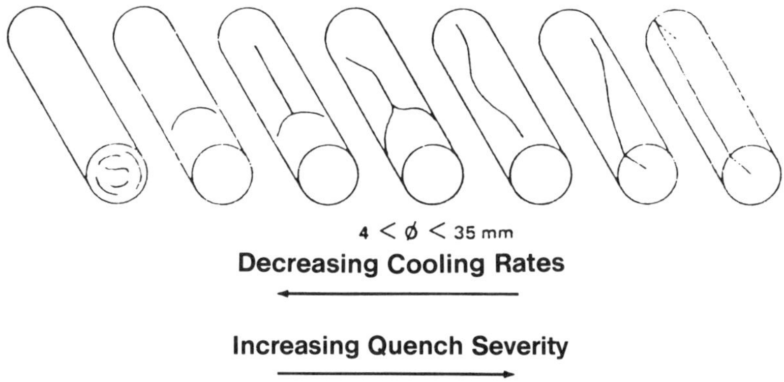

Fig. 11.40 Quench cracking morphology of a 60SC7 steel cylinder quenched in water

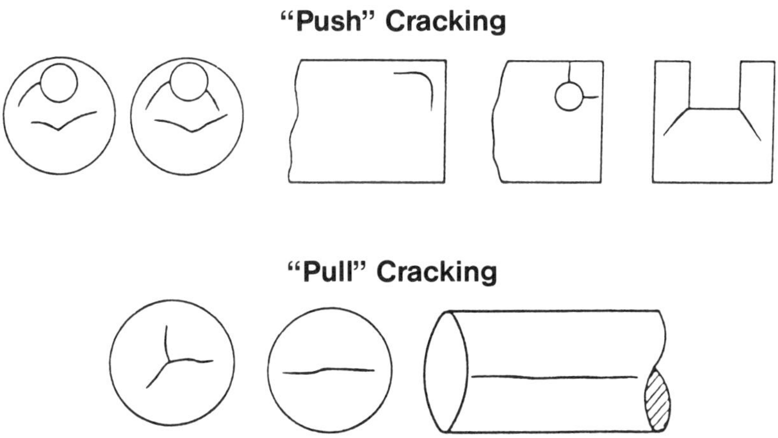

Fig. 11.41 Classification of quench cracking

There are two forms of cracking: pull-cracking and push-cracking (Fig. 11.41) (Ref 36). Pull-cracking is caused by a "pull" stress due to uneven surface cooling between the austenitizing temperature and M_s during quenching, and generally occurs where effective quenching has been delayed. Push-cracking is caused by a "push" stress due to uneven quenching within the body of the part between M_s and M_f, and occurs in rapidly cooled areas (Ref 36).

Cracking is often initiated by a "trigger," such as a notch (stress concentrator) in the part. Notches take two forms: geometric notches and notches such as grooves, keyways, and so on (Ref 36).

Examples of Favorable Stresses

The minimization or elimination of stress during quenching is not always desirable. For example, in the manufacture of carburized or induction-hardened gears and bearing parts, the tensile stresses created by bending and contact fatigue should be balanced (Ref 37, 38). Figure 11.42 shows the stress profile of induction-hardened AISI 51100 steel as a function of case depth (Ref 37). Very large compressive stresses are observed (–345 MPa, or 50 ksi) which are substantially higher than those produced by carburizing and quenching.

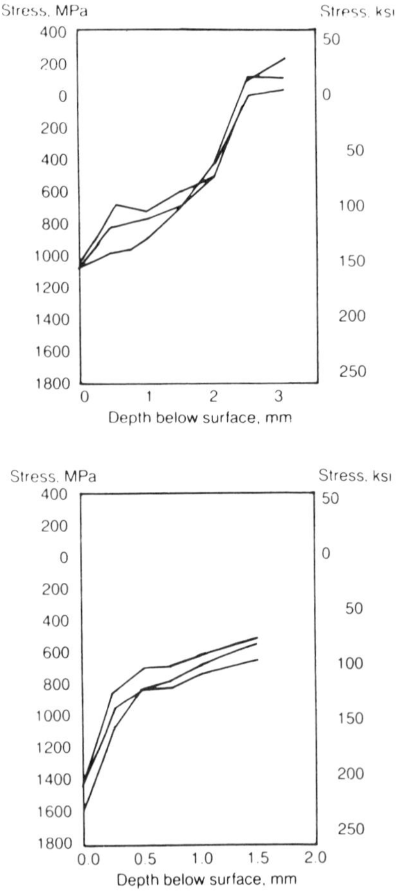

Fig. 11.42 Residual stress profile of a bearing race after rapid heating and cooling. (**a**) Outside diameter surface. (**b**) Inside diameter surface

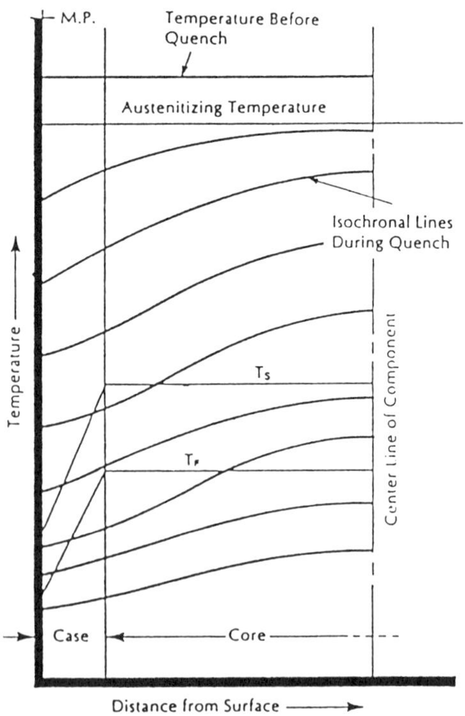

Fig. 11.43 Temperature distribution in steel parts during quenching after carburizing

Howes (Ref 38) published a diagram interrelating the temperature distribution as a function of case depth of a carburized steel part after quenching from the austenitizing temperature. The isochronal lines in Fig. 11.43 show how the surface cools faster than the center. The T_S and T_F values in Fig. 11.43 represent the start and finish temperatures of the martensitic transformation, respectively. These temperatures are depressed as the carbon concentration in the case increases. Thus, the carburized case will transform after the core transformation has occurred, resulting in a compressive stress at the surface.

Motoyama *et al.* have measured surface residual stresses for two carburized steels—S20C (Ref 39) and SCr22 (Ref 40)—as a function of the surface carbon content. The residual stress and carbon content correlations for these two steels are shown in Fig. 11.44 and 11.45, respectively. These data show that when the cooling rate was high (e.g., water quench), the temperature difference between the center and the surface was large. Under these conditions, the transformation started near the outer layer and then proceeded to the carburized surface and uncarburized core, producing very large transformation stresses.

The magnitude of the formation of surface compressive stresses is a function of the quenchant temperature and the alloy. For example, the surface compressive stresses increase with quenching temperature for a conventional, unaccelerated quench oil for both carburized AISI 4130 and 1526 (Fig. 11.46 and 11.47, respectively) (Ref 41). The higher compressive stresses in the carburized steels quenched in hot oil were attributed to a reduced temperature gradient between the surfaces. This would facilitate the formation of even greater amounts of martensite in the core relative to the surface. The relatively lower amounts of compressive stresses in the carburized AISI 1526 compared with 4130 lead to greater amounts of bainite formation at the expense of martensite in the 1526 alloy.

Saboury (Ref 42) performed a statistically designed process study on the effect of increasing oil quenchant temperature on the distortion of carburized AISI 4118 helical gears. The effects of normalization and increased carburization temperatures were also examined. The work was performed in a pusher furnace using an endothermic gas. An accelerated quench oil (Houghtoquench K) was used at temperatures varying from 49 to 74 °C (120 to 165 °F). When performed, the normalization cycle consisted of slow heating to 927 °C (1700 °F) under a controlled atmosphere and holding for 1 h, followed by air cooling.

Carburization was performed at either 900 or 927 °C (1650 or 1700 °F). The carbon content was 0.2% at the core and 0.6% at the surface. Reducing the temperature to 900 °C (1650 °F) increased the carburization time by 40%. The results of this study are shown in Fig. 11.48 (Ref 42). It was concluded that although decreasing the carburization temperature and normalizing reduced distortion, increasing the quenchant temperature was more effective.

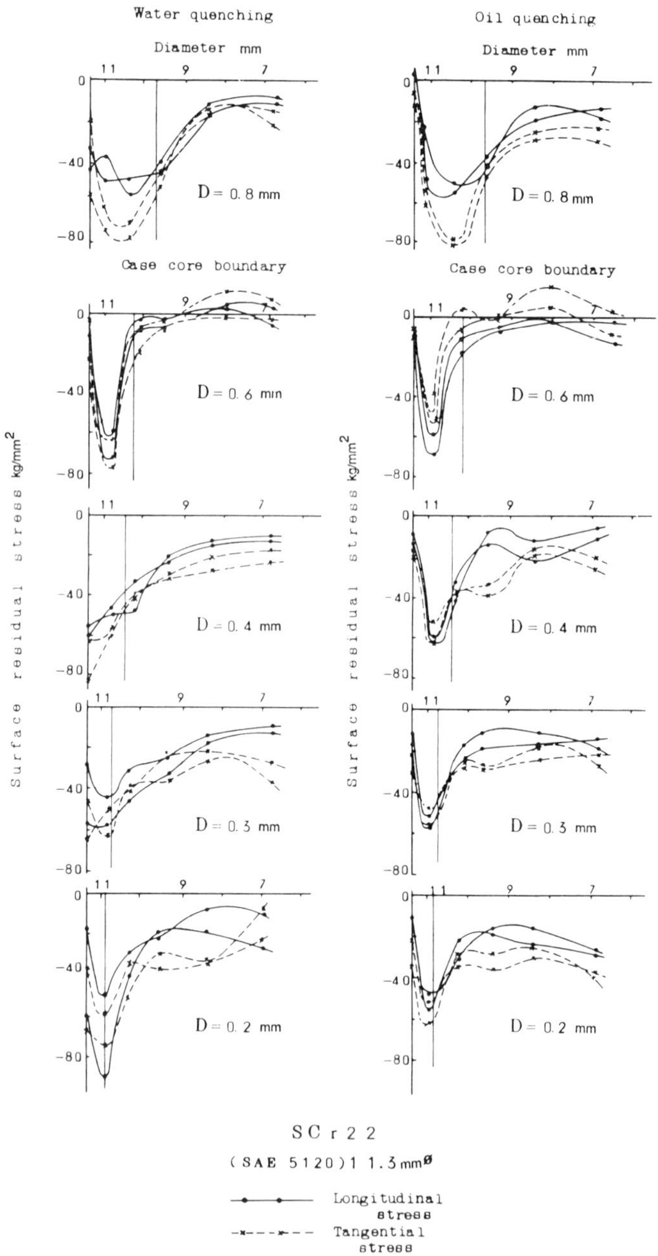

Fig. 11.44 Comparison of residual stresses of carburized S20C steel quenched in water and oil

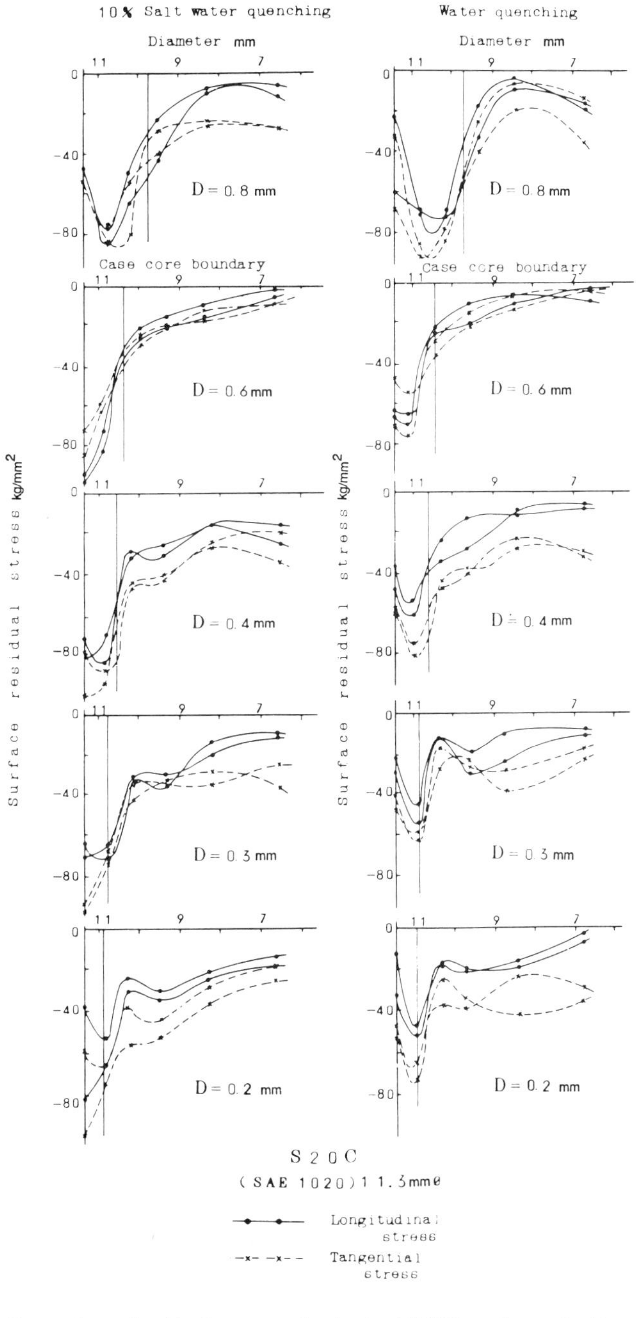

Fig. 11.45 Comparison of residual stresses of carburized SCr22 steel quenched in water and oil

Measurement of Temperature Gradients

Temperature-time cooling curves can be readily obtained by using appropriately instrumented probes or parts. For example, an AISI type 304 cylindrical

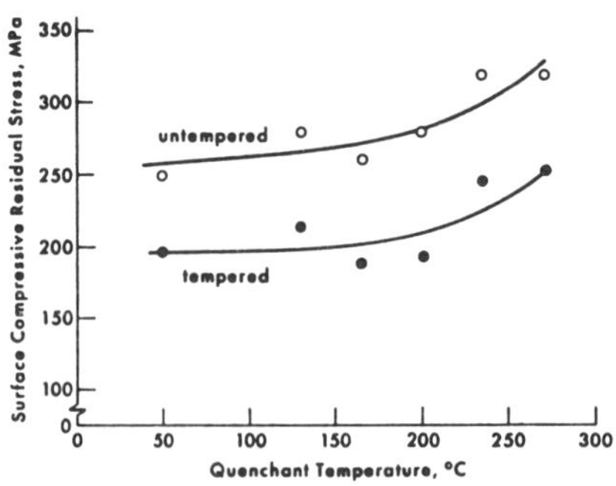

Fig. 11.46 Effect of quenchant temperature on residual stress in carburized AISI 4130 steel

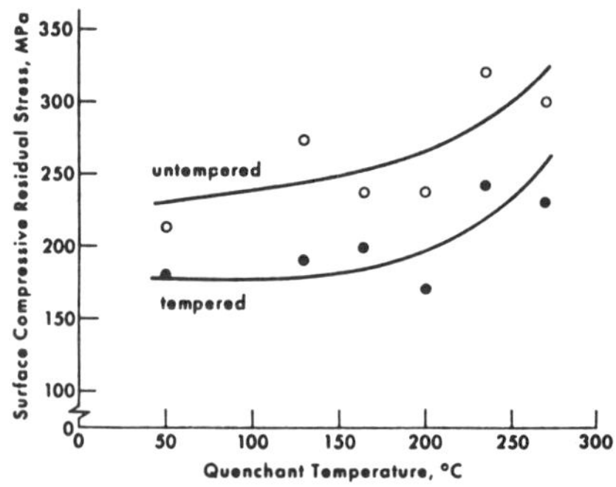

Fig. 11.47 Effect of quenchant temperature on residual stress in carburized AISI 1526 steel

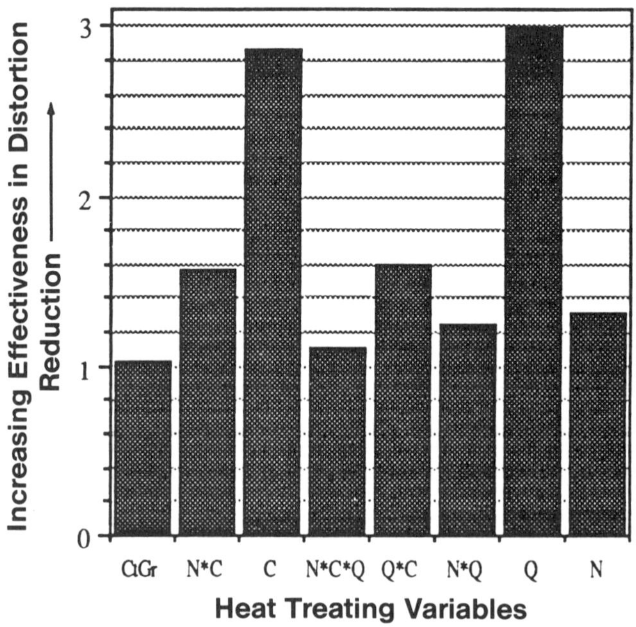

Fig. 11.48 Relative effects of quenchant temperature, carburization time, and normalizing on distortion of carburized AISI 4118 helical gears. N, normalized; C, lower carburization temperature; Q, increased quenchant temperature

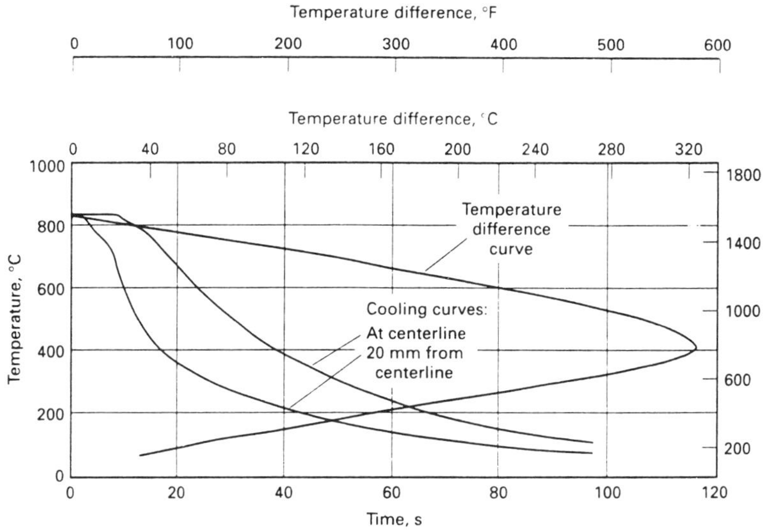

Fig. 11.49 Cooling curves and temperature difference curve for a 50 mm (2 in.) probe quenched in 32 °C (90 °F) water

stainless steel probe equipped with two thermocouples, one at the geometric center and the other at a position 20 mm (0.8 in.) from the centerline, was quenched into 32 °C (90 °F) water flowing at 0.5 m/s (1.6 ft/s). The cooling curve, showing the temperature difference between the two thermocouples, is depicted in Fig. 11.49, where the difference between the centerline temperature and the near-surface temperature is plotted as a function of near-surface temperature. The temperature difference for this process was a maximum when the near-surface temperature was 420 °C (790 °F). At this position, a maximum temperature difference of approximately 320 °C (575 °F) was obtained. This relatively high temperature gradient, which is near the M_S temperature of many steels, coupled with the loss of ductility during martensite formation, demonstrates why steel cracking, particularly with higher-hardenability steels, is often obtained with a "cold" water quench.

Similar near-surface and centerline cooling curves were obtained with a 50 mm (2 in.) AISI type 304 stainless steel probe that was quenched in an agitated fast oil. The cooling curves shown in Fig. 11.50 indicate that the maximum temperature gradient of approximately 210 °C (380 °F) occurred at a near-surface temperature of approximately 575 °C (1065 °F), which is substantially above the M_S temperature of many steels. The reduced maximum temperature gradient that occurs at a higher temperature relative to the data shown in Fig.

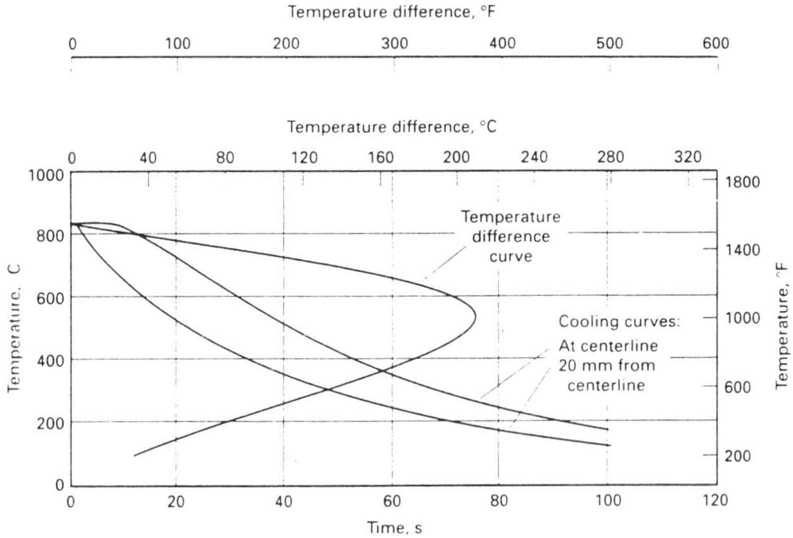

Fig. 11.50 Cooling curves and temperature difference curve for a 50 mm (2 in.) probe quenched in an accelerated oil

11.49 shows why a quench oil offers substantially greater protection from distortion and cracking than water.

The interrelationship between the centerline and surface (or near-surface) cooling curves and thermal stresses is illustrated in Fig. 11.51 (Ref 13) for a steel that does not undergo a transformation (an austenitic stainless steel). These data indicate that initially the surface cools more rapidly than the core, which means that the surface undergoes temperature-dependent shrinkage sooner than the core. Thus, there are tensile stresses at the surface and compressive stresses at the core. This gives rise to a plastic deformation at the core (see the shaded area in Fig. 11.51). After passing through point W, the core shrinks faster than the surface, at which point it produces an elastic reduction of surface stresses (curve C) until there is a stress conversion and the core goes into tension.

Of equal importance to the maximum temperature difference and the temperature where this occurs are the relative heat-transfer characteristics of the quenchant. Liscic and Letin (Ref 43) studied the relative difference in the quenching behavior of cold water and a conventional oil (see Fig. 3.63; these data were obtained with the probe shown in Fig. 3.10). Figure 3.63 shows that water produces a greater thermal gradient at a lower temperature and shorter time than the oil (which is consistent with the data discussed above). Additionally, water produces higher heat flux and heat-transfer coefficients, which means that even greater differences in the thermal gradients for complex shapes with thick

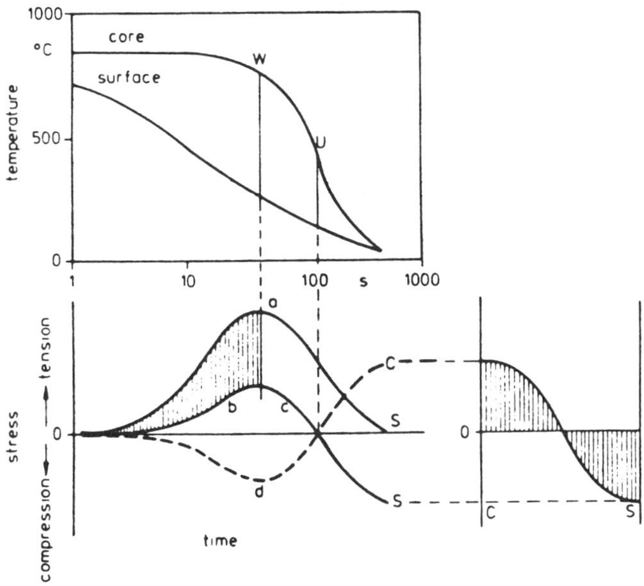

Fig. 11.51 Generation of thermal stress with no transformational behavior

or thin cross sections will be observed for water than for oil, leading to a greater tendency for distortion and cracking.

It is also of interest to facilitate temperature uniformity across the surface during the quench. This behavior can also be studied with probes or parts equipped with multiple thermocouples. An example of a probe that has been reported for this work is shown in Fig. 3.11.

Methods of Stress Measurement

In the previous discussion, it was shown that the propensity for distortion and cracking is dependent on both thermal- and transformation-induced stresses. Some of the methods that have been used to measure stress include Sach's hole-drilling (bore-out) procedure (Ref 44, 45), x-ray diffraction (Ref 44), bending-deflection methods (Ref 45), the "plate-cutout" method (Ref 46), Barkhausen noise level (Ref 47), neutron diffraction (Ref 48), energy-dispersive neutron diffraction (Ref 49), scanning tunneling microscopy (Ref 50), and ultrasonics (Ref 51).

The section on "Definition of Stress" at the beginning of this chapter provided a basic summary of the equations used to calculate stress from strain measurements, as stress is never measured directly. The remainder of this section will describe the fundamental principles of the calculation of steel stress using the two most common procedures for strain measurement: Sach's bore-out method and x-ray diffraction.

Sach's bore-out method is the classical technique for measuring residual stress. This procedure involves boring a hole in stages and measuring the reduction in longitudinal and circumferential strains at the outer surface caused by the residual stress-relieving hole-drilling process. Sach's bore-out method assumes that:

- The metal has a constant bulk modulus and Poisson's ratio.
- The residual stresses are distributed with rotational symmetry.
- The specimen is sufficiently long (or thick) to prevent binding.

Calculations of longitudinal stress (σ_r), circumferential stress (σ_θ), and radial stress (σ_r) are shown by Eq 11.12 to 11.14, respectively (Ref 45):

$$\sigma_z = E'\left[(A - A_o)\frac{d\lambda}{dA} + \lambda\right] \quad \text{(Eq 11.12)}$$

$$\sigma_\theta = -E'\left[(A - A_o)\frac{d\theta}{dA} + \frac{A + A_o}{2A}\theta\right] \quad \text{(Eq 11.13)}$$

$$\sigma_r = -E'\left(\frac{A - A_o}{2A}\theta\right) \quad \text{(Eq 11.14)}$$

where $E' = E(1 - r^2)$; $A_1 = \pi r^2$, where r_1 is the external radius of the cylinder; $A = \pi r^2$, where r is the internal radius of the cylinder; $\Lambda = \lambda + \gamma\theta$; λ is the change in longitudinal surface strain due to the removal of an inner layer; θ is the corresponding change in circumferential strain; and $\theta = \theta + \gamma\lambda$.

For a cylindrical specimen, a hole is progressively drilled at the center, and strain gages are attached to the unrestrained surfaces. Figure 11.52 illustrates data obtained by Sach's method for internal stresses in a brass tube.

The Sach's bore-out procedure is perhaps the most well-known method of stress measurement; however, it has a number of disadvantages (Ref 44, 45):

- It is slow and relatively expensive.
- Care must be taken to ensure that plastic deformation does not occur during hole drilling.
- Strain gage corrections for drift during measurement must be made.

- It can only measure final stress and thus cannot be readily applied during cooling.
- It results in damage to the part or specimen being tested.

X-ray diffraction is perhaps the most common alternative to the Sach's bore-out method (Ref 52). This procedure involves irradiating a steel sample

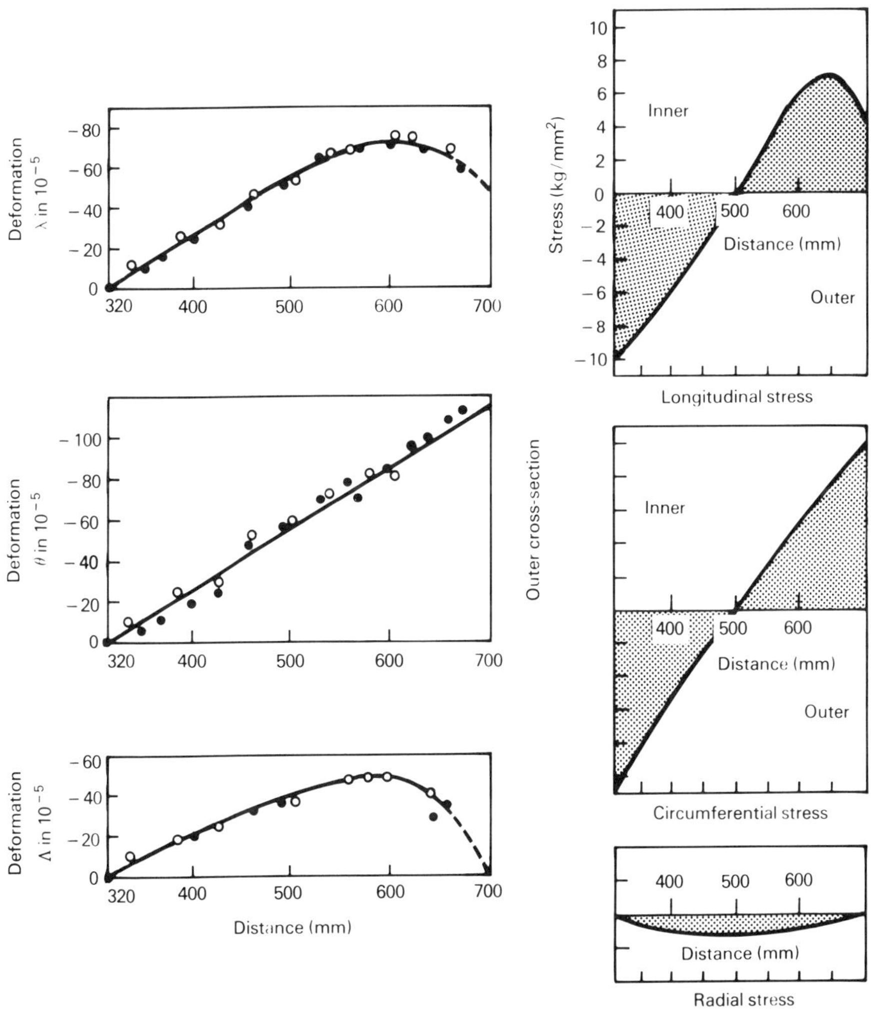

Fig. 11.52 Sach's internal stress data for a brass tube

with x-rays. The particular steel alloy analyzed will have a characteristic diffraction pattern that is dependent on the crystal structure of the iron and alloying elements present (see Fig. 11.7). The spacing between the lattice points, or *d*-spacing, can be calculated from the diffraction pattern.

The interplanar lattice *d*-spacing is calculated from the x-ray diffraction data using Bragg's equation:

$$n\lambda = 2d \sin \theta \qquad \text{(Eq 11.15)}$$

where n is an integer, λ is the wavelength of the x-ray beam, d is the spacing between reflecting planes, and θ is the angle of incidence of the beam with the sample. When a load is applied to the sample, there will be a perturbation in the *d*-spacing. Thus, changes in the measured diffraction patterns (Δd) are related to the lattice strain ($\Delta d/d$). The strain ($\Delta d/d$) is calculated from the diffraction data using:

$$\frac{\Delta d}{d} = -\cot \theta \left(\frac{\Delta 2\theta}{2}\right) \qquad \text{(Eq 11.16)}$$

There are a number of experimental procedures for calculating residual stress from diffraction data. The most common method is the $\sin^2 \psi$ technique. In this technique, the specimen is positioned and irradiated. Changes in the diffraction angle (ψ) are related to the interplanar spacing (d) and thus to strain ($\Delta d/d$). The change in interplanar spacing is determined by measuring d with different applied stresses. The relationship among $\Delta d/d$, $\sin^2 \psi$, and stress (σ) is described by:

$$\sigma = \frac{d-d_o}{d_o} \times \frac{E}{1+\nu} \times \frac{1}{\sin^2 \psi} \qquad \text{(Eq 11.17)}$$

The *d*-spacing is obtained from the Bragg equation (Eq 11.15). If a $\Delta d/d$ versus $\sin^2 \psi$ plot is constructed (e.g., Fig. 11.53), the stress can be calculated from the slope of the straight line as follows:

$$\text{Slope} = \frac{\sigma(1+\nu)}{E} \qquad \text{(Eq 11.18)}$$

Equation 11.19 is obtained by rearranging Eq 11.18 to solve for σ:

$$\sigma = \text{Slope} \times \left[\frac{E}{(1+\nu)}\right] \qquad \text{(Eq 11.19)}$$

Because Poisson's ratio (ν) is known and the bulk modulus (E) is also known, the stress (σ) can be readily calculated. The remainder of this section offers comments and guidelines for the measurement of stress by x-ray diffraction (Ref 53). As stated previously, it is recommended that the engineering modulus not be obtained from the

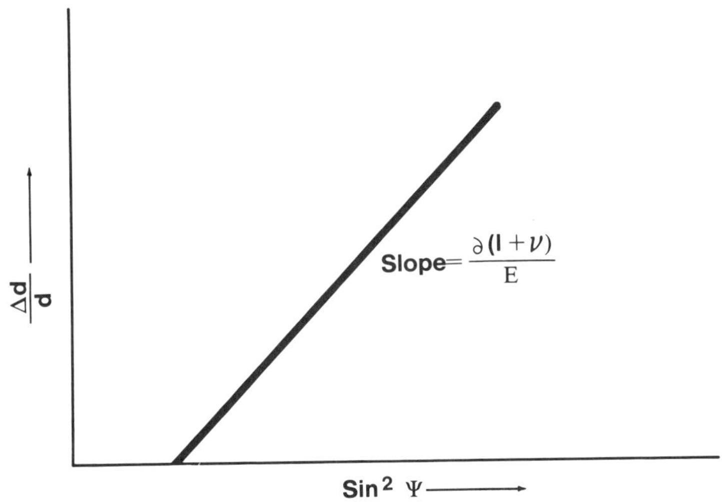

Fig. 11.53 Calculation of stress by the $\sin^2 \psi$ technique

literature. The modulus should be determined by x-ray measurement as part of the x-ray analysis experiment.

Rough surfaces (e.g., a fairly large grain size), may produce substantial errors. Measurement of the stress at a point other than the top of a rough spot may result in low value due to a partial or full relaxation of stresses at that point.

Possible sources of x-ray measurement errors include:

- Error in peak position
- Time between determinations (some materials, such as aluminum, may readily undergo stress relief by aging)
- Sample anisotropy
- Grain size
- Flat versus round (flat plates are preferable)

If x-ray analysis is to be performed under contract, the following questions should be asked for subsequent comparison between laboratories:

- What is the x-ray beam size?
- Is a standard collimated x-ray beam or a Sollier focus being used?

- What technique is going to be used: SET (single-exposure technique) or DET (double-exposure technique)?
- What is the previous experience and area(s) of specialty of the laboratory?
- How are differences in grain size or preferred orientation effects accounted for?
- What is the spatial resolution in the samples?

A "rule of thumb" dictates that measurements of a round bar should be taken at a distance from the ends of at least twice the diameter of the bar to minimize any possible end effects. If the specimen is a plate, the rule of thumb is that measurements should be taken at a distance from the edge of at least twice the thickness of the plate to minimize edge effects. More than one measurement generally is required along the length of the specimen to ensure that the maximum value is taken.

If a $\sin^2 \psi$ technique is being used, ask to see the plot and ask for the R_0 error. The plot should be linear. A $\sin^2 \psi$ plot should be requested if the part being tested is a new material and is suspected to have significant surface texture or preferred orientation. Aluminum is particularly susceptible in this regard.

Because stress changes with time and position, it is important to obtain the measurements in a standard and well-calibrated way as soon as possible after the stress-generating process (e.g., quenching). Usually the error that can be expected between instruments is ±35 MPa (5 ksi). The measurements should be standardized against a powder or a four-point bend test. More than one measurement at the same spot should be made to test for reproducibility.

Highly hardened steel is difficult to analyze because the x-ray peaks are typically very broad. Curve deconvolution techniques are usually employed. If curve deconvolution is used, each peak should be calibrated. Peak breadth may yield additional information. Be sure to have the analyst provide peak breadth data for each sample.

If a SET procedure is used, ask for repeated measurements. Better yet, the analysis should be viewed in person. A copy of the $\sin^2 \psi$ plot should be obtained.

Tests for Cracking and Distortion

The previous section of this chapter described various tests used to measure cracking and distortion. In some cases, these tests involved the quenching of actual parts (e.g., bearing races and gears); in others, standard test specimens were used. This section will summarize some of the more frequently encountered testing procedures for the evaluation of cracking and distortion, including the use of solid bars, plates with holes, notched C-rings, Navy C-rings, modified Navy C-rings, slotted disks, finned tubes, and the wedge test.

Solid Bars. The simplest test that can be run is to quench a round bar into the quenchant of interest and then examine it for cracks. Examples of these tests were shown previously in Fig. 11.28 and 11.40. Another variation of this test is to quench solid plates or rectangular bars. Kubota (Ref 54) performed cracking tests on 16 × 16 × 40 and 20 × 20 × 40 mm (0.6 × 0.6 × 1.6 and 0.8 × 0.8 × 1.6 in.) rectangular bars. When performing such tests, it is important to differentiate actual quench cracking from cracking that may occur during sample preparation.

Moerdijk (Ref 55) performed distortion tests on a series of quenchants, including water, oils, and aqueous polymers, by measuring the deformation of a "step" bar (Table 11.8). Similar studies could be performed by measuring the "running" distortion of a long bar/unit length.

Plates With Holes. A variation of the solid bar test is to evaluate the distortion of a plate with holes or slots (Ref 56). In these tests, the perturbations of both the outside plate and the hole dimensions are measured after quenching. Because the stresses operating on each of these dimensions are different, shapes such as this can be used to model more complex parts.

Table 11.8 Deformation of steel step bars after quenching

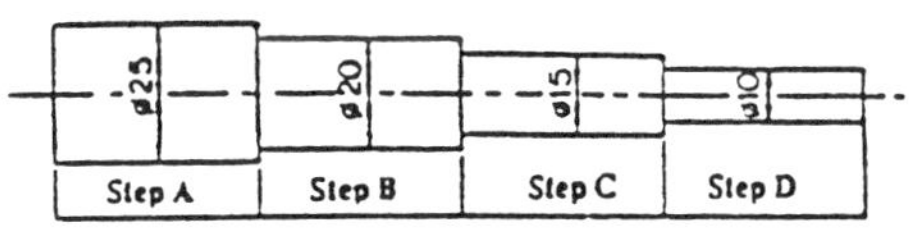

Stepped Bar Test

Quenchant	Concentration(a), %	Deformation, mm Ck45	C35 BKD	42CrMo4
Aquaquench 251	10	0.59	0.22	0.11
	20	0.24	0.26	0.17
UCON® Quenchant A	10	0.61	...	0.15
	20	0.10	...	0.10
Serviscol	10	0.22	0.56	0.40
	20	0.15	0.11	0.27
Martensal In at B	10	0.16	0.35	0.17
	20	0.16	0.20	0.08
Water	...	1.19	0.60	1.28
Isodur 220	...	0.09	0.14	0.16
Durixol	...	0.12	0.15	0.09
Isomax 166	...	0.05	0.18	0.18

(a) These are likely concentrations used. The original paper cited values of 1 and 2%, which are not reasonable for most of these steels.

C-Rings and Slotted Disks. One of the oldest standard tests for evaluating quench distortion is the so-called Navy C-ring test (Ref 57-59). Examples of

Table 11.9 Dimensional changes in Navy C-ring specimens in three vigorously agitated quench media

Types of quench	Dimensional change, 0.0001 in., at: A	B	C	D	E
Conventional oil at 60 °C (140 °F)......	83	95	80	240	30
Martempering in salt at:					
205 °C (400 °F)................................	52	60	50	0	10
245 °C (475 °F)................................	45	30	1	–20	10

A.

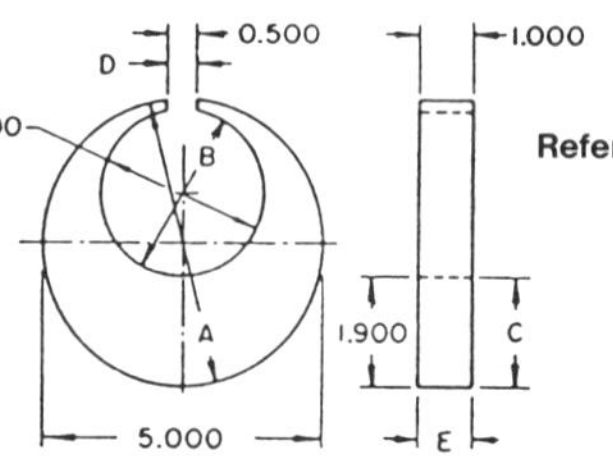

B.

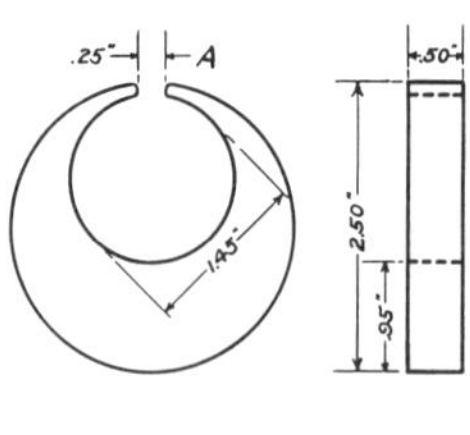

C.

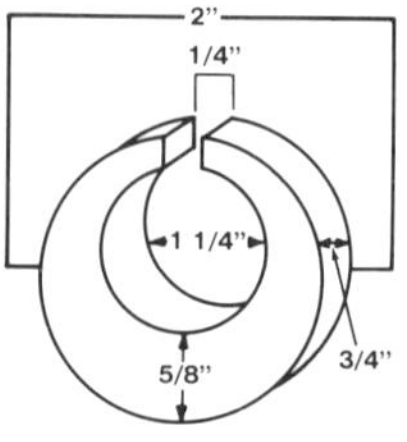

Fig. 11.54 Examples of C-ring test specimens used for quench distortion studies. **(a)** Ref 57. **(b)** Ref 58. **(c)** Ref 59

Table 11.10 Relative distortion with different quenchants

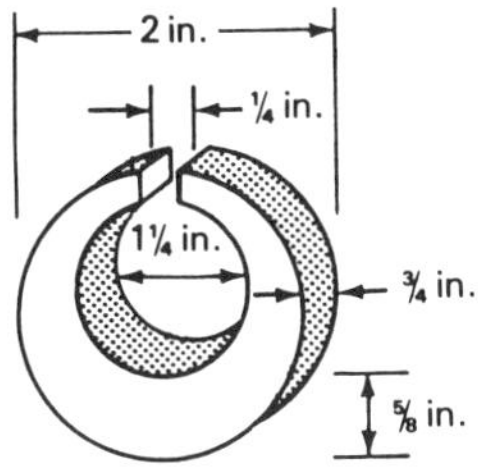

SAE steel	Quenchant	Dimensional change(b,c), in. Gap opening	Outside diameter
1045	Petroleum oil (65 °C, or 150 °F)	+0.0005	−0.0002
	UCON® Quenchant A(a) (40 °C, or 100 °F)	+0.0007	−0.0001
	Water (25 °C, or 80 °F)	+0.0003	+0.0003
4140	Petroleum oil (65 °C, or 150 °F)	+0.0016	+0.0006
	UCON® Quenchant A(a) (40 °C, or 100 °F)	+0.0017	+0.0006
	Water (25 °C, or 80 °F)	Cracked	Cracked
8620(d)..............	Petroleum oil (65 °C, or 150 °F)	+0.0007	+0.0002
	UCON® Quenchant A(a) (40 °C, or 100 °F)	+0.0008	+0.0002
	Water (25 °C, or 80 °F)	+0.0018	+0.0007

(a) This is a relatively fast aqueous PAG polymer quenchant even at 20%. UCON® Quenchant A was used at 20% with the same agitation level used for oil. **(b)** The dimensional changes for water were caused by increased martensite content. **(c)** Microstructural analysis indicated that the martensite contents of the UCON® Quenchant A and oil-quenched specimens were equivalent. **(d)** Carburized to 0.8% C to a case depth of 1 mm (0.04 in.).

some reported test specimens are shown in Fig. 11.54; illustrative distortion data are presented in Table 11.9 (Ref 57) and Table 11.10 (Ref 59).

A modified Navy C-ring test has also been reported (Ref 60). The notched test specimen (Fig. 11.55) imparts greater crack sensitivity, as shown in Table 11.11 along with some illustrative test results taken from Ref 60. A variant of the modified Navy C-ring test is the ASTM notched C-ring test (Ref 61). The test specimen and dimensions are shown in Fig. 11.56. Still another variant is the slotted disk test utilized by Kakiuchi (Ref 62). This test specimen is simply a relatively thin 8 mm (0.3 in.) disk with a 3 mm (0.12 in.) outside diameter, a 15.5 mm (0.6 in.) inside diameter, and a 1 mm (0.04 in.) slot or gap (see Fig. 11.57).

One relatively simple experimental procedure for measuring the approximate hoop stress in a tube is the split-ring test, which is illustrated in Fig. 11.58 (Ref 63). In this procedure, a section from the quenched tube is cut. A slit is cut lengthwise in the tube after the outside diameter (D_O) and tube thickness have been carefully measured. The distance that the tube separates at the slit (x) is then measured. The hoop stress (σ_{HS}) is calculated as follows:

$$\sigma_{HS} = \frac{Et}{1-\nu^2}\left[\frac{1}{D_o} - \frac{1}{(x/\pi + D_o)}\right] \qquad \text{(Eq 11.20)}$$

where E is Young's modulus, ν is Poisson's ratio, and t is the wall thickness.

Table 11.11 Modified C-ring test results for AISI 1085 steel in various quench media

	Hardness, HRC			Dimensional change, in.		
Quenchant	Maximum	Minimum	Deviation	Gap A	Diameter B	Diameter C
Water	67.0	63.0	4.0	0.0134	0.0102	0.0116
Fast oil	66.0	63.0	3.0	0.0021	0.0032	0.0024
Conventional (slow) oil	65.5	43.0	22.5	0.0022	0.0038	0.0038

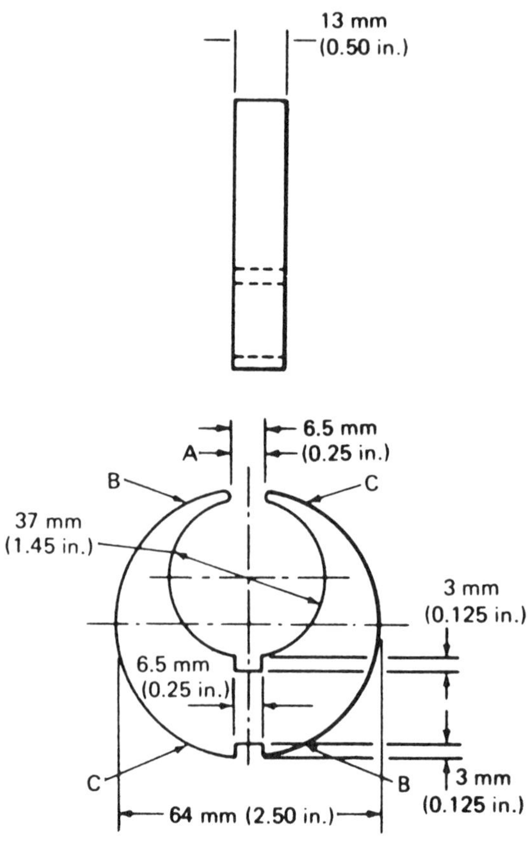

Fig. 11.55 Modified Navy C-ring distortion test specimen

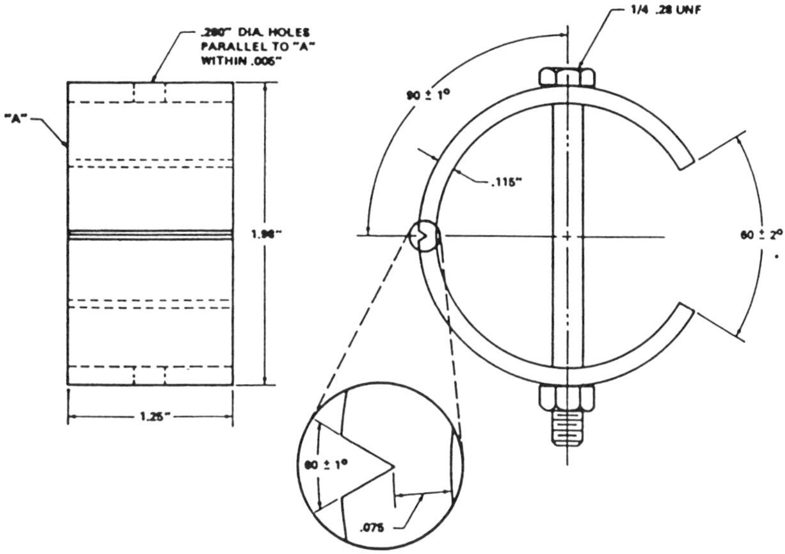

NOTE 1—Notch root radius to be 0.0030 ± 0.0005 in. (0.076 ± 0.013 mm).
NOTE 2—Notch surface finish to be 32 or better.
NOTE 3—Tolerances (unless otherwise specified):

X.XX ± 0.01 in. (0.2 mm)
X.XXX ± 0.001 in. (0.02 mm)

Fig. 11.56 Notched ASTM C-ring distortion test specimen

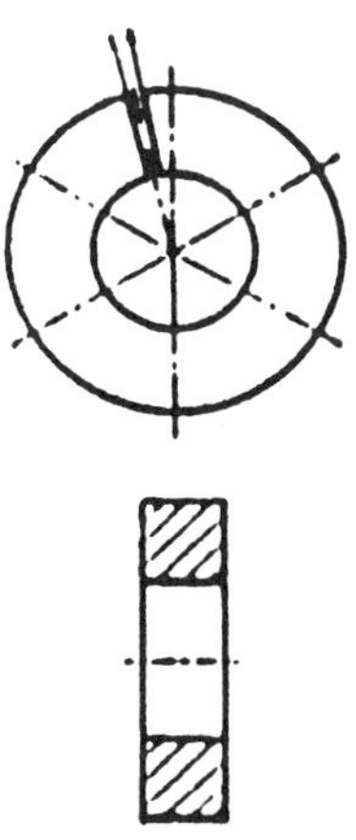

Fig. 11.57 Slotted disk distortion test specimen

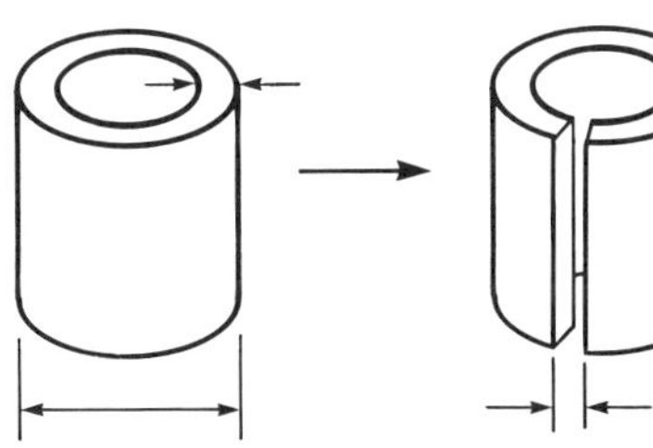

Residual Hoop Stress (psi) - ζHS

$$\zeta HS = \frac{Et}{1 - \nu^2} \quad \frac{1}{D_o} - \left[\frac{1}{(\chi/\pi + D_o)}\right]$$

Fig. 11.58 Split ring test procedure

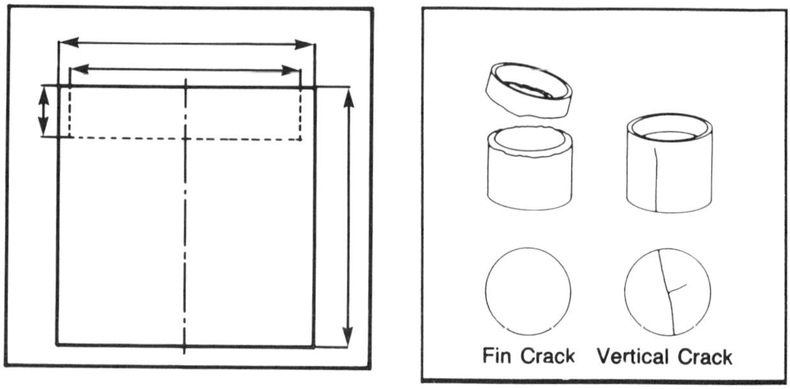

Fig. 11.59 Finned tube specimen

Finned Tubes. The distortion of hollow tubes has been studied (see Fig. 11.28). However, Mikita *et al.*, (Ref 64-68) varied this test by drilling a large hole in a solid block of the steel of interest to create a ring on one end of the block (see Fig. 11.59). In this test, the appearances of two types of cracks, fin and vertical, were quantified. The susceptibility toward fin cracking can be increased by cutting a V-notch in the fin (Ref 65).

Mikita *et al.* determined the relative cracking frequency for various quenchants using a limited number of test specimens. For each quenchant, the test specimen was repeatedly quenched until a crack appeared. So-called $1/n$ values, which are the inverse of the number of quenches required for cracking (n), were used to characterize the quenchants (Ref 64).

References

1. H.E. Boyer, *Atlas of Stress-Strain Curves*, ASM International, 1987, p 1-2
2. R.T. Von Bergen, in *Quenching and Distortion Control*, G.E. Totten, Ed., ASM International, 1992, p 275-282
3. R. Kern, *Heat Treat.*, April 1985, p 38-42
4. R. Kern, *Heat Treat.*, Feb 1971, p 1-4
5. R. Kern, *Heat Treat.*, March 1985, p 41-45
6. R.W. Edmonson, *Met. Treat.*, Dec 1969-Jan 1970, p 3-19
7. R.R. Blackwood and L.M. Jarvis, *Ind. Heat.*, March 1991, p 28-31
8. K.E. Thelning, *Steel and Its Heat Treatment*, Butterworths, London, 1985, p 584
9. T. Kunitake and S. Susigawa, *Sumitomo Search*, May 1971, p 16-25
10. H. Fujio, T. Aida, and Y. Masumoto. *Bull. Jpn. Soc. Mech. Eng.*, Vol 20, 1977, p 1655-1662
11. D.M. Nakhimov and D.I. Bron, *Metalloved. Term. Obrab. Met.*, Vol 7, 1966, p 16-22

12. Yu.A. Geller and V.P. Brimene, *Steel USSR*, July 1971
13. P. Mayr, *Residual Stresses in Science and Technology*, Vol 1, Garmisch-Partenkirchen, West Germany, 1986; DGM Informationsgesellschaft mbH, Adenaueralle 21, D-6370 Oberursel 1, West Germany, 1987 (Met. A, 8808-72-0399), p 57-77
14. D.V. Nelson, R.E. Ricklefs, and W.P. Evans, Paper No. 7102.83, Society of Automotive Engineers, X-Ray Fatigue Division, SAE Fatigue and Evaluation Committee, 23-24 Sept 1969
15. M. Wildau and H. Hougardy, *Härt. Tech. Mitt.*, Vol 42 (No. 5), 1987, p 269-277
16. K. Okamura and H. Kawashima, *32nd Jpn. Cong. Mater. Res.*, Dec 1988, p 323-329
17. F.M.B. Fernandes, C. Basso, S. Denis, and A. Simon, *Calc. Internal Stresses Heat Treat. Metall. Mater.*, Vol 2, 1984, p 275-297
18. G.K. Kuzmin, *Met. Sci. Heat Treat.*, Vol 28 (No. 1-2), 1986, p 87-91
19. H. Fujio, T. Aida, and Y. Masumoto, *Bull. Jpn. Soc. Mech. Eng.*, Vol 20, 1977, p 1051-1058
20. *ASM Handbook*, Vol 4, ASM International, 1991, p 76-81
21. M. Hetenyi, *Handbook of Experimental Stress Analysis*, John Wiley & Sons, 1950
22. Palmer *et al.*, *Tool Steel Simplified*, 4th ed., Chilton, 1978
23. Gulf Research and Development, product literature
24. V.A. Murzin, L.I. Koshkareva, and N.G. Polumordvinova, *Met. Sci. Heat. Treat. Met.*, Vol 2, 1990, p 101-105
25. C. Pavee, *Trait. Therm.*, Vol 191, 1985, p 53-56
26. K. Kakiuchi, *Netsu Shori*, Vol 29 (No. 3), 1988, p 125-130
27. Y. Toshioka, *Proc. 4th Conf. Irish Durability and Fracture Committee*, Queens University Press, Belfast, 1986
28. K. Kakiuchi, *Netsu Shori*, Vol 29 (No. 4), 1988, p 186-190
29. I.G. Uzlov and N.I. Danchenko, *Metalloved. Term. Obrab. Met.*, Vol 5, 1971, p 54-56
30. *Sourcebook on Heat Treating*, Vol I, *Materials and Processes*, American Society for Metals, 1975
31. Yu.P. Gul', V.S. Chmeleva, and V.V. Kirichenko, *Metalloved. Term. Obrab. Met.*, Vol 9, 1989, p 2-6
32. R.K. Zhelokhovtseva, *Steel USSR*, Vol 15, 1985, p 238-239
33. H.B. Bochum, *Radex Rundsch.*, Vol 1, 1989, p 40-57
34. F. Moreaux, A. Simon, and G. Beck, *J. Heat Treat.*, Vol 1 (No. 3), 1980, p 50-56
35. G. Beck, *Mém. Étud. Sci. Rev. Métall.*, June 1985, p 269-282
36. S. Owaku, *Netsu Shori*, Vol 30 (No. 2), 1990, p 63-67
37. D. Milam, *Heat Treat.*, Dec 1987, p 31-35
38. M.A.H. Howes, "The Effect of Heat Treatment on Metallurgical Structure and Gear Performance," presented at Conference on Advanced Gear Processing and Manufacturing, Society of Manufacturing Engineers, 17-19 Sept 1991
39. M. Motoyama and H. Horisawa, Paper No. 710281, Society of Automotive Engineers, X-Ray Fatigue Division, SAE Fatigue Design and Evaluation Committee, 24-25 Sept 1968
40. M. Motoyama, R.E. Ricklefs, and J.A. Larson, Paper No. 750050, Automotive Engineering Congress and Exposition, Society of Automotive Engineers, 24-28 Feb 1975
41. G. Krauss, in *Quenching and Distortion Control*, G.E. Totten, Ed., ASM International, 1992, p 181-191
42. S. Saboury, in *Quenching and Distortion Control*, G.E. Totten, Ed., ASM International, 1992, p 259-264

43. B. Liscic and T. F. Letin, *J. Heat Treat.*, Vol 57, 1988, p 115-124
44. K. Sakai, *Tetsu-to-Hagané* (*J. Iron Steel Inst. Jpn.*) Vol 60 (No. 12), 1974, p 1591-1598
45. A.A. Denton, *Metall. Rev.*, Vol 11, 1966, p 1-23
46. S. Nishimura, Y. Morita, and K. Tokimasa, *Bull. Jpn. Soc. Mech. Eng.*, Vol 18 (No. 116), 1975, p 116-122
47. K. Tiitto, A.S. Wojtas, W.J.P. Vink, and G. den Ouden, *Practical Applications of Residual Stress Technology*, ASM International, 1991, p 55-59
48. J.H. Root, R.R. Hosbons, and T.M. Holden, *Practical Applications of Residual Stress Technology*, ASM International, 1991, p 83-86
49. H.J. Prask and C.S. Choi, *Practical Applications of Residual Stress Technology*, ASM International, 1991, p 87-93
50. R. Wienendanger, M. Ringger, L. Rosenthaler, H.R. Hidber, P. Oelhafen, and H.J. Guntherodt, *Z. Phys. Chem. Neue Folge Bd.*, Vol 157, 1988, p 139-143
51. D.R. Allen and C.M. Sayers, *Ultrasonics*, July 1984, p 179-188
52. C.O. Ruud, in *Quenching and Distortion Control*, G.E. Totten, Ed., ASM International, 1992, p 193-198
53. C.O. Ruud, Pennsylvania State University, personal communication, 1990
54. T. Kubota, *Netsu Shori*, Vol 25 (No. 1), 1985, p 34-37
55. W.A.J. Moerdijk, *Adv. Mater. Proc.*, Vol 137 (No. 3), 1990, p 19-28
56. N. Hirose, S. Yamamoto, and K. Shinji, p 268-273
57. *Metals Handbook*, 8th ed., Vol 2, *Heat Treating, Cleaning and Finishing*, American Society for Metals, 1964, p 41
58. H.J. French, *The Quenching of Steels*, American Society for Steel Treating, 1930, p 133
59. "UCON Quenchant A—The Fast and Safe Way to Quench Steel Alloys," product information bulletin, Tenaxol, Inc., Dec 1972
60. H.E. Boyer and P.R. Cary, *Quenching and Control of Distortion*, ASM International, 1988, p 39
61. ANSI/ASTM F 519-77, ASTM specification, Sept 1978
62. K. Kakiuchi, *Netsu Shori*, Vol 30C (No. 1), 1990, p 40-44
63. H. Walton, Torrington Co., personal communication, 1992
64. Y. Mikita, I. Nakabayashi, and K. Sukamaki, *Heat Treat.*, Dec 1989, p 21-24
65. Y. Mikita and I. Nakabayashi, *Nippon Kikai Gakkai Ronbunshu (A-hen)*, Vol 53 (No. 489), 1987, p 884-889
66. Y. Mikita and I. Nakabayashi, *Nippon Kikai Gakkai Ronbunshu (A-hen)*, Vol 53 (No. 490), 1987, p 1007-1011
67. Y. Mikita, I. Nakabayashi, N. Ohga, and K. Ohsaka, *Nippon Kikai Gakkai Ronbunshu (A-hen)*, Vol 53 (No. 496), 1987, p 2211-2215
68. Y. Mikita, I. Nakabayashi, K. Sakamaki, and Siltori, *Nippon Kikai Gakkai Ronbunshu (A-hen)*, Vol 55 (No. 513), 1989, p 1116-1120

Index

A

B

C

D

E

F

G

H

I

J

K - L

M

N

O

P

Q

R

S

T

U

V

W

X - Y - Z